Progress in Drug Research
Fortschritte der Arzneimittelforschung
Progrès des recherches pharmaceutiques
Vol. 33

Progress in Drug Research
Fortschritte der Arzneimittelforschung
Progrès des recherches pharmaceutiques
Vol. 33

Edited by / Herausgegeben von / Rédigé par
Ernst Jucker, Basel

Authors / Autoren / Auteurs
David M. Warburton · Melvin J. Silver and Giovanni Di Minno ·
Anthony C. Allison and Simon W. Lee · Zell A. McGee, Gary L.
Gorby and Wanda S. Updike · Esteban Domingo · Erik T.
Michalson and Jacob Szmuszkovicz · George Kunos · H. Stähelin
and W. von Wartburg · Ranjan P. Srivastava and A. P. Bhaduri ·
Paul D. Hoeprich · Judit Ovádi · Hansjakob Müller · J. L. Stanford

1989

Birkhäuser Verlag
Basel · Boston · Berlin

The publisher cannot assume any legal responsibility for given data, especially as far as directions for the use and the handling of chemicals are concerned. This information can be obtained from the manufacturers of chemicals and laboratory equipment.

This work is subject to copyright. All rights reserved, whether the whole or part of the material is concerned, specifically those of translation, reprinting, re-use of illustrations, broadcasting, reproduction by photocopying machine or similar means, and storage in data banks. Under § 54 of the German Copyright Law where copies are made for other than private use a fee is payable to 'Verwertungsgesellschaft Wort', Munich

© 1989 Birkhäuser Verlag Basel
ISBN 3-7643-2306-X
ISBN 0-8176-2306-X

Contents · Inhalt · Sommaire

Nicotine: an addictive substance or a therapeutic agent?	9
David M. Warburton	
Aspirin as an antithrombotic agent	43
Melvin J. Silver and Giovanni Di Minno	
The mode of action of anti-rheumatic drugs. 1. Anti-inflammatory and immunosuppressive effects of glucocorticoids	63
Anthony C. Allison and Simon W. Lee	
The use of neutrophils, macrophages and organ cultures to assess the penetration of human cells by antimicrobials	83
Zell A. McGee, Gary L. Gorby and Wanda S. Updike	
RNA virus evolution and the control of viral disease	93
Esteban Domingo	
Medicinal agents incorporating the 1,2-diamine functionality	135
Erik T. Michalson and Jacob Szmuszkovicz	
Adrenergic receptor research: Recent developments	151
George Kunos	
From podophyllotoxin glucoside to etoposide	169
H. Stähelin and A. von Wartburg	
Emerging concepts towards the development of contraceptive agents	267
Ranjan P. Srivastava and A. P. Bhaduri	
Chemotherapy for systemic mycoses	317
Paul D. Hoeprich	
Effects of drugs on calmodulin-mediated enzymatic actions	353
Judit Ovádi	
The significance of DNA technology in medicine	397
Hansjakob Müller	
Immunotherapy for leprosy and tuberculosis	415
J. L. Stanford	
Index · Sachverzeichnis · Table des matières, Vol. 33	449
Index of titles · Verzeichnis der Titel · Index des titres, Vol. 1–33	457
Author and paper index · Autoren- und Artikelindex · Index des auteurs et des articles, Vol. 1–33	467

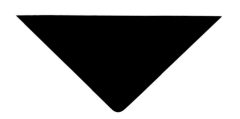

Foreword

In the first years of the existence of this series of monographs, during the so-called "Golden Age" of drug research, the majority of the papers published were mainly concerned with the traditional domains of drug research, namely chemistry, pharmacology, toxicology and preclinical investigations. The series' aim was to give coverage to important areas of research, to introduce new active substances with therapeutic potential and to call attention to unsolved problems.

This objective has not changed. The table of contents of the present volume makes evident, however, that the search for new medicines has become increasingly complex, and additional, new disciplines have entered the research arena. The series now includes reviews on biochemical, biological, immunological, physiological and medicinal aspects of drug research. Researchers actively engaged in the various scientific fields forming the entity of drug research can benefit from the wealth of knowledge and experience of the respective authors, and will be assisted in their endeavour to discover new pharmaceutical agents. Those simply wanting to keep abreast of new developments in the complex, multi-discipline science can turn to the "Progress in Drug Research" volumes as an almost encyclopaedic source of information without having to consult the innumerable original publications.

Volume 33 contains 13 reviews, a subject index, an index for the close to 400 articles published in the series so far, and an author and titles index for all 33 volumes.

I should like to thank all the authors for their willingness to prepare the reviews and for sharing their knowledge with the readers. Thanks are also due to L. Koechlin, H. P. Thür and A. Gomm of Birkhäuser Publishers for their most valuable help in the preparation of this volume.

Basel, October 1989   Dr. E. JUCKER

Vorwort

Die Gründung dieser Monographien-Reihe fiel in das «goldene Zeitalter» der Arzneimittelforschung. Eine nicht mehr zu überblickende Zahl von chemischen, pharmakologischen und klinischen Publikationen manifestierte die Suche nach neuen Medikamenten, und der aktive Forscher lief Gefahr, den Überblick zu verlieren. Die «Fortschritte der Arzneimittelforschung» hatten denn auch zum Zweck, in größeren Referaten wichtige Forschungsgebiete zusammenfassend darzustellen, neue Wirkstoffe vorzustellen und auf ungelöste Probleme hinzuweisen.

An dieser Zielsetzung hat sich nichts geändert; wie das Inhaltsverzeichnis des vorliegenden 33. Bandes jedoch illustriert, hat eine Verschiebung der Akzente in Richtung Biochemie, Biologie, Immunologie, Physiologie und Medizin stattgefunden. Damit verbunden ist gezielteres Forschen und vertieftes Verständnis der Wirkungsweise neuer Arzneimittel. Die Vielzahl der an dieser Forschung beteiligten Disziplinen dokumentiert auch die Komplexität dieses Arbeitsgebietes.

Der 33. Band der «Fortschritte» enthält 13 Übersichtsreferate, ein Stichwortverzeichnis des Bandes, einen Artikel- sowie Autoren- und Artikelindex aller bisher erschienenen 33 Bände. Weit über 300 Artikel der Reihe und die ihnen angegliederten unzähligen Hinweise auf die Originalliteratur sowie die erwähnten Verzeichnisse erlauben müheloses Auffinden der relevanten Angaben.

Der aktive Forscher wird direkten Nutzen für seine eigenen Arbeiten ziehen und Anregungen für neue Untersuchungen erhalten. Die Fülle des bisher publizierten Materials erfüllt indessen noch einen anderen Zweck: Wer sich über die Entwicklungen der Arzneimittelforschung rasch, umfassend und zuverlässig informieren will, wird in den «Fortschritten» eine nützliche, fast enzyklopädische Stütze finden, ohne sich in der Flut der Originalpublikationen umsehen zu müssen.

Nach diesen einleitenden Ausführungen verbleibt mir nur noch die angenehme Pflicht, den Autoren dieses und auch der vorangegangenen Bände für ihre große Arbeit zu danken. Dank gebührt auch den Mitarbeitern des Birkhäuser Verlages, vor allem Frau Koechlin und den Herren Thür und Gomm; ohne ihren großen persönlichen Einsatz wäre die Herausgabe der «Fortschritte» nicht möglich gewesen.

Basel, Oktober 1989   Dr. E. JUCKER

Nicotine: An addictive substance or a therapeutic agent?

By David M. Warburton
Department of Psychology, University of Reading, Building 3, Earley Gate, Whiteknights, Reading RG6 2AL, United Kingdom

1	Introduction	11
2	Pharmacokinetics	11
2.1	Absorption	11
2.1.1	Transdermal administration	11
2.1.2	Oral administration	12
2.1.3	Gastric absorption	12
2.1.4	Nasal administration	13
2.1.5	Administration by inhalation	13
2.2	Distribution	14
2.2.1	Whole body	14
2.2.2	Brain	14
2.3	Metabolism	15
2.3.1	Clearance	15
2.3.2	Metabolites	16
2.4	Summary	17
3	Nicotine and the control of body weight	17
3.1	The evidence	17
3.2	Mechanism	18
3.2.1	Physical activity	18
3.2.2	Caloric intake	18
3.2.3	Metabolic activity	19
3.2.3.1	Adrenal catecholamines	19
3.2.3.2	Adrenal glucocorticoids	20
3.2.4	Summary	20
4	Analgesia	21
4.1	The evidence	21
4.2	A mechanism of analgesia	22
5	Mood control	22
5.1	Evidence from laboratory studies	22
5.2	Evidence from nicotine substitution	23
5.3	Mechanisms of mood control	24
5.3.1	Brain release of catecholamines	24
5.3.2	Brain release of indoleamines	24
5.4	Summary	24
6	Cognitive enhancement	25
6.1	The evidence for improved attention	25
6.1.1	Focussed attention	25
6.1.2	Divided attention	27
6.1.3	Summary of attention studies	27
6.2	Information storage	28
6.2.1	Immediate memory	28
6.2.2	Delayed recall	28

6.2.3	Information retrieval		29
6.3	Mechanism of cognitive enhancement		30
6.3.1	Localization of action		30
6.3.2	Neurochemical action of nicotine at the cortex		30
6.3.3	Cortical actions of nicotine in people		31
6.3.3.1	Cortical desynchronization		32
6.3.3.2	Event-related potentials		32
6.4	Therapeutic implications of cognitive enhancement		33
7	Dangers of nicotine use		34
7.1	Duration of action		34
7.2	Toxicity		35
7.3	Desensitization and tolerance		35
7.4	Withdrawal symptoms		36
8	Conclusions		37
	References		38

1   **Introduction**

Recently, the Surgeon-General of the United States pronounced that nicotine use is an addiction like heroin and cocaine use, and that it is irrational and driven by withdrawal symptoms [1]. From another point of view, it can be argued that a more useful model for understanding nicotine use is a functional model [2, 3]. A functional model argues that nicotine use is not an irrational behaviour but is adopted because it enables the user to control their bodily state, especially their psychological state. Nicotine use is maintained because it satisfies these needs and it can be seen as a resource that the person uses.
This chapter will consider the potential of nicotine as a therapeutic agent, other than for use in smoking cessation. In addition, it will provide an explanation for smoking behaviour.

2   **Pharmacokinetics**

A number of reviews of the pharmacokinetics of nicotine have appeared recently [4, 5] and so this aspect of nicotine use will only be outlined.

2.1   Absorption

Nicotine is absorbed from the skin, mouth, stomach, nose and lungs and the amount absorbed from most sites depends on the amount of free base at that site. When the pH is 5.35, about 0.4 % of the nicotine is present as the free base, while at pH 8.5, 85 % of the nicotine is present as the free base.

2.1.1   Transdermal administration

It has been known for many years that nicotine could be absorbed through the skin [6]. A patch containing 9 mg of nicotine base produced saliva levels of 50 ng/ml and the levels were over 20 ng/ml for over 4 h [7]. Russell [4] tested doses of 10 mg, 20 mg and 30 mg and found slowly accumulating plasma levels which reached over 50 ng/ml after 6 h with the 30 mg patch. Marked skin irritation persisted for several days after removal of the patch.

2.1.2 Oral administration

Nicotine base is readily absorbed by the buccal membrane but the amount of free nicotine base depends on pH and so the amount of nicotine absorbed orally depends on pH. Beckett and Triggs [8] found that, from 1.2 mg of nicotine base, about 6 % was taken up at pH 5.5 and 25 % at pH 8.5. Animal studies with nicotine solutions have shown that a carotid nicotine level of 100 ng/ml can be achieved at pH 6 in the mouth but this increases to 500 ng/ml at pH 8 [9]. Unpublished research by Dr M.A.H. Russell and Dr K. Wesnes, showed that buccal absorption from alkaline tablets containing 1.5 mg nicotine gave venous levels of 6.0 ng/ml at pH 6 and 10.5 ng/ml at pH 9.

Buccal absorption is part of the process of nicotine intake from nicotine gum [5]. Of the nominal doses of 2 mg and 4 mg in the nicotine polacrilax gum, 1 mg remains in the gum and is not available for uptake. From the remaining nicotine in each piece about 0.9 mg is absorbed buccally. The remainder is swallowed for gastric absorption from a 2 mg, i. e. 0.2 mg, while about 1 mg is expectorated from the 4 mg leaving 1.0 mg which is swallowed. Consequently, the plasma levels of around 8 ng/ml from 2-mg nicotine gum [4] must represent buccal absorption from 30 min of chewing.

2.1.3 Gastric absorption

Nicotine uptake from the stomach is small in normal circumstances. Nicotine will dissolve in the saliva where pH ranges of between 5.6 and 7.6 would give about 6 % to 20 % free nicotine base, but the stomach is acidic. Travell [10] showed that nicotine was rapidly absorbed from a cat's stomach when the solution pH was between 7.8 and 8.6 but not when the solution was acidic, with a pH between 1.2 and 4.2.

It was mentioned earlier that about 1 mg of nicotine is swallowed from a piece of 4-mg nicotine gum [5]. Since the buccal absorption from both 2 mg and 4 mg gum is the same, the 18 ng/ml plasma nicotine levels from 4 mg gum [4] can be subdivided into 8 ng/ml from buccal absorption and 10 ng/ml from gastric absorption, if the estimates by Benowitz [5] of disposition are correct.

2.1.4 Nasal administration

Nicotine has been administered nasally, historically in the form of snuff, and more recently by nasal spray [4]. Tobacco leaves have a nicotine content between 0.2 % and 5 % of the dry weight. Russell has described the time course of plasma nicotine that resulted from nasal snuff taking by an experienced user [11]. Uptake of nicotine from the nasal mucous membrane was extremely rapid and nicotine concentrations of over 20 ng/ml were found in blood samples from a forearm vein. Nasal nicotine solution has been developed on the basis of this pharmacokinetic information. A 2-mg dose of nicotine produces peak plasma nicotine levels of 14 ng/ml after 7.5 min, i.e. much more rapidly than the previously described methods. It matches closely the absorption rate for inhalation.

2.1.5 Administration by inhalation

The most common method of taking nicotine by inhalation is in cigarette smoke. Each puff of cigarette smoke gives a mouth level of 150–250 μg of nicotine [12]. The pH of the mainstream smoke ranges between 5.5 and 6.2 for flue-cured cigarettes and so very little nicotine is absorbed orally from cigarette smoke (pH 5.5 to 6.2), perhaps as little as 30 % [12]. Some nicotine is absorbed nasally during smoke manipulation but the absorbed amount is small in comparison with the uptake that results from inhalation.

During inhalation, the smoke aerosol passes down the bronchi and into the alveoli. Particles of cigarette smoke are an ideal size (0.01 – 2 μg) for penetration into the alveoli and absorption occurs through the thin alveolar membrane into the pulmonary capillaries. It is estimated that more than 90–95 % of inhaled nicotine is absorbed [9, 13]. Nicotine diffuses so rapidly across the alveolar membrane and the velocity of blood flow through the capillaries is so slow that equilibrium is probably reached between alveolar nicotine and capillary nicotine, ensuring maximum uptake. On the basis of the estimate of between 150 and 250 μg mouth level of nicotine from each puff, over 100 μg would be taken up during each inhalation from a medium delivery cigarette, giving a total lung dose of over 1.0 mg nicotine per cigarette.

After absorption into the pulmonary capillaries, nicotine does not bind to plasma protein and so all the nicotine is available for biological activity. The nicotine-loaded blood leaves the lungs, via the pulmonary veins and passes through the left atrium of the heart into the left ventricles. From there the nicotine is pumped out into the aorta from which the large arteries branch off.

The time course of nicotine in human plasma after smoking has been studied most extensively. Smokers puffed ten times on a cigarette and plasma samples were taken every 5 min from an indwelling needle in a forearm vein. There was a rapid increase in plasma nicotine with each puff with irregularities in the ascent profile from the puff by puff boluses of nicotine. Peak venous nicotine levels of 15.5 to 38.4 ng/ml were reached at the end of the cigarette and then corresponded to about one fifth or one sixth of the carotid artery levels.

Nicotine decay is not smooth either, and Russell [14] argues that the irregularities represent nicotine redistribution and recycling.

2.2 Distribution

In the first part the distribution in the whole body will be considered and then the brain distribution will be outlined in the second part.

2.2.1 Whole body

As the partition coefficient is less than one, very little storage of nicotine in fatty tissues results. Instead, whole body autoradiograms [15] show a pattern of distribution after intravenous injection that corresponds to the blood supply, with the highest levels of radioactivity in the liver (4 % of the injected dose), about the same levels in the kidneys as in the brain (2 % of the injected dose) and less in the stomach.

2.2.2 Brain

From the point of view of the therapeutic use of nicotine, a significant distribution artery is the carotid, which leads directly to the brain, so that some absorbed nicotine passes directly, unmetabolised, from lung to brain within 10 s. About a fifth of the blood from the heart ascends in the carotid artery so that a fifth of the absorbed nicotine passes to the brain [16].

Nicotine is soluble in lipid (partition coefficient of 0.4) which suggests that it will freely pass through the blood-brain barrier. Rat studies have compared the percentage of nicotine remaining in the brain 15 s after a rapid intracarotid injection with tritiated water as a standard. Ninety percent of the tritiated water is taken up on the first pass through the brain. With tritiated water as 100 %, the uptake of nicotine is 131 % [17]. Thus, virtually all the nicotine that is delivered to the brain leaves the blood, since the volume of brain tissue to which it can distribute is so much larger than the capillary plasma volume, i.e. the amount of nicotine entering the brain is proportional to the cardiac output to the brain, i.e. about 250 μg nicotine per cigarette. As a result of this efficient uptake of nicotine, effective doses can be obtained with relatively low blood levels which minimises the risk of toxicity to other organs in the body.

Whole body autoradiograms of mice, given intravenous doses of 14 μg C-nicotine, reveal a high accumulation of nicotine in the grey matter (unmyelinated tissue) with much smaller quantities in the white matter. This distribution occurs because nicotine's partition coefficient of 0.4 allows good penetration of the blood-brain barrier, but does not result in depot fat storage. Microautoradiograms after 14 μg C-nicotine and 3 μg H-nicotine show radioactivity in cortical cells, high levels in molecular and pyramidal cells of the hippocampus, the molecular layer of the cerebellum, the nuclei of the hypothalamus and the brain stem [15]. This pattern of nicotine distribution throughout the brain allows wide scope for neurochemical interaction.

The time course of nicotine distribution in mouse brain shows that the maximum concentration is reached within one minute of an intravenous injection. The level then decreases rapidly to about 50 % in 5 min and 1 % after 1 h [18]. Similarly, Schmiterlow found a rapid nicotine decrease from 3.93 μg (per gram of brain tissue) at 5 min to 0.71 μg at 20 min, and 0.10 μg at 1 h [15]. The brain does not metabolise nicotine but the drug washes out quickly from the brain and so gives a short duration of action.

2.3   Metabolism
2.3.1   Clearance

The pathways of metabolism have been discussed at length in numerous publications, recently by Benowitz [5]. First, it is clear from the time course that nicotine is metabolised very efficiently by the liver, this limits nicotine's duration of action in the body. Following smoking or an intravenous infusion, nicotine distributes rapidly into tissues, after which it is eliminated, with a terminal half-life of about 2 h. Total clearance averages about 1300 ml/min and is highly variable between individuals. The relatively long terminal half-life in the presence of rapid clearance can be explained by the large volume of distribution, which averaged 183 l [5].

Nicotine is metabolised in the liver and kidneys and nicotine levels in mouse liver falls from 8.12 μg (per gram of tissue) at 5 min to 0.76 μg at 20 min, i.e. about 90 % is metabolised in 15 min. At 20 min the radioactivity in the liver and kidneys is mainly due to the metabolites of nicotine [15].

2.3.2 Metabolites

Complete characterization of the metabolism and disposition of a drug requires the determination of the rate of metabolism, elucidation of the structures of the various metabolites, quantitative assessment of the various metabolic pathways, and determination of the the percentage of the dose that is unchanged and the percentage of each metabolite. At the present time we can account for only about 25 % of ingested nicotine as urinary metabolites.

About 4 % of the nicotine intake is converted to nicotine N-oxide, which is largely, if not entirely, excreted in the urine without further metabolism. However, the major metabolic route is probably hydroxylation (insertion of a carbonyl group in the pyrolidine ring) to form cotinine. About 70 % of nicotine is converted to cotinine. In contrast to nicotine N-oxide, cotinine is extensively metabolised, only about 10–15 % being excreted unchanged in urine. Several metabolites of cotinine have been reported, but no quantitative data of these metabolites has been published. Cotinine N-oxide is a minor metabolite and its urinary excretion accounts for only about 3 % of nicotine ingested by smokers. However, recent experiments [19] indicate that trans-3'-hydroxycotinine is a major metabolite, its urinary concentrations exceeding cotinine concentrations by 2-3-fold. Consequently, a substantial portion of nicotine intake may be accounted for as urinary excretion of 3'-hydroxycotinine.

In contrast to nicotine, cotinine has a half-life of around 20 h, and blood concentrations are relatively stable throughout the smoking day. Nicotine concentrations are much more dependent upon sampling time with respect to smoking. The correlation of cotinine blood concentrations and daily intake of nicotine was only 0.53. This is probably due to individual differences in the metabolism of nicotine to cotinine and rate of cotinine metabolism. Certainly, the rate of nicotine elimination is highly variable, with clearance rates varying as much as 4-fold among cigarette smokers [5].

An important aspect of the conversion of nicotine is the fact that the metabolites appear to be inactive or virtually inactive. Thus, no compounds of demonstrated pharmacological action are produced and the pharmacological effects are determined almost completely by the action of nicotine alone.

2.4 Summary

Studies of nicotine pharmacokinetics have revealed it to be a substance which is absorbed most efficiently from the lungs. It readily enters and is quickly eliminated from the brain, and is rapidly metabolised to inactive metabolites. This pharmacokinetic pattern allows a brief duration of action and the possibility of central nervous action with minimum side effects from actions on the rest of the body. In the next major section we will consider the actions of nicotine on the body, especially the nervous system. The crucial questions are whether these changes can be extrapolated meaningfully to people to suggest therapeutic uses for nicotine.

3 Nicotine and the control of body weight

Smokers claim that they smoke to prevent weight gain [20]. First, we will consider the evidence for this claim.

3.1 The evidence

It seems clear that smokers do weigh less than non-smokers. Eight studies compared body weights of smokers and non-smokers. Two studies [21, 22] have reported no relationship between level of smoking and weight, but six have reported a curvilinear relationship [23–28]. Thus, non-smokers had the greatest body weight, moderate smokers had the lowest body weight and heavy smokers had body weight approaching those of non-smokers. When cigarette smokers quit smoking, they gain weight [20, 29, 30]. Smokers who smoke heavily tend to gain the most weight following smoking cessation.

Research on nicotine gum supports the hypothesis that nicotine is responsible. In a nicotine substitution study, Emont and Cummings [31] reported a significant negative relationship between the number of pieces of nicotine gum chewed per day and weight gain for those who had been heavy smokers. Abstainers were much less likely to gain weight when they consistently used nicotine gum. Abstinent subjects who regularly used the gum gained less than 1 kg at a follow-up 6 months later. In contrast, the infrequent gum users gained around 3 kg.

Animal studies have demonstrated that these effects are not some sort of "psychological" effect. Uniformly, nicotine administration results in decreased weight gains in young animals or weight loss in older ones. When dosing is discontinued, body weight gains occur which are greater than those of controls [32–40].

3.2 Mechanisms

Possible mechanisms of action could be increased caloric intake, increased physical activity or metabolic changes. Stamford and colleagues [41] evaluated changes in dietary intake, physical activity, and resting metabolic rate in women who had abstained from smoking for 48 days.

3.2.1 Physical activity

A review [42], which examined the relationship between smoking and physical activity, concluded that smoking and physical activity are negatively associated, but the association was weak. From the Stamford study [41], it is clear that changes in physical activity do not play a role in either differences in body weight between smokers and non-smokers or the weight gain associated with smoking cessation.

3.2.2 Caloric intake

After smoking cessation, the mean daily caloric consumption of abstainers increased by 227 kcal, which accounted for 69 % of the variance in post-cessation weight gain [41]. Smokers not allowed to smoke reported the greatest preference for sweet foods [33]. He also found that smokers, who were allowed to smoke during testing, consumed less sweet food than smokers who were not allowed to smoke, or non-smokers. Animal research provided striking confirmation of the importance of nicotine. The consumption of sweet foods by male rats is spectacularly decreased by nicotine [33, 43]. However, nicotine did reduce bland food intake in female rats. Nicotine had a greater effect on body weight of female rats than of male rats [38, 37].

3.2.3 Metabolic activity

There are a number of reports in the literature on animals that have documented nicotine-induced reductions in body weight without a concomitant reduction in food intake [36, 40], which would argue for a metabolic change. Metabolism increases as the result of acute nicotine administration and immediate effects of smoking [44, 45]. Possible explanations for the increased metabolism are changes in the secretion of the adrenal medulla and the adrenal cortex. Both increases in catecholamines and glucocorticoids have been reported after smoking.

3.2.3.1 Adrenal catecholamines

Smoking a single cigarette elevates both plasma noradrenaline and adrenaline within 5 min of starting the cigarette and they peak at, or soon after, the maximum plasma nicotine level is reached. The peak plasma concentration for noradrenaline is 330–350 pg/ml and for adrenaline it is 100–140 pg/ml [46]. The catecholamines have metabolic effects; they accelerate delivery of glucose from the liver, reduce glucose clearance from the circulation and, as a consequence, the plasma levels of glucose rise. In addition, the catecholamines stimulate lipolysis, increasing the delivery of free fatty acids to the liver. In the liver they may also act on hepatic metabolic processes to stimulate ketogenesis, which fuels those cells which do not depend on carbohydrates, and supplies glycerol for use in glucose synthesis. The increased free fatty acids will also reduce glucose uptake and use by cells with flexible metabolic requirements.
Cryer [46] has used graded infusions of noradrenaline and adrenaline to discover the threshold levels for haemodynamic and metabolic changes. The adrenaline thresholds were 50–100 pg/ml for increased heart rate, 75–125 pg/ml for systolic pressor effects and lipolysis and

150–200 pg/ml for detogenesis and glycolysis changes. Smoking produces suprathreshold elevations of adrenaline but not noradrenaline.

3.2.3.2 Adrenal glucocorticoids

The adrenal cortex secretes glucocorticoids, of which cortisol and corticosterone are the most important. The glucocorticoids, like the catecholamines, have marked metabolic effects. They are essential for the mobilisation of tissue proteins and for transferring the derived amino acids to the liver. The liver glycogen stores are limited and amino acids provide the major source of energy supply. In the liver, the glucocorticoids promote synthesis of enzymes that are used in gluconeogensis. As well as conversion of amino acids to glucose, they also speed up extensive mobilisation of depot lipid reserves, inhibit lipid synthesis and reduce glucose catabolism. Together these effects elevate plasma concentrations of glucose.

However, no changes in glucocorticoids occur after the nicotine from one cigarette. Kershbaum had to make habitual smokers smoke four unspecified cigarettes in 30 min to obtain significantly increased plasma glucocorticoids [47]. Kershbaum mentions that there was some release (range 22 to 46 %) when "Two cigarettes were smoked by each of several subjects" but no further details were given. In a second study [48], habitual smokers were given eight, 2.5-mg nicotine cigarettes to smoke in a 2-h session. It was only after the third cigarette in half an hour that cortisol had increased. The best conclusion that can be drawn from these studies is that high nicotine doses from exaggerated smoking rates can elevate plasma glucocorticoids. An alternative interpretation is that the rapid smoking was aversive and stressed the subjects and released glucocorticoids.

3.2.4 Summary

Nicotine does produce weight loss. Most of this does result from reduced caloric intake, while there is none from increased physical activity. The remainder of the effect is a result of enhanced metabolism. Smoking doses of nicotine may produce some release of glucocorticoids from the adrenal cortex, then these hormones, combined with the adrenaline release, increase metabolic rate and reduce weight.
An additional consequence will be that the hormones will make avail-

able energy sources for use by the brain and the rest of the body. In this way, nicotine would alleviate mental and muscle fatigue. Studies of carbohydrate and lipid use in physical work of different intensities have found that during exhausting work there was a depletion of carbohydrates, but subjective symptoms of fatigue disappeared within 15 min of ingesting glucose which increased the blood-sugar levels again [49]. The central nervous system has low reserves of glucose and depends on blood carbohydrate for its supply of energy. In man, about 60 % of the liver glucose output supplies brain metabolism, and so exhaustion will be reduced by nicotine.

4   Analgesia

Several studies have used pain thresholds as dependent variables in assessing the effects of smoking and nicotine on anxiety. However, the anxiolytic interpretation of increased pain thresholds has been questioned [50]. Instead, we will consider the evidence that nicotine may produce analgesia per se.

4.1   The evidence

Two studies tested the effects of smoking cigarettes of different nicotine yield on electric shock endurance and both reported elevated endurance thresholds in subjects who smoked relative to those who did not. There was greater endurance after a high nicotine cigarette compared with after a low nicotine cigarette, suggesting that nicotine was important for the effect [51, 52]. Other studies [53] have used the length of time that individuals are willing to endure pain associated with immersion of a hand or foot in ice water (the cold-pressor test). These studies also showed that smoking and snuff increase endurance in this test. However, it must be pointed out that several studies have failed to find increased shock thresholds associated with smoking [54, 55]. From these studies, it is unclear whether nicotine reduced pain thresholds or whether smoking deprivation increased them. However, Pomerleau et al. [53] also found that smoking a nicotine-containing cigarette nearly doubled the pain awareness thresholds in the cold pressor test. Pain tolerance thresholds were also elevated, although the increase was not as pronounced. In three of the five subjects, the effects of smoking on pain-awareness and tolerance thresholds were paralleled by changes in perceived pain, as assessed by the McGill Pain Questionnaire. Pomerleau et al. [53] argue that pain awareness thresholds may be a more accurate measure of the antinociceptive effects of nicotine than pain tolerance. The important issue was that the subjects were only deprived of cigarettes for half an hour, i.e. minimal deprivation.

4.2 A mechanism of analgesia

The mechanism for the analgesic effect is not known at the present time. We do know that nicotine induces release of the endogenous opioid beta-endorphin [56]. Peripheral analgesia is produced by beta-endorphin and the release of this peptide may account for nicotine's capacity to reduce pain.

5 **Mood control**

Smoking motive surveys have given evidence that smoking is perceived as providing negative affect reduction [57–66] and positive affect enhancement [57–60, 63]. It is found that women scored higher on negative-affect reduction [59, 61, 62]. These opinions suggest that smokers feel that smoking improves their mood and that this is an important reason for smoking [67].
Studies suggest that stress is antecedent to smoking [68, 69]. One prospective study [70] showed that measures of anger, restless sleep, and Type A personality were significantly related to onset of smoking. Two other prospective studies have found associations between anxious, aggressive, and generally neurotic personality traits in childhood and the tendency toward smoking later in life [71, 72].

5.1 Evidence from laboratory studies

Smoking a cigarette decreases both the subjective stress response and smooths out the body's stress response when subjects watched a mutilation film [73]. Smoking reduced the subjective annoyance to loud noise as well as the physiological stress response to noise [74]. In a study of smoking during gambling, smoking helped smokers to be less angry when they knew that they were being cheated by another person [75].

Pomerleau, Turk and Fertig [53] found that anxiety, as measured by the Profile of Mood States Questionnaire, could be generated using anagrams. The anxiety was markedly decreased in all subjects more after smoking nicotine cigarettes than after non-nicotine cigarettes. The subjects were minimally deprived of cigarettes for half an hour prior to the study. This finding supports the hypothesis that nicotine from smoking can reduce anxiety without abstinence from smoking.

During smoking smokers feel calmer, more relaxed, more contented, friendlier and happier, after a cigarette and are in a much better mood in general when smoking [3, 76]. These improvements in mood follow the same pattern as the increases in plasma nicotine during smoking. Non-nicotine cigarettes produced no improvements. When a second cigarette was given 30 min later (i.e. minimal deprivation), there were similar increases [3], as Pomerleau et al. [53] found.

Evidence for the importance of nicotine, rather than the smoking ritual and oral gratification, was obtained by Murphree [77]. Nicotine was infused until the heart rate increased by 15 beats per min, i.e. equivalent to a smoking dose. Murphree asked his subjects to fill in a mood adjective checklist before and after each infusion. He found mood improved and that there was no evidence that there were any differences in the responses of non-smokers and smokers. The mood shifts are similar to those obtained with smoking and give some support to the idea that nicotine is the agent responsible for the mood changes.

5.2 Evidence from nicotine substitution

Other major evidence, that nicotine is the agent in cigarette smoke which is responsible for mood changes, comes from nicotine replacement in smoking cessation studies. At least four studies have examined whether 2-mg nicotine gum can replace smoking in terms of mood changes. All aspects of mood have been changed significantly in abstaining smokers by nicotine gum, in comparison with placebo gum. Abstinent smokers felt less anxious and less tense when given 2 mg of nicotine gum than when they were given placebo gum [78], but others have found no difference between nicotine gum and placebo [79, 80]. On the anger scales, nicotine substitution resulted in abstinent smokers feeling less annoyed than when they were given placebo [80], also less hostile [80], and less irritable [79–81]. On other dimensions of mood, a 2-mg dose of nicotine gum reduced depression [79, 81], and increased sociability [81]. In general, nicotine gum was able to replace the effects of smoking on mood fluctuations [80].

5.3 Mechanisms of mood control

Mood is believed to be controlled by the central release of catecholamines and indoleamines.

5.3.1 Brain release of catecholamines

Studies of the effect of nicotine on the brain catecholamines, dopamine and noradrenaline have been carried out by Lichtensteiger and his co-workers [82], by Fuxe and his colleagues [83] and Westfall [84]. These studies have given the same general picture; there is increased activity in catecholamine neurones, although the changes are rather small [85].

5.3.2 Brain release of indoleamines

Changes in the concentration and turnover of serotonin have been found after doses of nicotine. A dose of 1 mg/kg of nicotine, intraperitoneally injected into mice, markedly increased the levels of serotonin in the mesencephalon and diencephalon within 15 min but not in the cortex [86]. In the same study, serotonin's major metabolite, 5-hydroxyindoleacetic acid, was also increased but there was a decreased serotonin turnover rate of 20 % and increased serotonin turnover time (+ 300 %). These data are not simple to interpret but clearly with this moderately large dose there are increased serotonin levels in some brain regions, probably due to increased synthesis and decreased release but, in spite of the decreased release, the large intraneuronal accumulation results in some leakage which elevates the level of 5-hydroxyindoleacetic acid.

5.4 Summary

We know that nicotine does act on the brain systems which are involved in mood control. The biochemical effects may seem weak, but smokers claim that they are important for them. "Smoking to calm oneself" is the major motive for smoking.

6   Cognitive enhancement

In this section, the effect of nicotine on information processing will be considered. The two aspects are attention and memory.
One aspect of information processing involves the selection of the relevant stimuli from all the possible inputs and maintenance of that selection. This cognitive mechanism is often called attention or concentration. We will consider this next.
Surveys have indicated that the majority of smokers (86 % in a smoking clinic population, 59 % among hospital workers and 86 % of students) believe that smoking helps them to concentrate [65, 66] and the question arises whether nicotine is responsible for these effects of smoking.

6.1   The evidence for improved attention

There are a number of attention tests. Focused attention tasks require subjects to attend to one source of information to the exclusion of others. Divided attention tasks require subjects to divide their monitoring between two or more sources of information.

6.1.1   Focussed attention

Vigilance tasks involve focused attention. In these tasks attention has to be directed to an input for long periods of time and the subject is required to detect and respond to brief, infrequent changes in the input. During a typical vigilance session, the detection rate decreases, a change called the vigilance decrement. In a study of smoking and visual vigilance, smoking cigarettes at 20-min intervals helped subjects maintain their detection of targets in the 80-min vigilance task [87]. In contrast, detection decreased for non-smokers and smokers who were not allowed to smoke. There was no significant difference between deprived smokers and non-smokers in their performance, i.e. no evidence of a withdrawal effect. Similar results were obtained in auditory vigilance studies [87–89]. In the Tong et al. and Mangan studies the smokers were only deprived for 3 h, i.e. minimally deprived.
Nicotine is the essential ingredient of cigarette smoke for producing the effect, because nicotine tablets prevented the vigilance decrement [90]. The nicotine tablets produced the same effects in non-smokers,

light smokers and heavy smokers. A number of attention tests involve picking out various sequences of numbers or letters from an array. Williams [91] tested the effects of smoking on smokers deprived for 10 h with a test of this kind which involved crossing out each letter E on sheets of randomly ordered letters arranged in lines of 30 letters. A highly significant improvement in performance of the letter cancellation was produced by smoking, in comparison with sham smoking. This finding was replicated in a study with 22 women and 26 men [92].

A computer version of this test is the Bakan task [93] in which a series of digits are presented at a rapid rate and subjects are required to detect certain specified three-digit sequences. Measures of both the speed and the accuracy of detection are made. Performance on this rapid visual information processing task by smokers deprived for 10 h after smoking was improved in both speed and accuracy above baseline levels, whereas either not smoking, or smoking herbal, nicotine-free cigarettes, resulted in a decline in speed and accuracy below baseline levels [94, 95]. Analyses of performance while smoking the cigarette show puff-by-puff improvement in both speed and accuracy as each bolus of nicotine enters the brain. The maximum level of improvement was of the order of 15 %. The improvement in both speed and accuracy is important because it shows that there is no speed and accuracy trade-off, i.e. an overall improvement in processing efficiency.

Evidence that nicotine was responsible comes from a study in which doses of 0.5 mg, 1.0 mg and 1.5 mg of nicotine were given to non-smokers for buccal absorption [96]. They then performed the rapid visual information processing task. The highest dose of nicotine (1.5 mg) produced a performance improvement in non-smokers which closely resembled that produced by smoking in smokers. The 1.0-mg dose had a lesser effect and 0.5 mg was without effect. This finding provides strong evidence that nicotine plays the major role in the improvements in focused attention tasks that are produced by smoking. The improvement in the performance of non-smokers argues against the improvement being only a reversal of nicotine withdrawal.

Another type of focused attention task has looked at performance under conditions of distraction. A task which has been used in nicotine research to examine distraction effects on attention is the Stroop Test. The Stroop Test uses three sets of display, a list of colour words printed in black, a set of colour patches, and a list of colour words with the words printed in incongruent colours (e.g. the word yellow which was

printed in blue). Colour naming the incongruently printed colour words takes much longer than naming the colour patches and the time difference between the two conditions is the Stroop effect. This score indicates the subject's ability to focus attention on a relevant stimulus dimension of print colour and ignore the irrelevant semantic one.
The effects of doses of nicotine (0.1 mg and 2.0 mg) administered buccally were studied on the Stroop performance of both smokers deprived for 10 h and non-smokers [87]. Both doses of nicotine reduced the size of the Stroop effect in this study by 16 %. No differences were found between the baselines of the deprived smokers and non-smokers or in the amount of improvement produced by nicotine at the time of the peak heart rate. Thus, there was no evidence of tolerance to the effects of nicotine in smokers for this type of performance.

6.1.2   Divided attention

A study of divided attention [97], used a test which was based on the rapid visual information task [95]. Subjects were presented with digits at a rate of 50 per min in both the visual and auditory modality, a different sequence for each modality. The detection of sequences in both modalities improved significantly after smokers deprived for 10 h smoked a cigarette, in comparison with not smoking; an improvement of 7 %. Smoking a cigarette also prevented the increase in reaction times that occurred in the non-smoking condition.
In a test which was designed to simulate driving [98], combined central guiding with peripheral visual monitoring. The task lasted for 2.5 h and the measure of performance reported was the percentage of the peripheral visual signals which were missed during the session. Monitoring performance was maintained by smoking, in contrast to the large increase in the percentage of missed signals when not smoking. The performance was similar for non-smokers and smokers who were deprived for 20 h.

6.1.3   Summary of attention studies

The clear summary of this data is that nicotine is improving a general attentional processing capacity and is not modifying one specific type. In the next section, a second type of cognitive function will be considered.

6.2 Information storage

Information storage consists of input of the information, registration of the information in immediate, shorter-term memory and consolidation of the information in longer-term memory. However, information storage can only be demonstrated by the effective retrieval of the information from storage. Thus, it is important to distinguish a drug's effect on storage from an effect on retrieval.

6.2.1 Immediate memory

The first study, that showed that smoking modified immediate memory, came from Andersson and Hockey [99]. In the study, heart rate was increased in the smokers who smoked and so it can be inferred that nicotine was absorbed and was the agent that was responsible for the changes. This study suggests that smoking may enhance storage of information but only of information that is thought to be relevant by the subjects. One interpretation of the finding is an attentional one; the subjects who smoke filter out irrelevant information.

Another study on immediate memory [100] used smokers who were deprived of cigarettes for 3 h. The immediate recall scores showed that the nicotine cigarette group had significantly poorer recall than the non-nicotine cigarette group. Unfortunately, the investigators used a rapid smoking rate with forced inhalation, which could have resulted in too much nicotine for some subjects.

In contrast to the previous data, Peeke and Peeke [101], found an effect of smoking on immediate memory in smokers deprived for 2 h. The high nicotine cigarette resulted in improved immediate recall. The low nicotine cigarette was less effective.

In agreement with the last study, when testing was given once just after the input, it has been found that memory is improved. After smoking a cigarette at their own pace, the subjects were shown a list of words and asked to write down as many as they could [102]. There was better immediate recall after smoking in comparison with not smoking.

6.2.2 Delayed recall

The effects of smoking a cigarette on learning were tested [103]. Cigarettes improved retention in the paired-associate learning, with task difficulty apparently having little relevance. In a second study, Mangan and Golding [104] looked at the effects on memory of smokers deprived for 1 h and those smoking a single cigarette after 1 h of deprivation but immediately after learning a set of words. After one month subjects who smoked a nicotine cigarette were better than non-smokers in their recall of words.

In a careful series of studies, Peeke and Peeke [101] tested the effects of smoking one cigarette on memory in four experiments, in which the smokers were only deprived for 2 h, i.e. minimally deprived. In one study, pre-learning smoking resulted in improved recall but there was no effect of smoking after learning, as Mangan and Golding [104] had found. In another study, a cigarette did produce improved recall on both immediate and delayed recall tests, and the low-nicotine (0.4 mg) cigarette was less effective.

6.2.3   Information retrieval

A state-dependent design enables researchers to distinguish a retrieval change from a storage effect. In this type of study, one group of subjects learns after a dose of compound while a second group learns after a placebo or nothing. For the recall test, both groups are divided so that half of each group are tested with the agent of learning and half are switched to the other condition.

In a recognition study [102], smokers who were deprived of cigarettes for over 10 h were given a 1.3-mg nicotine cigarette, or nothing, immediately before serial presentation of a set of Chinese ideograms. They were divided into four groups – a quarter who did not smoke prior to learning or recall; a quarter who did not smoke prior to learning but had a cigarette prior to recall; those who had a cigarette prior to learning and recall; and those who had a cigarette prior to learning but none prior to recall. Subjects who smoked prior to learning had significantly better recognition scores than the subjects who did not smoke in the first part of the experiment. There was no effect of smoking on recall performance itself.

Another study used 1.5-mg nicotine tablets in the state-dependent design in which smokers were deprived for over 10 h [102]. After the tablet, the subjects listened to 48 words and then did successive subtractions for 1 min to prevent rehearsal. As stated earlier, immediate recall was improved. One hour later, the subjects were given either nicotine or placebo tablets, depending on their group. They were asked to recall as many of the words as they could in another 10-min free recall test. Long-term recall was significantly better when subjects had taken nicotine prior to learning but not when taken prior to retrieval. A significant interaction term gave evidence for a state-dependent effect of nicotine and showed that nicotine was facilitating the input of information to storage but had no direct effect on retrieval.

6.3   Mechanism of cognitive enhancement

In this section, we will examine the possible mechanisms for nicotine enhancing information processing. This information can be obtained about the mechanism of action of drugs from the sites where the drug binds, although drug binding does not necessarily indicate biological interaction.

6.3.1   Localization of action

Nicotine is found in the reticular formation, lateral and medial geniculate, caudate nucleus, ventrobasal complex of the thalamus, hippocampus, cerebellum, inferior colliculus and the Betz cells of the deep pyramindal layer of the cerebral cortex. Cortical cells and caudate nucleus cells clearly have muscarinic receptors which were relatively insensitive to nicotine while acetylcholine receptors in the geniculate nuclei, ventrobasal thalamus, hippocampus and reticular formation nuclei were sensitive to both nicotinic and muscarinic drugs. Cholinergic inhibitory neurones, with mixed nicotinic and muscarinic receptors, have been found at the cortex in layers II, III and IV of the primary sensorimotor auditory and visual areas [105]. Thus, there is great scope for modification of the sensory information coming in from the environment.

6.3.2   Neurochemical action of nicotine at the cortex

"Smoking" doses of nicotine (e.g. 20 μg/kg in the cat) produce excitation of cortical cells [106, 107] and release of acetylcholine at the cortex [107]. In the study of Kawamura and Domino [106] blood pressure was kept constant with drugs, so that the effect was not due to vascular changes. It has been found that nicotine depletes whole brain acetylcholine in the rat [86]. There is no evidence that nicotine modifies acetylcholine synthesis [108], but there is strong evidence for increased free acetylcholine at the cortex after a "smoking" dose [107] which is consistent with increased release from the nucleus basalis of Meynert. Essman [86] found evidence of a decrease of acetylcholine in synaptic vesicles and in bound acetylcholine at the neocortex which suggests that acetylcholine was being released from storage by nicotine. However, there was no increase in the free acetylcholine pool concentration which argues for increased release of the unbound transmitter and subsequent inactivation by acetylcholinesterase. The phenomenon of increased release at the cortex would be explained if nicotine enhanced presynaptic release mechanisms in cortical tissue but there is only weak *in vitro* evidence [109]. Thus, we are left with the hypothesis that it is increased activity in the cholinergic neurones to the cortex which produces the *in vivo* depletion.

Cortical acetylcholine release and cortical excitation can be produced by stimulation of the mesencephalic reticular formation and this phenomenon can be reduced in one hemisphere by unilateral destruction of this region ipsilaterally [110]. In a neuropharmacological analysis of the effects of "smoking" doses of nicotine after destruction of the midbrain [106, 111], 20 μg/kg of nicotine produced cortical desynchronization in cats with a caudal midbrain transection at the junction of the pons, after bilateral lesions in the tegmental region of the midbrain. However, nicotine in doses up to five times the 20 μg/kg "smoking" dose did not activate the cortex. Clearly, nicotine's action on the cortex depends on an intact tegmental region. The ventral tegmental region of the mesencephalic reticular formation activates a cholinergic pathway which projects to the cortex and there is good evidence that it terminates on the pyramidal Betz cells at the sensory cortex and produces electrocortical arousal [112].

6.3.3   Cortical actions of nicotine in people

In this section we will consider the action of nicotine on the human cortex. The first part will consider the effects of nicotine on cortical desynchronization and the second part will outline nicotine's action on event related potentials.

6.3.3.1 Cortical desynchronization

Many human studies have shown that smoking increases the amount of cortical desynchronization in the form of an upward shift in dominant alpha frequency, i.e. more 10 – 12 Hz [113], less total alpha activity and more beta activity [77, 114]. Thus, these human studies show that smoking produces cortical desynchronization just as nicotine does in animal studies. In a study correlating performance with electrocortical activity, Warburton and Wesnes [115] found that both cigarettes and nicotine tablets increased the dominant alpha frequency (11.5 – 13.5 Hz] and beta activity (13.5 – 20 Hz). It is significant that Murphree [77] was unable to find any differences in the cortical desynchronization of smokers, who smoked at least 10 cigarettes per day, and non-smokers produced by an intravenous nicotine dose of 1.86 mg, a "smoking" dose. In other words, the smokers responded in the same way as non-smokers and there was no evidence of tolerance to nicotine's effect on electrocortical activity. It must be emphasised that the shifts produced by nicotine are within normal limits, i.e. they are indistinguishable from those that occur when a person is concentrating hard. There is certainly no evidence of the EEG abnormalities that occur with stimulants, like amphetamine.

6.3.3.2 Event-related potentials

Event-related potentials, are the complex electrical changes recorded at the same time as a physical or mental event. The potentials' components provide a continuous record of events occurring in the brain during psychological processes. Of particular interest are the endogenous components whose characteristics are partially independent of the physical characteristics of stimuli. The major endogenous component which has been identified is the P_3, whose latency ranges from 275 – 600 ms. The P_3 varies according to the task requirements and experimental instructions [116].
An experiment [117] enabled an analysis to be made of the effects of smoking on the ERPs to correct target detections in a rapid information processing task. Good performance depended on subjects maintaining their concentration throughout the 20-min session. Smoking produced a large decrease in the latency of P_3, 20-30 ms decrease. A reduction in P_3 latency is interpreted as a decrease in the time taken to

evaluate a stimulus. Analysis of the attentional performance showed that smoking increased the targets detected and decreased the reaction time. This improvement of performance fits neatly with the P_3 changes which are indicative of more efficient neural processing of the stimulus. The occurrence of the P_3 wave depends on the completion of certain stimulus evaluation processes [116] and the reduced latency of the P_3, indicates quicker stimulus evaluation, i.e. more efficient processing. It is important that cholinergic antagonists like scopolamine have the opposite effect [118].

These data argue for nicotine affecting cortical mechanisms involved in information processing [112]. The enhancement results from increased activity in the pathways which project to cortical sensory neurones from the nucleus basalis of Meynert. Nicotine maintains the cortex in a desynchronized state by driving the ascending cholinergic pathways and so reducing fluctuations in cortical arousal and the concomitant variations in information processing that occur in the undrugged state.

6.4 Therapeutic implications of cognitive enhancement

Recent research has linked Alzheimer's Disease, in which memory loss is a cardinal symptom, with damage to the ascending cholinergic pathways from the nucleus basalis of Meynert to the cortex [119]. It is of interest to know if nicotine has any effect on patients who are in the early stages of the disease. It is conceivable that by driving the ascending pathway with nicotine, it would be possible to ameliorate some of their information processing deficits. In a study at the Institute of Psychiatry in London, the effects of subcutaneous doses of nicotine on information processing performance was examined using patients with senile dementia of the Alzheimer type [120]. The comparison group was a set of age-matched elderly subjects.

Doses of 0.4 mg, 0.6 mg and 0.8 mg of nicotine were given subcutaneously and the patient's performance in a simplified version of the rapid visual information processing task was looked at. The normal elderly performed very well on the simple task, while the patients with senile dementia of the Alzheimer type had poor performance. After nicotine, there was a dose-related improvement in performance in the detection of signals in the rapid visual information processing task and they approached the performance of the normal elderly. The equivalent data for the reaction time showed that nicotine also produced improvements in reaction times in the patients. A third task was critical flicker fusion. Nicotine produced a dose-related improvement in the frequency with which the patients with senile dementia of the Alzheimer type saw the lights as fused. Higher resolution of flashes is interpreted as improved cortical functioning. As a result of this improved cortical functioning, there is an improvement in attention in patients with senile dementia of the Alzheimer type.

In these patients, it was presumed that nicotine was driving the remaining ascending pathways to the cortex or, alternatively, that nicotine was acting on presynaptic nicotinic receptors on the cholinergic neurones at the cortex and producing release of acetylcholine. Recently, there has been biochemical evidence to suggest the existence of these presynaptic nicotinic receptors [109]. Whatever the mechanism, these data give hope that nicotine, or a compound like nicotine, which acts on these cholinergic pathways, may provide one method of ameliorating the deficits in senile dementia of the Alzheimer type.

7 Dangers of nicotine use

Comparisons with reference substances are an essential part of the assessment of any therapeutic agent so that the relative costs and benefits can be estimated and decisions made about the agent's usefulness. In the case of nicotine, which has stimulant and sedative properties, some reference agents are the psychostimulants, amphetamine and caffeine, and the sedatives, diazepam and alcohol.

The comparison takes the form of an analysis of both the pharmacokinetic and pharmacodynamic features of these substances which could constitute risk to the user.

7.1 Duration of action

An important aspect of pharmacokinetics, which is also related to personal control of a person's psychological state, is duration of action. Psychological needs vary and most people want coping techniques for specific occasions rather than for chronic states.

Nicotine has a half-life of only 120 min after smoking and 70% is cleared from the plasma on each pass through the liver [121] in comparison to the elimination half-life of diazepam [122]. In addition, diazepam has active metabolites that further prolong their action while cotinine, the major metabolite of nicotine, seems to be virtually inactive.

Clearly nicotine fits the pharmacokinetic specifications of a compound which a person can use to exert control over their psychological state on an hour-by-hour basis.

7.2 Toxicity

Almost all substances have toxic effects if taken in large doses for a sufficiently long period of time. Acute doses of nicotine give rise to many undesirable effects, such as sweating, tremor, nausea, vomiting, abdominal pain, diarrhoea, palpitation, fatigue and headache [123]. In experienced nicotine users, these effects are not reported which suggests that the person can learn to titrate the dose to avoid the adverse effects. In the long term, there is no evidence that people suffer intellectual impairment during use.

7.3 Desensitization and tolerance

An important issue is the extent to which a decreased response to nicotine develops with repeated use. It is often found that the effect of a drug gradually diminishes when it is given continuously or repeatedly; *desensitization* and tachyphylaxis are terms used to describe decreased responsiveness when it develops rapidly, often as quickly as a few minutes.

Human studies have shown desensitization of the heart rate response when nicotine is infused intravenously [121]. In the same study, they also measured skin temperature during infusion and found no evidence of desensitization of the skin temperature response.

Given the differences in desensitization for these two physiological systems, what are the changes in psychological systems with short-interval dosing with nicotine or with chronic dosing?

One piece of evidence against the occurrence of desensitization with nicotine comes from the studies with minimal deprivation, although zero deprivation is unsuitable because of the consequent chances of a nicotine "overdose". Performance improvements have been found in vigilance tests with deprivation intervals of 1 h (Revell, Wesnes and Warburton, unpublished), 2 h [89] and 3 h [88]. On the mood tests, improvements have been found with 0.5-h deprivation [3, 53]. In other words, desensitization does not occur for the mood and performance effects of nicotine, which are major motives for smoking.

The question of tolerance can be studied by comparing the effects of nicotine on smokers and non-smokers. In a vigilance experiment the effects of nicotine tablets on non-smokers, light smokers and heavy smokers were studied [90]. There was no difference in the effect of nicotine on the performance of light and heavy smokers or non-smokers. Exposure to nicotine had not affected the response. Similar results were obtained with another test, the Stroop Test [87]. In a study of Murphree [77], nicotine was infused into smokers and non-smokers. The effect of nicotine on alpha activity showed an alpha blockade and the alpha activity remained suppressed until the end of the 20-min experiment. The effect of nicotine was the same for deprived smokers and non-smokers. This absence of tolerance explains why smokers do not have to keep on escalating their nicotine dose to achieve the desired amounts of sedation or stimulation. This contrasts markedly with the dramatic tolerance which develops for heroin [124].

7.4 Withdrawal symptoms

The consequence of some types of tolerance is that there will be withdrawal or abstinence symptoms when drug use is stopped, especially when physiological adaptation has occurred. According to DSM IIIR [125], the essential feature of a withdrawal syndrome is the development of a substance-specific syndrome that follows the cessation of, or reduction in, intake of a psychoactive substance that the person previously used regularly. The syndrome that develops varies according to the psychoactive substance the person was using. The nicotine withdrawal syndrome includes craving for nicotine; irritability, frustration or anger; anxiety; difficulty concentrating; restlessness; decreased heart rate; and increased appetite or weight gain. However, DSM IIIR [125] pointed out: – "In any given case it is difficult to distinguish a withdrawal effect from the emergence of psychological traits that were suppressed, controlled or altered by the effects of nicotine or from a behavioural reaction (e.g. frustration) to the loss of a reinforcer." (Pp 150) These changes are more properly called abstinence symptoms. The functional model explains why individuals experience abstinence symptoms when they stop nicotine use and why the symptoms vary between individuals, depending on the function that it served for that individual. After heroin withdrawal, the syndrome includes lacrimation, rhinorrhoea, pupillary dilation, piloerection, sweating, diarrhoea, yawning, mild hypertension, tachycardia, fever and insomnia. The symptoms and signs of opioid withdrawal may be precipitated by the abrupt cessation of opioid administration after a one- or two-week period of continuous use or by administration of a narcotic antagonist (e.g. naloxone or nalorphine) after therapeutic doses of an opioid given four times a day for as short a period of time as three or four days. Depending on the observer and the environment, there may be complaints, pleas, demands and manipulations, all directed toward the goal of obtaining more opioids [124].

8   Conclusions

Nicotine, as used by the smoker, is a low risk substance in terms of acute and chronic toxicity. Its rapid uptake into the brain allows smokers to control their psychological state at will so that they can obtain either stimulation or sedation to help them to cope with situations. Nicotine's specific action on the brain produces neural changes which are within normal limits and so nicotine acts in a normal way to produce stimulation and sedation. Nicotine is absorbed very efficiently, enters the brain very quickly and a smoking dose is metabolised quickly which allows for brief duration of action. The rapid absorption and rapid metabolism make this substance suitable for hour-by-hour self-medication because of the personal control that can be exercised. In this respect nicotine is superior to other compounds for medication.
Nicotine taken in smoking doses seems to be relatively safe for healthy adults to use. On the evidence available at the moment it acts specifically on only a subset of the acetylcholine pathways in the body producing few side effects and none of these interfere with the normal functioning of body and mind. Even after a lifetime of use, the chronic toxicity of nicotine appears to be low.
Nicotine ist not like other so-called "addictive" substances with which the Surgeon-General's Report compared it [1]. Unlike heroin and cocaine, tolerance to the *psychological* effects of nicotine does not occur, and there is not a stereotyped pattern of withdrawal symptoms. There are however, abstinence symptoms which arise from the person missing the functions that nicotine served.

References

1 U.S. Dept. of Health and Human Services: Nicotine Addiction: A Report of the Surgeon-General. DHHS Publication, No. (CDC) 88-8406, U.S. Dept. of Health and Human Services, Office of the Assistant Secretary for Health. Office on Smoking and Health, Rockville, MD. (1988).
2 D.M. Warburton: in Tobacco Smoking and Nicotine. W.R. Martin, G.R. Van Loon, E.T. Iwamoto and L. Davis (Eds.) Plenum Publishing Corporation, New York. pp. 51–61 (1987).
3 D.M. Warburton, A. Revell and A.C. Walters: in The Pharmacology of Nicotine. M.J. Rand and K. Thurau (Eds.) IRL Press, Oxford. pp. 359–373 (1988).
4 M.A.H. Russell: in Nicotine Replacement. O. Pomerleau and C. Pomerleau (Eds.), Liss, New York. pp. 63–94 (1988).
5 N.L. Benowitz: in The Pharmacology of Nicotine. M.J. Rand and K. Thurau (Eds.) IRL Press, Oxford. pp. 3–18 (1988).
6 S.H. Gehlbach, W.A. Williams and Freeman: Arch. Envir. Hlth 45, 111–114 (1979).
7 J. Rose: in Pharmacologic Treatment of Tobacco Dependence. J.K. Ockene (Ed.), Institute for the Study of Smoking Behavior and Policy, Cambridge, MA. pp. 158–166 (1986).
8 A.H. Beckett and E.J. Triggs: J. Pharm. Pharmac. 19, 315 (1967).
9 A.K. Armitage and D.M. Turner: Nature 226, 1231–1232 (1970).
10 J. Travell: Ann. NY. Acad. Sci. 90, 13 (1960).
11 M.A.H. Russell, M.J. Jarvis and C. Feyerabend: Lancet 1, 474 (1980).
12 A.K. Armitage: in Smoking Behavior: Motives and Incentives. W.L. Dunn, Jr. Ed. Winston (Wiley & Sons), Washington, D.C. pp. 83–91 (1973).
13 A.J. Von Artho and K. Grob: Z. Präventivmed. 9, 14 (1964).
14 M.A.H. Russell: in Research Advances in Alcohol and Drug Problems. R.J. Gibbins, Y. Isreael, H. Kalant, R.E. Popham, W. Schmidt and R.G. Smith (Eds.) pp. 1–48, Wiley, New York (1976).
15 C.G. Schmiterlow, E. Hanson, G. Andersson, L. Appelgren and P.C. Hofman: Ann. NY. Acad. Sci. 142, 2 (1967).
16 W.H. Oldendorf: in Psychopharmacology in the Practice of Medicine. M.E. Jarvik (Ed.) Appleton-Century-Crofts, New York. pp. 167–175 (1977).
17 W.H. Oldendorf, S. Hyman, L. Braun and S.Z. Oldendorf: Science 176, 984 (1972).
18 T. Stalhandske, P. Slanina, H. Tjaelve, E. Hansson and C.G. Schmiterlow: Acta Pharmac. 27, 363 (1969).
19 P. Jacob, N.L. Benowitz and A.T. Shulgin: Pharmac. Biochem. Behav. 30, 249 (1988).
20 N.E. Grunberg: Advances in Behavioural Medicine, vol. 2. JAI Press Inc. Connecticut (1986a).
21 E. Bjelke: Br. J. prev. soc. Med. 25, 192 (1971).
22 J. Kopczynski: Epid. Rev. 26, 452 (1972).
23 D.M. Albanes, Y. Jones, M.S. Micozzi and M.E. Mattson: Am. J. pub. Hlth 77, 439 (1987).
24 J. Hjermann, A. Helgeland, I. Holme, P.G. Lund-Larsen and P. Leren: Acta med. scand. 200, 479 (1976).
25 H.S. Holcomb and J.W. Meigs: Archs Envir. Hlth 25, 295 (1972).
26 D.R. Jacobs and S. Gottenborg: Am. J. pub. Hlth 71, 391 (1981).
27 T. Khosla and C.R. Lowe: Br. med. J. 4, 10 (1971).
28 J.E. Lincoln: J. Am. med. Ass. 214, 1121 (1970).
29 J. Rodin and J.T. Wack: Behavioral Health. A Handbook for Health Enhancement and Disease Prevention, pp. 671. John Wiley & Sons, New York (1984).

30 J.T. Wack and J. Rodin: Am. J. clin. Nutr. *35*, 366 (1982).
31 S.L. Emont and K.M. Cummings: Addict. Behav. *12*, 151 (1987).
32 D.J. Bowen, S.E. Eury and N.E. Grunberg: Pharm. Biochem. Behav. *25*, 1131 (1986).
33 N.E. Grunberg: Addict. Behav. *7*, 317 (1982).
34 N.E. Grunberg: Br. J. Addict. *80*, 369 (1985).
35 N.E. Grunberg: Psychopharmac. Bull. *22*, 875 (1986b).
36 N.E. Grunberg, D.J. Bowen and D.E. Morse: Psychopharmacology *83*, 93 (1984).
37 N.E. Grunberg, D.J. Bowen and S.E. Winders: Psychopharmacology *90*, 101 (1986).
38 N.E. Grunberg, S.E. Winders and K.A. Popp: Psychopharmacology *91*, 221 (1987).
39 E. McNair and R. Bryson: Pharm. Biochem. Behav. *18*, 341 (1983).
40 M.D. Schechter and P.G. Cook: Eur. J. Pharm. *38*, 63 (1976).
41 B.A. Stamford, S. Matter, R.D. Fell and P. Papenek: Am. J. clin. Nutr. *43*, 486 (1986).
42 S.N. Blair, D.R. Jacobs and K.E. Powell: Pub. Hlth Rep. *100*, 172 (1985).
43 N.E. Grunberg, D.J. Bowen, V.A. Maycock and S.M. Nespor: Psychopharmacology *87*, 198 (1985).
44 S. Robinson and D.A. York: Int. J. Obesity *10*, 407 (1986).
45 H. Schievelbein, G. Heinemann, K. Loeschenkohl, C. Troll and J. Schlegel: in Smoking Behaviour. R.E. Thornton (Ed.) Churchill Livingstone, Edinburgh (1978).
46 P.E. Cryer: N. Engl. J. Med. *303*, 436 (1980).
47 A. Kershbaum, D.J. Pappajohn, S. Bellet, M. Hirabayashi and H. Shafiiha: J. Am. med. Ass. *203*, 113 (1968).
48 W.W. Winternitz and D. Quillen: J. clin. Pharm. *17*, 389 (1977).
49 E.H. Christensen and O. Hansen: Skand. Arch. Physiol. *81*, 160 (1939).
50 D.G. Gilbert: Psychol. Bull. *86*, 643 (1979).
51 P.D. Nesbitt: Doctoral Dissertation. Thesis No. 31/04-A. Universitiy Microfilms International (1969).
52 B. Silverstein: J. pers. soc. Psychol. *42*, 946 (1982).
53 O.F. Pomerleau, D.C. Turk and J.B. Fertig: Add. Behav. *9*, 265 (1984).
54 J. Milgrom-Friedman, R. Penman and R. Meares: Clin. exp. Pharmac. Physiol. *10*, 161 (1983).
55 S.M. Shiffman and M.E. Jarvik: Add. Behav. *9*, 95 (1984).
56 O.F. Pomerleau and C.S. Pomerleau: Neurosci. Biobehav. Rev. *8*, 503 (1984).
57 S.S. Tomkins: Am. J. pub. Hlth *56*, 17 (1966).
58 S.S. Tomkins: in Smoking, Health and Behaviour. E. Borgatta and R. Evans (Eds.). Aldine, Chicago. pp. 165–186 (1968).
59 F.F. Ikard, D.E. Green and D.A. Horn: Int. J. Addict. *4*, 649 (1969).
60 A.C. McKennell: Br. J. soc. clin. Psychol. *9*, 8 (1970).
61 C.D. Frith: Brit. J. soc. clin. Psychol. *10*, 73 (1971).
62 F.F. Ikard and S. Tomkins: J. Abnorm. Psychol. *81*, 172 (1973).
63 R.W. Coan: J. pers. soc. Psychol. *26*, 86 (1973).
64 A.C. McKennell: Research Paper *12*. Tobacco Research Council, London (1973).
65 M.A.H. Russell, J. Peto and U.A. Patel: Royal Statist. Soc. A. *137*, 313 (1974).
66 D.M. Warburton and K. Wesnes: in Smoking Behaviour. Thornton, R.E. (Ed.). Churchill-Livingstone, Edinburgh, 19–43 (1978).
67 H. Ashton and R. Stepney: in Smoking: Psychology and Pharmacology. Cambridge University Press, Cambridge (1982).
68 T.A. Wills: in Coping and Substance Use. S. Shiffman and T.A. Wills (Eds.) Academic Press, Orlando. pp. 67–94 (1985).

69 T.A. Wills: Hlth Psychol. *5*, 503 (1986).
70 C.C. Seltzer and F.W. Oechsli: Chron. Dis. *38*, 17 (1985).
71 N. Cherry and K. Kiernan: Br. J. prev. Med. *30*, 123 (1976).
72 J.V. Lerner and J.R. Vicary: Drug Educ. *14*, 1 (1984).
73 D.G. Gilbert and R.L. Hagen: Add. Behav. *5*, 247 (1980).
74 P.P. Woodson, R. Buzzi, R. Nil and K. Battig: Psychophysiol. *23*, 272 (1986).
75 D.R. Cherek: Psychopharmacology *75*, 339 (1981).
76 G. Mangan and J. Golding: in Smoking Behaviour: Physiological and Psychological Influences. R.E. Thornton (Ed.) Churchill-Livingstone, Edinburgh. pp. 208–228 (1978).
77 H.B. Murphree: in Electrophysiological Effects of Nicotine. A. Remond, A. Izard and C. Izard (Eds.). Elsevier-North Holland Biomedical Press, Amsterdam. pp. 227–344 (1979).
78 J.R. Hughes, D.K. Hatsukami, R.W. Pickens, D. Krahn, S. Malin and G. Luknic: Psychopharmacology *83*, 82 (1984).
79 M.J. Jarvis, M. Raw, M.A.H. Russell and C. Feyerabend: Br. Med. J. *285*, 537 (1982).
80 N.G. Schneider: in The Pharmacologic Treatment of Tobacco Dependence: Proceedings of the World Congress, Nov. 4–5, 1985. J.K. Ockene (Ed.). Institute for the Study of Smoking Behavior and Policy, Cambridge, MA. pp. 233–248 (1986).
81 R.J. West, M.J. Jarvis, M.A.H. Russell, M.E. Curruthers and C. Feyerabend. Br. J. Add. *79*, 215 (1984).
82 W. Lichtensteiger: in Biological Cellulaire des Processes Neurosecretoires Hypothalamiques. J.C. Vincent and C. Kornon (Eds.) Colloques Internationaux du C.N.R.S., Paris. pp. 179–191 (1979).
83 K. Fuxe, L. Agnati, P. Eneroth, J.-A. Gustafsson, J. Hokfelt, A. Lofstrom, B. Skett and P. Skett: Med. Biol. *55*, 148 (1977).
84 T.C. Westfall: Neuropharmacology *13*, 693 (1974).
85 M.J. Rand and K. Thurau (Eds.): The Pharmacology of Nicotine. IRL Press, Oxford (1988).
86 W.B. Essman: in Drugs and Cerebral Function. W.L. Smith (Ed.) Charles C. Thomas, Springfield USA. pp. 46–84 (1971).
87 K. Wesnes and D.M. Warburton: in Smoking Behaviour: Physiological and Psychological Influences. R.E. Thornton (Ed.) Churchill-Livingstone, Edinburgh. pp. 131–147 (1978).
88 J.E. Tong, G. Leigh, J. Campbell and D. Smith: Br. Psychol. *68*, 365 (1977).
89 G.L. Mangan: Gen. Psychol. *106*, 77 (1982).
90 K. Wesnes, D.M. Warburton and B. Matz: Neuropsychobiology *9*, 41 (1983).
91 D. Williams: Psychol. *71*, 83 (1980).
92 D. Williams, P.R. Tata and J. Miskella: Addict. Behav. *9*, 207 (1984).
93 P. Bakan: Br. Psychol. *50*, 325 (1959).
94 K. Wesnes and D.M. Warburton: Neuropsychobiology *9*, 223 (1983).
95 K. Wesnes and D.M. Warburton: Psychopharmacology *82*, 338 (1984a).
96 K. Wesnes and D.M. Warburton: Psychopharmacology *82*, 147 (1984b).
97 D.M. Warburton and A.C. Walters: in Smoking and Human Behaviour. A. Ney and A. Gale (Eds.). Wiley & Sons, Chichester, pp. 223–237 (1988).
98 C. Tarriere and F. Hartemann: Ergonomics *525*, 530 (1964).
99 K. Andersson and G.R.J. Hockey: Psychopharmacology *52*, 223 (1977).
100 J.P. Houston, N.G. Schneider and M.E. Jarvik: Am. Psychiat. *135*, 220 (1978).
101 S.C. Peeke and H.V.S. Peeke: Psychopharmacology *84*, 205 (1984).
102 D.M. Warburton, K. Wesnes, K. Shergold and M. James: Psychopharmacology *89*, 55 (1986).

103 G.L. Mangan: Gen. Psychol. *108,* 203 (1983).
104 G.L. Mangan and J.F. Golding: Psychology *115,* 65 (1983).
105 K.J. Kellar, B.A. Giblin and A. Martino-Barrows: in The Pharmacology of Nicotine. M.J. Rand and K. Thurau (Eds.). IRL Press, Oxford, pp. 193–207 (1988).
106 H. Kawamura and E.F. Domino: Int. Neuropharmac. *8,* 105 (1969).
107 A.K. Armitage, G.H. Hall and C.M. Sellers: Pharmacology *35,* 152 (1969).
108 P.D. Hrdina: Drug Metab. Rev. *3,* 89 (1974).
109 P.P. Rowell and D.L. Winkler: Neurochemistry *43,* 1593 (1984).
110 G.G. Celesia and H.H. Jasper: Neurology *16,* 1053 (1966).
111 E.F. Domino: Ann. N.Y. Acad. Sci. *142,* 216 (1967).
112 D.M. Warburton: Br. med. Bull. *37,* 121 (1981).
113 H. Hauser, B. Schwartz, G. Roth and R. Bickford: Clin. Neurophysiol. *10,* 576 (1958).
114 V.J. Knott: Electrophysiological Effects of Nicotine. A. Remond and C. Izard (Eds.) Elsevier-North Holland Biomedical Press, Amsterdam (1979).
115 D.M. Warburton and K. Wesnes: Electrophysiological Effects of Nicotine. A Remond and C. Izard (Eds.). Elsevier-North Holland Biomedical Press, Amsterdam (1979).
116 E. Donchin: in Cognitive Psychophysiology. E. Donchin (Ed.) Lawrence Erlbaum, Hillsdale, New Jersey (1984).
117 J.A. Edwards, K. Wesnes, D.M. Warburton and A. Gale: Addict. Behav. *10,* 113 (1985).
118 E. Callaway, R. Halliday, H. Naylor and G. Schechter: Psychopharmacology *85,* 133 (1985).
119 S.E. Robinson: in Central Cholinergic Mechanisms and Adaptive Dysfunctions. M.M. Singh, D.M. Warburton, H. Lal. (Eds.). Central Plenum Press, New York (1985).
120 B. Sahakian, G. Jones, R. Levy, J. Gray and D.M. Warburton: Br. Psychiat. 1988 in press.
121 N.L. Benowitz, P. Jacob, R.T. Jones and J. Rosenberg: Pharmac. exp. Ther. *221,* 368 (1982).
122 D.D. Breimer: in Sleep Research. R.G. Priest, A. Pletscher and J. Ward (Eds.). MTP Press, Lancaster, UK (1979).
123 A.J. Cohen and F.J.C. Roe: in Monograph on the Pharmacology and Toxicology of Nicotine. Tobacco Advisory Council, London (1981).
124 H.P. Rang and M.M. Dale: in Pharmacology. Churchill Livingstone, Edinburgh (1987).
125 American Psychiatric Association. Diagnostic and Statistical Manual of Mental Disorders. American Psychiatric Association, Washington (1987).

Aspirin as an antithrombotic agent

By Melvin J. Silver and Giovanni Di Minno
Cardeza Foundation for Hematologic Research, Thomas Jefferson
University, Philadelphia, PA 19107, USA, and Second Medical School,
Naples University, Naples, Italy

1	Original realization of the antihemostatic properties of aspirin and recognition of its potential antithrombotic properties	44
1.1	Involvement of platelets in hemostasis and thrombosis	45
2	Mechanism of action of aspirin as an antithrombotic agent	46
2.1	Platelet adhesion	46
2.2	Platelet aggregation	46
3	Clinical trials of aspirin in the prophylaxis of thrombosis	47
3.1	Coronary heart disease, atherosclerosis and platelets	48
3.1.1	Prevention of myocardial infarction	48
3.1.2	Unstable angina	49
3.2	Thrombosis complicating coronary angioplasty	50
3.3	Coronary artery by-pass graft disease	51
3.4	Ischemic cerebrovascular disease	51
3.5	Peripheral vascular disease	52
3.6	Venous thrombosis	52
3.7	Prosthetic heart valves	54
3.8	Prosthetic vascular grafts	54
3.9	Arteriovenous shunts	55
4	Dosage – timing – controlled slow-release formulations	55
5	Use of aspirin in combination with other drugs	58
6	Concluding remarks	58
	References	59

1   **Original realization of the antihemostatic properties of aspirin and recognition of its potential antithrombotic properties**

In retrospect, it appears that the earliest glimmers that aspirin might have antithrombotic properties were inherent in anecdotal reports of gastrointestinal bleeding in some patients, such as arthritics, who were consuming large amounts of aspirin. These reports were followed by controlled studies in 1938 which showed gastric hermorrhage [1] or, in 1958, occult blood in the feces, after the ingestion of aspirin [2].

In 1961, Grossman et al. [3] reported fecal blood loss in humans who had received 3 g per day of aspirin either by ingestion or *intravenously*, thus indicating that this deleterious effect of aspirin might not be due solely to a direct effect on the gastric mucosa but could also involve systemic factors. The possible systemic effect was investigated further by Blatrix [4] who showed in 1963 that aspirin prolongs the bleeding time in normal humans after doses of 40 mg/kg/day. This was confirmed in 1966 by Quick [5] who reported prolonged bleeding times in normal humans 2 hours after the ingestion of 1.3 g of aspirin but no change in bleeding time after the ingestion of similar amounts of sodium salicylate. This appears to be the first indication that the ability of aspirin to increase bleeding was specific for acetylsalicylic acid and not a non-specific effect of all salicylates. Prolongation of the bleeding time by aspirin was confirmed by Mielke [6], using a new standardized Ivy bleeding time technique. Thus, it was clear that ingested aspirin could cause a mild antihemostatic defect in some individuals. However, the mechanism for this was unknown.

Two seminal reports in 1967 and 1968 by Weiss and Aledort [7] and Weiss, Aledort and Kochwa [8] clearly showed that the antihemostatic effect of aspirin was related to its ability to inhibit the aggregation of blood platelets and was associated with inhibition of the release of nucleotides from the platelets. In addition, they showed that these inhibitory effects were persistent, lasting for 4 to 7 days after a single dose of 1.8 g of aspirin, and that platelet aggregation was inhibited when aspirin was added to platelet rich plasma *in vitro*. Similar effects were reported by O'Brien [9] and Zucker and Peterson [10] in 1968.

Thus, in the late 1960's, while most thinking was focused on the antihemostatic effects of high doses of aspirin, Weiss and Aledort called attention to the possibility that aspirin might have antithrombotic properties [7]. Such thinking was later supported by similar conclusions based on retrospective epidemiological data in the reports of the Boston Collaborative Drug Surveillance Program [11] and Jick and Miettinen [12] in 1974 and 1976, respectively. Looking at the incidence of regular use of aspirin in hundreds of patients who had had a myocardial infarct and thousands of control subjects who had not had an infarct they concluded that there was a "negative association between regular aspirin ingestion and non-fatal infarction which is consistent with the hypothesis that aspirin confers some degree of protection" and that "this evidence calls for further evaluation of aspirin as a potential preventive agent for thrombotic disease" [11]. There was now a clear indication for the need for careful clinical trials. These are described later.

1.1   Involvement of platelets in hemostasis and thrombosis

Important factors in both hemostasis and thrombosis are now recognized to include blood platelets which adhere to areas of exposed subendothelium on the wall of an injured blood vessel and accumulate by aggregation to form the physiological platelet plug which is important in the arrest of bleeding. Under pathological conditions, thrombi may be formed on an injured vessel wall, which can occlude, and so inhibit the flow of blood in small vessels. Also, pieces of thrombi may break off and travel downstream as thromboemboli to block the flow of blood in smaller vessels. Thus, adhesion of platelets to an injured vessel wall and aggregation of platelets to each other are important factors in both hemostasis and thrombosis. In addition, the blood vessel wall to which the platelets adhere, is an important component of the hemostatic and thrombotic process. For example, vasoconstriction or vasodilation may inhibit or facilitate the flow of blood in a vessel and so contribute to the beneficial or deleterious effects of hemostasis or thrombosis. It is thus important to understand the mechanism(s) whereby aspirin can influence these processes and determine the optimal dose to obtain a minimal antihemostatic effect with maximal antithrombotic activity.

2   Mechanism of action of aspirin as an antithrombotic agent
2.1   Platelet adhesion

It has been shown in an experimental, *in vivo* model in rabbits that platelets adhere, form pseudopods and spread over the surface of exposed subendothelium on a damaged peripheral artery and that aspirin (8 mg/kg) significantly inhibits this adhesion [13]. This suggests that part of the antithrombotic effect of aspirin may be related to its ability to inhibit the adhesion of platelets to exposed subendothelium. In fact, it has been shown (see below) that aspirin does inhibit thrombosis and the accumulation of platelets at sites of anastomosis on vein grafts and on sites of angioplasty. Platelet accumulation, of course, includes both adhesion and aggregation. However, in these studies the methods employed did not allow for dissociation or quantitation of inhibition of adhesion of platelets in contradistinction to accumulation of platelets by aggregation. More sophisticated methods are needed to dissociate adhesion from aggregation of platelets so that the ability of aspirin and other drugs to inhibit each of these processes can be determined quantitatively in humans.

2.2   Platelet aggregation

The way in which aspirin might inhibit platelet aggregation and so manifest its antithrombotic properties began to be elucidated when Smith and Willis [14] reported in 1971 that aspirin could inhibit the synthesis of prostaglandins (PG) by washed human platelets when they were stimulated by thrombin. However, their data were restricted to the finding that aspirin could inhibit the formation of prostaglandins E_2 and $F_{2\alpha}$ by platelets. Since there was no evidence that PGE_2 or $F_{2\alpha}$ could induce platelet aggregation, there was still no direct cause and effect linkage between inhibition of PG synthesis and inhibition of platelet aggregation by aspirin. At that time, it was not known whether platelets could form other prostaglandins or related metabolites of arachidonic acid. It remained to be shown that stimulated platelets were indeed capable of synthesizing a prostaglandin or related substance which could indeed induce platelet aggregation. In addition, if this turned out to be true, a further remaining question was "What is the precise mechanism by which aspirin might inhibit the synthesis of such substances or by which it might interfere with the ability of stimulated platelets to aggregate with each other?"

In 1974, evidence was educed that cyclic endoperoxides of arachidonic acid were formed during platelet aggregation and this suggested that these substances could induce platelet aggregation [15–17]. Then, Hamberg et al. [18] clearly showed that the cyclic endoperoxides prostaglandin G_2 and H_2 can induce platelet aggregration. In addition, it was shown that the newly discovered thromboxane A_2 could both induce platelet aggregation and cause vasoconstriction [19]. These inducers of platelet aggregation were shown to be derived from arachidonic acid released from phospholipids when platelets were stimulated by thrombin [20, 21]. This is reviewed in detail elsewhere [22]. The mechanism whereby aspirin inhibits both platelet aggregation and synthesis of the metabolites of arachidonic acid which induce platelet aggregation was shown to be via irreversible acetylation of the platelet cyclooxygenase enzyme [23].

3 Clinical trials of aspirin in the prophylaxis of thrombosis

Two reports which appeared over 30 years ago *suggested* the clinical potential of aspirin as an antithrombotic agent. In a review of 131 arthritic patients at autopsy it was found that the mortality related to stroke was only 2 %, while that related to myocardial infarction was 4 %. These frequencies are considerably lower than anticipated for those subjects. Since such patients were normally treated with aspirin, it was suggested that this might explain the low frequency of thrombotic episodes in these patients [24]. In the other study, there was a 6-year follow-up of 8,000 men who had taken 325 or 650 mg aspirin per day with no reported cases of myocardial infarction during that period [25].
Based on data from a variety of later clinical trials, we have assembled information on current evidence for the effectiveness of aspirin in the prophylaxis or treatment of various thrombotic episodes (Table 1), and correlate these observations with current knowledge on the role of platelets in various ischemic events. It is beyond the scope of this chapter to consider in detail all the trials of aspirin related to the prevention of thrombosis. For this the reader is referred to several excellent, recent reviews [26–31].

Table 1
Indications for the use of aspirin in the prevention or treatment of thrombosis

1. Coronary artery disease (unstable angina and myocardial infarction)
2. Transient cerebral ischemic attacks, stroke
3. Aortocoronary by-passes
4. Arterial thromboembolism
5. Thrombosis complicating coronary angioplasty
6. Peripheral vascular disease
7. Venous thrombosis (hip operations)
8. Prosthetic heart valves
9. Prosthetic vascular grafts
10. Arteriovenous shunts
11. In combination with thrombolytic agents after myocardial infarction

3.1 Coronary heart disease, atherosclerosis and platelets

Atherosclerosis of coronary arteries is a critical determinant of ischemic coronary heart disease, and a variety of data [26–31] suggest a major role for platelets in the pathogenesis of coronary atherosclerosis. Asymptomatic, raised, fatty streaks appear in the arteries of most young people. For reasons which are only beginning to be elucidated, a complex series of events occur in some of these streaks which leads to the progressive formation of atheromatous plaques. Some of these plaques undergo fissuring [32] and platelets contacting these abnormal surfaces adhere, aggregate and form thrombi. This may result in an acute ischemic event associated with unstable angina, myocardial infarction and sudden cardiac death [32]. In other cases, thrombi formed become organized and lead to growth of the plaques, thus triggering repeated chronic, ischemic events such as angina pectoris [31]. Knowledge of the mechanisms by which these events (as well as similar events in peripheral arteries) occur should provide the rationale(s) for the use of drugs that modify platelet function and are effective in the prophylaxis of arterial thrombosis.

3.1.1 Prevention of myocardial infarction

Angiographic studies confirm that acute transmural myocardial infarctions are often associated with thrombotic blockage of a diseased coronary artery and occluding thrombi are present in over 90 % of patients with transmural myocardial infarction [33, 34]. Another common cause of myocardial infarction is coronary vasospasm [35], and it has been documented that prostaglandin endoperoxides/thromboxane A_2

may induce vasospasm. Obviously, use of drugs that inhibit platelet function and suppress the synthesis of platelet prostaglandin endoperoxide/thromboxane A_2 should be effective in preventing myocardial infarction. However, the results of some early controlled trials were inconclusive with respect to the ability of these drugs to affect *reinfarction* and/or mortality *in patients who survive a first myocardial infarction*. On the other hand, early observations from a hospital-based case control study [12] suggested that regular ingestion of aspirin might protect against myocardial infarction. This conclusion was supported by the observation that regular ingestion of aspirin reduced cardiovascular mortality in male veterans with symptomatic coronary or cerebral vascular disease [36, 37]. Furthermore, in two double-blind clinical trials, it was found that aspirin was effective in protecting against myocardial infarction as well as cerebral ischemia [38, 39]. In spite of the fact that other clinical trials challenged these conclusions [40, 41], two large prospective trials on the primary prevention of myocardial infarction were planned. In these trials, British and American middle-aged physicians were recruited to evaluate the effectiveness of a single dose of 325 mg aspirin (ingested daily or every second day) in the primary prevention of myocardial infarction, stroke and death. The preliminary results of the American study showed a 47 % reduction in the incidence of myocardial infarction in aspirin users [42], while the British study showed no effect [43]. However, when considered together, the two primary prevention studies indicated a 33 % reduction in nonfatal myocardial infarction [44]. A slightly increased risk of stroke associated with cerebral hemorrhage was noted in these studies, and this needs to be considered by physicians prescribing long-term treatment with aspirin.

The results of *all* randomized trials of prolonged antiplatelet treatment in the secondary prevention of myocardial reinfarction have recently been assembled and analyzed. From the data on about 29,000 patients evaluated, it was found that a daily dose of 300–325 mg aspirin reduced the incidence of reinfarction or stroke by 30 % and the incidence of vascular mortality by 15 % [45].

3.1.2   Unstable angina

Patients with unstable angina and recent chest pain have been reported to have high thromboxane B_2 levels in the coronary sinus [46, 47] and intracoronary platelet aggregates [48, 49]. These two observations are consistent with the concept that platelet activation occurs during episodes of angina, and that inhibition of arachidonate metabolism might reduce the risks involved. Studies in which aspirin was administered support this hypothesis. In a report by Lewis et al. [50], the administration of 324 mg per day of buffered aspirin during the first 3 months after hospitalization for unstable angina caused a reduction of 50 % in the risk of non-fatal myocardial infarction as well as in the risk of mortality. Both these differences were highly significant ($p < 0.05$). Cairns et al. [51] extended these observations to show long-term benefit with aspirin in unstable angina. These authors reported that oral administration of 325 mg aspirin, four times daily, for an average of 19 months, starting 8 days after the diagnosis of unstable angina, reduced the incidence of non-fatal myocardial infarction as well as cardiac death by 50 % ($p < 0.005$). The incidence of side effects such as gastric discomfort, nausea, vomiting and gastric bleeding was greater in the latter study probably because of the much higher dose of aspirin employed. Similar beneficial effects are likely to be achieved with much smaller doses. However, this needs to be tested.

3.2 Thrombosis complicating coronary angioplasty

Acute coronary thrombosis is a possible complication of percutaneous transluminal coronary angioplasty. Platelet activation leading to thrombosis may occur following balloon-induced vascular injury. It is therefore reasonable to hypothesize that aspirin may diminish the risk of thrombosis in these cases. In fact, it was shown in 1984 that aspirin inhibits platelet deposition at sites of angioplasty [52]. More recently, Barnathan et al. [53], in a retrospective study involving 263 patients undergoing angioplasty showed that aspirin administered before percutaneous transluminal angioplasty "is associated with a decreased incidence and significance of acute coronary thrombosis complicating" the angioplasty. In the cases studied, the dosage of aspirin and the timing relative to the time of the procedure varied considerably and this merits further study. In a recent report [54] in which patients were given a combination of high dose aspirin plus dipyridamole, the authors concluded that the "six month rate of restenosis after successful angioplasty" was not reduced, but the incidence of transmural myocardial infarction, during or soon after angioplasty, was markedly reduced. Restenosis after coronary angioplasty has been reviewed recently [55].

3.3 Coronary artery by-pass graft disease

Coronary artery by-passes are currently used in patients with atherosclerotic occlusion of coronary arteries. Their effectiveness is related to their ability to increase blood flow to ischemic myocardial areas. Therefore, the continued patency of these grafts is essential for their long-term effectiveness. It is well documented [26–31] that one year after performing a coronary artery by-pass, 15–35 % of the grafts are occluded. The major factors contributing to their occlusion are thrombosis and proliferation of smooth muscle cells in the intima. Platelets contribute to both these processes through their ability to adhere to injured subendothelium, aggregate with each other, release intracellular components (such as platelet-derived growth factor which can induce proliferation of smooth muscle cells) and to form and release thromboxane A_2 [22, 26–31]. Furthermore, platelets release serotonin and $PGF_{2\alpha}$, other substances known to affect coronary vascular tone, as well as ADP and ATP, nucleotides known to produce direct effects on smooth muscle cells [22, 26–31, 56]. A clear benefit from antiplatelet drugs has been demonstrated, provided the treatment is started before or during the first day after surgery. When initiated 24–36 h before surgery, 1 g aspirin plus 225 mg dipyridamole per day reduced the incidence of graft occlusion by 50–70 % [56–58]. Similar results may be achieved using lower doses of aspirin [59]. In contrast, when the treatment is started 48 h after surgery is completed, its effectiveness is markedly reduced [26–31], an observation consistent with the concept that the treatment is effective against early mural thrombosis rather than late proliferation. Many patients with coronary artery by-pass grafts are taking aspirin. For example, in a follow-up of such patients Simons and Simons [60] reported that 81 % were taking aspirin.

3.4 Ischemic cerebrovascular disease

Under the umbrella of ischemic cerebrovascular disease, three different entities are recognized: transient ischemic attacks (TIAs), reversible ischemic neurological deficiencies and completed strokes. As many as 25–40 % of patients with TIAs experience a completed stroke within 5 years with 15 % of these occuring during the first year [61]. The atherosclerotic origin of these events as well as the involvement of platelets in their pathogenesis is supported by the facts that atheromas of carotid or vertebrobasilar arteries may be detected in 50–80 % of patients with TIAs [26–31], and that embolic material has been observed in the retinal vessels during episodes of amaurosis fugax [62]. These observations are currently thought to be correlated, and reversible cerebral ischemic episodes are interpreted as resulting from emboli formed by platelets and fibrin at the level of carotid or vertebrobasilar artery plaques. These emboli are then swept into the cerebral circulation [63], where they occlude small arteries and cause cerebral ischemic events. The effectiveness of aspirin in inhibiting fatal thromboembolic episodes has been clearly demonstrated in two animal models [64, 65].

Other explanations for the pathogenesis of TIAs must also be considered. In about 15–20 % of cases, TIAs derive from emboli formed on diseased heart valves or walls [66]. In a very few cases, platelet emboli are not detectable and heart valves and walls appear to be normal. The obvious explanation of these TIAs is that they are caused by vasospasms occurring at sites of stenosis [29]. It is to be noted that prostaglandin $F_{2\alpha}$ and thromboxane A_2 formed and released from activated platelets are both vasoconstrictors and could possibly induce such vasospasms.

The question of proper design of protocols to answer specific questions concerning the efficacy of aspirin has recently been reviewed [67]. A number of studies have evaluated the ability of aspirin or other antiplatelet drugs to protect patients with TIAs from the occurrence of further TIAs as well as more severe cerebral ischemic events [26–31]. The majority of them support the conclusions of three big studies: the Canadian Cooperative Study, the American Canadian Cooperative Study Group and the Study of Fields et al. [66, 68–70]. All these trials showed a 30–45 % reduction of stroke or death in males taking about 1 g of aspirin daily (325 mg/three-four times daily). In addition, the data also showed a substantial reduction in the incidence of TIAs. Consistent with these findings, studies focused primarily on the secondary prevention of myocardial infarction [28] also showed the effectiveness of aspirin in preventing stroke. This has been further supported by the recent trial of the UK Medical Research Council in which it was shown that a single dose of 300 mg aspirin per day is as effective as higher doses in preventing completed strokes in patients with TIAs [71]. The question as to whether still lower doses of aspirin are as effective as higher doses in the prevention of cerebrovascular disease is being addressed by two studies, the Swedish and the Danish low-dose trials, whose results on the prevention of TIAs and strokes using 50–60 mg of aspirin per day will be forthcoming*. It would be unfortunate if such doses proved to be ineffective because they were lower than the effective threshold dose which is most likely about 200 mg per day in divided doses. See section 4 below. The question of whether or not females respond to aspirin treatment as well as males is still unanswered. The number of females studied in past trials is relatively small and there have been findings indicating similarities or differences in responses in various single trials. This issue will be resolved when enough females have been included in appropriate studies to allow for statistical validity of the conclusions. In the meantime, there would seem to be no logic in withholding aspirin from female patients where it is otherwise indicated.

3.5 Peripheral vascular disease

Very few epidemiological data are available on the effectiveness of aspirin in the prevention of thrombosis and/or the progression of atherosclerosis in the arteries of the lower limbs in patients with peripheral vascular disease. In a trial [72] in patients with symptomatic peripheral vascular disease, the combination of aspirin (330 mg three times daily) plus dipyridamole (75 mg three times daily) significantly inhibited the progression of peripheral vascular occlusions. Aspirin alone had a somewhat lesser effect. However, aspirin alone was shown to be effective in inhibiting adhesion of platelets to an injured peripheral artery in an experimental, *in vivo* model in rabbits [13]. This subject has been reviewed recently [28].

---

* When this article was in the galley stage, the results of a Danish trial of very low dose aspirin (most patients received 50 mg/day) on 150 patients who had recently undergone carotid endoarterectomy were published (G. Boysen et al.: Stroke *19*, 1211, 1988). They concluded that "there was no significant effect of very-low-dose aspirin" in their trial concerned with the combined effects of transient ischemic attack, stroke, acute myocardial infarction and vascular death.

Since this was almost predictable, based on the material that we present on pages 14, 15 and 17 of this article, it would seem wise to initiate a similar trial using a dosage schedule of 300 mg/day contained in a well-controlled slow release preparation.

3.6 Venous thrombosis

In addition to their role in arterial thrombosis and atherosclerosis, several lines of evidence *suggest* a possible role for platelets in the pathogenesis of venous thrombosis [26–31]. When aspirin was administered in such cases, there appeared to be a strong tendency (not reaching statistical significance) toward prevention of venous thrombosis in patients with hip operations, but it was not at all effective in the prevention of venous thrombosis in general surgery [72–76].

3.7 Prosthetic heart valves

In a recent 10–19 year, follow-up study of patients with prosthetic heart valves, it was found that about 1/4 of them experienced at least one episode of thromboembolism, especially when anticoagulation was not properly carried out and atrial fibrillation was present [77]. There has been a review of papers published on this subject since 1979 [78]. The thromboembolic risk has been considerably reduced in recent years by use of improved types of prosthetic valves. However, the problem still remains an important one. When administered in combination with anticoagulants, aspirin (1 g/day or 250 mg twice daily) and/ or dipyridamole (400 mg) is more effective than the anticoagulant alone in preventing thromboembolic episodes in these patients [77–83]. The combination of anticoagulant and aspirin greatly increases the incidence of gastrointestinal bleeding. Therefore, the use of aspirin plus anticoagulants is not recommended for patients with prosthetic heart valves.

3.8 Prosthetic vascular grafts

Vascular grafts often fail because of thrombus formation on the luminal surface of the graft, and several reports agree that the junction of the graft with the blood vessel is the locus where thrombosis takes place most often [26–31]. The frequency of occlusion is very high in vascular grafts with internal diameter of less than 5 mm. A controlled clinical study [84] showed that aspirin, in combination with dipyridamole, reduces platelet deposition on prosthetic femoropopliteal grafts. In another study, it was shown that perioperative treatment with aspirin/dipyridamole inhibited platelet accumulation at the site of carotid endarterectomy [85]. The authors suggested that such treatment might reduce the risk of operative stroke and the long-term risk of carotid restenosis.

3.9 Arteriovenous shunts

In patients with chronic uremia, hemodialysis is made possible by arteriovenous shunts of silicone rubber cannulae, and the thrombotic occlusion of these cannulae is a major problem for patients undergoing chronic hemodialysis [26–31]. The platelet thrombus that forms in these shunts is localized at the vein-shunt junction [86]. Low daily doses of aspirin (160 mg) have been shown by Harter et al. [87] to reduce the incidence of thrombotic occlusions in arteriovenous shunts in patients on long-term hemodialysis. The effectiveness of low dose aspirin may be related to platelet dysfunction present in some patients on chronic hemodialysis [88]. Therefore, further studies correlating platelet function and the prevention of thrombosis are needed to strengthen the argument for the effectiveness of low-dose aspirin in the prevention of thrombosis in uremic patients with arteriovenous shunts.

4 Dosage – timing – controlled slow-release formulations

Besides its inhibitory effects on the cyclooxygenase of platelets, aspirin has also been shown to inhibit the synthesis by endothelial cells of the vasodilator and antiaggregating agent prostacyclin. It has been hypothesized that this would limit its usefulness as an antithrombotic agent. Such thinking presumes that possible inhibition of prostacyclin synthesis by aspirin *in vivo* could contribute to a thrombotic episode. There is actually no good evidence from retrospective or prospective reports that this is so. To the contrary, the real precaution that users of aspirin must consider is possible induction or enhancement of bleeding in the gastrointestinal tract or cerebral hemorrhage leading to stroke. In spite of this, there have been a number of studies aimed at finding an appropriate dosage of aspirin to maximize inhibition of platelet cycloogenase while minimizing its inhibition of prostacyclin by endothelial cells. Since many decades of wide experience with aspirin indicate that the thrombotic potential of aspirin, especially in low doses, is negligible, the dosage studies related to effects on prostacyclin synthesis by the cells of the vessel wall seem futile and not worth pursuing further.

On the other hand, it is worthwhile to find the optimal dosage and timing of administration for a maximal prophylactic, antithrombotic effect in normal individuals and individuals who may be at greater risk because of previous thrombotic episodes or possible genetic disposition. The high doses (often several grams per day) employed in many early studies were obviously chosen because they were handed down to clinicians interested in the antithrombotic properties of aspirin from clinicians concerned with its anti-inflammatory or analgesic properties. It was also a matter of convenience. The standard pill available in the United States contained 325 mg while in European Countries it was generally 500 mg. With our present knowledge of effective doses to inhibit platelet cyclooxygenase activity and platelet aggregation *in vivo*, we realize the high doses previously employed were unnecessarily high and should be reduced. To do this rationally, it is important to have knowledge of the *pharmacokinetics* of aspirin, and this has been studied and reviewed in some detail [89–92].

The relevant kinetic information regarding orally ingested aspirin may be summarized as follows: The drug is partially hydrolyzed by esterases in the gastrointestinal tract to salicylate and acetate. Thus, presystemic metabolism begins here. Absorption and systemic availability vary depending on whether a regular aspirin tablet or a sustained release or enteric-coated preparation has been ingested. The latter 2 generally are associated with lower systemic availability. Intact aspirin which enters the circulation is rapidly hydrolyzed by esterases in blood, lungs and liver. Aspirin itself has a half life of about 15 min in blood plasma *in vivo* while that of salicylate is about 2.5 h. The acetyl group has been shown to bind to platelet cyclooxygenase as well as other platelet proteins and albumin and hemoglobin. Since systemic availability is low, it is possible that the major effect of aspirin on platelet cyclooxygenase occurs in the presystemic circulation. Salicylate has been shown to interfere with the acetylation of platelet cyclooxygenase by aspirin [92].

In addition, when considering dosage in patients, or compliance in clinical trials, it must be borne in mind that there are at least several hundred over-the-counter preparations which contain varying amounts of aspirin [93, 94].

The above information may be helpful in understanding and interpreting data related to the prophylactic use of aspirin as an antithrombotic agent in normal subjects and those considered at high risk for a thrombotic episode.

A very important consideration, which has frequently been overlooked in past publications on this subject, is the following: while aspirin permanently acetylates and inactivates the cyclooxygenase of platelets that it contacts, little intact aspirin is present in the circulation 1 h after the ingestion of an aspirin tablet. In addition, over a 24 h period, while the concentration of circulating aspirin is negligible, the numbers of new platelets entering the circulation is significant. It has been reported that 10 % of platelets turn over every day [95], and we have demonstrated the entrance of about 5 to 6 % of newly formed circulating platelets in as little as 12 to 15 h after a single dose of aspirin [96]. Since only 2 to 3 % of platelets that have not contacted aspirin may produce enough thromboxane to cause platelet aggregation when acting synergistically with tiny amounts of other naturally occurring aggregating agents, we hypothesized that administration of aspirin four times a day would be enough to *continually* abolish the synthesis of thromboxane [96]. As a matter of fact, it was found that platelet thromboxane synthesis could be continually suppressed by a dose as low as 25 mg of aspirin four times a day in *normal subjects.* However, in *patients* with diabetic angiopathy neither 25 nor 100 mg four times a day was sufficient to continually suppress platelet thromboxane synthesis. However, we also found that the rate of entry of new platelets into the circulation of the diabetic patients was about twice that of normals. (In addition, it is possible that there may be abnormalities in the pharmacokinetics of aspirin in these patients.) These findings indicate that dosage schedules of aspirin based on studies in normal subjects are not applicable to patients with diabetic angiopathy (and perhaps other patients with a high rate of entry of new platelets into the circulation) and that *doses of aspirin administered at relatively long intervals allow for a temporal hiatus* in which new platelets entering the circulation, and that have not contacted aspirin, may aggregate and secrete in response to naturally occurring combinations of aggregating agents *in vivo.* Thus, in such patients *continual* suppression of platelet cyclic endoperoxide/thromboxane synthesis may greatly enhance the antithrombotic potential of aspirin. Since long-term, very frequent, oral dosing with aspirin tablets (e.g., every 2–3 hours) is not practical, carefully controlled, "low dose, slow-release preparations might be the ideal long term approach for the prevention of thrombosis in patients with a high rate of entry of new platelets into the circulation" [96].

5   Use of aspirin in combination with other drugs

Over the years, there have been many reports of the use of aspirin with other drugs, especially dipyridamole, as antithrombotic modalities. In general, there is little evidence that any of the combinations are constistently better than can be achieved with aspirin alone. However, there has been a successful recent advance in the use of aspirin in combination with thrombolytic agents. An important approach to the patient with myocardial infarction associated with blockage of coronary arteries with thrombi involves the use of thrombolytic agents. These agents, when properly administered, can lyse thrombi and reduce mortality of patients. However, their use is associated with a reocclusion rate of about 20 % often resulting in reinfarction [97]. To see whether this could be overcome by using aspirin, a randomized trial involving over 17,000 patients each assigned to one of four groups – intravenous streptokinase, oral aspirin, both or neither – was instituted by Isis-2 (Second International Study of Infarct Survival) Collaborative Group [98]. It has been shown that platelet activation occurs in association with fibrinolytic therapy, and that this activation can be suppressed by aspirin [99]. In the Isis-2 study, patients in the streptokinase group received 1.5 MU of 'Streptase' and those in the placebo group received an albumin solution intravenously; those in the oral aspirin group received 162.5 mg in enteric-coated tablets and those in the related placebo group received enteric-coated starch tablets daily for one month. Amongst the patients on aspirin alone, there was a significant reduction in both non-fatal reinfarction and non-fatal stroke with no significant increase in cerebral hemorrhage. Amongst those receiving streptokinase alone, there was an excess of non-fatal reinfarction. Those treated with both streptokinase and aspirin had significantly fewer reinfarctions, strokes or deaths than those on placebo or streptokinase alone.

6 Concluding remarks

In 1972, a prophetic proposal was made by Lee Wood [100]. "I suggest that men over the age of twenty and women over the age of forty should take one aspirin tablet (0.325 g) a day on a chronic, long-term basis in the hope that this will lessen the severity of arterial thrombosis and atherosclerosis. Exceptions to this would be people with bleeding disorders, aspirin allergy, uncontrolled hypertension, and those with a history of bleeding lesions of the gastrointestinal tract or other organ system." He concluded by saying ". . . , to me the rationale for this regimen seems sound, the risks small and the possible benefits enormous." Based on the accumulated data of the last 15 years, we would make a similar proposal with modification only as to dose and type of aspirin preparation. Long-term prophylaxis of thrombosis for apparently normal individuals in future years, probably will involve ingestion once a day of 100 mg aspirin in a carefully, controlled slow-release form (making available about 25 mg every 6 h). For the population with increased risk for a thrombotic episode, the dose will probably be somewhat higher (perhaps 300 mg per day with release of 50 mg every 4 h). Surely, all indications are that "the risks are small and the possible benefits enormous."

References

1. A. H. Douthwaite and G. A. M. Lintott: Lancet 2, 1222 (1938).
2. L. T. F. L. Stubbe: Brit. Med. J. 2, 1062 (1958).
3. M. I. Grossman, K. K. Matsumoto and R. J. Lichter: Gastroenterology 40, 383 (1961).
4. C. Blatrix: Nouv. Revue fr. Hémat. 3, 346 (1963).
5. A. J. Quick: Am. J. med. Sci. 252, 265 (1966).
6. C. H. Mielke, Jr., M. M. Kaneshiro, I. A. Maher, J. M. Weiner and S. I. Rapaport: Blood 34, 204 (1969).
7. H. J. Weiss and L. M. Aledort: Lancet II, 495 (1967).
8. H. J. Weiss, L. M. Aledort and S. Kochwa: J. clin. Invest. 47, 2169 (1968).
9. J. R. O'Brien: Lancet I, 779 (1968).
10. M. B. Zucker and J. Peterson: Proc. Soc. exp. Biol. Med. 127, 547 (1968).
11. Boston Collaborative Drug Surveillance Group: Br. med. J. 1, 440 (1974).
12. H. Jick and O. S. Miettinen: Br, med. J. 1, 1057 (1976).
13. M. J. Silver, C. M. Ingerman, A. W. Sedar and M. Smith: Exp. molec. Path. 41, 141 (1984).
14. J. B. Smith and A. L. Willis: Nature 231, 235 (1971).
15. J. B. Smith, C. Ingerman, J. J. Kocsis and M. J. Silver: J. clin. Invest. 53, 1468 (1974).
16. A. L. Willis and D. C. Kuhn: Prostaglandins 4, 127, (1973).

17 M. Hamberg and B. Samuelsson: Proc. natl. Acad. Sci. USA *71*, 3400 (1974).
18 M. Hamberg, J. Svensson and B. Samuelsson: Proc. natl. Acad. Sci. USA *71*, 3824 (1974).
19 M. Hamberg, J. Svensson and B. Samuelsson: Proc. natl. Acad. Sci. USA *72*, 2994 (1975).
20 T. Bills, J. B. Smith and M. J. Silver: Biochim. biophys. Acta *424*, 303 (1976).
21 T. Bills, J. B. Smith and M. J. Silver: J. clin. Invest. *60*, 1 (1977).
22 M. J. Silver: Adv. Pharmacol. Chemotherapy *18*, 1 (1981).
23 G. J. Roth and P. Majerus: J. clin. Invest. *56*, 624 (1975).
24 S. Cobb, F. Anderson and W. Bauer: N. Engl. J. Med. *249*, 553 (1953).
25 L. L. Craven: Miss. Valley Med. J. *78*, 213 (1956).
26 G. De Gaetano, C. Cerletti, E. Dejana and J. Vermylen: Drugs *31*, 517 (1986).
27 L. A. Harker: Circulation *73*, 206 (1986).
28 A. S. Gallus: Clin. Haemat. *15*, 509 (1986).
29 J. Webster and A. S. Douglas: Blood Rev. *1*, 9 (1987).
30 J. A. Oates, A. Garret, R. A. Branch, E. K. Jackson, H. R. Knapp and L. J. Roberts: N. Engl. J. Med. *319*, 689 (1988).
31 V. Fuster, L. Badimon, J. Badimon, P. Adams, V. Turitto and J. H. Chesebro: Thrombosis and Haemostasis, p. 349. Leuven University Press, Leuven 1987.
32 M. J. Davies and A. C. Thomas: Br. Heart J. *53*, 363 (1985).
33 M. A. DeWood, J. Spores, R. Notske, L. T. Mouser, R. Burroughs, M. S. Golden and H. T. Lang: N. Engl. J. Med. *303*, 897 (1980).
34 K. P. Rentrop, H. Blanke, K. R. Karsch, H. Kaiser, H. Kostering and K. Leitz: Circulation *63*, 307 (1981).
35 A. Maseri, A. L'Abbate, G. Baroldi, S. Chierchia, M. Marzilli, A. M. Ballestra, S. Serveri, O. Parodi, A. Biagini, A. Distante and A. Pesola: N. Engl. J. Med. *299*, 1271 (1978).
36 J. A. Blakely and M. Gent: Platelets, Drugs and Thrombosis, p. 284, S. Karger Publisher, Basel 1975.
37 R. Heikinheimo and K. Jarvinen: J. Am. ger. Soc. *19*, 403 (1971).
38 M. G. Bousser, E. Eschwege, M. Haguenau, J. M. Lefaucconier, K. Thibult, and P. J. Touboul: Stroke *14*, 5 (1983).
39 P. S. Sorensen, H. Pedersen, J. Marquardsen, H. Petersson, A. Heltberg, N. Simonsen, O. Munch and L. A. Andersen: Stroke *14*, 15 (1983).
40 E. C. Hammond and L. Garfinkel: Br. med. J. *2*, 269 (1975).
41 C. H. Hennekens, L. H. Karlson and B. Rosner: Circulation *58*, 35 (1978).
42 The Steering Committee of the Physicians Health Study Research Group Preliminary Report: N. Eng. J. Med. *318*, 262 (1988).
43 R. Peto, R. Gray et al.: Br. med. J. *296*, 313 (1988).
44 C. H. Hennekens, R. Peto and G. B. Hutchison: N. Engl. J. Med. *318*, 923 (1988).
45 Antiplatelet Trialist Collaboration: Br. med. J. *296*, 320 (1988).
46 R. I. Lewy, L. Wiener, P. Walinsky, A. M. Lefer, M. J. Silver and J. B. Smith: Circulation *61*, 1165 (1980).
47 D. J. Fitzgerald, L. Roy, F. Catella and G. A. Fitzgerald: N. Engl. J. Med. *315*, 983 (1986).
48 C. I. Sherman, F. Litvack and W. Guredfest: N. Engl. J. Med. *315*, 913 (1986).
49 M. J. Davies, A. C. Thomas, P. A. Knapman and J. R. Hangartner: Circulation *78*, 418 (1986).
50 H. D. Lewis, J. W. Davis, D. G. Archibald, W. E. Steinke, J. E. Smitherman Doeherty, H. W. Schnape, M. M. Le Winter, E. Linares, J. M. Pouget, S. C. Harwal, E. Chesler and H. DeMots: N. Engl. J. Med. *309*, 396 (1983).

51 J. Cairns, M. Gent, J. Singer, K. J. Finnie, G. M. Froggatt, D. A. Holder, G. Jablonsky, W. J. Kostik, L. J. Melendez, M. G. Myers, L. D. Sackett, B. J. Sealey and P. H. Tanser: N. Engl. J. Med. *313*, 1369 (1985).
52 D. A. Cunningham, B. Kumar, B. A. Siegal, L. A. Gilula, W. G. Totty and M. J. Welch: Radiology *151*, 487 (1984).
53 E. S. Barnathan, J. S. Schwartz, L. Taylor, W. K. Laskey, J. P. Kleaveland, W. G. Kussmaul and J. W. Hirshfeld, Jr.: Circulation *76*, 125 (1987).
54 L. Schwartz, M. G. Bourassa, J. Lesperance, H. E. Aldridge, M. B. F. Kazim, V. A. Salvatori, M. Henderson, R. Bonan and P. R. David: N. Engl. J. Med. *318*, 1714 (1988).
55 W. McBride, R. A. Lange and L. D. Hillis: N. Engl. J. Med. *318*, 1735 (1988).
56 L. Campeau, M. Enjalbert, J. Lesperance, M. G. Bourassa, P. Kwiterovich, Jr., S. Wacholder and A. Sniderman: N. Engl. J. Med. *311*, 1329 (1984).
57 J. E. Mayer, W. G. Lindsay, W. Casteneda and D. M. Nicoloff: Ann. Thor. Surg. *31*, 204 (1981).
58 J. H. Chesebro, I. P. Clements, V. Fuster, L. R. Elverback, H. C. Smith, W. T. Bardsley, R. L. Frye, D. R. Holmes, R. E. Vlietrstra, J. R. Pluth, R. B. Wallace, F. J. Puga, T. A. Orszulak, J. M. Piehler, H. V. Schaff and G. K. Danielson: N. Engl. J. Med. *307*, 73 (1982).
59 R. L. Lorenz, M. Webber, J. Kotzur, K. Thiesen, C. V. Schacky, W. Meister, B. Reichardt, P. C. Weber: Lancet *I*, 1261 (1984).
60 L. A. Simons and J. Simons: Med. J. Australia *146*, 573 (1987).
61 D. G. Sherman and R. G. Hart: J. Am. Coll. Cardiol. *8*, 88b (1986).
62 R. W. Ross Russell: Lancet *II*, 789 (1968).
63 H. J. M. Barnett: Med. Clins N. Am. *63*, 649 (1979).
64 M. J. Silver, W. Hoch, J. J. Kocsis, C. M. Ingerman, J. B. Smith: Science *183*, 1085 (1974).
65 G. DiMinno and M. J. Silver: J. Pharmac. exp. Ther. *225*, 57 (1983).
66 W. S. Fields, N. A. Lemak, R. F. Frankowski and R. J. Hardy: Stroke *8*, 541 (1977).
67 M. Gent: Stroke *18*, 541 (1988).
68 Canadian Cooperative Study Group: N. Engl. J. Med. *299*, 53 (1978).
69 W. S. Fields, N. A. Lemark, R. F. Frankowski, R. J. Hardy and R. H. Bigelow: Circulation *62*, 90 (1980).
70 American-Canadian Cooperative Study Group: Stroke *16*, 406 (1985).
71 UK-TIA Study Group: Br. Med. J. *296*, 316 (1988).
72 H. Hess, A. Mietaschk and G. Diechsel: Lancet *I*, 415 (1985).
73 E. W. Salzmann, W. H. Harris and R. W. De-Sanctis: N. Engl. J. Med. *284*, 1287 (1971).
74 Medical Research Council Report of the Steering Committee: Lancet *II*, 441 (1972).
75 G. P. Clagett, P. Schneider, C. B. Rosoff and E. W. Salzmann: Surgery *77*, 61 (1975).
76 W. H. Harris, E. W. Salzmann, C. Athanasoulis et al.: J. Bone Joint Surg. *56*, 1552 (1974).
77 V. Fuster, C. W. Pumphrey, M. D. McGoon, J. H. Chesebro, J. R. Pluth and D. C. McGoon: Circulation *66*, 157 (1982).
78 L. H. Edmunds, Jr.: Ann. thor. Surg. *44*, 430 (1987).
79 R. Altman, F. Boullon, J. Rouvier, R. Raca, L. de la Fuente and R. Favaloro: J. thorac. cardiovasc. Surg. *72*, 127 (1976).
80 J. Dale, E. Myhre, O. Storstein, H. Stormorken and L. Efskind: Am. Heart J. *94*, 101 (1977).
81 Groupe de Recherche PACTE: Cour *9*, 915 (1978).
82 J. H. Chesebro, V. Fuster, L. R. Elveback, D. C. McGoon, J. R. Pluth, F. J. Puga, R. B. Wallace, G. K. Danilson, T. A. Orszulak, J. M. Piehler and H. V. Schaff: Am. J. Cardiol *51*, 1537 (1983).

83 C. K. Mok, J. Boey, R. Wang, T. K. Chan, K. L. Cheung, P. K. Lee, J. Chow, R. P. Ng and T. F. Tse: Circulation 77, 1059 (1985).
84 M. D. Goldmann, D. Simpson, R. J. Hawker, H. C. Norcott and C. N. McCollum: Ann. Surg. 198, 713 (1983).
85 J. M. Findlay, W. M. Lougheed, F. Gentili, P. M. Walker, M. F. X. Glynn and S. Houle: J. Neurosurg. 63, 693 (1985).
86 J. L. Ritchie, A. Lindner, G. W. Hamilton and L. A. Harker: Nephron 31, 333 (1982).
87 H. R. Harter, J. W. Burch, P. W. Majerus, N. Stanford, J. A. Delkmez, C. B. Anderson and C. A. Weerto: N. Engl. J. Med. 301, 577 (1979).
88 G. Di Minno, J. Martinez, M. L. McKean, J. De la Rosa, J. F. Burke and S. Murphy: Am. J. Med. 79, 552 (1985).
89 A. K. Pedersen and G. A. Fitzgerald: N. Engl. J. Med. 311, 1206 (1984).
90 A. K. Pedersen and G. A. Fitzgerald: Circulation 72, 1164 (1985).
91 C. Cerletti, R. Latini, A. Del Maschio, F. Galletti, E. Dejana and G. de Gaetano: Thromb. Haemostasis 53, 415 (1985).
92 C. Cerletti, M. Livio and G. de Gaetano: Biochim. biophys. Acta 714, 122 (1982).
93 E. R. Leist and J. G. Banwell: N. Engl. J. Med. 291, 710 (1974).
94 J. Selner: N. Engl. J. Med. 292, 372 (1975).
95 L. H. Harker and C. A. Finch: J. clin. Invest. 48, 963 (1969).
96 G. DiMinno, M. J. Silver, A. M. Cerbone and S. Murphy: Blood 68, 886 (1986).
97 D. H. Schaer, A. M. Ross and A. G. Wasserman: Circulation 76 Suppl. II, 57 (1987).
98 Isis-2 (Second International Study of Infarct Survival) Collaborative Group: Lancet II, 349 (1988).
99 D. J. Fitzgerald, F. Catalla, L. Roy and G. A. Fitzgerald: Circulation 77, 142 (1988).
100 L. Wood: Lancet II, 532 (1972).

The mode of action of anti-rheumatic drugs. 1. Anti-inflammatory and immunosuppressive effects of glucocorticoids

By Anthony C. Allison and Simon W. Lee
Syntex Research, 3401 Hillview Avenue, Palo Alto, CA 94304, USA

1	Introduction: the lipocortin hypothesis	64
2	Production of IL-1	65
2.1	Structure	66
2.2	Immunoassay	66
2.3	Measurement of IL-1 messenger RNAs	67
2.4	Control of IL-1 formation	67
2.5	Production of IL-1 by RA synovia	68
2.6	Production of IL-1 by gingival tissue in periodontal disease	69
3	Pro-inflammatory and catabolic effects of IL-1	70
3.1	Effects on endothelial cells	70
3.2	Induction of prostaglandin formation	70
3.3	Induction of metalloproteinase secretion by chondrocytes	71
3.4	Induction of procollagenase secretion	71
3.5	Induced secretion of tissue plasminogen activator	71
3.6	Bone demineralization	71
3.7	Effects of IL-1 on hematopoiesis	72
3.8	The role of IL-1 in late allergic reactions	72
4	Antagonism by glucocorticoids of the formation of Il-1	73
5	Inhibition by glucocorticoids of the effects of Il-1	74
6	IL-1 induces the release of glucocorticoids	75
7	Antagonism by IL-1 of the effects of glucocorticoids	76
8	Acute-phase protein synthesis	76
9	Immunosuppressive effects of glucocorticoids	77
10	Side effects of glucocorticoids	78
	Acknowledgment	79
	References	79

1   Introduction: the lipocortin hypothesis

Although glucocorticoids are among the most potent and widely used anti-inflammatory agents, their mode of action is poorly understood. The most popular theory has been that glucocorticoids induce the formation of a group of proteins, collectively termed lipocortins, that inhibit phospholipase A_2 activity, thereby decreasing the release of arachidonic acid and the production of pro-inflammatory prostaglandins and leukotrienes [1, 2].

Lipocortins are a family of proteins with sequence homology and certain common properties [reviews 3, 4]. Several lipocortins have been cloned and expressed: lipocortin I [5], lipocortin II [6] and lipocortins III and V [7]. Recombinant human lipocortin I combines with an antibody raised against purified lipocortin and inhibits leukotriene-induced thromboxane release from isolated perfused guinea pig lung in an analogous fashion [8], showing that the molecules have the same properties. Independently a family of proteins termed calpactins, because they bind calcium, phospholipid and actin, have been defined, cloned and expressed. Lipocortin I has a relative molecular mass (M_r) of 35,000 (p35) and an amino acid sequence identical with that of calpactin II [9] while lipocortin II (p36) has an amino acid sequence identical with that of calpactin I [10]. The p35 is a physiological substrate for epidermal growth factor receptor tyrosine kinase while p36 is a major substrate for pp60v-src-mediated phosphorylation [3, 4].

There are several reasons for doubting whether lipocortins/calpactins represent an important mechanism by which glucocorticoids exert their effects. First, p35 and p36 are abundant proteins: in fibroblasts p36 accounts for 0.1 to 0.4 % of total cell protein [4]. It would be expected that the concentration of an inducible regulatory protein would be much lower.

Second, p35 and p36 are not inducible by glucocorticoids. It was claimed by Wallner et al. [5] that dexamethasone increased the level of p35 messenger RNA (mRNA) in peritoneal macrophages but did not increase the level of p35 as determined by immunoblotting. In subsequent studies [11, 12] glucocorticoids had no effect on p35 or p36 mRNA or protein levels in a variety of cell types including U-937 promonocytes and pulmonary alveolar macrophages. Furthermore, all the proteins in this family lack an N-terminal signal peptide and are rather highly charged. There is no good experimental evidence that they are secreted, even following treatment with steroids [4]. Even if p35 and/or p36 is secreted by a novel mechanism it is not clear how these proteins could from the outside of the cell inhibit phospholipase A_2 activity on the cytoplasmic face of the plasma membrane.

Third, p35 and p36 inhibit phospholipase A_2 activity indirectly. Phospholipase A_2 requires Ca^{2+} for activity, and in the presence of Ca^{2+} p35 and p36 bind to phospholipids and limit access to the enzyme [13,14]. If assays with limiting concentrations of labelled bacterial phospholipids are used, apparent inhibition by p35 and p36 is observed, but at high phospholipid concentrations approaching those present in the cell no inhibitory effect of p35 or p36 is found [13].

Fourth, although it has been claimed that bacterially expressed p35 has anti-inflammatory effects in a lung effusion assay [8], p35 purified from placenta does not show such activity in a paw edema assay [15]. For all these reasons the role of lipocortins as mediators of the anti-inflammatory effects of glucocorticoids is in doubt. Although there may be other proteins with properties more closely matching those originally postulated for lipocortin, alternative or supplementary mechanisms by which glucocorticoids might exert anti-inflammatory effects deserve consideration. We propose that inhibition of the production and effects of the two known varieties of interleukin-1 (IL-α and IL-β) contributes to the anti-inflammatory and immunosuppressive effects of glucocorticoids. In fact, the antagonism is mutual: IL-1 in low concentrations can inhibit some of the effects of glucocorticoids. This balance between pro-inflammatory effects of IL-1 and anti-inflammatory effects of glucocorticoids is a basic mechanism regulating host responses to infection and injury. When the balance is disturbed, inflammatory responses can persist. Therapeutic administration of potent glucocorticoid analogs can restore the balance, but at the expense of various side effects, some of which can now be explained at a molecular level.

2 Production of IL-1

It is widely accepted that IL-1 is a central mediator in the pathogenesis of rheumatoid arthritis (RA), periodontal disease and other diseases with chronic inflammatory pathogenesis. IL-1 also appears to be a mediator of acute allergic reactions. Immunoassays for the two known varieties of IL-1 and the corresponding messenger RNAs (mRNAs) provide a secure foundation for analyses of the role of IL-1 in RA and other inflammatory conditions. They are also required for analyses of the effects of drugs, such as glucocorticoids, and procedures, such as total lymphoid irradiation, that can improve the clinical status of patients.

2.1 Structure

Cloning investigations [16–18] have demonstrated that there are two human IL-1 genes producing in several cell types primary translation products with 271 and 269 amino acids for IL-1α and IL-1β, respectively (M_r about 31K). The primary gene product of IL-1α is biologically active whereas that of IL-1β is not; both are cleaved to biologically active forms of 17.5K. The 17.5K IL-1 β molecule is also known from its isoelectric point as the pI7 form, whereas the 17.5K IL-1α molecule is also termed the pI5 form. Human IL-1β has only 26 % amino acid homology with human IL-1α, but the homology of the nucleic acid bases is greater (45 %), suggesting that the two genes arose by duplication of a common precursor in the distant past. Nevertheless IL-1α and IL-1β bind to the same receptor on several cell types, which has recently been cloned and expressed [19]. It is not surprising that IL-1α and IL-1β share many biological activities, in which they are usually about equipotent.

2.2 Immunoassay

IL-1 has usually been measured by bioassays, especially the murine thymocyte co-mitogenic (lymphocyte-activating factor or LAF) assay. However, the assay is ambiguous because IL-1α and IL-1β are both active, and factors other than IL-1 (e.g. IL-6 and IL-7) can have thymocyte co-stimulatory activity. Moreover other mediators, such as PGE_2 (which are often present in synovial fluid or supernatants of synovial cell cultures) and drugs such as glucocorticoids, which can inhibit the response, and are not easily eliminated by dialysis. For these reasons we have developed IL-1α and IL-1β immunoassays. Our assay for IL-1β is a two-site ELISA using two high-affinity monoclonal antibodies against non-overlapping epitopes with co-operative binding [20]. It can detect IL-1β in concentrations down to 15 pg/ml in serum, other biological fluids, culture media, and released from cells by Triton X-100 lysis. This is close to the sensitivity of most bioassays. We have also developed an immunoassay of comparable sensitivity and specificity for IL-1α. A radioimmunoassay for IL-1β[20] is less sensitive than our assay.

2.3 Measurement of IL-1 messenger RNAs

Because we wished to analyze the expression of IL-1 genes in tissues such as RA synovia, as well as the molecular biology of glucocorticoid and other drug effects on various cell types, we improved methods for extracting messenger RNAs (mRNAs) so as to minimize degradation by ribonuclease and applied dot-blot hybridization with complementary DNA probes for IL-1α and IL-1β to measure the corresponding mRNAs [22].

2.4 Control of IL-1 formation

We have used human peripheral blood monocytes or U-937 cells in culture to study the control of IL-1 formation [22]. The U-937 is a hu-

Figure 1
Production of IL-1β messenger RNA in the human promonocytic cell line U-937 induced by phorbol myristate acetate and lipopolysaccharide. At the times indicated, dexamethasone (10μM) was added. After 1-h levels of IL-1β mRNA rapidly fall [from data in reference 22].

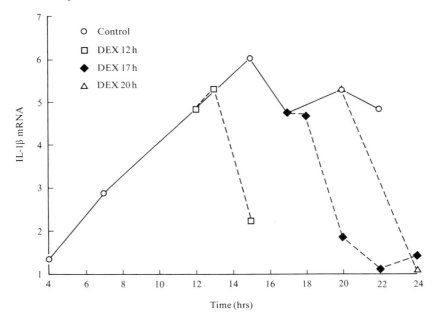

man promonocytic cell line. In both cell types levels of IL-1α and IL-1β mRNAs are barely detectable. When monocytes are stimulated with lipopolysaccharide (LPS), substantial amounts of IL-1β mRNA and lesser amounts of IL-1α mRNA are found (ratio about 15 to 1). Hence the control is at the level of transcription of the IL-1α and IL-1β genes. U-937 cells respond maximally to LPS only when they are primed with phorbol-12-myristate-13-acetate (PMA, see fig. 1). Monocytes do not require such priming and cycloheximide does not inhibit IL-1β gene transcription. Beginning about 2 h after stimulation, increased levels of IL-1β are detectable in the cells by immunoassay. In human peripheral blood monocytes stimulated with LPS, a substantial amount of IL-1β is formed. Relatively large pools of IL-1β accumulate within cells; the IL-1 is readily demonstrable by immunofluorescence in the cytoplasm [23, 24]. Western blots show that both the 31K precursor form and the 17.5K biologically active form of IL-1β are present within the cells and in the extracellular medium [24]. In fact, the mechanism of secretion of IL-1 is not understood since the molecule lacks the typical leader sequence found in most secreted proteins.

2.5 Production of IL-1 by RA synovia

Several groups of investigators found IL-1 by bioassays in RA synovial fluid [25, 26]. We have consistently observed IL-1β mRNA and IL-1α mRNA in RA synovial tissue (fig. 2), as well as IL-1β in supernatants of RA synovial tissue cultures [27] (fig. 3). IL-1β is released first, the PGE_2. The delayed release of PGE_2 is presumably due in part to escape

Figure 2
Messenger RNA extracted from rheumatoid arthritis synovial tissue hybridized with probes for IL-1α, IL-1β and the proto-oncogenes H-ras, c-fos and c-myc. Substantial amounts of mRNAs for c-fos and c-myc are present, and both IL-1α and IL-1β genes are expressed [from ref. 27].

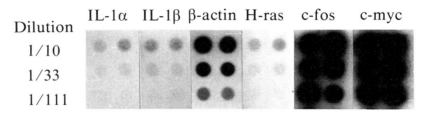

Figure 3
Release from cultured rheumatoid arthritis synovial tissue of IL-1β measured by immunoassay, of lymphocyte-activating activity and of PGE_2 [from ref. 27].

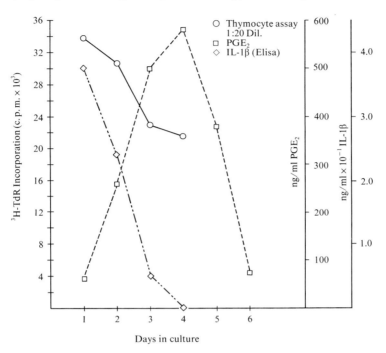

from control of drugs used by the patient and in part to IL-1-induced production of PGE_2 by synovial cells in culture. The amount of IL-1 produced (several nanograms per ml of culture fluid) is sufficient to induce inflammatory changes as well as cartilage proteoglycan degradation and bone erosion. We have also found that total lymphoid irradiation (a treatment decreasing the signs and symptoms of RA) reduces production of IL-1 by RA synovial tissue cultures without affecting the concentrations of rheumatoid factors in the circulation or synovial fluid [28].

2.6   Production of IL-1 by gingival tissue in periodontal disease

We have also analyzed the role of IL-1 in the pathogenesis of periodontal disease, a common inflammatory disorder induced by products of bacteria in dental plaque, which is the major cause of tooth loss in adults in North America, Europe and Japan [29]. Lipopolysaccharides (LPS) of oral *Bacteroides* spp. were found to induce the formation of IL-1β by cultured human monocytes as efficiently as LPS of *Escherichia coli*. Gingival tissues taken from patients with progressive periodontal disease showed substantial amounts of IL-1α and IL-1β mRNA, and immunoassays showed IL-1β in nanogram per ml amounts in most gingival fluid samples; some showed also IL-1α in comparable amounts. It is likely that IL-1β is the major mediator of alveolar bone erosion and consequent tooth loss in periodontal disease.

3 Pro-inflammatory and catabolic effects of IL-1

Although IL-1 has been reported to have many biological effects, some of these are no longer accepted, for example induction of synthesis of acute-phase proteins discussed below. A brief review will therefore be given of pro-inflammatory and catabolic effects of IL-1 which are well established.

3.1 Effects on endothelial cells

IL-1 acts on endothelial cells in culture to increase adhesion of neutrophils, monocytes and lymphocytes [30, 31]. This is a stage in the recruitment of cells from the circulation into inflammatory sites, including the joints in RA, thereby initiating or perpetuating inflammatory reactions.

3.2 Induction of prostaglandin formation

IL-1 acts on synovial fibroblast-type cells [32] and on chondrocytes [33] to induce the production and release of PGE_2 and on endothelial cells to induce production of prostacyclin (PGI_2) [34]. In chondrocytes this action of IL-1 is associated with increased phospholipase A_2 activity and release of the enzyme from the cells [33]. Whether this is due to activation of enzyme already present, to induced formation of phospholipase, or to both of these effects, has not yet been established. PGE_2 and PGI_2 are vasodilators and co-mediators of edema and pain, which are among the manifestations of inflammation.

3.3 Induction of metalloproteinase secretion by chondrocytes

IL-1 induces in chondrocytes the formation of a neutral metalloproteinase that can degrade cartilage proteoglycan. Human monocyte-derived IL-1 and recombinant human IL-1α and IL-1β are potent inducers of metalloproteinase secretion by isolated rabbit chondrocytes [35]. Since tumor-necrosis factor (TNF-α) shares some properties of IL-1, it is interesting that recombinant human TNF-α did not induce prostaglandin or neutral metalloproteinase release by chondrocytes [35]. In chondrocytes activated by IL-1 we have found that indomethacin does not influence metalloproteinase secretion or proteoglycan degradation, suggesting that prostaglandins do not mediate this effect [36].

3.4 Induction of procollagenase secretion

The metalloproteinase just described also cleaves procollagenase to active collagenase that can degrade collagen in cartilage, bone and loose connective tissue. IL-1 induces the release of procollagenase by synovial fibroblast-type cells [32]. Thus IL-1 induces the synthesis and release of both types of enzymes required for breakdown of connective tissue matrix and fibers. This could explain cartilage erosion in sites of highest concentrations of IL-1 adjacent to proliferating pannus in RA and alongside the joint space in septic arthritis.

3.5 Induced secretion of tissue plasminogen activator

IL-1 stimulates plasminogen activator activity in human synovial fibroblast-type cells [37]. Plasminogen activator is a serine proteinase which cleaves plasminogen to form the less specific proteinase plasmin. This activation has been correlated with inflammation, as well as cell migration and tissue remodelling. The IL-1 induced increase in plasminogen activator activity was found to be inhibited by indomethacin, suggesting that prostaglandins mediate the effect.

3.6 Bone demineralization

Following studies with IL-1-containing supernatants [38], it has been shown that recombinant human IL-1α and IL-1β are able to induce the release of calcium from organ cultures of calvaria and long bones [39]. In this assay IL-1β is active in concentrations less than 10^{-11}M, more active than IL-1α; IL-1β is the most potent known inducer of demineralization, about one thousand times as active on a molar basis as parathyroid hormone. Recombinant TNF-α also induces demineralization although it is less potent than IL-1; the two have synergistic effects, as do TNF-β (lymphotoxin) and PGE$_2$ with IL-1 [39]. By inducing demineralization of bone and degrading intercellular matrix and collagen fibers, IL-1 could contribute to joint erosion in RA and loss of alveolar bone in chronic inflammatory periodontal disease. IL-1 induced loss of calcium from bone cultures is inhibited by drugs blocking cyclooxygenase.

3.7 Effects of IL-1 on hemopoiesis

Recent studies have shown IL-1 to have major effects on hemopoiesis. IL-1 acts on stromal cells in the bone marrow to induce the production of G-CSF and GM-CSF [40]. IL-1 acts synergistically with these factors in stimulating the proliferation of granulocyte-monocyte precursors; IL-1 is, in fact, identical to a factor previously termed hemopoietin-1.
In addition, we have found that IL-1α and IL-1β (but not TNF-α) antagonize effects of erythropoietin on late erythroid precursors [41]. We suggest that IL-1 production may be a factor in the pathogenesis of hypoplastic anemia in RA and systemic lupus erythematosus.

3.8 The role of IL-1 in late allergic reactions

Subramanian and Bray [42] reported that IL-1 induces histamine release from human basophils. In collaborative experiments [43] we have not found that IL-1 induces histamine release from human basophils by itself but it does exert a synergistic effect with anti-IgE on histamine release. These findings are relevant to the pathogenesis of the late phase of allergy in the skin, the lung and the nose. The reaction occurring within minutes of exposure to antigen is termed the early response. In some allergic persons there is also a late response several hours after the early reaction. This can occur in the absence of further exposure to antigen, but small doses of antigen increase the severity of the late reaction. The late response is correlated with inflammation following antigen exposure and may be the counterpart of clinical disease [44].

The early phase has been attributed to release of mediators (including prostaglandin D_2) from mast cells and the late phase to release of mediators (histamine and TAME esterase, but not PGD_2) from basophils [45]. Thus following the early phase there is recruitment of circulating basophils into the reaction site. We have developed a model for this recruitment (J. Kowalski and A. C. Allison, unpublished). IL-1 acting on endothelial cells and monoclonal IgE and antigen (DNP-BSA) acting on a rat basophil cell line induce greater attachment of the basophil cells to the endothelial cells than either one alone. Thus it can be postulated that during the early phase of acute allergy IL-1 is released. This participates in the recruitment of blood basophils and acts as a co-factor with IgE and antigen in the late-phase release of histamine and other mediators from the recruited basophils.

Systemic glucocorticoid therapy inhibits the late but not the early response, although prophylaxis with topical steroids can reduce also the early response [46]. Thus the late response is more sensitive to control by glucocorticoids than the early response. Since IL-1 production is inhibited by glucocorticoids (see below) this may, at least in part, explain the efficacy of the drugs on the late response.

4   Antagonism by glucocorticoids of the formation of IL-1

Unlike polypeptide hormones, which usually interact with cell surface receptors, steroids enter cells by passive diffusion through the plasma membrane and within the cell bind high-affinity protein receptors which have specificity for different classes of steroids [47, 48]. The binding of the steroid leads to activation of the receptor-steroid complex, which is thought to involve a conformational change in the protein. The steroid-bound receptor molecules pass from the cytoplasm to the nucleus, bind to enhancer regions of several genes and act as specific transcriptional regulators to induce or repress the transcription of those genes [47, 48].

Glucocorticoids have been reported to inhibit the release of IL-1 by mouse peritoneal macrophages stimulated with lipopolysaccharide (LPS) [49]. Because inhibition of IL-1 formation by glucocorticoids has important implications for immunopharmacology, we have analyzed the phenomenon at the molecular level in the human promonocytic cell line U-937 [22] and in cultured human peripheral blood monocytes. In U-937 cells dexamethasone and other glucocorticoids were found in therapeutic concentrations to inhibit selectively the transcription of IL-1α and IL-1β genes and to induce the degradation of IL-1β mRNA (fig. 1). In monocytes no effect on transcription was observed, but glucocorticoids decreased IL-1β mRNA stability. Release of IL-1β, as measured by immunoassay, was inhibited by glucocorticoids. All these effects of glucocorticoids were blocked by the steroid receptor antagonist RU-486. The formation and stability of other mRNAs, for β-actin and the proto-oncogene *c-fos*, were unaffected by dexamethasone. The effect of glucocorticoids on mRNA stability requires the synthesis of protein, possibly a ribonuclease with activity restricted by the primary or secondary structure of mRNA. This represents an unexpected controlling mechanism for steroids, although in other situations regulation of protein synthesis by effects on mRNA stability is known [50].

5   Inhibition by glucocorticoids of the effects of IL-1

Binding of IL-1α or IL-1β to specific receptors on target cells [19] also induces expression of several genes, and this effect is blocked by glucocorticoids in therapeutic concentrations. We have analyzed in detail one situation: the induction by IL-1 of expression of the IL-6 gene in

Figure 4
Levels of IL-6 mRNA compared with β-actin mRNA in cultured synoviocytes. IL-1α or IL-1β (1 ng/ml) induce a high level of IL-6 mRNA expression at 6 h which is maintained to 24 h. Dexamethasone (10^{-6} M, Dex) completely suppresses IL-6 mRNA expression at 6 h but there is a low level of expression at 24 h.

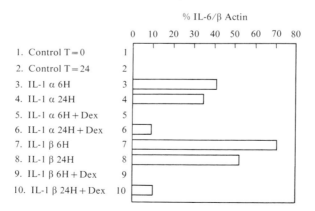

connective tissue-type cells cultured from synovial tissue of patients with rheumatoid arthritis (fig. 4). In the uninduced synoviocytes mRNA for IL-6 is not detectable. Within 6 hours of exposure to IL-1α or IL-1β, mRNA for IL-6 is formed, and maintained at a maximal level to 24 h. IL-6 is released from RA synovial tissue and cell cultures and induces synthesis of immunoglobulin in human peripheral blood B-lymphocytes activated by antibody against surface membrane immunoglobulin or pokeweed mitogen. Indeed, IL-6 appears to be the principal factor produced by RA synovial tissue that induces differentiation of B-lymphocytes; this effect in supernatants of RA synovial cultures is inhibited by antibody against IL-6. It has long been known that RA synovial tissue produces IgM, including rheumatoid factor (IgM anti-IgG), and IgG at a high rate both in culture and *in vivo* [51, 52]. Glucocorticoids strongly antagonize the induction by IL-1 of IL-6 gene expression in RA synovial cells (fig. 4).

Analogous situations are the induction by IL-1 in synovial cells of procollagenase [32] and tissue plasminogen activator [37] and decreases of the steady-state levels of the corresponding mRNAs in these cells by dexamethasone [53, 54]. IL-1 increases formation of PGE_2 in several cell types [32, 33] and of PGI_2 in endothelial cells [34], whereas glucocorticoids inhibit the production of PGE_2 in several cell types [55] and PGI_2 in endothelial cells [56]. It will be interesting to ascertain whether these effects of glucocorticoids are due to inhibition of expression of phospholipase A_2 and cyclooxygenase genes in sensitive target cells.

6 IL-1 induces the release of glucocorticoids

IL-1 formation is a central feature of inflammatory responses, required for defense against bacteria and other invaders. However, the formation and activity of IL-1 must be regulated, and glucocorticoids provide one feedback control mechanism. In mice injected with IL-1, plasma corticosterone is considerably increased [57–59], and this would suppress further IL-1 production. The mechanism by which IL-1 mediates this effect has recently been analyzed [58, 59]. IL-1 induces the secretion from the hypothalamus, by a prostaglandin-independent mechanism, of corticotrophin-releasing factor, which passes through the portal circulation to the anterior pituitary and releases ACTH. IL-1 does not directly induce the secretion of ACTH by pituitary cells or of glucocorticoids by adrenal cortical cells.

Figure 5
Inhibition by IL-1β of the induction of alkaline phosphatase in endothelial cells [ref. 61].

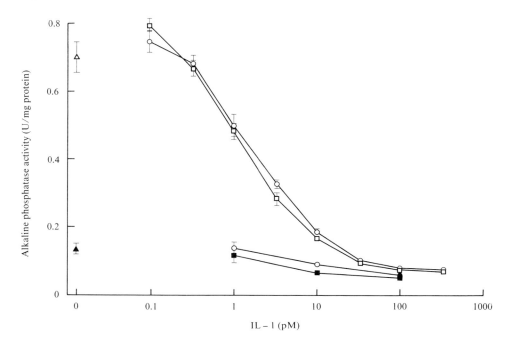

7 Antagonism by IL-1 of the effects of glucocorticoids

The antagonism can also be effected in the opposite direction: IL-1 decreases the induction by glucocorticoids of phosphoenolpyruvate carboxykinase in liver cells [60] and of alkaline phosphatase in endothelial cells [61] (fig. 5). The latter effect is not mediated by a detectable change in the number or affinity of glucocorticoid receptors. However, induction by glucocorticoids of tyrosine aminotransferase in rat hepatoma cells is much less sensitive to inhibition by IL-1, so the antagonism may be selective (M. Mulkins and A. C. Allison, unpublished).

8 Acute-phase protein synthesis

In addition to the effects described above, glucocorticoids act synergistically with an inflammatory mediator or mediators to induce the synthesis of acute-phase proteins by hepatocytes [62]. At least some of

Figure 6
Diagram showing the antagonistic effects of IL-1 and glucocorticoids. Stimulatory effects are shown by solid arrows, inhibitory effects by interrupted arrows.

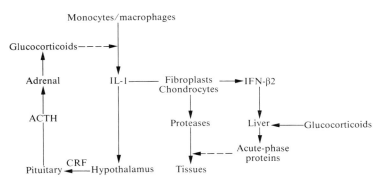

these proteins (e.g. α_2-macroglobulin and α_1-proteinase inhibitor) inhibit protease activity and can be considered anti-inflammatory and anti-catabolic. It was formerly believed that IL-1 increases the synthesis of acute-phase proteins by hepatocytes. However it has more recently been shown that IL-6, also known as interferon-$\beta 2$, and not IL-1, induces the synthesis of acute-phase proteins by these cells [63]. In several cell types IL-1 stimulates the formation of IL-6 [64 and see above], so it seems likely that the alleged *in vivo* effects of IL-1 in inducing acute-phase proteins *in vivo* may be due to a cascade of mediators, including IL-6.

Thus glucocorticoids may not act synergistically with IL-1 to induce synthesis of acute-phase proteins by hepatocytes, which would contradict to the rule of antagonism (or at least independent action) of these mediators proposed in this chapter. Instead IL-1 induces the release of both glucocorticoids and IL-6; these act synergistically to increase the formation and release of acute-phase proteins, which can oppose some effects of IL-1 and thereby constitute part of a feedback control system (fig. 6).

9 Immunosuppressive effects of glucocorticoids

The preceding section has been concerned with the anti-inflammatory effects of glucocorticoids. The inhibition of IL-1 formation by glucocorticoids also contributes to their immunosuppressive effects. Different subsets of lymphocytes vary in sensitivity to glucocorticoids, which in these cells selectively suppress transcription of some genes but not others. Proliferation of B-lymphocytes and their differentiation into antibody-forming cells are relatively resistant to glucocorticoids [65, 66]. Clones of helper T-lymphocytes are, in general, more sensitive to glucocorticoids than clones of cytotoxic T-lymphocytes [67]. In mice helper lymphocytes cloned in IL-2 can be divided into a T_H and a T_H2 subset [68]. T_H cells activated by antigen produce IL-2 and IFN-γ whereas T_H2 cells activated by antigen produce IL-4. The T_H2 subset is more sensitive than T_H1 to glucocorticoids, which strongly decrease levels of mRNAs for IL-4 and IL-5 in these cells [69 and F. Lee, personal communication 1988]. As described above, glucocorticoids strongly inhibit the expression of the IL-6 gene induced by IL-1 (fig. 4). IL-6 is a major B-cell differentiation factor produced by T-lymphocytes as well as several other cell types. In contrast glucocorticoids do not decrease levels of IL-2 receptor and β-chain T-cell receptor mRNAs, which are expressed in the absence of antigenic stimulation, and they do not inhibit expression of the preproencephalin gene, which requires antigenic stimulation. Glucocorticoids also inhibit class II major histocompatibility gene expression in mouse macrophages and B-lymphocytes [70].

As expected from the molecular biological findings, glucocorticoids antagonize some helper effects on antibody formation [71]. This effect of the drugs can be overcome by the addition of cytokines [72], later shown to be IL-1 (R. I. Mishell, personal communication, 1988). This is evidence that in the models studied inhibition of IL-1 production by glucocorticoids is the limiting factor in helper effects on antibody formation.

Glucocorticoids also suppress granulopoiesis, and this effect can be reversed by addition of IL-1 [73]. Thus inhibition of IL-1 formation contributes not only to anti-inflammatory effects of glucocorticoids but also to their effects on the immune system and hematopoiesis.

10 Side effects of glucocorticoids

Major side effects of glucocorticoids are on carbohydrate metabolism, which complicates the use of these drugs in diabetics. Induction of phosphoenolpyruvate carboxykinase and in the liver [60] and other regulatory mechanisms [47] explain these effects. A side effect of continued usage is skin thinning, which is attributable to decreased transcription of the procollagen type 1 gene [74]. Other important side effects of glucocorticoids, such as loss of calcium from bones, have not yet been explained in molecular terms.

Acknowledgments

We are grateful to the following colleagues who contributed to various aspects of our research reviewed in this paper: H. Chan, E. Eugui, J. Kenney, J. Kowalski, M. Masada, Y. Amano, Y. Nawata, A.-P. Tsou and R. Waters.

References

1 R. J. Flower and G. J. Blackwell: Nature *278*, 456 (1979).
2 F. Hirata, E. Schiffman, K. Venkatsubramanian, D. Solomon and J. Axelrod: Proc. natl Acad. Sci. USA *77*, 2533 (1980).
3 M. R. Crompton, S. E. Moss and M. J. Crumpton: Cell *55*, 1 (1988).
4 T. Hunter: Adv. exp. Med. Biol. *234*, 169 (1988).
5 B. P. Wallner, R. J. Mattaliano, C. Hessian, R. L. Cate, R. Tizard, L. V. Sinclair, C. Foeller, E. P. Chow, J. L. Browning, K. L. Ramchandran, and R. B. Pepinsky: Nature *320*, 72 (1986).
6 K.-S. Huang, B. P. Wallner, R. J. Mattaliano, R. Tizard, C. Burne, A. Fey, C. Hession, P. McGray, L. K. Sinclair, E. P. Chow, J. L. Browning, K. L. Ramchandran, J. Yang, J. E. Smart and R. B. Pepinsky: Cell *46*, 191 (1986).
7 R. B. Pepinsky, R. Tizard, R. J. Mattaliano, L. K. Sinclair, G. T. Miller, J. L. Browning, E. P. Chow, C. Burne, K.-S. Huang, D. Pratt, L. Wachter, C. Hession, A. Z. Frey and B. P. Wallner: J. biol. Chem. *263*, 10799 (1988).
8 G. Cirino, R. J. Flower, J. L. Browning, L. K. Sinclair and R. B. Pepinsky: Nature *238*, 270 (1987).
9 C. J. Saris, B. F. Tack, T. Kristensen, J. R. Glenny Jr and T. Hunter: Cell *46*, 201 (1986).
10 R. Kaplan, M. Jaye, W. H. Burgess, D. D. Schlaepfer and H. T. J. Haigler: Biol. Chem. *263*, 8037 (1988).
11 M. Brönnegaard, O. Anderson, D. Edwall, J. Lund, G. Nonstedt and J. Carlstedt-Duke: Molec. Endocr. *2*, 732 (1988).
12 C. M. Isacke, R. A. Lindberg and T. Hunter: Molec. cell. Biol. in press, 1988.
13 F. F. Davidson, E. A. Dennis, M. Powell and J. R. Glenney Jr: J. biol. Chem. *262*, 1698 (1987).
14 H. T. Haigler, D. D. Schlaepfer and W. H. Burgess: J. biol. Chem. *262*, 6921 (1987).
15 M. D. Hollenberg, J. K. Northrup, K. A. Valentine-Braun, L. K. Johnson and D. L. Severson: Clin. Res. *35*, 639a (1987).
16 P. E. Auron, A. C. Webb, J. L. Rosenwasser, S. F. Mucci, A. Rich, S. M. Wolff and C. A. Dinarello: Proc. natl Acad. Sci. USA *81*, 7907 (1984).
17 C. J. March, B. Mosley, A. Larsen, D. P. Cerretti, G. Braedt, V. Price, S. Gillis, C. S. Henney, S. R. Kronheim, K. Grabstein, P. J. Conlon, T. P. Hopp and D. Cosman: Nature *315*, 641 (1985).
18 V. Gubler, A. O. Chua, A. S. Stern, C. P. Hellmann, M. P. Vitek, T. M. Dechiara, W. R. Benjamin, K. J. Collier, M. Kukovitch, P. C. Familletti,

C. Fielder-Nagy, J. Jensen, K. Kaffka, P. L. Kibian, D. Stremlo, B. H. Wittreich, D. Woehle, S. B. Mizel and P. T. Lomedico: J. Immun. *136,* 2492 (1986).
19 J. E. Sims, C. J. March, D. Cosman, M. B. Widmer, H. R. MacDonald, C. J. McMahan, C. E. Grubin, J. M. Wignall, J. L. Jackson, S. M. McCall, D. Friend, A. R. Alpert, S. Gillis, D. L. Urdal and S. K. Dower: Science *241,* 585 (1988).
20 J. S. Kenney, M. P. Masada, E. M. Eugui, B. De Lustro, M. A. Mulkins and A. C. Allison: J. Immun. *138,* 4236 (1987).
21 P. J. Lisi, C.-W. Chu, G. A. Koch, S. Endres, G. Lonnemann and C. A. Dinarello: Lymphokine Res. *6,* 229 (1987).
22 S. W. Lee, A. P. Tsou, H. Chan, J. Thomas, K. Petrie, E. M. Eugui and A. C. Allison: Proc. natl. Acad. Sci. USA *85,* 1204 (1988).
23 E. K. Bayne, E. A. Rupp, G. Limjuco, J. Chin and J. A. Schmidt: J. exp. Med. *163,* 1267 (1986).
24 D. J. Hazuda, J. C. Lee and P. R. Young: J. biol. Chem. *263,* 8473 (1988).
25 A. Fontana, H. Hengartner, E. Weber, K. Fehr, P. K. Grob and G. Cohen: Rheumat. Int. *2,* 49 (1982).
26 D. D. Wood, E. J. Ihrie, C. A. Dinarello and P. L. Cohen: Arthr. Rheum. *26,* 975 (1983).
27 A. C. Allison: In Immunopathogenetic Mechanisms of Arthritis, p.211. Eds. J. Goodacre and W. C. Dick (Eds). MTP Press, Lancaster (1988).
28 J. S. H. Gaston, S. Strober, J. J. Solovera, D. Gaudow, N. Lane, D. Schurman, R. T. Hoppe, R. C. Chen, E. M. Eugui, J. H. Vaughan and A. C. Allison: Arthr. Reum. *31,* 21 (1988).
29 R. C. Page: Int. Dent. J. *36,* 153 (1986).
30 M. P. Bevilacqua, J. S. Pober, M. E. Wheller, R. S. Cotran and M. A. Gimbrone Jr: J. clin. Invest. *76,* 2003 (1985).
31 D. E. Cavender, D. O. Haskard, B. Joseph and M. Ziff: J. Immun. *136,* 203 (1986).
32 J.-M. Dayer and S. Demczuk: Springer Semin. Immunopath. *7,* 387 (1984).
33 J. Chang, S. C. Gilman and A. J. Lewis: J. Immun. *136,* 1283 (1986).
34 V. Rossi, F. Brevario, P. Ghezzi, E. Dejana and A. Mantovani: Science *229,* 174 (1985).
35 J. Schnyder, T. Payne and C. A. Dinarello: J. Immun. *138,* 496 (1987).
36 R. L. Smith, A. C. Allison and D. J. Schuurman: Connect. Tiss. Res. in press (1988).
37 T. Leizer, B. J. Clarris, P. E. Ash, J. van Damme, J. Saklatvala and J. A. Hamilton: Arthr. Rheum. *30,* 562 (1987).
38 M. Gowen, D. D. Wood, E. J. Ihrie, M. K. B. McGuire and R. G. G. Russell: Nature *306,* 378 (1983).
39 P. Stashenko, F. E. Dewhirst, W. J. Peros, R. Kent and J. M. Ago: J. Immun. *138,* 1464 (1987).
40 D. Pennick, G. Yang, L. Gemmell and F. Lee: Blood *69,* 682 (1987).
41 J. C. Schooley, B. Kullgren and A. C. Allison: Br. J. Haemat. *67,* 11 (1987).
42 N. Subramanian and M. A. Bray: J. Immun. *138,* 271 (1987).
43 W. A. Massey, T. Randall, S. M. MacDonald, S. Gillis. A. C. Allison, A. Kagey-Sobotka and L. M. Lichtenstein: FASEB J. *2,* A1251 (1988).
44 G. J. Gleich, J. Allergy clin. Immun. *70,* 160 (1982).
45 R. M. Naclerio, D. Proud, A. G. Tokias, N. F. Adkinson Jr. D. A. Meyers, A. Kagey-Sobotka, M. Plant, P. S. Norman and L. M. Lichtenstein: New Engl. J. Med. *313,* 65 (1985).
46 V. Pipkorn, D. Proud, L. M. Lichtenstein, A. Kagey-Sobotka, P. S. Norman and R. M. Naclerio: New Engl. J. Med. *316,* 1506 (1987).
47 K. R. Yamamoto: A. Rev. Genet. *19,* 209 (1985).
48 S. Green and P. Chambon: Nature *324,* 615 (1986).

49 D. S. Snyder and E. R. Unanue: J. Immun. *129*, 1803 (1982).
50 T. Morris, F. Marashi, L. Weber, E. Hickey, D. Greenspan, J. Bonner, J. Stein and G. Stein: Proc. natl Acad. USA *83*, 981 (1986).
51 R. M. Werrick, P. E. Lipsky, E. Marbon-Arcos, J. J. Maliakkol, D. Edelbaum and M. Ziff: Arthr. Rheum. *28*, 742 (1985).
52 A. J. Slivinski and N. J. Zwaifler: J. Lab. clin. Med. *76*, 304 (1970).
53 C. E. Brinckerhoff, I. M. Plecinska, L. A. Sheldon and G. T. O'Connor: Biochemistry *25*, 6378 (1986).
54 R. L. Medcalf, R. I. Richards, R. J. Crawford and J. A. Hamilton: EMBO J. *5*, 2217 (1986).
55 M. D. Mitchell, B. R. Carr, J. I. Mason and G. R. Simpson: Proc. natl Acad. Sci. USA *79*, 7547 (1982).
56 D. J. Crutchley, U. S. Ryan and J. W. Ryan: J. Pharm. exp. Ther. *233*, 650 (1985).
57 B. M. R. N. J. Woloski, G. M. Smith, W. J. Meyer III, G. M. Fuller and J. E. Blalock: Science *230*, 1035 (1985).
58 H. Besedovsky, A. del Rey, E. Sorkin and C. A. Dinarello: Science *233*, 652 (1986).
59 R. Sapolsky, A. Rivier, K. Yamamoto, P. Plotsky and W. Vale: Science *238*, 522 (1987).
60 M. R. Hill, R. D. Smith and R. E. McCallum: J. Immun. *137*, 858 (1986).
61 M. Mulkins and A. C. Allison: J. Cell Physiol. *133*, 539 (1988).
62 A. Koj, J. Gauldie, E. Regoeczi, D. N. Sauder and G. D. Sweeney: Biochem. J. *224*, 505 (1984).
63 J. Gauldie, C. Richards, D. Harnish, P. Lansdorf and H. Baumann: Proc. natl Acad. Sci. USA *84*, 7251 (1987).
64 P. B. Sehgal, L. T. May, I. Tamm and J. Vilcek: Science *235*, 731 (1987).
65 T. C. Cupps and A. S. Fauci: Immun. Rev. *65*, 133 (1982).
66 T. Paavonen: Scand. J. Immun. *21*, 63 (1985).
67 A. Kelso and A. Munck: J. Immun. *133*, 784 (1984).
68 H. M. Cherwinski, J. H. Schumacher, K. D. Brown and T. R. Mossmann: J. exp. Med. *124*, 1669 (1987).
69 J. Culpepper and F. Lee: In Molecular Cloning and Analysis of Lymphokines p. 275. Eds. D. R. Webb and D. V. Goeddel. Academic Press, Orlando (1987).
70 V. M. McMillan, G. J. Dennis. L. H. Glimcher, F. D. Finkelman and J. H. Mond: J. Immun. *140*, 2549 (1988).
71 L. M. Bradley and R. I. Mishell: Eur. J. Immun. *12*, 91 (1982).
72 L. M. Bradley and R. I. Mishell: Cell Immun. *73*, 115 (1982).
73 R. I. Mishell, D. A. Lee, K. H. Grabstein and L. B. Lachman: J. Immun. *128*, 1614 (1982).
74 D. Cockayne, K. M. Sterling Jr, S. Shull, K. P. Mintz, S. Illeyne and K. R. Cutroneo: Biochemistry *25*, 3202 (1986).

The use of neutrophils, macrophages and organ cultures to assess the penetration of human cells by antimicrobials

By Zell A. McGee, Gary L. Gorby and Wanda S. Updike

The Center for Infectious Diseases, Diagnostic Microbiology and Immunology, University of Utah, School of Medicine, Salt Lake City, Utah 84132, USA

1	Introduction	84
2	PMN Leukocytes	86
3	Macrophages	88
4	Organ cultures	89
5	Summary	91
	References	91

1   **Introduction**

The knowledgeable physician, choosing an antimicrobial to aid in the management of an infectious disease, does not simply pick a broad-spectrum antimicrobial, which may not be the best antimicrobial for the infecting microorganism, but rather goes through a three-step thought process wherein 1) he or she diagnoses the *disease* in the patient; and 2) on the basis of historical, physical and laboratory information, predicts the *most likely organism* to be causing the disease in this particular patient and then; 3) chooses the *best antimicrobial* for the most likely organism or organisms. Once the list of most likely organisms has been generated from appropriate clinical and laboratory data, the choice of the best antimicrobial is systematically made on the basis of the factors listed in Table 1.

Table 1
Major factors to be considered in choosing the *best antimicrobial* in a particular patient.

a)	Assure that the candidate antimicrobial is likely to be active against the most likely infecting organism.
b)	From the drugs likely to be active against the most likely organism, choose drugs which are delivered to the site of the infection.
c)	Choose a bactericidal drug if that is necessary for the particular type of infection (e.g. endocarditis) or the particular host (e.g. neutropenia).
d)	Choose drugs whose complications, though they be infrequent, do not especially threaten the particular patient.
e)	Of the drugs remaining, determine which is the least expensive.

In this article, the authors will focus on Step b in Table 1, determination of whether the drug is delivered to the site of the infection – for example, in meningitis, delivery of certain antimicrobials (e.g. aminoglycosides) into the CSF is quite difficult [1]. Problems of comparable or greater difficulty are encountered when the site of the infection is the intracellular space. We have long appreciated the fact that many bacterial infections are primarily intracellular (see Table 2). The term "intracellular" has been used to connote the residence of the bacterial pathogens within host cells, usually "professional phagocytes" such as macrophages or cells of the reticuloendothelial system. Traditional intracellular parasites are listed in Table 2; however, this more traditional view of intracellular pathogens is much too limited. Studies using organ cultures of various mucosal tissues have shown that a wide variety of human pathogens attach to mucosal epithelial cells and induce

Table 2
Traditional and newly-appreciated intracellular pathogens.

Traditional:
- Brucella species (e.g. abortus, suis)
- Mycobacteria species (e.g. tuberculosis, leprae, etc.)

- *Listeria monocytogenes*
- Legionella species (e.g. pneumophila, micdadei)
- Rickettsia species (e.g. rickettsii, akari, etc.)
- Chlamydia species (e.g. trachomatis, psittaci, etc.)

Newly appreciated [2]:
- *Neisseria gonorrhoeae*
- *Neisseria meningitidis*
- *Haemophilus influenzae*
- entero-invasive *E. coli*, Shigella, Salmonella

---

their own internalization and transport across the epithelial barrier by means of a process which has been designated "parasite-directed endocytosis" [2]. The pathogens that participate in the process whereby they are endocytosed by endothelial cells, phagocytic cells and cells of mucosal epithelia are listed in Table 2. Whereas the studies documenting the intracellular residence of *N. gonorrhoeae,* and *H. influenzae* in human epithelial cells have been performed in human mucosal organ cultures [3–5], studies of naturally occurring disease in humans [6, 7] or experimental animals [8] have verified the observations made in organ cultures.

The ability of antimicrobials to penetrate human cells differs greatly, depending on the type of antimicrobial, some entering cells easily, others, in low concentrations only or not at all. Penetration of antimicrobials into human cells does not appear to depend substantially on the type of human cell. Most studies of cellular penetration of antimicrobials have been performed using human polymorphonuclear leucocytes (PMN's) or macrophages.

In treating bacterial infections, one often focuses concern primarily on extracellular microbes in the bloodstream or in anatomic foci of infection where they are relatively accessible to antimicrobials. However, because microbes may become inaccessible to antimicrobials once phagocytized by polymorphonuclear leukocytes (PMN's) or macrophages, but are not necessarily killed by these cells and they may emerge after the extracellular antimicrobial is gone, one's concern should also be focused on the intracellular phase of the infection.

There are a number of disorders in which PMN's have defective killing mechanisms that make the host more susceptible to certain infecting microbes [9] and there are some infecting microbes (e.g. *Salmonella typhi* and *Brucella abortus*) that have adapted to intracellular survival and multiplication while evading killing by PMN's [10, 11]. It is under these conditions that the intracellular concentrations of antibiotics assume new importance in determining the efficacy of antimicrobial therapy.

Even in the setting of normally functioning PMN's and infecting organisms that are susceptible to neutrophil-mediated killing, the intracellular penetration of antimicrobial agents may hasten the demise of phagocytized bacteria [12].

2 PMN Leukocytes

To effectively design antimicrobial regimens that can eradicate bacteria within PMN's, one must consider the ability of various antimicrobials to penetrate PMN's. A summary of antimicrobial penetration of PMN's is provided in Table 3. Briefly, Table 3 indicates that roxithromycin, erythromycin, clindamycin, tetracycline, ciprofloxacin, trimethoprim/sulfamethoxazole, ethambutol, rifampin, isoniazid, chloramphenicol, lincomycin and metronidazole achieve concentrations inside neutrophils equal to or up to 40 times greater than the extracellular concentration of the antimicrobial.

However, the intracellular penetration of an antimicrobial does not guarantee that it will be able to augment the killing of bacteria that are residing within host PMN's. To do so requires: 1) that the antibiotic be delivered to the intracellular location of the bacteria, which in most cases is the phagosome or endocytic vacuole; and 2) that the antimicrobial be able to function within the intraphagosomal acidic environment.

Many studies of cellular penetration of antimicrobials only show that the antimicrobial reaches the cytoplasm of cells but do not incisively demonstrate that the antimicrobial reaches the true site of infection, the phagosome or endocytic vacuole.

Some antibiotics, especially erythromycin, rifampin, ciprofloxacin, clindamycin, trimethoprim/sulphamethoxazole are concentrated in neutrophils and might be expected to augment killing of intracellular bacteria or assist defective neutrophils in generating bactericidal activ-

Table 3
Penetration of antibiotics into PMN leukocytes expressed as a Cellular Extracellular (C:E) ratio.

Antibiotic	C:E Ratio*	Citation
Roxithromycin	34	[13]
Erythromycin	4.4	[14]
	18.0–24.0	[15]
	13.32	[16]
Clindamycin	40	[17]
	40	[18]
	30	[19]
	11.08	[16]
Tetracycline	7.1	[20]
Ciprofloxacin	5.5	[21]
Trimethoprim	9–13	[13]
	4.0	[19]
	4.1	[20]
Sulfamethoxazole	1.7	[20]
Ethambutol	4.83	[16]
Rifampin	2.43	[16]
	2.2	[12]
Isoniazid	1.5	[16]
Chloramphenicol	2.63	[16]
Lincomycin	1.78	[16]
Metronidazole	1.0	[13]
Gentamicin	0.84	[16]
Imipenem	3.3–.09	[13]
Penicillin	0.7	[15]
	0.4	[16]
	0.3	[20]
Cephalexin	0.57	[16]
Cefazolin	<0.01	[16]
Cefamandole	<0.01	[16]
Cefotaxime	<0.3	[13]

* Lower values indicate less penetration and lower intracellular concentrations.

ity. Only rifampin, ciprofloxacin and trimethoprim/sulfamethoxazole seem to have bactericidal activity within neutrophils [12, 15, 19] whereas clindamycin, erythromycin and roxithromycin seem to possess only bacteriostatic activity within leukocytes [15, 17, 23]. It is important to treat intracellular infections and infections in the setting of defective host neutrophils with antimicrobials that penetrate neutrophils well. Further research is needed to better define the role of intracellularly active antibiotics in a wide variety of infections.

In a broader context, it is also very important that antibiotics not affect adversely crucial neutrophil functions such as chemotaxis, phagocytosis, respiratory burst and degranulation. For example, at therapeutic levels, cefoperazone, a beta lactam unlikely to penetrate human cells, has been shown to inhibit neutrophil chemotaxis [22] and the phagocytic function of PMN's and macrophages as well [23]. This drug also compromises another major host defense system by wiping out normal flora, so that its substantial risks must be weighed against its potential benefits.

3   Macrophages

Many of the problems of delivering antimicrobials into PMN's also apply to delivering antimicrobials into macrophages and the patterns

Table 4
Relative penetration of antibiotics into macrophages.

Antibiotic	Relative macrophage penetration	Reference
Ampicillin	low	[25]
Cefazolin	low	[26]
Penicillin G	low	[26]
Cefamandole	moderate	[26]
Gentamicin	moderate	[25, 26]
Isoniazid	moderate	[26]
Streptomycin*	moderate	[27, 28]
Sulfamethoxazole-trimethoprim	moderate	[29]
Tetracycline	moderate	[26, 29]
Ciprofloxacin	high	[29, 30]
Chloramphenicol	high	[26]
Ethambutol	high	[26]
Ofloxacin	high	[29, 31]
Pefloxacin	high	[29]
Rifampin	high	[25, 26, 29]
Clindamycin**	very high	[26]
Erythromycin**	very high	[26, 29]

*   Significant accumulation in macrophages over a period of days.
** Active transport of antimicrobial into macrophages with intracellular concentration greater than 10 times extracellular concentration.

of antimicrobial penetration are predictably parallel. As will be noted by comparing Table 3 with Table 4, the penicillins and cephalosporins penetrate phagocytic cells poorly, whereas erythromycin, clindamycin, rifampin and ciprofloxacin penetrate well.

4   Organ cultures

Organ cultures, such as those of the fallopian tube mucosa [32] and nasopharyngeal mucosa [33] offer a number of advantages over PMN's and macrophages for the study of antimicrobial penetration of human cells: 1) the nonciliated, low columnar cells of the fallopian tube mucosa and the nasopharyngeal mucosa endocytose a variety of bacteria [2], but do not appear to have lysosomes or lysosomal enzymes, so that they do not kill bacteria within the endocytic vacuoles and any killing effect that results with the administration of an antimicrobial is likely to be the result of the antimicrobial acting alone rather than in concert with lysosomal enzymes; 2) organ cultures provide endocytosing cells in their usual cellular milieu, rather than isolated phagocytic cells attached to synthetic surfaces; 3) organ cultures can be maintained in serum-free medium [32], so that the problem of binding of antimicrobials to serum proteins is obviated.

In one study of the penetration of human mucosal epithelial cells by antimicrobials, Schlech et al. [34] showed that if human fallopian tube mucosa in organ culture was infected with gonococci and treated with penicillin 18 to 20 hours later ("attachment phase"), when the organisms were attached to the surface of the epithelium, but prior to "invasion", homogenization of the tissues and plate counts revealed that all the organisms were dead. In contrast, if treatment with penicillin was delayed until 48 hours after infection ("invasion phase") when the organisms were in endocytic vacuoles at the base of the epithelial cells (see Fig. 1), substantial numbers of gonococci remained viable when the tissues were homogenized and plate counts were performed [34]. However, if the tissues were treated with rifampin during the "invasion" phase, all the organisms were killed.

Thus, organ cultures provide a convenient and easily managed model for testing the penetration of antimicrobials into human cells. In contrast to many studies of antimicrobial penetration, the organ culture model of Schlech et al. [34] does not simply test the entry of drugs into the cytoplasm, which has little relevance to human disease, but the

Figure 1
Human fallopian tube mucosa in organ culture 48 hours after infection with *Neisseria gonorrhoeae*. Note entire thickness of the epithelium with large vacuoles packed with gonococci near the base of the nonciliated cells. The organisms have dark chromatin and occur as diplococci, indicating their viability.

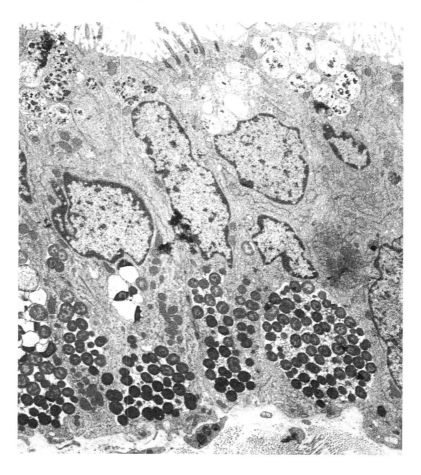

model allows assessment of antimicrobial penetration into endocytic vacuoles, where pathogenic bacteria reside in the course of infectious diseases. Thus, the organ culture model may be more predictive of antimicrobial efficacy in intracellular infections than are studies performed using PMN's or macrophages.

5 Summary

Recent studies of pathogenic mechanisms of bacteria have revealed that in addition to traditional intracellular parasites such as *Brucella* spp. and Listeria, many other human pathogens reside inside cells during disease proceses. Thus, studies of how well antimicrobials are delivered to and perform within phagocytic or endocytic vacuoles have become of increasing importance. Whereas studies of penetration of PMN neutrophils and macrophages have indicated whether antimicrobials penetrate the cytoplasm of human cells, studies in organ cultures can reveal whether antimicrobials enter phagosomes or endocytic vacuoles where the bacteria actually reside. Such information is probably more predictive of antimicrobial efficacy in naturally occurring intracellular infections than are data from studies with PMN's or macrophages.

References

1. A. B. Kaiser and Z. A. McGee: N. Engl. J. Med. *293*, 1215 (1975).
2. Z. A. McGee, G. L. Gorby, P. B. Wyrick, R. Hodinka and L. H. Hoffman: Rev. infect. Dis. *10*, S 311 (1988).
3. Z. A. McGee and R. G. Horn: in: Microbiology 1979, pp. 158–161. Ed. D. Schlessinger. American Society for Microbiology, Washington D.C. 1979.
4. D. S. Stephens, L. H. Hoffman and Z. A. McGee: J. infect. Dis. *148*, 369 (1983).
5. M. M. Farley, D. S. Stephens, M. H. Mulks, M. D. Cooper, J. V. Bricker, S. S. Mirra and A. Wight: J. infect. Dis. *154*, 752 (1986).
6. M. E. Ward and P. J. Watt: J. infect. Dis. *126*, 248 (1977).
7. C. E. Pollock and L. W. Harrison: Gonococcal Infections. Oxford University Press, London 1912.
8. A. L. Smith, M. C. Roberts, J. E. Haas, T. L. Stull and P. M. Mendelman: in: Bacterial Meningitis, pp. 11–21. Eds. M. A. Sande, A. L. Smith and R. K. Root. Churchill Livingstone, New York 1985.
9. R. K. Root, A. S. Rosenthal and D. J. Balestra: J. clin. Invest. *51*, 649 (1972).
10. R. E. Kossack, J. Schadelin, R. L. Guerrant, P. Densen and G. L. Mandell: Clin. Res. *26*, 28 A (1978).
11. D. L. Kreutzer, L. A. Dreyfus and D. C. Robertson: Infect. Immun. *23*, 737 (1979).
12. G. L. Mandell: J. clin. Invest. *52*, 1673 (1973).
13. W. L. Hand, N. K. King-Thompson and J. W. Holman: Antimicrob. Agents Chemother. *31*, 1553 (1987).
14. G. A. Dette and H. Knothe: J. Antimicrob. Chemother. *18*, 73 (1986).
15. M. F. Miller, J. R. Martin, P. Johnson, J. T. Ulrich, E. J. Rdzok and P. Billing: J. infect. Dis. *149*, 714 (1984).
16. R. C. Prokesch and W. L. Hand: Antimicrob. Agents Chemother. *21*, 373 (1982).
17. C. S. F. Easmon and J. P. Crane: Br. J. exp. Path. *65*, 725 (1984).

18 M. S. Klempner and B. Styrt: J. infect. Dis. *144*, 472 (1981).
19 R. F. Jacobs and C. B. Wilson: Pediatr. Res. *17*, 916 (1983).
20 F. K. Gmunder and R. A. Seger: Pediatr. Res. *15*, 1533 (1981).
21 C. S. F. Easmon, J. P. Crane and A. Blowers: J. Antimicrob. Chemother. *18* (suppl. D), 43 (1986).
22 A. Fieta, F. Sacchi, C. Bersani, F. Grassi, P. Mangiarotti and G. G. Grassi: Antimicrob. Agents Chemother. *23*, 930 (1983).
23 H. Ohnishi, H. Kosuzume, H. Inaba, M. Okura, H. Mochizuki, Y. Suzuki and R. Fujii: Antimicrob. Agents Chemother. *23*, 874 (1983).
24 R. Anderson, C. E. J. Van Rensburg, G. Joone and P. T. Lukey: J. Antimicrob. Chemother. *20* (suppl. B), 57 (1987).
25 M. C. Lobo and G. L. Mandell: Proc. Soc. exp. Biol. Med. *142*, 1048 (1973).
26 J. D. Johnson, W. L. Hand, J. B. Francis, N. King-Thompson and R. W. Crown: J. Lab. clin. Med. *95*, 429 (1980).
27 U. T. Chang: Appl. Microbiol. *17*, 750 (1969).
28 P. F. Bonventre and J. G. Imhoff: Infect. Immun. *2*, 89 (1970).
29 J. L. Vilde', E. Dournon and P. Rajagopalan: Antimicrob. Agents Chemother. *30*, 743 (1986).
30 R. B. Fitzgeorge: J. infect. Dis. *10*, 189 (1985).
31 A. Saito, K. Sawatan, Y. Fukuda, M. Nagasawa, H. Koga, A. Tomonaga, H. Nakagato, K. Fujita, Y. Shigeno, Y. Suzuyama, K. Yamaguchi, K. Izumikawa and K. Hara: Antimicrob. Agents Chemother. *28*, 15 (1985).
32 Z. A. McGee, A. P. Johnson and D. Taylor-Robinson: Infect. Immun. *13*, 608 (1976).
33 D. S. Stephens, L. H. Hoffman and Z. A. McGee: J. infect. Dis. *148*, 369 (1983).
34 W. F. Schlech, III and Z. A. McGee: Clin. Res. *27*, 778 A (1979).

RNA virus evolution and the control of viral disease

By Esteban Domingo*
Department of Biology, University of California San Diego, La Jolla, California 92093, USA

* On leave from Centro de Biologia Molecular (CSIC–UAM), Canto Blanco, 28049-Madrid, Spain.
Dedicated to the memory of Prof. Ferran Calvet i Prats who died in Barcelona on the 16th of June, 1988.

1	Introduction: the dynamics of RNA genomes	94
2	RNA versus DNA mutation rates, mutation frequencies and fixation of mutations	96
3	Genetic heterogeneity of RNA viruses	100
4	Population equilibrium and quasi-species	103
5	Phenotypic heterogeneity	105
6	Resistance of RNA viruses to antiviral agents	105
6.1	Drugs that act at an early stage of the virus infectious cycle	106
6.2	Drugs that act at an intracellular step of virus development	110
6.3	Additional antiviral strategies	112
6.4	Combination therapy	113
7	Vaccines	114
7.1	Virus variation and vaccine design	116
7.2	Is there a viable strategy?	119
8	Error-prone polymerase subsets, hypermutability, and error catastrophe. Is there a limit to RNA virus variation?	120
9	Conclusion	122
10	Summary	123
	Acknowledgments	124
	References	124

Abbreviations

AIDS, acquired immune deficiency syndrome; DI, defective interfering; ds RNA, double-stranded RNA; FMDV, foot-and-mouth disease virus; HIV, human immunodeficiency virus; IV, influenza virus; M, matrix; MV, measles virus; PV, polio virus; RSV, respiratory syncytial virus; SSPE, subacute sclerosing panencephalitis; s/s/yr, substitutions per site per year; ts, temperature sensitive; VSV, vesicular stomatitis virus.

1 Introduction: the dynamics of RNA genomes

RNA molecules display several features – some of which have been recognized only very recently – that contribute to their potential for modification and evolution. Cellular transcripts as well as autonomous RNA genetic elements may possess catalytic activity [1–3]. The following examples demonstrate the catalytic ability of RNA molecules, an activity which in the past was associated only with proteins. The active site of ribonuclease P – an enzyme that processes precursor tRNA molecules – resides in RNA [4,5]. Excision of the intervening sequence and subsequent splicing of *Tetrahymena* rRNA [6,7] is a protein-independent process. Regarding autonomous genetic elements, replication of viroids and certain plant virus satellites involves some protein-free RNA modification steps [8–12]. The catalytic potential of RNA includes hydrolysis of substrate RNA molecules *in trans* [13,14] and addition of nucleotides to preexisting chains [15–17]. Thus, biological catalysis, embodied in polynucleotide chains, adds to the potential for generating novel molecules in evolving RNA populations. This has strengthened the belief in a central role of RNA or of RNA-like molecules in the self-organization of a genetic memory and the early evolution of life on earth [18–20].
Present day RNA-replicating elements have developed other mechanisms of genetic variation to ensure their adaptability. Molecular recombination, initially thought to occur mainly in DNA, has now been shown to play an important role in several positive stranded RNA viruses [21]. [An RNA virus is positive stranded when the polarity of virion RNA coincides with that of mRNA; if virion RNA is of opposite polarity to mRNA the virus is negative stranded.]. Early observations on RNA recombination were made by selecting progeny picornaviruses from mixed infections with two parents harboring distinguishable selectable markers [22–24]. Direct proof of covalent linkage between two different parental molecules to yield a recombinant virus was obtained with foot-and-mouth disease virus (FMDV) [25] (review in [21]). RNA recombination occurs in animal and plant viruses [26] and in phage [27]. It may involve homologous, very closely related genomes, or very divergent molecules such as a cellular tRNA and viral RNA [28]. The molecular events leading to those different kinds of recombinant molecules are not known at present [21]. Poliovirus (PV) recombinants occur at high frequency *in vivo*. In a child fed all three

Sabin vaccine serotypes, the majority of novel antigenic variants shed during a 50-day period were type 2-type 3 intertypic recombinants [29]. In the coronavirus mouse hepatitis virus RNA, recombination is frequent in tissue culture [30] and in the brain [31], suggesting that it plays a role in natural evolution and pathogenesis of coronaviruses [31].

Recombination is distinguished from reassortment, exchanges of entire genome segments which occur among viruses with segmented genomes such as the influenza viruses (IV). New pathogenic influenza viruses have arisen by reassortment between human viruses and viruses from an animal reservoir (reviews in [32,33]). When it involves segments that encode the immunologically relevant surface antigens hemagglutinin or neuraminidase, the process is known as antigenic shift. Reassortments and RNA recombination events may occur simultaneously, as shown with rotaviruses, segmented double-stranded (ds) RNA viruses that cause diarrheal disease in infants. In chronically infected, immunodeficient children, rotaviruses with atypical genomic profiles were found; segment 11 was missing and additional ds RNA bands consisting of concatemers of segment-specific sequences were observed [34]. Rotaviruses with rearranged genome segments do reassort, and rearranged ds RNA can replace normal RNA segments functionally, and can also be replaced by normal RNA segments [35,36]. Reassortment and recombination permit large evolutionary jumps in RNA by bringing together genes or gene segments initially present in distinct ecological niches. Mutations that arise and are unfit in one genetic context can become viable or even advantageous in another context.

Probably the most widespread mechanism of RNA variation is mutation. Elements with RNA as genetic material (RNA viruses, retroviruses, viroids and satellites) or that utilize RNA as a replication intermediate (hepadnaviruses such as hepatitis B, retroposons, retrotransposons) mutate at rates estimated at about one million-fold higher than their host cells [37]. Because of their high mutability and their tolerance to accept change while remaining functional, RNA genomes are extremely heterogeneous collections of molecules (section 3). This structure for RNA viruses is relevant to the development of resistance to antiviral agents (section 6) and to difficulties encountered in the use of some anti-viral vaccines (section 7). Because of these and other implications for human and animal disease, current results on mutation rates and frequencies for RNA viruses, (section 2), as well as the meaning of the extremely heterogeneous *(quasi-species)* nature of viral populations (section 4) are also reviewed. The subject has drawn increasing attention, as seen by the recent articles that have covered theoretical [37–43,49] as well as practical [37,41,44–48] implications of RNA genome evolution.

2 RNA versus DNA mutation rates, mutation frequencies and fixation of mutations

Mutation rate is the frequency of a misincorporation event during a single replicative round of nucleic acid synthesis. For the discussion that follows, it is important to distinguish *mutation rate* from *mutation (or mutant) frequency*, which is the proportion of genomes with a mutated residue in an RNA population. *Rate of fixation of mutations* is the number of mutations which per unit time become dominant among replicating genomes. *Fixation* may also refer to new dominant genomes during infections in cell culture (even during the development of a single plaque on a cell monolayer) or in host organisms, without a time factor being necessarily quoted. Some published values do not conform to the definitions given here, making comparisons of mutation rates and frequencies between viral systems difficult. This point is discussed in depth in ref. [49].

Several measurements suggest that mutation rates for cellular DNA are in the range of 10^{-7} to 10^{-11} substitutions per nucleotide per replication [50–53]. Mutations in DNA may arise by base pairing of rare tautomers of the usual bases, purine-purine mispairs with the free nucleotide substrate in the *syn* configuration, and by depurination of DNA, among other mechanisms [51–55]. The fidelity of copying is a result of the inherent accuracy of nucleotide incorporation and the subsequent proofreading step [56]. If this latter activity is suppressed, as in some mutant DNA polymerases, or by using homopolymeric templates, error levels of 10^{-3} to 10^{-4} per nucleotide are attained [51–53]. DNA hypermutability has been described in immunoglobulin gene segments [57–59], and in certain shuttle plasmid vectors during their replication in mammalian cells [60]. Transformed cells show increased genetic instability [61–63], that probably causes tumor cell heterogeneity [64,65], an important property for invasiveness and metastatic potential [61,65]. The basis of DNA genetic instability is not understood, with transposition of mobile elements and mutational events being probably involved. It has been proposed that transient DNA hypermutability may have contributed to accelerated evolution at certain times, thus determining punctuated rather than gradual evolution [66–68].

Rates of fixation of mutations in cellular genes (estimated by comparing homologous genes from organisms which diverged at times suggested by the fossil record) are in the range of 10^{-8} to 10^{-9} substitutions per nucleotide site per year (s/s/yr) [69–72]. Up to five-fold differences have been estimated among phylogenetic groups, the slowest values being for some bird lineages and primates (the so-called "hominoid slowdown" [70]) and the fastest for rodents, sea urchins and Drosophila [71]. These variations have been attributed to the different generation times among species or to the development of more efficient repair mechanisms [70,71,73].

Mutation rates and frequencies for RNA viruses have been estimated to be 10^{-3} to $< 10^{-7}$ substitutions per nucleotide and genome doubling, using a variety of procedures (recent reviews in [44,49]). An extracistronic mutant (A (-40) → G) of phage Qβ prepared by site-directed mutagenesis [74] reverted at a rate of 10^{-3} to 10^{-4} per RNA doubling [75]. The transition G (-40) → A was quantitated by direct chemical analysis of the proportion of wild type and mutant sequence in the evolving population [74]. This was possible because the mutant sequence replicated less efficiently than its wild type counterpart [74]. Transversions G(-40) → C or G(-40) → U could possibly occur, but never be revealed in the analyses because they did not endow the genome with a measurable selective advantage over the mutant RNA. Because of the high mutation rates, populations of phage Qβ are genetically heterogeneous [76] (see section 3).

Steinhauer and Holland have developed a procedure to detect nucleotide sequences present at very low levels in RNA populations [77]. The method uses the absolute specificity of ribonuclease T1 for G residues. Selected RNA segments with a G residue flanked by sequences with only A, U and C are protected from ribonuclease (RNase) hydrolysis by hybridization with a complementary deoxyoligonucleotide. The protected RNA segment is then digested with RNase T1 and an "error oligonucleotide" is obtained when the molecules do not include the G residue. The presence of mutated bases is confirmed by sequencing. The proportion of "error oligonucleotide" relative to the two "consensus", shorter oligonucleotides permits a calculation of the mutation frequency at the G site [77]. A total of 57 clonal RNA preparations of vesicular stomatitis virus (VSV) wild type and temperature sensitive *(ts)* mutants, with different passage histories have consistently provided error frequencies of 1×10^{-4} to 1×10^{-3} for G sites from the N, M and L genes and 5'-extracistronic region of the genome [78]. In a different experimental approach, the frequency of revertants of *amber* nonsense mutants of VSV was 10^{-3} to 10^{-4} [78a].

Measurements with other RNA viruses have yielded, however, lower error frequencies. The *ts* character of a Sindbis virus mutant reverted at a frequency $< 5 \times 10^{-7}$ [79]. It is not clear, however, that the phenotypic change was due to a single substitution. An *amber* mutant in a serine codon (position 28) in the poliovirus 3D (polymerase) gene reverted to wild type with a frequency of 2.5×10^{-6} [80]. The significance of this value is unclear because the selection of amber mutants involved small, slowly growing plaques, and early-arising wild type revertants could not have been scored as components of an amber plaque (small plaque) population. However, it is possible that this site has a low mutation rate.

Dougherty and Temin have constructed vectors that contain sequences from the avian retrovirus spleen necrosis virus plus several selectable markers [81]. Based on the expression of neomycin resistance in one of such vectors, it was estimated that the total mutation frequency (point mutations, additions and deletions) after a single round of virus replication was 5×10^{-3} [81]. In a subsequent study, the rate of a transition $A \rightarrow G$ (from an *amber* UAG codon to wild type UGG) was estimated at 2×10^{-5} per replication cycle [82]. It is not possible at present to exclude that other substitutions occurred at the same residue and were selected against [82]. The base insertion rate was about 10^{-7} per base pair per replication cycle [81].

Palese and colleagues have carried out repetitive sequencing of viral genes from clonal viral preparations derived from single plaques [83]. The rate of fixation of mutations was estimated in 1.5×10^{-5} for influenza NS gene and $< 2.1 \times 10^{-6}$ for polio VP1 segment [83]. In an extension of the same type of calculation, progeny from a single virion of Rous sarcoma virus was analyzed at several genomic sites by denaturing gradient gel electrophoresis [84]. The frequency of mutation was about 1×10^{-4} [84], a value that agrees with other estimates for Rous sarcoma virus [85].

The fidelity of RNA- and DNA-dependent RNA polymerases has been measured using enzyme preparations in cell-free systems. For several retroviral reverse transcriptases, misincorporation frequencies ranged from 10^{-3} to 10^{-5} per nucleotide [86–90]. For PV RNA polymerase, Ward et al. [91] measured error frequencies of 7×10^{-4} to 5×10^{-3} using synthetic templates. All those measurements are influenced by the nature of the enzyme and of the templates, ionic composition of the reaction buffer, relative nucleoside-triphosphate concentrations, etc. [51,53], and it is not possible to assess at present to what extent they reflect error rates *in vivo*.

The limitations of several of the above measurements have been discussed by Eigen and Biebricher [49], who have compared values derived for DNA and RNA genomes. One difficulty is the evaluation of the relative fitness of the mutant molecules generated. Until many RNA sites for different RNA viruses, strains and isolates are analyzed using the same procedure, it will not be possible to reach conclusions on constancy or variation of mutation rates for different viruses or sites on a genome. It is even possible that a value for one site deviates from those at other sites, the latter being significant for one biological activity of the virus. For example, for Sindbis virus the lowest mutation rate so far calculated for an RNA virus ($<5 \times 10^{-7}$, ref. [79]) was reported. Yet, antigenic variants were found at frequencies of $10^{-3.5}$ to 10^{-5} [92], not far from values for other viruses [93].

There is no basis to support the notion that the replication of some groups of RNA viruses (including retroviruses and HIV) is more error-prone than the replication of others. In keeping with theoretical concepts of Eigen and colleagues [43,49], RNA viruses may derive a selective advantage by maintaining their replication fidelity near the "error threshold" for stable information (section 8). In this situation, mutation rates would not differ by more than ten-fold [49,93,94], perhaps with the bias towards higher fidelity for larger, unsegmented genomes because of their increased information content within a single genomic molecule [43,49].

The rates of fixation of mutations during propagation of RNA viruses in nature may reach exceedingly high values nearing 10^{-2} s/s/yr. An interesting comparative figure is the 10^{6}-fold higher rate for the viral v-*mos* gene relative to its cellular counterpart c-*mos* [95]. The NS gene of influenza A virus showed a uniform rate of evolution of 2×10^{-3} s/s/yr [96], a value similar to that of the neuraminidase N2 gene [97]. Rates for the *env* and *gag* genes of the AIDS virus were 1×10^{-2} to 3×10^{-3} and 1.85×10^{-3} to 3.7×10^{-4} s/s/yr [98,99]. Gebauer et al. [100] sequenced the

VP1 gene of foot-and-mouth disease virus (FMDV) during an inapparent, persistent infection of cattle that was established with plaque-purified virus [101]. They measured an evolution rate of 9 x 10^{-3} to 7.4 x 10^{-2} s/s/yr in a single animal [100], a range similar to that estimated during an episode of acute disease [102], or for lentiviruses [102 a]. In contrast, a long-term conservation of the VP1 coding region of FMDV was evidenced by sequence comparison of two viruses isolated over a 29-year interval [103]. An extreme case of long-term conservation has been observed by Gibbs and colleagues [42] with a strain of turnip yellow mosaic tymovirus that suggest at most 1 % nucleotide variation during the past 12000–15000 years. That the same virus shows a dual potential for rapid variation and for long-term conservation has been clearly demonstrated by Holland and his colleagues using VSV [37,45,104–110]. Rapid evolution was driven by defective-interfering (D.I.) particles during high multiplicity passages or in the course of persistence in cell culture [104–109]. In the absence of selective pressures, the same or very similar average sequence can be maintained for many generations [110] (see also section 4).

Again, from the data presently available no bias towards higher rates of fixation of mutations or increased potential for long-term conservation of sequences for any group of virus is apparent.

3   Genetic heterogeneity of RNA viruses

A considerable wealth of evidence indicates that populations of RNA viruses are genetically heterogeneous. By fingerprinting RNA from individual clones of phage $Q\beta$ [76] it was estimated that, assuming a random distribution of the mutations found, each infectious $Q\beta$ genome differs in one to two positions form the "average" or "consensus" sequence (see section 4). Many field isolates and laboratory-adapted populations of RNA viruses have proven genetically heterogeneous. This is also true for clonal pools, derived from a single infectious unit. Multiple variants are present in a single infected organism, as shown for FMDV [111,112], subacute sclerosing panencephalitis (SSPE) virus 113–115] and more recently for IV [116] and the AIDS viruses [117,118], among others. Independent isolates of the same virus are, in general, genetically distinct (recent review in [44] and Table 1 for selected examples).

Table 1
Heterogeneity of RNA genome populations

Virus	Type of population	References
Phage Qβ	Clonal populations	76
Vesicular stomatitis virus	Field isolates	119–121
Measles virus	Clinical isolates	122
Subacute sclerosing panencephalitis virus	Among viral molecules of an infected brain	113–115
Avian paramyxoviruses	Field isolates	123, 124
Lymphocytic choriomeningitis virus	Comparison of strains	125, 126
Bunyaviruses	Comparison of strains	127, 128
Influenza virus	Clonal populations	83, 129
	Infected human	116
Poliovirus	Clinical isolates	29, 130–135
	Clonal population in vivo	135a
Foot-and-mouth disease virus	Field isolates	136
	Clonal populations	137, 138
	Clonal populations in vivo	100
Enterovirus 70	Clinical isolates	139
Coxsackievirus A10	Clinical isolates	140
Drosophila C	Natural isolates	141
Hepatitis A	Clinical isolates	142
Venezuelan equine encephalomyelitis	Field isolates	143
Western equine encephalomyelitis	Field isolates	144
Sindbis	Field isolates	145
St. Louis encephalitis	Clinical isolates	146
Dengue	Clinical isolates	147
Yellow fever	Clinical isolates	148
Murray Valley encephalitis	Clinical isolates	149
Murine hepatitis	Comparison of strains	150, 151
Tobacco mosaic virus	Comparison of stocks and strains	152, 153
Reovirus	Natural isolates	154, 155
Rous sarcoma virus	Natural isolate	156
AIDS viruses	Clinical isolates	117, 118
Hepatitis B	Chronic carrier	157
Yeast killer elements	One isolate	158
Viroids	Natural isolates	159–162a

Figure 1 depicts homogeneous and extremely heterogeneous genome populations. Lines represent genomic molecules, sprinkled with mutations in populations B and C, but not in A. It is noteworthy that the sets A and B share the same "consensus" or "average" sequence (the one that includes in each position the residue most represented in the set of molecules). However, A and B differ in the nature of the individual genomes that comprise the population. Since individual molecules cannot be sequenced without prior amplification, the heterogeneous nature of population B can only be revealed after biological or molecular cloning (arrow in Fig. 1). A new distribution C with a "consensus" or

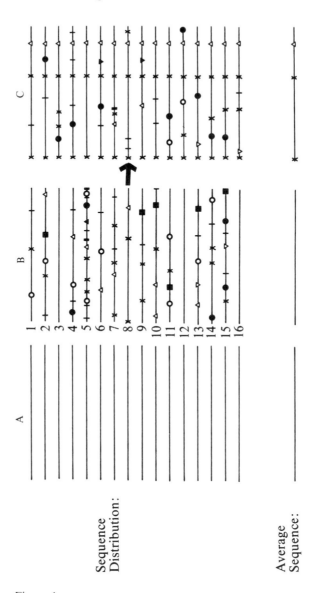

Figure 1
Diagram of homogeneous (A) and heterogeneous (B and C) genome populations. Each line is an RNA molecule, and symbols on the lines are mutations. The average sequence contains in each position the most frequent nucleotide in the sequence distribution. Note that in spite of the heterogeneity in B (average of 4 mutations per genome) this distribution has the same average than A. The arrow represents the amplification of molecule 8 from distribution B to yield a new distribution C, with a new average. Events such as plaque purification of a virus on a cell monolayer or transmission of a single infectious virion from one infected to a susceptible host are equivalent to the generation of C.

"average" sequence different from B is generated. Molecules 3, 12 and 16 in set B coincide with the "average", but they are a minority in the population. Sampling of nucleotide sequences has indicated that RNA viruses consist of pools of variants as in sets B or C. The shape of the distributions (average number of mutations per genome, standard deviations from the mean, etc.) are largely unknown, but the available evidence suggests that those parameters vary for different viruses and different populations.

4   Population equilibrium and quasi-species

There are two remarkable features of RNA genome distributions not reflected in Figure 1: (i) the extremely large population size, and (ii) the variations in the composition of the nucleotide sequence distribution with time. Measurements of the number of infectious viral particles in several organisms have yielded 10^9 to 10^{12} per infected individual (reviewed in [41]). Given the size of RNA viral genomes (3000 to 30000 residues), taking a range of heterogeneity for clonal pools of one to ten mutations per genome (several experimental measurements are included in Table 1) it can be calculated that all possible single and double mutants, as well as decreasing proportions of triple, quadruple, etc. mutants, are *potentially* present in most natural viral populations [41,76,93]. The proportion of a variant in an evolving population will depend on the rate at which it arises as well as on its fitness relative to the other variants (present or arising) in the population. The studies of Weissmann and colleagues with phage $Q\beta$ [74–76] led to the proposal that "a $Q\beta$ phage population is in a dynamic equilibrium with viable mutants arising at a high rate on the one hand, and being strongly selected against on the other. The genome of $Q\beta$ phage cannot be described as a defined unique structure, but rather as a weighted average of a large number of different individual sequences" [76]. At least some aspects of such a description apply to animal and plant RNA viruses as well ([37,41,44,110] and Table 1). This has lead to the proposal of the "population equilibrium model" for RNA genomes [93]. Little is known on the mechanisms that maintain equilibrium: What is the proportion of neutral or quasi-neutral variants relative to disadvantaged variants? What are the chances of generating advantageous combinations of mutations in constant or changing environments? Several such questions are now under investigation. That a relatively stable equilib-

either that a large number of mutations – above the range expected to be represented in a *quasi-species* – are required or, more trivial, that the drug acts in a rather unspecific manner [171].

Antiviral agents may interact with virions and prevent their infectivity by inhibiting an early step in infection such as attachment to the host cell, penetration or uncoating. Others interfere with the intracellular viral nucleic acid or protein syntheses, protein processing, virion assembly or release from cells. The mechanism of action of a wide range of antiviral agents and the problem of development of resistance have been reviewed recently [172,173]. Here we will discuss selected antivirals for which some quantitation of detection of resistant variants has been made and the molecular mechanisms of drug action are understood at least to some extent.

6.1   Drugs that act at an early stage of the virus infectious cycle

Amantadine (1-aminoadamantane) (Fig. 2) and rimantadine (α-methyl-1-adamantane methylamine) (Fig. 3) are used for the prophylaxis and treatment of IV type A infections [174–177]. Resistant mutants have been obtained in tissue culture upon passage of the virus in the presence of the drug [178–180], in animals [181,182] and from humans subject to treatment [185–187]. The isolated variants tested showed cross-resistance to the two drugs, suggesting that both act by the same or a very similar mechanism. Resistance maps in the matrix (M) gene [180,187–189], although a possible influence of other gene products has not been excluded [187,190,191]. The surface antigen hemagglutinin may be indirectly involved as a result of its interaction with protein M2 [191]. Different human IV A strains vary in their sensitivity to these drugs [179,187]. For most human IV strains, an early stage in the infection is inhibited. For avian viruses, however, the late assembly step appears to be the target. Perhaps critical interactions between M2 and hemagglutinin that occur both during penetration and assembly are affected to a different extent in the two steps, depending on the viral strain [190]. Amantadine-resistant mutants of IV A show single amino acid substitutions in M2 [188,191]. Rimantadine resistance was associated with the corresponding genome segment 7 in reassortant viruses generated with a resistant clinical isolate [190]. Nucleotide sequencing showed the single amino acid changes Ala/30→Val or Ser/31 →Asn of M2 to be associated with the resistant phenotype [190]. Since the

Figure 2
Amantadine (1-aminoadamantane).

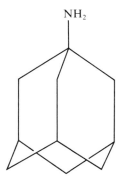

Figure 3
Rimantadine (α-methyl-1-adamantane methylamine).

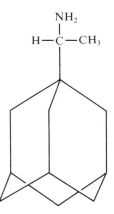

variants replicated efficiently at least in cell culture, the results suggest that amantadine- and rimantadine-resistant mutants should occur frequently in IV populations. Indeed, their proportion in various IV preparations was 1×10^{-3} to 4×10^{-4} [179,180]. Among children treated with rimantadine, 27 % shed resistant viruses [190]. Thus, selection of resistant IV variants is likely to be a frequent event, and an important problem mainly during therapeutic use of these drugs, since selection is then exerted on a large pool of replicating genomes.

A number of drugs bind to the capsid of picornaviruses and inhibit their uncoating. Dichloroflavan (4', 6-dichloroflavan) (DCF) (Fig. 4), 4'-ethoxy-2'-hydroxy-4, 6' dimethochalcone (RO-09-0410) (Fig. 5), 2-[-(1,5,10,10a-tetrahydro-3H-thiazolo [3,4b] isoquinolin-3-ylindene) amino]-4-thiazole acetic acid (44-081 R.P.) (Fig. 6), disoxaril, 5-[7-[4-(4,5 dihydro-2-oxazolyl) phenoxy] heptyl]-3-methyl-isoxazole (WIN

51711) and its methyl derivative WIN 52084 (Fig. 7) and arildone (4-[6-(2-chloro-4-methoxyphenoxy) hexyl]-3, 5-heptanedione (Fig. 8) belong to this group. Crystallographic analysis has shown that WIN 51711 and WIN 52084 go to the hydrophobic interior of the VP1 β-bar-

Figure 4
Dichloroflavan (4',6-dichloroflavan).

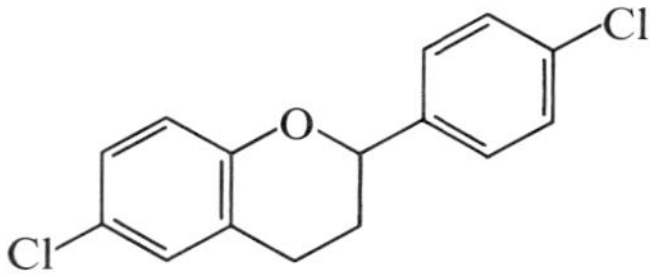

Figure 5
4'-ethoxy-2'-hydroxy-4,6' dimethochalcone (R0-09-0410).

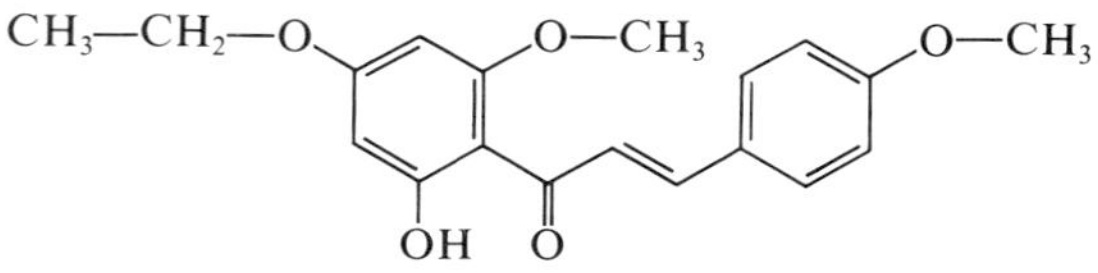

Figure 6
2-[-(1,5,10,10a-tetrahydro-3H-thiazolo[3,4b]isoquinolin-3-ylindene) amino]-4-thiazole acetic acid (44-081 R.P.).

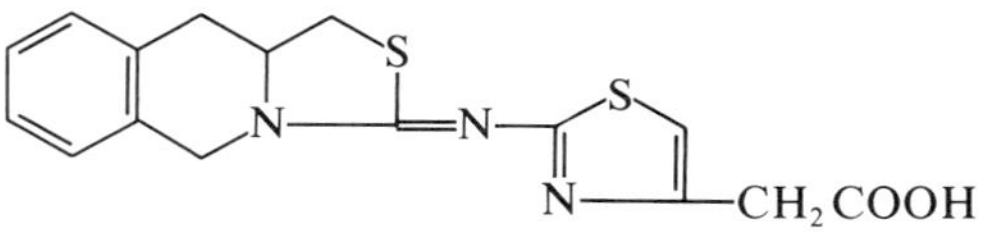

Figure 7
R = H: Disoxaril, 5-[7-[4-(4,5 dihydro-2-oxazolyl) phenoxy] heptyl]-3-methyl-isoxazole (WIN 51711). R = CH_3, WIN 52084.

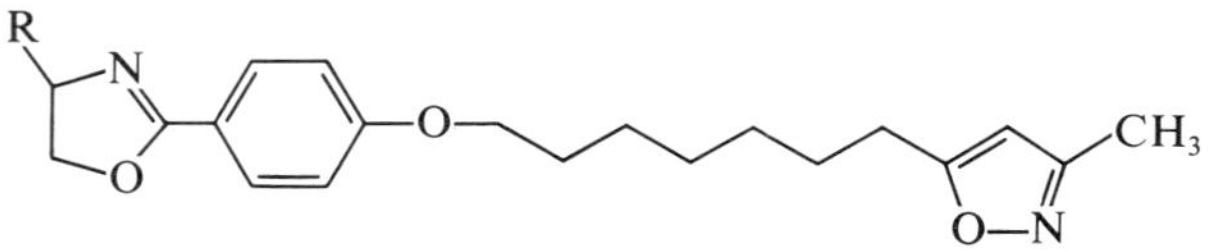

Figure 8
Arildone (4-[6-(2-chloro-4-methoxyphenoxy) hexyl]-3,5-heptanedione.

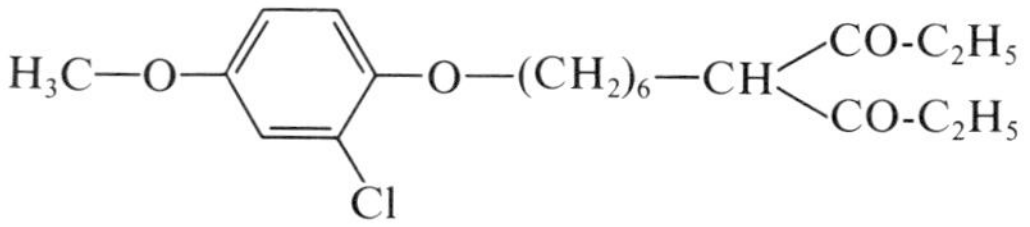

rel of human rhinovirus 14 [192]. Much emphasis has been put on the conservation of the general architecture of picornaviruses [193]. Since these drugs accommodate into a pore of the viral capsid, one would anticipate that most viral strains would be equally sensitive, and that resistant mutants would be rare. However, the antiviral activity of these agents varies up to 10^3-fold for different rhinovirus serotypes [194–197]. Moreover, low level resistance to disoxaril was found at frequencies of 10^{-3} to 10^{-4} and high level resistance at 10^{-5} (quoted in [197]). A rhinovirus 9 mutant resistant to the chalcone R0-09-0410 was found at a frequency of 10^{-5}, and grew less efficiently than wild type [198]. It is not known whether resistant strains would generally show a reduced fitness. Clinical trials have yielded poorer results than anticipated from cell culture assays due in part to insufficient drug concentration in the target tissues. This increases the chances of selecting variant viruses with low-level resistance *in vivo* [199].

There are more than a hundred serotypes of rhinovirus, suggesting an increased difficulty in finding effective, wide-spectrum anti-rhinovirus agents. However, 90 % of the serotypes interact with the same cell receptor [200]. A MAb directed to such receptor, effectively competed with virus binding [201], but prevention of infection or illness *in vivo* was not observed [202]. Interference with virus attachment and penetration has also been achieved with synthetic peptides that represent amino acid sequences from the viral surface needed either for receptor recognition or membrane fusion [203,204]. Choppin and colleagues sequenced the fusion (F) protein gene of a measles virus mutant resistant to fusion inhibiting polypeptides [205] and identified three amino acid changes located at the carboxy terminal half of F. It is not possible to know at what frequency the mutant arose since it was selected by repeated passage of single plaques in the presence of one oligopeptide (quoted in [205]). An alternative approach is to inhibit virus attachment by soluble receptor protein, as currently tested with CD4 to prevent HIV infectivity [206]. Many clinical trials using peptides and soluble receptor proteins are currently in progress with AIDS patients [207]. Little is known of the possibility of selecting variant viruses with altered receptor specificity, as shown previously for Coxsackie virus in cell culture [208]. Most viral receptors probably play a role in normal cell physiology [208a], and it is currently a concern that their blockade or manipulation may have undesirable side effects.

6.2 Drugs that act on an intracellular step of virus development

Some nucleoside analogs are very effective antiviral agents. Usually they are activated with the participation of virus-coded enzymes, and inhibit a viral nucleic acid polymerase. A classical antiherpes drug, acyclovir or 9-(2-hydroxy-ethoxymethyl) guanine, is converted into the monophosphate form by the viral thymidine kinase and then into the active triphosphate by cellular enzymes. The molecular basis and clinical importance of resistance to acyclovir have been recently reviewed [209]. A number of analogs inhibit the multiplication of RNA viruses. Ribavirin, (1-β-D ribofuranosyl) 1-H-1,2,4-triazole-3-carboxyamide) (Fig. 9) is a broad spectrum antiviral agent [210–212]. Ribavirin lowers the intracellular pool of GTP by inhibiting inosine monophosphate dehydrogenase [213], resulting in inhibition of viral polymerases, or capping of mRNA [214], or both. Many viruses that cause important diseases are effectively inhibited by ribavirin. For FMDV, the drug was ten-fold more effective in inhibiting viral replication during a persistent than during a lytic infection [215]. It has proven particularly successful when applied as an aerosol at the upper respiratory tract, for influenza and respiratory syncitial virus infections [216], and in some severe infections such as Lassa fever and Argentinian hemorrhagic fevers [217,218]. It has been difficult to derive ribavirin-resistant mutants [219]. The only example, to my knowledge, is a Sindbis virus (SV) mutant derived by Scheidel *et al.* [220]. By serial passage of SV in the presence of inhibitory concentrations of mycophenolic acid, mutants resistant to the drug, that showed cross-resistance to ribavirin, were obtained. They suggested that an altered viral enzyme – RNA polymerase or, more likely, RNA guanylyltransferase – was selected that

Figure 9
Ribavirin, (1-β-D ribofuranosyl) 1-$\underline{H}$-1,2,4-triazole-3-carboxamide.

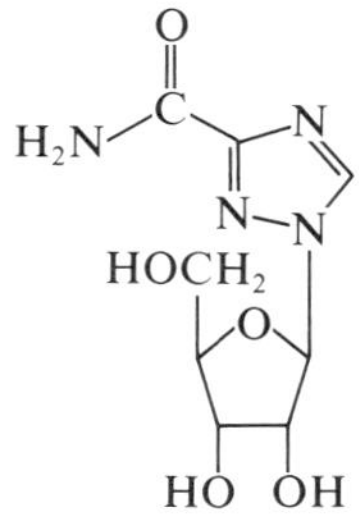

Figure 10
3'-azido-3'-deoxythymidine (AZT or retrovir).

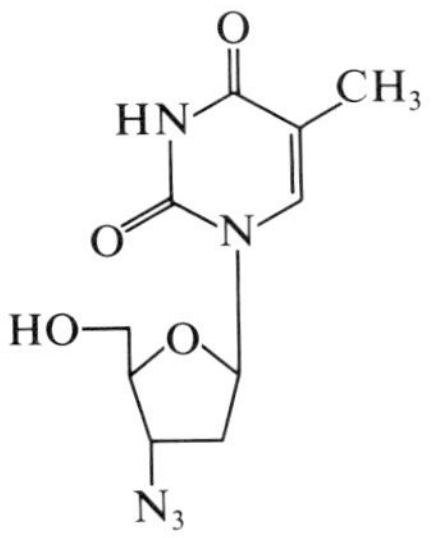

was functional at lower GTP concentrations than those required by the wild type enzyme [220].

The 2', 3'-dideoxynucleoside analogues are inhibitors of retrovirus replication [221]. They are converted into the 5'-triphosphate form by cellular kinases, and act as chain terminators during reverse transcription [222]. The analog 3'-azido-3'-deoxythymidine (AZT or retrovir) (Fig. 10) is currently being tested in AIDS patients, with promising results ([207,221] and references therein). Recently, AZT-resistant HIV variants have been isolated from AIDS patients [222a]. Selection of such mutants may be quite widespread, in view of the lengthy treatments required for this disease.

Many other nucleoside analogs are currently under investigation, and because they often affect the active sites of essential viral enzymes, it may be more difficult to select resistant variants and may prove active against a wider range of naturally occurring variants. It is encouraging that no differences were noted in the inhibitory activity of 20 nucleoside analogues (including ribavirin and 3-deazaguanine) on ten strains of IV types A and B and one isolate of type C [223].

Guanidine (Fig. 11) inhibits several animal and plant viruses, including picornaviruses. Its action on FMDV and PV is at the RNA synthe-

Figure 11
Guanidine.

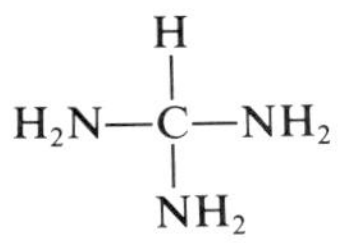

Figure 12
Enviroxime, anti-6-[(hydroxyimino)-phenyl-methyl]-1-[(-methylethyl) sulfonylimidazol-2-amine].

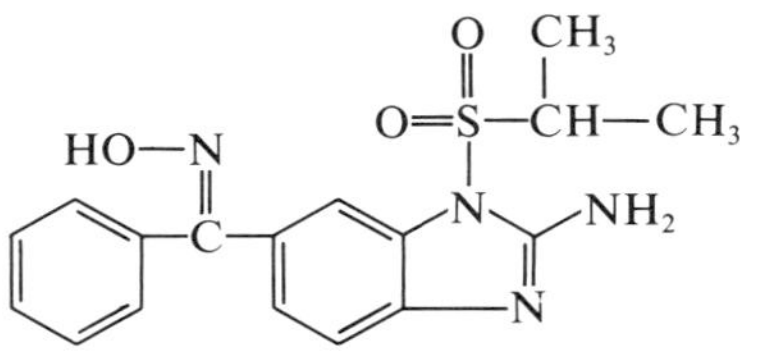

sis step, and after some conflicting results, there is now good evidence that mutations that confer guanidine resistance or dependence map in viral polypeptide 2C (or P34) [224,225]. PV type 1 mutants resistant to 2mM guanidine (there are several degrees of resistance to guanidine) showed one amino acid substitution, Asn/179 → Gly or Asn/179 → Ala in 2C, that involved two nucleotide substitutions in each case [226]. Two guanidine-dependent mutants, selected upon serial passage of PV in 2 mM guanidine, had two amino acid substitutions each: Asn/179 → Gly (the same observed in resistant mutants), Ile/227 → Met in one mutant, and Ile/142 → Val, Met/187 → Leu in another mutant [226]. The mutation frequencies of PV to guanidine resistance were 1.8×10^{-5} to 4×10^{-8} [226]. These values are compatible with multiple mutations causing the resistance phenotype, as indeed observed by genomic sequencing. Infectious cDNA clones that included the relevant mutations, yielded the expected phenotypes, thus indicating the direct involvement of 2C [227]. The role of 2C in the picornavirus replication complex is not known, and thus the molecular basis of guanidine action remains undefined [228].

Enviroxime, (anti-6-[(hydroxyimino)-phenyl-methyl]-1-[(-methylethyl) sulfonylimidazol-2-amine]) (Fig. 12) is effective in inhibiting replication of many rhinovirus serotypes [229]. Recent evidence suggests that the drug may inhibit the formation of the replication complex (quoted in ref. [197]) and no resistant mutants have been studied [219].

6.3   Additional antiviral strategies

Many compounds, in addition to those discussed in previous paragraphs, are presently under study as antiviral agents [172,173,230]. Among them, interferons, now available in large amounts via recombinant DNA expression systems, have generally given poor results in clinical trials [230], and more work is needed to understand the basis of their antiviral activity.

The expression of polynucleotides or viral polypeptides has induced antiviral activity in plant and animal cells. Short oligodeoxynucleotides or RNAs complementary to viral messengers ("antisense" RNA) have, in several instances, blocked virus replication (reviewed in [230]). Gene transfer techniques may allow the constitutive expression of such antiviral macromolecules in plants or animals. In addition to the possible selection of viral variants (a largely unexplored subject), it is uncertain whether living cells will maintain a stable expression of abnormal nucleic acids and/or proteins. The potential for cell variation is, at least on occasions, remarkable. The induction of multidrug resistance in cultured cells or in tumors is a pertinent example. Strategies for its reversal is key to antitumor therapy and an active field of research on its own (see also section 2).

6.4 Combination therapy

The early [171,231] and more recent studies on virus inhibitors summarized above (sections 6.1 to 6.3) demonstrate that the emergence and selection of virus variants resistant to antiviral agents is quantitatively significant, and a potential severe problem in medical practice. Frequent appearance of drug-resistant viruses is the expected consequence of the successive population equilibrium steps through which RNA viruses evolve (section 4). Even a very rare (infrequent) genome from the *mutant spectrum* of the *quasi-species,* when able to grow in the presence of an inhibitor of the replication of other members of the distribution, will be selected to form a new *quasi-species*. The process can be represented as the transition from distribution B to C depicted in Figure 1 (section 3). In this process, a new genomic distribution (not a single individual) will be selected that may adjust to be quite fit while maintaining the drug resistance trait, in the absence of the drug. Even wide-spectrum agents that profoundly affect biochemical parameters in the cell (example, depletion of the GTP pool by ribavirin) have proven capable of selecting resistant viruses, as illustrated by the isolation of a ribavirin-resistant mutant of SV [220] (section 6.2). We are dealing with inherently variable and indeterminate distributions of genomes, with ever-evolving mutant profiles (section 4). Such profiles are presently inaccessible to detailed analysis (except for computer simulations [165] and experiments with simple RNA molecules [166,167], section 4) since they would require the sequencing of thousands of viral genome molecules from several populations, a task that necessitates waiting for technical improvements. This indeterminacy of the profile of the distributions generates an uncertainty as to the proportion of drug-resistant mutants likely to arise in an evolving virus population. For viruses that can be grown in cell or tissue cultures, parallel serial passages in the presence of the drug should provide an experimental system to determine in a statistically reliable fashion the frequency of emergence of variants. It must be noted, however, that the results may not be relevant to viral multiplication *in vivo,* where the virus finds a different environment.

In light of the above considerations, the most adequate strategy for an antiviral therapy would be a *combination therapy* applied in cases of fatal illness or, exceptionally, a *combination prophylactic treatment* directed to groups at high risk of fatal illness. Indiscriminate, widespread use of antiviral agents should be avoided. Combination therapy has been discussed in several recent reviews [207,232,233]. It should involve non-antagonic, multiple drugs with an independent target site of their inhibitory action, or drug-interferon combinations [234] with synergistic activity. Synergisms and antagonisms are difficult to anticipate, as shown by the finding that ribavirin antagonizes AZT in its anti-HIV activity, apparently by inhibiting its phosphorylation [207]. Statistical considerations suggest that with an adequate combination therapy, the likelihood of selecting a variant with the multiple mutations required for multidrug resistance is many orders of magnitude lower than that of selecting for resistance to one drug. This point is testable in serial passage experiments in which the frequency of appearance of resistant viruses to one or several drugs can be compared.

7   Vaccines

The success of vaccination in controlling important viral diseases such as smallpox, poliomyelitis, mumps or measles [235–239] does not justify overlooking the problems encountered with current vaccines. Furthermore, for diseases of increasing concern, such as AIDS, vaccines are not available. The types of vaccines in use or under experimenta-

Table 2
Types of antiviral vaccines (from refs. [230, 238, 240, 241])

Inactivated, whole virus.
Live-attenuated (various mutants selected *in vivo* or produced by site-directed mutagenesis on cDNA copies of viral genomes; vaccinia or other recombinant viruses).
Synthetic (capsid proteins or structures; non-structural proteins [241]; oligopeptides).
Anti-idiotypic [240].

tion are shown in Table 2. Anti-idiotypic antibodies [240] provide reagents that may maintain the antigenic properties of complex protein conformations important in the immune response, and difficult to mimic with synthetic proteins or oligopeptides. Attenuated viruses such as vaccinia have been manipulated to include foreign viral antigens and considerable effort has been put in the development of such recombinant vaccines [230,242–244], not always successfully [245]. The chances of a vaccine inducing a protective immune response are higher when the response resembles that produced during a natural infection with the virus [238]. Obviously, a vaccine must be safe, stable, and available to a substantial proportion of individuals from the populations at risk [238].

During their replication, attenuated viruses undergo genetic variation (sections 2 to 4). PV serotype 2 and 3 vaccine strains can fix more than 100 mutations during replication in one or two individual humans [130]. A single nucleotide substitution at position 472 (from U in type 2 Sabin vaccine to C in mutants) occurred regularly during viral multiplication in the intestine of vaccinated persons [246]. This well-documented variability of PV (see also Table 1, section 3) has caused reversion of attenuated viruses to virulent forms and a number of cases of vaccine-associated poliomyelitis [247,248]. Among vaccinees and their contacts, the incidence of paralysis in different countries ranged from 0.13 to 2.29 cases per million doses of vaccine distributed [248]. Even in well-vaccinated communities, outbreaks of poliomyelitis have occurred among the unvaccinated individuals, as in China in 1982 [249]. The PVs isolated during an outbreak in Finland in 1984 were heterogeneous [135] and antigenically different from the previous isolates of PV types 1,2 or 3 [250]. The availability of infectious cDNA copies of the PV genome may permit the preparation of attenuated variants with low frequencies of reversion to virulence. However, this is proving a difficult task since the molecular basis of attenuation remains elusive. Recombinational analysis [251] and sequencing of PV variants [252] has indicated that several loci spread over the vaccine virus genomes may influence the attenuation phenotype. It has been suggested that infectious cDNA clones of poliovirus Sabin strains may be used as a repository of inocula for vaccine production [253], reducing the risk of reversion to virulence by repeated passage. Antigenic chimaeras of poliovirus have been suggested as potential new vaccines [254]. However, the progeny from a cDNA clone will also be a *quasi-species* distribution of genomes, with a "range" of phenotypes (sections 5,6). Thus, eradication of poliomyelitis [255], as inevitable as it may seem, will probably still require considerable research effort.

Reversion from attenuated to virulent forms occurred with *ts* and cold-adapted *(ca)* mutants of IV [230,256] and *ts* mutants of respiratory syncytial virus (RSV) [256]. Some vaccinees shed *ts* + RSV revertants [256]. The attenuated RSV were assayed since an inactivated vaccine not only did not induce protection, but aggravated the clinical response of children to subsequent RSV infection [257]. Encephalitis following measles, mumps or rubella vaccination occurs at a low rate of about one case per million doses of vaccine [239], and the few cases of SSPE in children without previous evidence of measles, may also have a vaccine origin [258]. These observations emphasize a likely role of variant viruses in chronic, degenerative disease [37]. Attenuated viruses share with their wild type counterparts the potential for initiating atypical infections (see also sections 8,9) and for triggering immunopathological responses. In 1976–1977, many cases of Guillain-Barré syndrome in the U.S. were attributed to vaccination with influenza virus A/New Jersey [259], but the mechanisms involved are not clear.

7.1   Virus variation and vaccine design

RNA virus variation in the field is increasingly recognized as one of the main obstacles for vaccine efficacy [37,41,47,93,136,260–263] but, curiously, this problem was very often overlooked in the initial design of synthetic vaccines. Most human and animal viruses exist as several serotypes, subtypes and antigenic variants. Even for viruses considered antigenically stable such as hepatitis A, RSV or rabies, use of MAbs has shown antigenic differences between isolates [93,230]. Whether these differences may cause vaccine failures or not depends on a complex set of parameters, some of which are discussed below. A useful model system for vaccine studies is provided by FMDV, an economically important pathogen that infects several animal species suitable for experimentation (mice, guinea pigs, swine, etc.). Seven serotypes and more than 65 subtypes of this virus were recognized by application of classical serological techniques (complement fixation and neutralization assays). A recent quantitation of the reactivity of field isolates of serotype C with MAbs directed at epitopes involved in neutralization of infectivity indicated an extensive antigenic heterogeneity, even among viruses of one disease outbreak [261,263]. Mateu *et al.* suggested that given the *quasi-species* nature of FMDV populations, each viral isolate may be not only genetically unique, but also antigenically unique when a large enough panel of MAbs is used to quantitate antigenic reactivities. Whole-virus inactivated vaccines prepared with each one of two viruses from the same FMD episode were able to protect swine against challenge with homologous virus, but only partially against heterologous virus [262]. In this case, amino acid substitutions fixed during the natural evolution of FMDV caused a discernible effect in vaccine potency. The same problem has recently become apparent when the genetic and antigen heterogeneity of the AIDS viruses has been recognized [98,99,117,118,264]. The conclusions on antigenic heterogeneity of FMDV and HIV are very similar to those reached previously by Prabhakar, Notkins and colleagues, comparing clinical isolates of Coxsackie virus B4 [265–267] and they may be common to many more pathogenic RNA viruses.

The spectrum of antigenic variants that at any one time will be represented in a viral population is indeterminate. For each epitope, there will be tolerance for certain amino acid substitutions. Some replacements will alter epitope reactivity. Not only variations fixed on the surface proteins, but also those in non-structural proteins may be immunologically relevant, since they may affect T-cell recognition of infected cells. The total number of epitopes involved in triggering the humoral and cellular responses leading to protection will influence the weight that changes in individual epitopes have in the ability of a virus to escape from an immune response directed to a related variant. If the equilibrium populations (section 4) remain stable, changes in consensus antigenic structure are unlikely and a vaccine may show long-lasting efficacy. A stable equilibrium may be maintained for many generations in spite of a remarkable genetic heterogeneity and high mutation rates (several examples in section 4). A fictional story might clarify the point. Assume phage $Q\beta$ had the nasty property of occasionally infecting humans causing an abortive but fatal infection, from which no viable particles can be rescued. Imagine that such a fatal disease would be amenable to control by an inactivated $Q\beta$ vaccine, with phage produced in *E. coli*, the sole productive host of $Q\beta$. Such a vaccine would maintain its efficacy from the times of Sol Spiegelman (see refs. [76,93] for an account of the passage history of $Q\beta$), in spite of the undisputed high mutability and heterogeneity of $Q\beta$ populations (section 3). Thus, factors that favor genetic disequilibrium in natural populations of viruses will increase the likelihood of antigenic drift. Again, disequilibrium may be caused by environmental changes and transmission bottlenecks (or founder effect), in which one genome is greatly amplified [41,111]. The ability of many viruses to infect different host organisms and different cells within one organism provides changes in environment. It is very likely that transmission bottlenecks have been a major driving force in generating HIV diversity [41,98].

Progress in the development of synthetic vaccines has been slow. Proteins or oligopeptides present the immune system with a more limited repertoire of relevant epitopes than the entire virus particle. Then, for statistical reasons, the problems summarized in previous paragraphs are aggravated: the level of immunity is lower and the chances of selecting variant viruses able to escape neutralization by the suboptimal immune response are increased. In the successful development of a hepatitis B vaccine (review in [268]) it became clear that viral particles or core structures afforded a better protection than an individual protein or synthetic peptides. In the course of studies to develop new anti-FMD vaccines, several protein or peptide constructions have been engineered that induce an enhanced response: amino acid bridging of two synthetic peptides [269], incorporation of a peptide epitope into viral core structures [260], and inclusion of foreign helper T-cell determinants [271]. It is likely, however, that important elements of the global response needed to afford protection are omitted by greatly optimizing particular epitopes. Very important new knowledge and progress has been derived from such efforts. Yet, it has not been possible to substitute whole virus-inactivated anti-FMD vaccines by synthetic vaccines.

7.2   Is there a viable strategy?

The antigenic diversity and instability of RNA viruses is best interpreted as a consequence of the *quasi-species* nature of viral populations and modifications in the equilibrium distributions of genomes (section 4). From the emerging data of crystal structures of antigen-antibody complexes [273], it appears that conservative amino acid substitutions (likely to be well represented in the viral *quasi-species* because they will hardly affect virion fitness) at key sites may diminish the affinity of antigens for antibodies. The factors likely to affect vaccine efficacy are: (i) the *rate of fixation* of amino acid replacements in virus-coded proteins (not directly the mutation rate as defined in section 2); (ii) the *number* of different epitopes involved in triggering the humoral and cellular immune response leading to protection; (iii) the *tolerance* of such epitopes for amino acid substitutions that result in a decreased affinity for antibodies or cells from the immune system. An added difficulty is the polymorphism of molecules of the major histocompatibility complexes through which most antigen-specific T-cells recognize antigens. The allelic form of such molecules influences which amino acid sequences are recognized by the immune system. The viral epitopes recognized as T-cell inducing determinants vary among individuals from one population [274,275]. This has led to some pessimistic views on the feasibility of vaccines based on T-cell inducing epitopes [274].

Since in spite of the above problems, some classical vaccines have been successful, the conclusion appears to be that the less epitope-specific, i. e., "the less synthetic" a synthetic vaccine looks to the immune system, the higher its chances of inducing protection. A new antiviral vaccine formulation should include: (i) a wide repertoire of epitopes recognized by B and T-cells; (ii) a number of variant amino acid sequences for the important and variable epitopes; (iii) complex, structured epitopes in the form of anti-idiotypic antibodies or capsid structures. In addition, the effects of carriers and adjuvants should be cautiously tested since they may have enhancing or suppressive effects on the immune response. Furthermore, in the process of conjugation to a carrier macromolecule, the antigens may be altered [276].

By ensuring a response to several viral epitopes, chances of selecting variants resistant to one or a family of antibodies are enormously reduced. This "multisite" response has an aim parallel, in principle, to combination therapy with antiviral drugs (section 6.4). Partial immunity or biased immune responses will favor selection of antigenic virus variants. It must be emphasized, however, that selection of mutants resistant to "multisite" responses cannot be excluded; simply, their probability of occurrence is lower (see also section 8). For synthetic vaccine development, the animal protection tests should use not only the homologous virus for challenge, but a collection of natural variants likely to be present (or to arise) in the natural population of the virus which spread is to be controlled (see ref. [136]).

When possible, and for groups at high risk of severe disease, a combination of prevention by vaccination and antiviral therapy has also been suggested. Much research is needed to develop any such strategies, and whether they ultimately will prove successful is an open question.

8   Error-prone polymerase subsets, hypermutability, and error catastrophe. Is there a limit to RNA virus variation?

A clustering of A → G transitions was observed by O'Hara et al. in RNA from a VSV DI [108]. They suggested that it might have been generated by error-prone viral polymerases. In some positions around the G residues selected for error oligonucleotide analysis (section 2) to measure mutation frequencies in VSV, remarkable heterogeneities were found [77,78]. Cattaneo et al. identified an extremely high level of U→C transitions in the M gene of viral molecules isolated from the brain of a child afflicted with measles inclusion body encephalitis [115]. Relative to a consensus sequence, about 50 % of the U residues within the M-coding segment were changed to C in that particular set of molecules. They suggested that a selective hypermutability event led to enhanced U→C transitions, but not to other substitutions [115]. Since viral polymerase genes are themselves subject to variation, polymerase subsets with decreased fidelity of template copying could generate highly altered molecules, that would allow a transient, accelerated evolution [108]. An alternative mechanism for the generation of hypermutated molecules is the temporary incursion of RNA replication into the error catastrophe zone, as defined by Eigen, Schuster and colleagues [49,164,277]. The maximum polynucleotide length (V_{max}) whose information may be stably replicated relates to the error rate per nucleotide (1-q̄, with q̄ being the average copying fidelity) and to the selectivity (or superiority) (δ_o) of the master sequence over its mutant spectrum. The threshold relation is $V_{max} < \ln \delta_o/1-\bar{q}$. Near the error threshold, a replicating RNA population has the advantage of an abundant mutant spectrum [49,164,277]. Occasional transitions into error catastrophe and the consequent "melting of information" may occur by either decreasing the fidelity of copying or by decreasing the selectivity of the master. The result may be either an abortive, dead-end infection, or the rescue of a new *quasi-species* if the brevity of the stay within the catastrophe zone allows it. A decrease in master superiority could be brought about by an environmental change, such as cell-to-cell (non receptor-mediated) invasion of brain cells by MV. If a new *quasi-species* is "rescued" in the new environment, completely different molecules modified at selected genomic locations may arise due to the different (not absence of) selective constraints. In some of the above examples, the bias in the type of mutations produced could be due to the participation of a double-stranded RNA unwinding activity [277a].

In slowly progressing brain disease, the many years elapsed between the initial infection and the final stages of a persistent infection do not permit excluding that frequent replication bottlenecks occurred in which only one or a few molecules succeeded in being amplified. Highly mutated molecules present at the very tail of the mutant spectrum could have been selected, the system having never entered the error catastrophe zone. Similar founder effects may explain the extremely high rates of fixation of mutations in prolonged, persistent infections involving low amounts of virus [100]. Hypermutational events cause abrupt changes in RNA and, along with recombination, permit greater evolutionary jumps than the continuum of high mutability. Is there a limit to the variation of RNA viruses? Are possible drug- or antibody-resistant mutants a limited set of genomes, or are we isolating only a few out of a hopelessly large number of representatives? In a recent phylogenetic analysis of VP1 of 15 isolates of FMDV, it was shown that fixation of amino acid replacements had occurred at a limited number of residues, in relation to that expected from a Poisson distribution of changes [278]. The comparison was among a limited number of relatively recent FMDV isolates. It is becoming apparent, however, that as new isolates are entered into the comparisons, changes are increasingly found in previously invariant positions. As a recent example, the hemagglutinin of IV type B of isolate B/Ann Arbor/1/86 showed six amino acid substitutions not found in 1940–1983 isolates, and each of the hemagglutinins compared had unique substitutions [279]. The same applies to the M protein of MV isolates [280]. Long-term evolution tends to preserve functional domains as defined by charge distributions and polypeptide foldings, rather than primary sequences. Thus, variations at one site on a viral polypeptide may demand "accompanying" changes at other sites to maintain functionality. Since several amino acids have similar chemical properties, the number of possibilities for compensating substitutions appears to be very large. An experimental analysis that supports this notion has come from the comparison of 31 reverse transcriptase sequences from different origins [281]. Only three out of about 300 residues were invariant in all the enzymes! With time, an enzyme that is essential for the replication cycle of retroviruses changed almost completely its primary sequence and yet maintained its function. Considering that even single, conservative amino acid substitutions may alter the behavior of an RNA virus (section 5), and that each infected organism includes distributions as shown in Figure 1B and C (section 3) (but with 10^9 to 10^{12} genomes with an ever-changing succession of distributions!), it may be concluded that the potential for genetic as well as for phenotypic variation of RNA viruses is indeed enormous.

9 Conclusion

RNA genetics deals with uncertainty and with probabilities [93]. Nucleotide sequencing is currently providing great insight into the detailed organization of the genetic material. Yet, RNA geneticists are becoming aware that each RNA virus molecule sequenced will most likely never be found again as an exact replica. This is becoming dramatically apparent in the sequencing of HIV [117,118, 281a]. However, it must be remembered that it is the very same phenomenon previously documented for many other RNA viruses, including retroviruses (sections 3 and 4). This individual indeterminacy of RNA genomes may be safely overlooked in some studies, but not in others. It will hardly affect studies such as the elucidation of three-dimensional structures by crystallographic methods or the gene expression strategy of a virus. Other important biological traits, however, are critically dependent on the fine, single-base residue or single amino acid residue make-up of the virus. Cell tropism and host-range, attenuation or virulence, affinity for antibodies, recognition by cells of the immune system, and resistance to antiviral agents are some of the important phenotypic traits critically dependent on single-residue variations [37,41,45,93]. It is for such biological properties that the uncertainty of relating a *defined* nucleotide sequence to an observed behavior becomes apparent [93]. RNA heterogeneity is directly relevant to viral pathogenesis. Infection of a susceptible organism can be viewed as a succession of invasions of cells and tissues modulated by a series of responses from the organism. It is a very adequate playground for the continuous generation of new genomic distributions, even though the same *average* genome may be repeatedly selected in the same tissue or organ, or intact organism or populations of organisms.

The above considerations make it very unlikely that RNA viruses evolve simply by a steady accumulation of mutations. Instead, unpredictable shifts of genomic distributions, along with very frequent transmission bottlenecks are the main driving evolutionary forces. It may be a matter of chance that the comparison of viral sequences belonging to isolates distant in time may appear as a gradual accumulation of mutations. It may also be the result of an averaging of rapid evolution events along with periods of stability, as predicted by the population equilibrium model [93] for RNA genomes (section 4) and by the punctuated equilibrium model of evolution generally. As a consensus distribution, a virus may remain relatively invariant for thousands of years [42] or undergo exceedingly fast change in *one* infected individual [100]. "New" diseases such as human hemorrhagic conjunctivitis [139], and AIDS or the adaptation of a morbillivirus to seals [282–284] can be viewed as examples of rapid genetic and biologic shifts in RNA genomes. As a consequence, dealing with the evolution of RNA viruses required a theoretical framework different from the classical phylogenetic methods consisting in the derivation of rooted trees. At present, the most adequate theoretical study is provided by the concepts of *quasi-species* as developed by Eigen, Schuster and colleagues (section 4).

In a certain sense, RNA viruses appear as "collections of entities waiting to be selected for", and perhaps that is why they have been successful parasites all along.

10 Summary

RNA viruses and other RNA genetic elements must be viewed as organized distributions of sequences termed *quasi-species*. This means that the viral genome is statistically defined but individually indeterminate [76]. Stable distributions may be maintained for extremely long time periods under conditions of population equilibrium. Perturbation of equilibrium results in rapid distribution shifts. This genomic organization has many implications for viral pathogenesis and disease control. This review has emphasized the problem of selection of viral mutants resistant to antiviral drugs and the current difficulties encountered in the design of novel synthetic vaccines. Possible strategies for antiviral therapy and vaccine development have been discussed.

Acknowledgments

This review was written while I was spending a sabbatical year in Dr. J.J. Holland's laboratory at UC San Diego, supported by a Fulbright Fellowship. I am indebted to J.J. Holland for his continuous encouragement to our group in Madrid to pursue research on RNA virus heterogeneity, and for comments, ideas, and unpublished results that have been included in the text. I also thank J.C. de la Torre, D.K. Clarke, E. Meier and D. Steinhauer for directing my attention to relevant literature, and for critically evaluating the manuscript. I thank L. Carrasco and M. E. Gonzalez (C.B.M., Madrid) for very valuable information on the action of antiviral agents.

While this review was in the press, Heinz et al. (J. Virol. *63* [1989] 2477) reported that human rhinovirus 14 mutants with high resistance to WIN 52084 (Fig. 7, section 6.1) occur at frequencies of 4×10^{-5} and those with low level resistance are found at frequencies of 1×10^{-3} to 4×10^{-4}, in excellent agreement with current estimates of mutation frequencies for RNA viruses (section 2).

References

1. T.R. Cech and B.L. Bass: A. Rev. Biochem. *55*, 599 (1986).
2. S. Altman, M. Baer, C. Guerrier-Takada and A. Vioque: Trends Biochem. Sci. *11*, 515 (1986).
3. G.M. McCorkle and S. Altman: J. chem. Educ. *64*, 221 (1987).
4. C. Guerrier-Takada, K. Gardiner, T. Marsh, N. Pace and S. Altman: Cell *35* 849 (1983).

5   C. Guerrier-Takada and S. Altman: Science 223, 285 (1984).
6   T.R. Cech, A.J. Zaug and P.J. Grabowski: Cell 27, 487 (1981).
7   K. Kruger, P.J. Grabowski, A.J. Zaug, J. Sands, D.E. Gottschling and T.R. Cech: Cell 31, 147 (1982).
8   H.D. Robertson, D.L. Rosen and A.D. Branch: Virology 142, 441 (1985).
9   G.A. Prody, J.T. Bakos, J.L. Buzayan, I.R. Schneider and G. Bruening: Science 231, 1577 (1986).
10 W.L. Gerlach, J.M. Buzayan, I.R. Schneider and G. Bruening: Virology 150, 172 (1986).
11 J.M. Buzayan, W.L. Gerlach and G. Bruening: Proc. natl Acad. Sci. USA 83, 8859 (1986).
12 G. Bruening, J.M. Buzayan, A. Hampel and W.L. Gerlach: RNA Genetics, Vol. II, Retroviruses, Viroids and RNA Recombination, pp. 127–145. Eds E. Domingo, J.J. Holland and P. Ahlquist. CRC Press Inc., Florida 1988.
13 A.J. Zaug, M.D. Been and T.R. Cech: Nature 324, 429 (1986).
14 J. Haseloff and W.L. Gerlach: Nature 334, 585 (1988).
15 A.J. Zaug and T.R. Cech: Science 231, 470 (1986).
16 T.R. Cech: Proc. natl Acad. Sci. USA 83, 4360 (1986).
17 M.D. Been and T.R. Cech: Science 239, 1412 (1988).
18 W. Gilbert: Nature 319, 618 (1986).
19 E.G. Nisbert: Nature 322, 206 (1986).
20 J.P. Ferris: Cold Spring Harbor Symp. Quant. Biol. 52, 29 (1987).
21 A.M.Q. King: RNA Genetics, Vol. II, Retroviruses, Viroids and RNA Recombination, pp. 149–165. Eds E. Domingo, J.J. Holland and P. Ahlquist. CRC Press Inc., Florida 1988.
22 P.D. Cooper: Comprehensive Virology, Vol. 9, Genetics of Animal Viruses, pp. 133–207. Eds H. Fraenkel-Conrat and R.R. Wagner. Plenum Press, New York 1977.
23 D. McCahon, W.R. Slade, R.A.J. Priston and J.R. Lake: J. gen. Virol. 35, 555 (1977).
24 C.R. Pringle: Virology 25, 48 (1965).
25 A.M.Q. King, D. McCahon, W.R. Slade and J.W.I. Newman: Cell 29, 921 (1982).
26 J.J. Bujarski and P. Kaesberg: Nature 321, 528 (1986).
27 A.V. Munishkin, L.A. Voronin and A.B. Chetverin: Nature 333, 473 (1988).
28 S.S. Monroe and S. Schlesinger: Proc. natl Acad. Sci. USA 80, 3279 (1983).
29 P.D. Minor, A. John, M. Ferguson and J.P. Icenogle: J. gen. Virol. 67, 693 (1986).
30 S. Makino, J.G. Keck, S.A. Stohlman and M.M.C. Lai: J. Virol. 57, 729 (1986).
31 J.G. Keck, G.K. Matsushima, S. Makino, J.O. Fleming, D.M. Vannier, S.A. Stohlman and M.M.C. Lai: J. Virol. 62, 1810 (1988).
32 R.G. Webster, W.G. Laver, G.M. Air and G.C. Schild: Nature 296, 115 (1982).
33 F.I. Smith and P. Palese: RNA Genetics, Vol. III, Variability of RNA Genomes, pp. 123–135. Eds E. Domingo, J.J. Holland and P. Ahlquist. CRC Press Inc., Florida 1988.
34 S. Pedley, F. Hundley, I. Chrystie, M.A. McCrae and U. Desselberger: J. gen. Virol. 65, 1141 (1984).
35 F. Hundley, B. Biryahwaho, M. Gow and U. Desselberger: Virology 143, 88 (1985).
36 B. Biryahwaho, F. Hundley and U. Desselberger: Arch. Virol. 96, 257 (1987).
37 J.J. Holland, K. Spindler, F. Horodyski, E. Grabau, S. Nichol and S. VandePol: Science 215, 1577 (1982).
38 D.C. Reanney: A. Rev. Microbiol. 36, 47 (1982).

39   C.K. Biebricher: Evolutionary Biology, pp. 1–52. Eds M.K. Hechet, B. Wallace and G.T. Prance. Plenum, New York 1983.
40   D.C. Reanney: The Microbe, Part. I. Viruses, pp. 175–195. Eds B.W.J. Mahy and J.R. Pattison. Cambridge University Press, Cambridge 1984.
41   E. Domingo, E. Martinez-Salas, F. Sobrino, J.C. de la Torre, A. Portela, J. Ortín, C. López-Galindez, P. Pérez-Breña, N. Villanueva, R. Nájera, S. VandePol, D. Steinhauer, N. DePolo and J.J. Holland: Gene *40*, 1 (1985).
42   A. Gibbs: J. Cell. Sci. Suppl. *7*, 319 (1987).
43   M. Eigen: Cold Spring Harbor Symp. Quant. Biol. *52*, 307 (1987).
44   E. Domingo, J.J. Holland and P. Ahlquist: RNA Genetics, Vol. III, Variability of RNA Genomes. CRC Press Inc., Boca Raton, Florida 1988.
45   J.J. Holland: Concepts in Viral Pathogenesis, pp. 137–143. Eds A.L. Notkins and M.B.A. Oldstone. Springer-Verlag, New York 1984.
46   D.B. Smith and S.C. Inglis: J. gen. Virol. *68*, 2729 (1987).
47   D. Steinhauer and J.J. Holland: A. Rev. Microbiol. *41*, 409 (1987).
48   J.H. Strauss and E.G. Strauss: A. Rev. Microbiol. *42*, 657 (1988).
49   M. Eigen and C.K. Biebricher: RNA Genetics, Vol. III, Variability of RNA Genomes, pp. 211–245. Eds E. Domingo, J.J. Holland and P. Ahlquist. CRC Press Inc., Florida 1988.
50   J.W. Drake: Nature *221*, 1132 (1969).
51   L.A. Loeb and T.A. Kunkel: A. Rev. Biochem. *52*, 429 (1982).
52   B. Singer and J.T. Kusmierek: A. Rev. Biochem. *52*, 655 (1982).
53   M. Fry and L.A. Loeb: Animal Cell DNA Polymerases. CRC Press Inc., Florida 1986.
54   M.D. Topal and J.R. Fresco: Nature (London) *263*, 285 (1976).
55   W. Saenger: Principles of Nucleic Acid Structure. Springer-Verlag, New York 1984.
56   A. Kornberg: DNA Replication. Freeman-Cooper, San Francisco, California 1980.
57   M. Wabl, P.D. Burrows, A. von Gabain and C. Steinberg: Proc. natl Acad. Sci. USA *82*, 479 (1985).
58   R.L. O'Brien, R.L. Brinster and U. Storb: Nature *326*, 405 (1987).
59   M. Siekevitz, C. Kocks, K. Rajewsky and R. Dildrop: Cell *48*, 757 (1987).
60   A. Razzaque, H. Mizusawa and H. Seidman: Proc. natl Acac. Sci. USA *80*, 3010 (1983).
61   M.A. Cifone and I.J. Fidler: Proc. natl Acad. Sci. USA *78*, 6949 (1981).
62   M.S. Grigorian, D.A. Kramerov, E.M. Tulchinsky, E.S. Revasova and E.M. Lukanidin: EMBO J. *4*, 2209 (1985).
63   J.C. de la Torre, E. Martinez-Salas, J. Diez and E. Domingo: J. Virol. *63*, 59 (1989).
64   E.C. Crouch, K.R. Stone, M. Block and R.W. McDivitt: Canc. Res. *47*, 6086 (1987).
65   G.L. Nicolson: Canc. Res. *47*, 1473 (1987).
66   H.L. Carson: Am. Nat. *109*, 83 (1975).
67   D.H. Erwin and J.W. Valentine: Proc. natl Acad. Sci. USA *81*, 5482 (1984).
68   Z. Agur and M. Kerszberg: Am. Nat. *129*, 862 (1987).
69   W.H. Li, C.I. Wu and C.C. Luo: Molec. biol. Evol. *2*, 150 (1985).
70   M. Goodman: Bioessays *3*, 9 (1986).
71   R. J. Britten: Science *231*, 1393 (1986).
72   C. Weissmann and H. Weber: Prog. nucl. Acid Res. molec. Biol. *33*, 251 (1986).
73   H. Temin: Canc. Res. *48*, 1697 (1988).
74   E. Domingo, R. A. Flavell and C. Weissmann: Gene *1*, 3 (1976).
75   E. Batschelet, E. Domingo and C. Weissmann: Gene *1*, 27 (1976).
76   E. Domingo, D. Sabo, T. Taniguchi and C. Weissmann: Cell *13*, 735 (1978).
77   D. Steinhauer and J.J. Holland: J. Virol. *57*, 219 (1986).

78   D. Steinhauer, J. C. de la Torre and J.J. Holland: J. Virol. *63*, 2063 (1989).
78a B.T. White and D.J. McGeoch: J. gen. Virol. *68*, 3033 (1987).
79   R.K. Durbin and V. Stollar: Virology *154*, 135 (1986).
80   J.M. Sedivy, J.P. Capone, U.L. Raj Bhandary and P.A. Sharp: Cell *50*, 379 (1987).
81   J.P. Dougherty and H.M. Temin: Molec. cell. Biol. *6*, 4387 (1987).
82   J.P. Dougherty and H.M. Temin: J. Virol. *62*, 2817 (1988).
83   J.D. Parvin, A. Moscona, W.T. Pan, J.M. Leider and P. Palese: J. Virol. *59*, 377 (1986).
84   J.M. Leider, P. Palese and F. Smith: J. Virol. *62*, 3084 (1988).
85   J.M. Coffin, P.N. Tsichlis, C.S.Barker and S. Voynow: Ann. N.Y. Acad. Sci. *354*, 410 (1980).
86   N. Battula and L.A. Loeb: J. biol. Chem. *249*, 4086 (1974).
87   K.P. Gopinathan, L.A. Weymouth, T.A. Kunkel and L.A. Loeb: Nature (London) *278*, 857 (1979).
88   J.D. Roberts, B.D. Preston, L.A. Johnston, A. Soni, L.A. Loeb and T.A. Kunkel: Molec. cell. Biol., *9*, 469 (1989).
88a J.D. Roberts, K. Bebenek and T.A. Kunkel: Science *242*, 1171 (1988)-
89   B.D. Preston, B.J. Poiesz and L.A. Loeb: Science, *242*, 1168 1988).
90   Y. Takeuchi, T. Nagumo and H. Hoshino: J. Virol. *62*, 3900 (1988).
91   C.D. Ward, M.A.M. Stokes and J.B. Flanegan: J. Virol. *62*, 558 (1988).
92   D.S. Stec, A. Waddell, C.S. Schmaljohn, G.A. Cole and A.L. Schmaljohn: J. Virol. *57*, 715 (1986).
93   E. Domingo and J. Holland: RNA Genetics, Vol. III, Variability of RNA Genomes, pp. 3–36. Eds E. Domingo, J.J. Holland and P. Ahlquist. CRC Press Inc., Florida 1988.
94   A. Villaverde, E. Martinez-Salas and E. Domingo: J. molec. Biol. *204*, 771 (1988).
95   T. Gojobori and S. Yokoyama: Proc. natl Acad. Sci. USA *82*, 4198 (1985).
96   D.A. Buonagurio, S. Nakada, J.D. Parvin, M. Krystal, P. Palese and W. Fitch: Science *232*, 980 (1986).
97   C. Martinez, L. del Rio, A. Portela, E. Domingo and J. Ortín: Virology *130*, 539 (1983).
98   B.H. Hahn, G.M. Shaw, M.E. Taylor, R.R. Redfield, P.D. Markham, S.Z. Salahuddin, F. Wong-Staal, R.C. Gallo, E.S. Parks and W.P. Parks: Science *232*, 1548 (1986).
99   F. Wong-Staal: RNA Genetics, Vol. III, Variability of RNA Genomes, pp. 147–157. Eds E. Domingo, J.J. Holland and P. Ahlquist. CRC Press Inc., Florida 1988.
100 F. Gebauer, J.C. de la Torre, I. Gomes, M.G. Mateu, H. Barahona, B. Tiraboschi, I. Bergmann, P. Augé de Mello and E. Domingo: J. Virol. *62*, 2041 (1988).
101 M.P. Costa Giomi, I.E. Bergmann, E.A. Scodeller, P. Augé de Mello, I. Gomes and J.L. La Torre: J. Virol. *51*, 799 (1984).
102 F. Sobrino, E.L. Palma, E. Beck, M. Dávila, J.C. de la Torre, P. Negro, N. Villanueva, J. Ortín and E. Domingo: Gene *50*, 149 (1986).
102a J.E. Clements, S.L. Gdovin, R.C. Montelaro and O. Narayan, Ann. Rev. Immunol. *6*, 139 (1988).
103 M.E. Piccone, G. Kaplan, L. Giavedoni, E. Domingo and E.L. Palma: J. Virol. *62*, 1469 (1988).
104 J.J. Holland, E.A. Grabau, C.L. Jones and B.L. Semler: Cell *16*, 494 (1979).
105 K.R. Spindler, F.M. Horodyski and J.J. Holland: Virology *119*, 96 (1982).
106 F.M. Horodyski, S.T. Nichol, K.R. Spindler and J.J. Holland: Cell *33*, 801 (1983).
107 P.J. O'Hara, F.M. Horodyski, S.J. Nichol and J.J. Holland: J. Virol. *49*, 793 (1984).

108 P.J. O'Hara, S.T. Nichol, F.M. Horodyski and J.J. Holland: Cell 36, 915 (1984).
109 N.J. DePolo, C. Giachetti and J.J. Holland: J. Virol. 61, 454 (1987).
110 D.A. Steinhauer, J.C. de la Torre, E. Meier and J.J. Holland: J. Virol. 63, 2072 (1989).
111 E. Domingo, M. Dávila and J. Ortín: Gene 11, 333 (1980).
112 D.J. Rowlands, B.E. Clarke, A.R. Carroll, F. Brown, B.H. Nicholson, J.L. Bittle, R.A. Houghten and R.A. Lerner: Nature 306, 694 (1983).
113 R. Cattaneo, A. Schmid, G. Rebmann, K. Baczko, V. ter Meulen, W.J. Bellini, S. Rozenblatt and M.A. Billeter: Virology 154, 97 (1986).
114 R. Cattaneo, A. Schmid, M.A. Billeter, R.D. Sheppard and S.A. Udem: J. Virol. 62, 1388 (1988).
115 R. Cattaneo, A. Schmid, D. Eschle, K. Baczko, V. ter Meulen and M.A. Billeter: Cell, 55, 255 (1988).
116 J.M. Katz and R.G. Webster: Virology 165, 446 (1988).
117 M.S. Saag, B.H. Hahn, J. Gibbons, Y. Li, E.S. Parks, W.P. Parks and G.M. Shaw: Nature 334, 440 (1988).
118 A.G. Fisher, B. Ensoli, D. Looney, A. Rose, R.C. Gallo, M.S. Saag, G.M. Shaw, B.H. Hahn and F. Wong-Staal: Nature 334, 444 (1988).
119 J.P. Clewley, D.H.L. Bishop, C.Y. Kang, J. Coffin, W.M. Schnitzlein, M.E. Reichmann and R.E. Shope: J. Virol. 23, 152 (1977).
120 S.T. Nichol: J. Virol. 61, 1029 (1987).
121 S.T. Nichol: J. Virol. 62, 572 (1988).
122 J.R. Stephenson and V. ter Meulen: Arch. Virol. 71, 279 (1982).
123 K. Nerome, M. Ishida, A. Oya and S. Bosshard: J. gen. Virol. 64, 465 (1983).
124 K. Nerome, M. Shibata, S. Kobayashi, R. Yamaguchi, Y. Yoshioka, M. Ishida and A. Oya: J. Virol. 50, 649 (1984).
125 F.J. Dutko and M.B.A. Oldstone: J. gen. Virol. 64, 1689 (1983).
126 P.J. Southern and M.B.A. Oldstone: RNA Genetics, Vol. III, Variability of RNA Genomes, pp. 159–170. Eds E. Domingo, J.J. Holland and P. Ahlquist. CRC Press Inc., Florida 1988.
127 J.F. Lees, C.R. Pringle and R.M. Elliot: Virology 148, 1 (1986).
128 C.S. Schmaljohn, J. Arikawa, S.E. Hasty, L. Rasmussen, H.W. Lee, P.W. Lee and J.M. Dalrymple: J. gen. Virol. 69, 1949 (1988).
129 S. Fields and G. Winter: Gene 15, 207 (1981).
130 O.M. Kew, B.K. Nottay, M.H. Hatch, J.H. Nakano and J.F. Obijeski: J. gen. Virol. 56, 337 (1981).
131 B.K. Nottay, O.M. Kew, M.H. Hatch, J.T. Heyward and J.F. Obijeski: Virology 108, 405 (1981).
132 P.D. Minor, G.C. Schild, M. Ferguson, A. MacKay, D.I. Magrath, A. John, P.J. Yates and M. Spitz: J. gen. Virol. 61, 167 (1982).
133 B.A. Jameson, J. Bonin, E. Wimmer and O.M. Kew: Virology 143, 337 (1985).
134 P.J. Hughes, D.M.A. Evans, P.D. Minor, G.C. Schild, J.W. Almond and G. Stanway: J. gen. Virol. 67, 2093 (1986).
135 A. Huovilainen, L. Kinnunen, M. Ferguson and T. Hovi: J. gen. Virol. 69, 1941 (1988).
135a E.J. Rozhon, A.K. Wilson and B. Jubelt: J. Virol. 50, 137 (1984).
136 E. Domingo, M.G. Mateu, M.A. Martinez, J. Dopazo, A. Moya and F. Sobrino: Applied Virology Research, Vol. 2, Virus Variation and Epidemiology. Eds E. Kurstak, R.G. Marusyk, F.A. Murphy and M.H.V. Van Regenmortel. Plenum Publ. Co., New York, in press (1989).
137 F. Sobrino, M. Dávila, J. Ortín and E. Domingo: Virology 123, 310 (1983).
138 J.C. de la Torre, E. Martinez-Salas, J. Diez, A. Villaverde, F. Gebauer, E. Rocha, M. Dávila and E. Domingo: J. Virol. 62, 2050 (1988).
139 N. Takeda, K. Miyamura, T. Ogino, K. Natori, S. Yamazaki, N. Sakwiai, N. Nakazono, K. Ishii and R. Kono: Virology 134, 375 (1984).

140 T. Kamahora, A. Itagaki, N. Hattori, H. Tsuchie and T. Kurimura: J. gen. Virol. 66, 2627 (1985).
141 J.P. Clewley, J.K. Pullin, R.J. Avery and N.F. Moore: J. gen. Virol. 64, 503 (1983).
142 M. Weitz and G. Siegl: Virus Res. 4, 53 (1985).
143 D.W. Trent, J.P. Clewley, J. France and D.H.L. Bishop: J. gen. Virol. 43, 365 (1979).
144 D.W. Trent and J.A. Grant: J. gen. Virol. 47, 261 (1980).
145 K. Olson and D.W. Trent: J. gen. Virol. 66, 979 (1985).
146 D.W. Trent, J.A. Grant, A.V. Vorndam and T.P. Monath: Virology 114, 319 (1981).
147 D.W. Trent, J.A. Grant, L. Rosen and T.P. Monath: Virology 128, 271 (1983).
148 V. Deubel, J.P. Digoutte, T.P. Monath and M. Girard: J. gen. Virol. 67, 209 (1986).
149 R.J. Coelen and J.S. MacKenzie: J. gen. Virol. 69, 1903 (1988).
150 M.C. Lai and S.A. Stohlman: J. Virol. 38, 661 (1981).
151 H. Wege, J.R. Stephenson, M. Koga, H. Wege and V. ter Meulen: J. gen. Virol. 54, 67 (1981).
152 P. Goelet, G.P. Lomonossoff, P.J.G. Butler, M.E. Akam, M.J. Gait and J. Karn: Proc. natl Acad. Sci. USA 79, 5818 (1982).
153 F. Garcia-Arenal, P. Palukaitis and M. Zaitlin: Virology 132, 131 (1984).
154 D.B. Hrdy, L. Rosen and B.N. Fields: J. Virol. 31, 104 (1979).
155 E.A. Wenske and B.N. Fields: RNA Genetics, Vol. III, Variability of RNA Genomes, pp. 101–121. Eds E. Domingo, J.J. Holland and P. Ahlquist. CRC Press Inc., Florida 1988.
156 J.L. Darlix and P.F. Spahr: Nucl. Acids Res. 11, 5953 (1983).
157 H. Okamoto, M. Imai, M. Kametani, T. Nakamura and M. Mayumi: Japan. J. exp. Med. 57, 231 (1987).
158 J.A. Bruenn and V.E. Brennan: Cell 19, 923 (1980).
159 J.E. Visvader and R.H. Symons: Nucl. Acids Res. 13, 2907 (1985).
160 H.J. Gross, G. Krupp, H. Domdey, M. Raba, P. Jank, C. Lossow, H. Alberty, K. Ramm and H.L. Sänger: Eur. J. Biochem. 121, 249 (1982).
161 F. Garcia-Arenal, V. Pallàs and R. Flores: Nucl. Acids. Res. 15, 4203 (1987).
162 P. Keese, J.E. Visvader and R.H. Symons: RNA Genetics, Vol. III, Variability of RNA Genomes, pp. 71–98. Eds E. Domingo, J.J. Holland and P. Ahlquist. CRC Press Inc., Florida 1988.
162a V. Pallás, I. Garcia-Luque, E. Domingo and R. Flores: Nucl. Acids Res. 16, 9864 (1988).
163 M. Eigen: Naturwissenschaften 58, 465 (1971).
164 M. Eigen and P. Schuster: The Hypercycle. A. Principle of Natural Self-Organization. Springer-Verlag, Berlin 1979.
165 W. Fontana and P. Schuster: Biophys. Chem. 26, 123 (1987).
166 C.K. Biebricher: Cold Spring Harbor Symp. Quant. Biol. 52, 299 (1987).
167 C.K. Biebricher and M. Eigen: RNA Genetics, Vol. I, RNA-Directed Virus Replication, pp. 1–21. Eds E. Domingo, J.J. Holland and P. Ahlquist. CRC Press Inc., Florida 1988.
168 G.N. Rogers, J.C. Paulson, R.S. Daniels, J.J. Skehel, I.A. Wilson and D.C. Wiley: Nature 304, 76 (1983).
169 C.W. Naeve, V.A. Hinshaw and R.G. Webster: J. Virol. 51, 567 (1984).
170 R.F. Siliciano, T. Lawton, C. Knall, R.W. Karr, P. Berman, T. Gregory and E.L. Reinherz: Cell 54, 561 (1988).
171 E.C. Herrmann and J.A. Herrmann: Ann. N.Y. Acad. Sci. 284, 632 (1977).
172 H.J. Field: Antiviral Agents: The Development and Assessment of Antiviral Chemotherapy, Vols. I and II. CRC Press Inc., Florida 1988.
173 E. De Clercq: Clinical Use of Antiviral Drugs. Martinus Nijhoff Publishing, Boston 1988.

174 W.L. Davies, R.R. Grunert, R.M. Haff, J.W. McGahen, E.M. Neumayer and M. Paulshock: Science *144*, 862 (1964).
175 R. Dolin, R.C. Reichman, H.P. Madore, R. Maynard, P.N. Linton and J. Webber-Jones: N. Engl. J. Med. *307*, 580 (1982).
176 T.A. Bektimirov, R.G. Douglas, R. Dolin, G.P. Galasso, V.F. Krylov and J.S. Oxford: WHO Bull. *63*, 51 (1985).
177 R. Dolin: Clinical Use of Antiviral Drugs, pp. 277–287. Ed. E. De Clercq. Martinus Nijhoff Publ., Boston 1988.
178 K.W. Cochran, H.R. Massaab, A. Tsudona and B.S.Berlin: Ann. N.Y. Acad. Sci. *130*, 432 (1965).
179 G. Appleyard: J. gen. Virol. *36*, 249 (1977).
180 M.D. Lubeck, J.L. Schulman and P. Palese: J. Virol. *28*, 710 (1978).
181 J.S. Oxford, I.S. Logan and C.W. Potter: Nature *226*, 82 (1970).
182 R.G. Webster, Y. Kawaoka, W.J. Bean, G.W. Beard and M. Brugh: J. Virol. *55*, 173 (1985).
183 H. Heider, B. Adamczyk, H.W. Presber, C. Schroeder, R. Feldblum and M.K. Indulen: Acta virol. *25*, 395 (1981).
184 R.M. Pemberton, R. Jennings, C.W. Potter and J.S. Oxford: J. antimicrob. Chemother. *18* (Suppl. B), 135 (1986).
185 C.B. Hall, R. Dolin, C.L. Gala, D.M. Markovitz, Y.Q. Zhang, P.H. Madore, F.A. Disney, W.B. Talpey, J.L. Green, A.B. Francis and M.E. Pichichero: Pediatrics *80*, 275 (1987).
186 J. Thomson, W. Fleet, E. Lawrence, E. Pierce, L. Morris and P. Wright: J. med. Virol. *21*, 249 (1987).
187 C. Scholtissek and G. Faulkner: J. gen. Virol. *44*, 807 (1979).
188 A.J. Hay, A.J. Wolfstenholme, J.J. Skehel and M.H. Smith: EMBO J. *4*, 3021 (1985).
189 A.J. Hay, M.C. Zambon, A.J. Wolfstenholme, J.J. Skehel and M.H. Smith: J. antimicrob. Chemother. *18* (Suppl. B), 19 (1986).
190 R.B. Belshe, M. Hall Smith, C.B. Hall, R. Betts and A.J. Hay: J. Virol. *62*, 1508 (1988).
191 R.A. Lamb, S.L. Zebedee and C.D. Richardson: Cell *40*, 627 (1985).
192 T.J. Smith, M.J. Kremer, M. Luo, G. Vriend, E. Arnold, G. Kamer, M.G. Rossmann, M.A. McKinlay, G.D. Diana and M.J. Otto: Science *233*, 1286 (1986).
193 M.G. Rossmann and R.R. Rueckert: Microbiol. Sci. *4*, 206 (1987).
194 H. Ishitsuka, Y. Ninomiya, C. Oshawa, T. Ohiwa, M. Fujiu, I. Umeda, H. Shirai and Y. Suhara: Current Chemotherapy and Immunotherapy, pp. 1083–1085. Eds P. Periti and G.G. Grassi. American Society for Microbiology, Washington, D.C. 1982.
195 M.J. Otto, M.P. Fox, M.J. Fancher, M.F. Kuhrt, G.D. Diana and M.A. McKinlay: Antimicrob. Agents Chemother. *27*, 883 (1985).
196 J.W.T. Selway: Progr. clin. biol. Res. *213*, 521 (1986).
197 S.J. Sperber and F.G. Hayden: Antimicrob. Agents Chemother. *32*, 409 (1988).
198 A.L.M. Ahmad, A.B. Dowsett and D.A.J. Tyrrell: Antiviral Res. *8*, 27 (1987).
199 D.A.J. Tyrell and W. Al-Nakib: Clinical Use of Antiviral Drugs, pp. 241–276. Ed. E. De Clercq. Martinus Nijhoff Publ., Boston 1988.
200 G. Abraham and R.J. Colonno: J. Virol. *51*, 340 (1984).
201 R.J. Colonno, P.L. Callahan and W.J. Long: J. Virol. *57*, 7 (1986).
202 F. Hayden. J. Gwaltney, Jr. and R. Colonno: Antiviral Res. *9*, 233 (1988).
203 C.D. Richardson and P.W. Choppin: Virology *131*, 518 (1983).
204 C.B. Pert, J.M. Hill, M.R. Ruff, R.M. Berman, W.G. Robey, I.O. Arthur, F.W. Ruscetti and W.L. Farrar: Proc. natl Acad. Sci. USA *83*, 9254 (1986).
205 J.D. Hull, D.L. Krah and P.W. Choppin: Virology *159*, 368 (1987).

206 R.A. Fisher, J.M. Bertonis, W. Meier, V.A. Johnson, D.S. Costopoulos, T. Liu, R. Tizarol, B.D. Walker, M.S. Hirsh, R.T. Schooley and R.A. Flavell: Nature *331,* 76 (1988).
207 M.S. Hirsch: Immunobiology and Pathogenesis of Persistent Virus Infections, pp. 262–277. Ed. C. Lopez. American Society for Microbiology, Washington, D.C. 1988.
208 K.J. Reagan, B. Goldberg and R.L. Crowell: J. Virol. *49,* 635 (1984).
208a M.S. Co, B.N. Fields and M.I. Greene: Concepts in Viral Pathogenesis, Vol. 2, pp. 126–131. Eds A.L. Notkins and M.B.A. Oldstone. Springer-Verlag, New York 1986.
209 C. Crumpacker: Clinical Use of Antiviral Drugs, pp. 207–222. Ed. E. De Clercq. Martinus Nijhoff Publ., Boston 1988.
210 R.W. Sidwell, J.H. Huffmann, G.P. Khare, L.B. Allen, J.T. Witkowski and R.K. Robins: Science *177,* 705 (1972).
211 R.W. Sidwell, L.N. Simon, J.T. Witkowski and R.K. Robins: Prog. Chemotherapy *2,* 889 (1974).
212 R.A. Smith and W. Kirkpatrick: Ribavirin: A Broad Spectrum Antiviral Agent. Academic Press, New York 1980.
213 E. De Clercq: The Scientific Basis of Antimicrobial Chemotherapy, pp. 155–184. Eds D. Green and F. O'Grady. Cambridge University Press, Cambridge 1985.
214 B.B. Goswami, E. Borek, D.K. Shalma, J. Fujitaki and R.A. Smith: Biochem. biophys. Res. Commun. *89,* 830 (1979).
215 J.C. de la Torre, B. Alarcón, E. Martinez-Salas, L. Carrasco and E. Domingo: J. Virol. *61,* 233 (1987).
216 V. Knight and B. Gilbert: Clinical Use of Antiviral Drugs, pp. 289–304. Ed. E. De Clercq. Martinus Nijhoff Publ., Boston 1988.
217 J.B. McCormick and S.P. Fisher-Hoch: Clinical Use of Antiviral Drugs, pp. 305–315. Ed. E. De Clercq. Martinus Nijhoff Publ., Boston 1988.
218 K.T. McKee, Jr., J.W. Huggins, C.J. Trahan and B.G. Mahlandt: Antimicrob. Agents Chemother. *32,* 1304 (1988).
219 H.J. Field: Antiviral Agents: The Development and Assessment of Antiviral Chemotherapy, Vol. I, pp. 127–149. CRC Press Inc., Florida 1988.
220 L.M. Scheidel, R.K. Durbin and V. Stollar: Virology *158,* 1–7 (1987).
221 J. Balzarini and S. Broder: Clinical Use of Antiviral Drugs, pp. 361–385. Ed. E. De Clercq. Martinus Nijhoff Publ., Boston 1988.
222 H. Mitsuya, K.J. Weinhold, P.A. Furman, M.H. St. Clair, S.N. Lehrman, R.C. Gallo, D. Bolognesi, D.W. Barry and S. Broder: Proc. natl Acad. Sci. USA *82,* 7096 (1985).
222a B.A. Larder, G. Darby and D.D. Richman: Science *243,* 1731 (1989).
223 S. Shigeta, K. Konno, T. Yokota, K. Nakamura and E. De Clercq: Antimicrob. Agents Chemother. *32,* 906 (1988).
224 K. Saunders and A.M.Q. King: J. Virol. *42,* 389 (1982).
225 K. Anderson-Sillman, S. Bartal and D.R. Tershak: J. Virol. *50,* 922 (1984).
226 S.E. Pincus, D.C. Diamond, E.A. Emini and E. Wimmer: J. Virol. 57, 638 (1986).
227 S.E. Pincus and E. Wimmer: J. Virol. *60,* 793 (1986).
228 S.E. Pincus, H. Rohl and E. Wimmer: Virology *157,* 83 (1987).
229 D.C. DeLong: Microbiology 1984, pp. 431–434. Eds L. Leive and D. Schlessinger. American Society for Microbiology, Washington, D.C. 1984.
230 D.O. White: Monographs in Virology, Vol. 16, Antiviral Chemotherapy, Interferons and Vaccines. Ed. J.L. Melnick. Karger 1984.
231 D. Sergiescu, F. Horodniceanu and A. Aubert-Combiescu: Progr. med. Virol. *14,* 123 (1972).
232 A.A. Newton: Antiviral Agents: The Development and Assessment of Antiviral Chemotherapy, Vol. I, pp. 23–66. Ed. H.J. Field. CRC Press Inc., Florida 1988.

233 M.J. Hall and I.B. Cuncan: Antiviral Agents: The Development and Assessment of Antiviral Chemotherapy, Vol. II, pp. 29–84. Ed. H.J. Field. CRC Press Inc., Florida 1988.
234 K.L. Hartshorn, E.G. Sandstrom, D. Neumeyer, T.J. Paradis, T.C. Chou, R.T. Schooley and M.S. Hirsch: Antimicrob. Agents Chemother. *30*, 189 (1986).
235 H.J. Parish: Victory with Vaccines. E. and S. Livingstone, Ltd., Edinburgh and London, 1968.
236 D. Salk: Rev. Infect. Dis. *2*, 228 (1980).
237 A.B. Sabin: J. Am. med. Assoc. *246*, 236 (1981).
238 A.S. Evans: Viral Infections of Humans: Epidemiology and Control. Plenum Medical Book Co., New York 1984.
239 A.R. Hinman: Rev. infect. Dis. *8* (4), 573 (1986).
240 A.G. Dalgleish and R.C. Kennedy: Vaccine *6*, 215 (1988).
241 C.A. Gibson, J.J. Schlesinger and A.D.T. Barret: Vaccine *6*, 7 (1988).
242 D. Panicali, S.W. Davis, R.L. Weinberg and E. Paoletti: Proc. natl Acad. Sci. USA *80*, 5364 (1983).
243 G.L. Smith, M. Mackett and B. Moss: Nature *302*, 490 (1983).
244 R.M. Chanock, R.A. Lerner, F. Brown and H. Ginsberg: Vaccines 87. Cold Spring Harbor Laboratory, 1987.
245 V. Deubel, R.M. Kinney, J.J. Esposito, C.B. Cropp, A.V. Vorndam, T.P. Monath and D.W. Strent: J. gen. Virol. *69*, 1921 (1988).
246 D.M.A. Evans, G. Dunn, P.D. Minor, G.C. Schild, A.J. Cann, G. Stanway, J.W. Almond, K. Currey and J.V. Maizel: Nature *314*, 548 (1985).
247 L.B. Schonberger, J.E. McGowan and M.B. Gregg: Am. J. Epidem. *104*, 202 (1966).
248 WHO Bulletin *53*, 319 (1976).
249 R.J. Kim-Farley, P. Lichfield, W.A. Orenstein, K.J. Bart, G. Rutherford, S.T. Hsu, L.B. Schonberger, K.J. Lui and C.C. Lin: Lancet *2*, 1322 (1984).
250 D.I. Magrath, D.M.A. Evans, M. Ferguson, G.C. Schild, P.D. Minor, F. Horaud, R. Crainic, M. Stenvik and T. Hovi: J. gen. Virol. *67*, 899 (1986).
251 T. Omata, M. Kohara, S. Kuge, T. Komatsu, S. Abe, B.L. Semler, A. Kameda, H. Itoh, M. Arita, E. Wimmer and A. Nomoto: J. Virol. *58*, 348 (1986).
252 N. La Monica, V.R. Racaniello and J.W. Almond: Vaccines 87, pp. 338–344. Eds R.M. Chanock, R.A. Lerner, F. Brown and H. Ginsberg. Cold Spring Harbor Laboratory, 1987.
253 M. Kohara, S. Abe, S. Kuge, B. Semler, T. Komatsu, M. Arita, H. Itoh and A. Nomoto: Virology *151*, 21 (1986).
254 K.L. Burke, G. Dunn, M. Ferguson, P.D. Minor and J.W. Almond: Nature *332*, 81 (1988).
255 A.R. Hinman, W.H. Foege, C.A. de Quadros, P.A. Patriarca, W.A. Orenstein and E.W. Brink: WHO Bull. *65* (6), 835 (1987).
256 R.M. Chanock and B.R. Murphy: Rev. infect. Dis. *2* (3), 421 (1980).
257 A.Z. Kapikian, R.H. Mitchell, R.M. Chanock, R.A. Shvedoff and R.A. Stewart: Am J. Epidemiol. *89*, 405 (1969).
258 J.E. Conte, Jr. and S.L. Barriere: Manual of Antibiotics and Infectious Diseases. Lea and Febiger, Philadelphia 1988.
259 L.B. Schonberger, D.J. Bregman, J.Z. Sullivan-Bolyai, R.A. Keenlyside, D.W. Ziegler, H.F. Retailliau, D.L. Eddins and J.A. Bryan: Am. J. Epidem. *110*, 105 (1979).
260 M.W. Cloyd, M.J. Holt: Virology *161*, 286 (1987).
261 M.G. Mateu, E. Rocha, O. Vicente, F. Vayreda, C. Navalpotro, D. Andreu, E. Pedroso, E. Giralt, L. Enjuanes and E. Domingo: Virus Res. *8*, 261 (1987).
262 M.A. Martinez, C. Carrillo, J. Plana, R. Mascarella, J. Bergada, E.L. Palma, E. Domingo and F. Sobrino: Gene *62*, 75 (1988).

263 M.G. Mateu, J.L. Da Silva, E. Rocha, D.L. de Brum, A. Alonso, L. Enjuanes, E. Domingo and H. Barahona: Virology *167*, 113 (1988).
264 D. Langley and R.E. Spier: Vaccine *6*, 3 (1988).
265 B.S. Prabhakar, M.V. Haspel, P.R. McClintock and A.L. Notkins: Nature *300*, 374 (1982).
266 B.S. Prabhakar, M.A. Menegus and A.L. Notkins: Virology *146*, 302 (1985).
267 S.R. Webb, K.P. Kearse, C.L. Foulke, P.C. Hartig and B.S. Prabhakar: J. med. Virol. *12*, 9 (1986).
268 A.J. Zuckerman: WHO Bull. *65*(6), 785 (1987).
269 R. DiMarchi, G. Brooke, C. Gale, V. Cracknell, T. Doel and N. Mowat: Science *232*, 639 (1986).
270 B.E. Clarke, S.E. Newton, A.R. Carroll, M.J. Francis, G. Appleyard, A.D. Syred, P.E. Highfield, D.J. Rowlands and F. Brown: Nature *330*, 381 (1987).
271 M.J. Francis, G.Z. Hastings, A.D. Syred, B. McGinn, F. Brown and D.J. Rowlands: Nature *330*, 168 (1987).
272 F. Brown: Vaccine *6*, 180 (1988).
273 R.A. Mariuzza, S.E.V. Phillips and R.J. Poljak: A. Rev. Biphys. Chem. *16*, 139 (1987).
274 R.H. Schwartz: Curr. Top. Microbiol. Immunol. *130*, 79 (1986).
275 F. Celada and E.E. Sercarz: Vaccine *6*, 94 (1988).
276 J.P. Briand, S. Muller and M.H.V. Van Regenmortel: J. immun. Methods *78*, 59 (1985).
277 J. Swetina and P. Schuster: Biphys. Chem. *16*, 329 (1982).
277a B.L. Bass, H. Weintraub, R. Cattaneo and M.A. Billeter: Cell *56*, 331 1989).
278 J. Dopazo, F. Sobrino, E.L. Palma, E. Domingo and A. Moya: Proc. nat. Acad. Sci. USA *85*, 6811 (1988).
279 J.S. Bootman and J.S. Robertson: Virology *166*, 271 (1988).
280 M.D. Curran and B.K. Rima: J. gen. Virol. *69*, 2407 (1988).
281 R.F. Doolittle, D.F. Feng, M.S. Johnson and M.A. McClure: The Quarterly Review of Biology, *64*, 1 (1989).
281a R. Weiss: Nature *338*, 458 (1989).
282 A.D.M.E. Osterhaus and E.J. Vedder: Nature *335*, 20 (1988).
283 B.W.J. Mahy, T. Barrett, S. Evans, E.C. Anderson and C.J. Bostock: Nature *336*, 115 (1988).
284 S.L. Cosby, S. McQuaid, N. Duffy, C. Lyons, B.K. Rima, G.M. Allan, S.J. McCullough, S. Kennedy, J.A. Smyth and F. McNeilly: Nature *36*, 115 (1988)

Medicinal agents incorporating the 1,2-diamine functionality

By Erik T. Michalson and Jacob Szmuszkovicz
Department of Chemistry, University of Notre Dame, Notre Dame, IN 46556, USA

1	Introduction	136
2	Antiarrhythmic agents	136
3	Hypotensive agents	137
4	Antipsychotics	138
5	Analgesics	139
6	Antianxiety agents	140
7	Anticonvulsants	141
8	Gastrointestinal agents	142
9	Local anesthetics	143
10	Anticancer agents	144
11	Additional examples	145
12	Conclusion	147
	References	148

Substituted 1,2,3,4-tetrahydroaminonaphthols (*1*) have been reported to be calcium channel blockers with hypotensive activity[9, 10]. Propranolol, pindolol and related 1,2-amino alcohols are important hypotensives. Interestingly, the sodium channel activity of propranolol is related to that of 1,2-diamines such as lidocaine and procainamide[11].

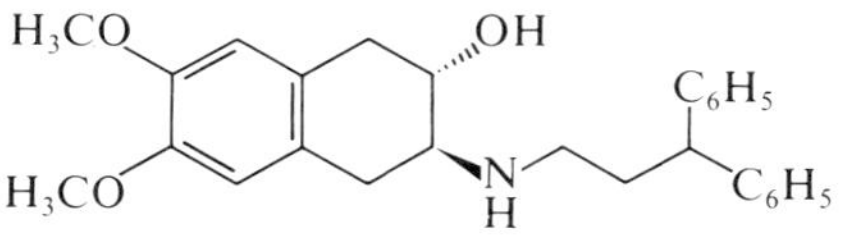

1

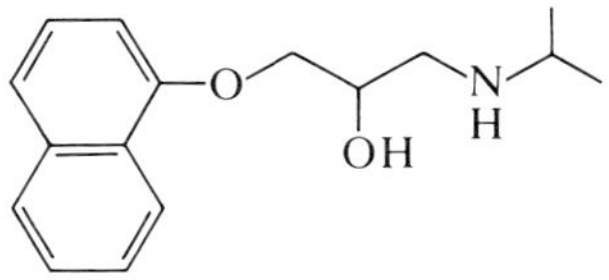

Propranolol

4 Antipsychotics

The orthopramides, which are substituted *o*-methoxybenzamides, are an important group of 1,2-diamines. These compounds are dopamine D-2 receptor antagonists which display antipsychotic, antimanic, neuroleptic and antiemetic activities. Prominent among these substituted benzamides are remoxipride, raclopride and sulpiride[12, 13]. The mechanism of action of the orthopramides is not well understood, but certain structural elements are clearly vital. The aromatic ring and its substituents, especially the 2-methoxy group, are essential for D-2 receptor affinity. The 2-methoxy group by itself, however, is not sufficient for activity, as other aromatic substituents are required[14]. Sulpiride is used in its racemic form for the treatment of psychosis. However, levosulpiride has been shown to be more active than either racemic or dextrosulpiride[15]. Levosulpiride displays outstanding antiemetic activity without extrapyramidal side effects, as well as neuroleptic activity.

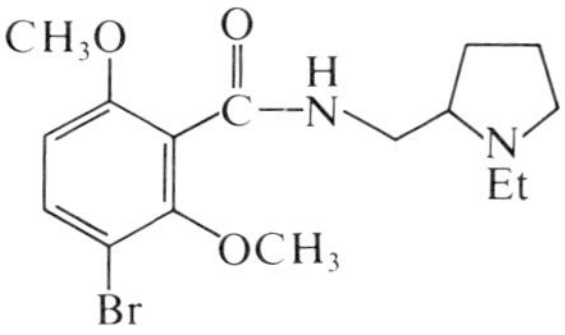

Remoxipride 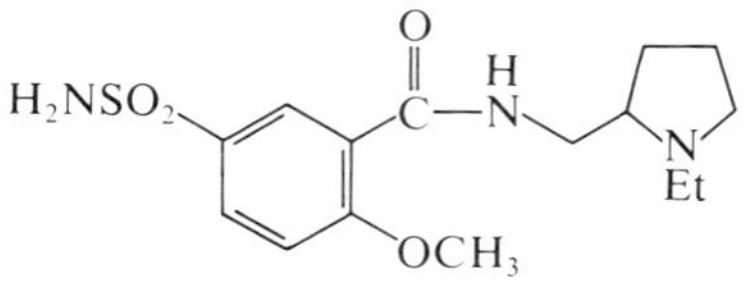   Sulpiride 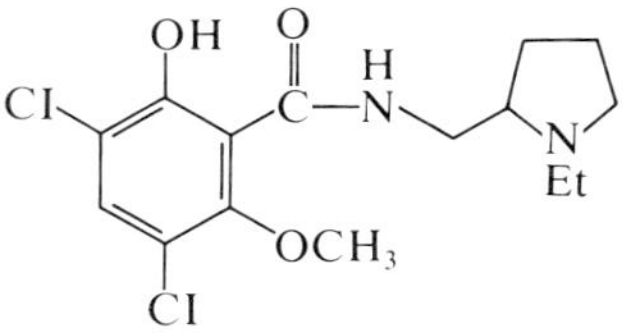   Raclopride

A group of substituted 2-phenylpyrroles, which are conformationally related to the benzamides, were reported to have good antipsychotic activity[16]. Two important representatives of the 2-phenylpyrroles are shown below (*2* and *3*). These new compounds, which possess vicinal diamine linkages, showed improved oral absorption and good dopamine D-2 receptor antagonist activity.

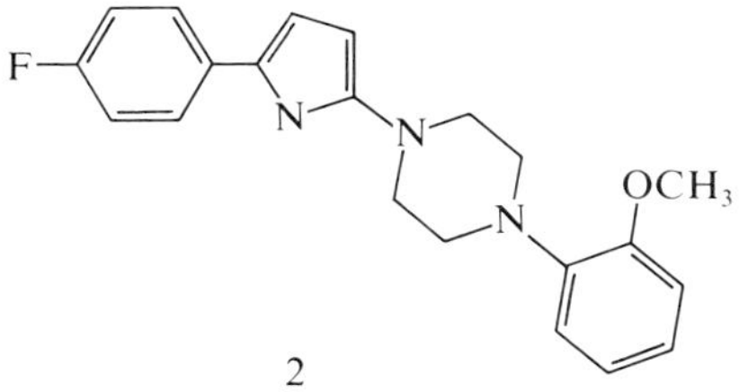

5   Analgesics

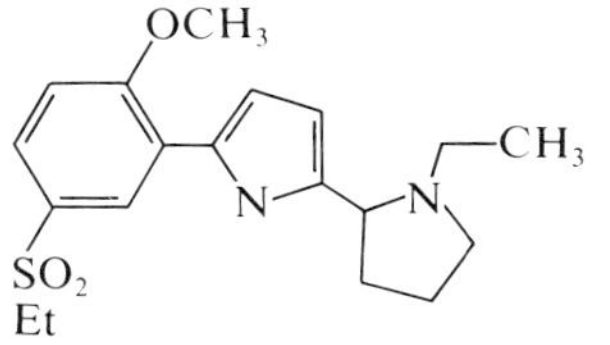

In the area of analgesics, several 1,2-diamines have been shown to be effective agents. A series of N-aryl-propionamide derivatives having good analgesic activity was discovered by Wright[17]. Two important members of this class of compounds were diampromid and phenampromid.

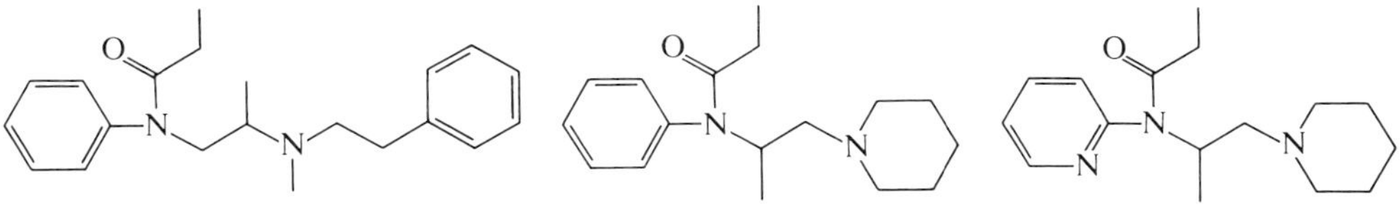

Diampromid   Phenampromid   Propiram

Both of these compounds were reported to have analgesic activities similiar to those of morphine, meperidine, and codeine. A large number of propionyl anilides were prepared and the structure-activity relationships for these N-aryl-propionamides were extensively studied.
Very few substitutions of the N-phenyl ring of the propionyl anilides have been investigated. A meta-methoxy group was the lone modification attempted, and it reduced the analgesic activity substantially as compared to the unsubstituted compound.
A successful variation of the propionyl anilide structure was found in the compound propiram, in which a 2-pyridinyl ring replaced the N-

seizures, as demonstrated in the electroshock test in mice. It is a selective opioid anticonvulsant which lacks the analgesic activity of U-50,488. Interestingly, the structurally similar kappa opioid agonist U-50,488 displays anticonvulsant activity very similar to that of U-50,494[25].

U-54,494

8 Gastrointestinal agents

A novel series of 1,2-diamines, namely N-(1-azabicyclo[2.2.2]oct-3-yl) benzamides, have been made which display gastric prokinetic and antiemetic activities[26]. It is of interest that the prominent member of this class of benzamides, zacopride, shows no neuroleptic or dopamine D-2 receptor antagonist activity. Substituted o-methoxybenzamides (orthopramides) are typically dopamine D-2 antagonists.

An important 1,2-diamine is the glutaramic acid derivative proglumide. It is a cholecystokinin (CCK) antagonist which is used in the treatment of peptic ulcers[27, 28]. Recently, a series of 4-benzamido-N,N-dialkyl-glutaramic acid derivatives related to proglumide was prepared, and their anti-CCK activity was studied[29, 30]. It was found that some of these compounds, such as the 3,4-dichlorobenzamide derivative 6 had significantly higher affinity for the CCK binding site than proglumide.

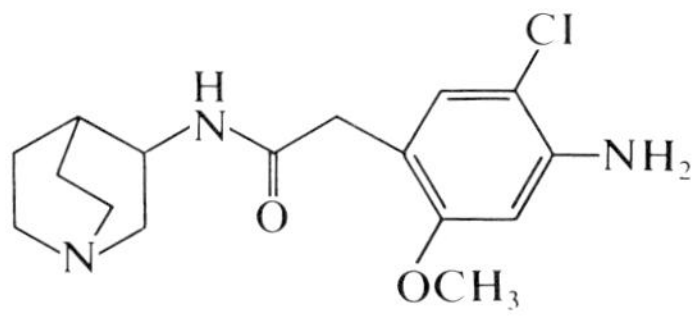

Zacopride

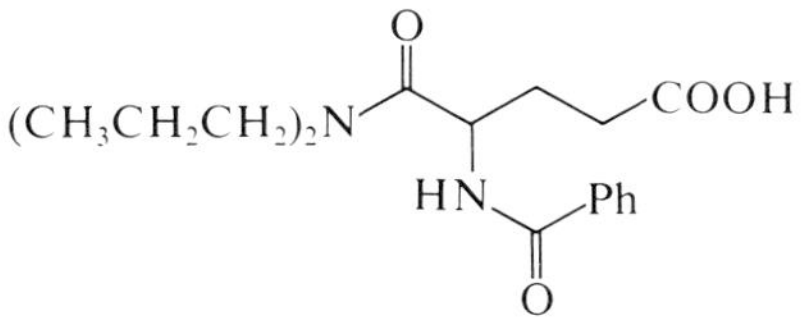

Proglumide

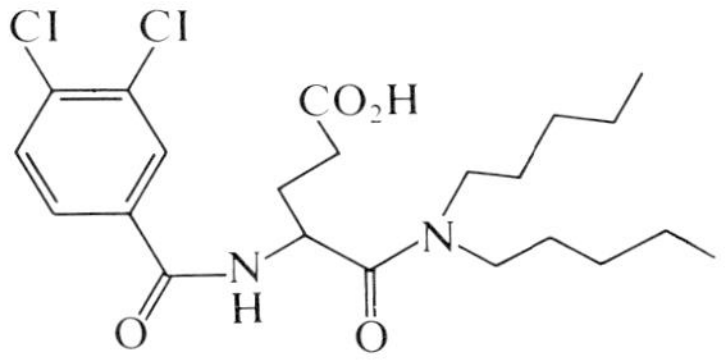

6

9 Local anesthetics

Many local anesthetic agents including lidocaine, etidocaine, bupivacaine, mepivacaine, dibucaine and prilocaine incorporate the 1,2-diamine linkage as a structural feature. An excellent review of this class of compounds is available[31, 32]. Interestingly, some of the above-mentioned vicinal diamino local anesthetics are known inhibitors of acylCoA-cholesterol acyltransferase (ACAT), which is the predominant enzyme of cholesterol esterification in atheromatous arteries[33]. The order of inhibitory potency of these compounds is dibucaine > lidocaine > procaine.

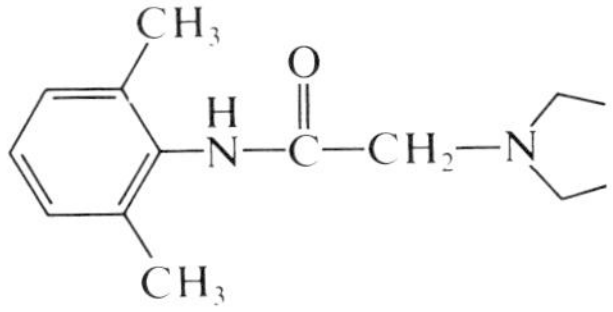

Lidocaine

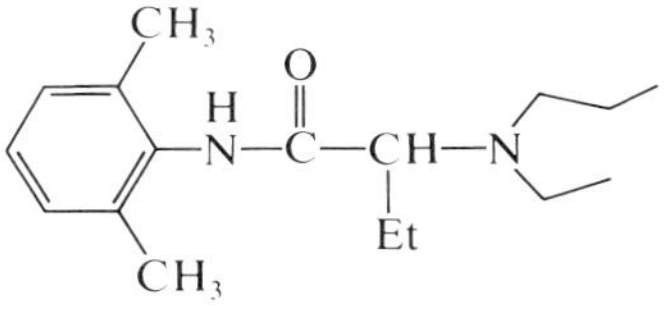

Etidocaine

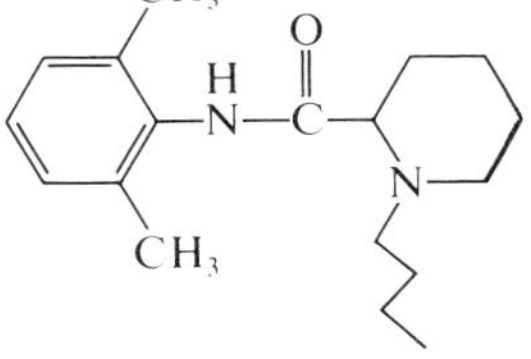

Bupivacaine

Mepivacaine   Dibucaine   Prilocaine

10 Anticancer agents

Cisplatin (cis-diaminedichloroplatinum [II]), which was discovered by Rosenberg, is clinically useful against solid tumors[34]. It has inspired the synthesis of many other platinum diamine complexes in a search for drugs having greater activity and less toxicity. Interestingly, the trans platinum complexes are inactive against neoplastic diseases. It is known that platinum antitumor agents interact with chromosomal DNA and other cellular macromolecules. The platinum anti-cancer agents have been recently reviewed[35, 36].

cisPlatin

Subsequent organoplatinum antitumor compounds have shown less nephrotoxicity and broader antitumor activity than cisplatin. Excellent examples of 1,2-diamines in this important area are 1,2-diaminocyclohexane platinum (II) complexes. These compounds show less nephrotoxicity, greater antitumor activity and less cross-resistance with cisplatin. The labile ligands (X) of the 1,2-diaminocyclohexane platinum (II) complexes greatly influence the water solubility of the complexes. Consequently, modification of these compounds has involved substitution of these labile ligands as well as the non-exchangable amine ligands. Recently, 1,2-dihydroxy-4,5-diaminocyclohexane platinum (II) complexes have been prepared and evaluated for antitumor activity[37].

It was anticipated that the hydroxyl groups would increase the water solubility and thus decrease toxicity because of improved kidney excretion. However, these 1,2-dihydroxy-4,5-diaminocyclohexane platinum (II) complexes were found to be less active than cisplatin or the unsubstituted 1,2-diaminocyclo-hexane platinum (II) complexes.

11 Additional examples

At this point, several medicinally important examples of 1,2-diamines remain to be discussed. Compounds of interest not falling into the previously mentioned categories will be highlighted in the following section.

A new oral antidiabetic agent possessing a 1,2-diamino configuration was recently reported[38]. It is the clinical candidate MTP-1307, which is a thiopyrano-pyrimidine. This series of compounds shows antidiabetic activity in genetically diabetic KK mice.

MTP-1307   Oxotremorine   U-48,753

The novel antidepressant, U-48,753, which contains a trans-1,2-diaminocyclopentane moiety[19, 39], represents a departure from the known antidepressant templates.

Oxotremorine, an ethinylogous 1,2-diamine, is a representative of the cholinergic receptor agonist class of compounds[40].

The H_1-antihistamines tripelennamine, pyrilamine and antazoline all display vicinal diamine subunits.

Tripelennamine   Pyrilamine   Antazoline

Lysergic acid, a representative of the tetracyclic ergot alkaloids (ergoline series), and serotonin, a neurotransmitter, are additional examples of vinylogous 1,2-diamines.

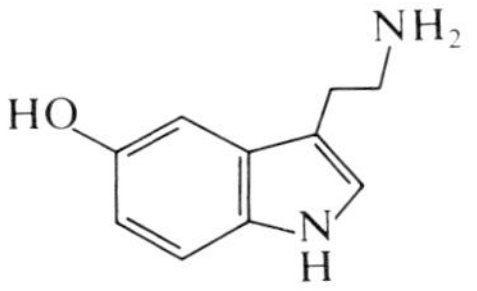

Serotonin

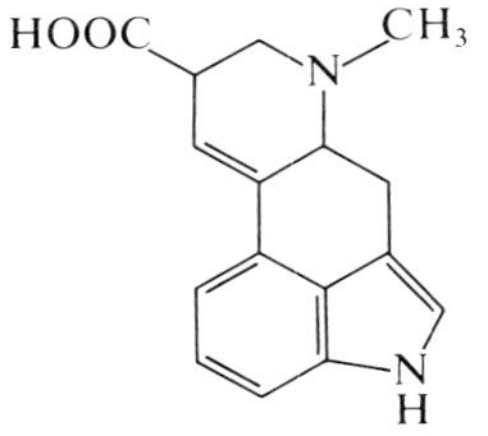

Lysergic acid

Mebrofenin is a substituted 1,2-diamine useful in visualizing the liver and bile excretion patterns[41].
Promethazine is a phenothiazine derivative which displays potent antihistaminic and antiemetic activities. It is also a good sedative and hypnotic. The 1,2-diamino side chain is apparently responsible for the unique properties of promethazine. Phenothiazine derivatives having three carbon side chains, as in the case of chlorpromazine, are usually antipsychotics showing little or no antihistaminic activity.

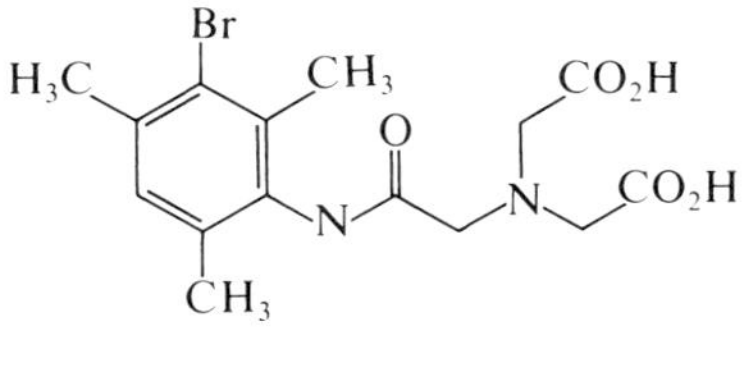

Mebrofenin

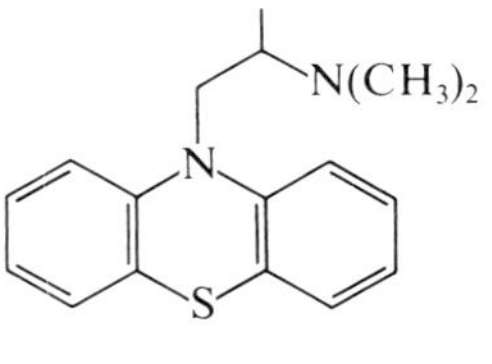

Promethazine

A great variety of 1,2-diamine derivatives are found in the area of antischistosomal drugs. For example, the tricyclic 1,2-diamine hycanthone is an antitumor agent in mice, as well as a potent antischistosomal drug. Hycanthone and the related compound lucanthone have been shown to intercalate DNA. However, the antischistomal effect of these drugs is not fully understood. In addition, several tetrahydroisoquino-

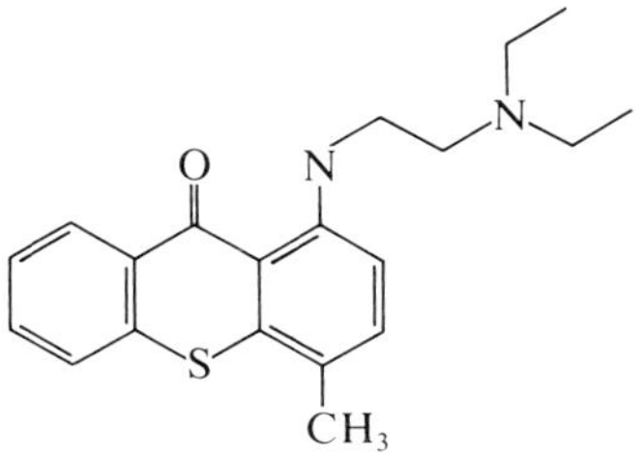

Lucanthone 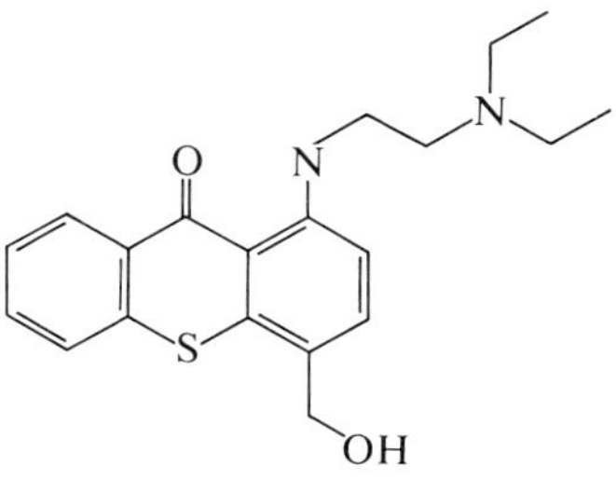   Hycanthone

line derivatives having 1,2-diamine side chains are also active antischistosomal agents. Oxamniquine is the leading representative of this class of tetrahydroisoquinolines. An excellent review of the chemotherapy of schistosomiasis is available [42].

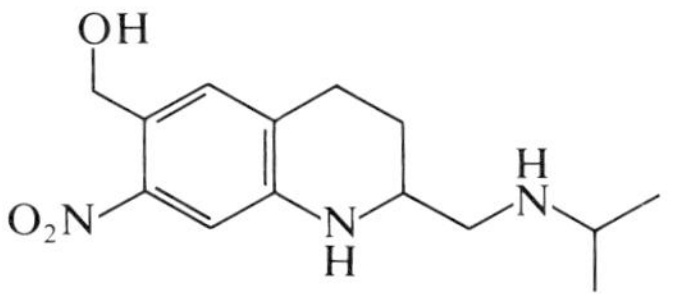

Oxamniquine 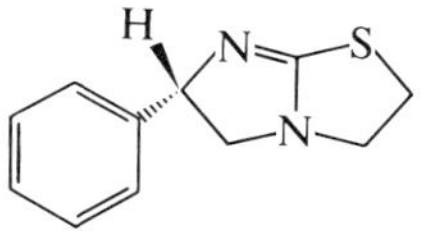   Levamisole

Levamisole is an example of an important 1,2-diamine in the anthelmintic area.

12 Conclusion

This review offers a collection of medicinal agents which contain the 1,2-diamine moiety. Not reviewed in this article, however, are a number of medicinal agents containing other 1,2-dinucleophiles (e.g. 1,2-amino alcohols). The possibility of a relationship between the 1,2-diamine moiety and other 1,2-dinucleophiles is an intriguing question. Furthermore, the amino acids serine and threonine contain a 1,2-aminoalcohol functionality and play a key role in a variety of enzymatic processes. Also, the neurotransmitters acetylcholine and norepinephrine, and also sphingosine are 1,2-aminoalcohol derivatives.
Interestingly, some of the physiologically important diamines such as spermidine and spermine have 1,3- and 1,4- rather than 1,2-diamine functionalities.

The requirement of specific amphiphilic drug-membrane interactions may help to explain why many medicinal agents rely on 1,2-diamine and 1,2-aminoalcohol moieties.

The reader should be aware of the significant differences between the NCH_2CH_2N and $NCH_2CH_2CH_2N$ moieties, as seen for example in promethazine on one hand, and the tricyclic compounds imipramine and chlorpromazine on the other.

It would be interesting to find out what happens in various physiological processes when the 1,2-aminoalcohol and 1,3- and 1,4-diamine functionalities are replaced by 1,2-diamines. Perhaps evolution has this item on its program during the next fifty million years.

Drug design is entering an era in which specific drug-receptor interactions will be emphasized in order to produce efficacious compounds without serious side effects. Today's medicinal chemist should be thoroughly conversant with the few generalizations that are emerging in drug design. The relationship between peptide transmitters and modulators, peptidomimetic agents and the incorporation of the x,y-dinucleophile functionality into molecules should be added to the armamentarium of the medicinal chemist, particularly in terms of their interaction with the receptor structure.

Quoting a recent provocative article by Susan Iversen: "The possibility of developing selective cholinergic agonists and antagonists for the different receptors in different tissues and areas of the brain opens a new era of cholinergic pharmacology" [43].

Acknowledgement

Our work in this area has been supported by The Upjohn Company.

References

1 D. C. Harrison: Drugs *31*, 93 (1986).
2 R. N. Fogoros: Am. Family Physn *33*, 236 (1986).
3 J. J. Hanyok, M. S. S. Chow and J. Kluger: Pharm. Int. 175 (1986).
4 Cardiovascular Drugs, p. 309. Ed. J. A. Bristol. Wiley, 1986.
5 H. E. Kendall: Drugs of Today *23*, 399 (1987).
6 S. L. Chase and G. E. Sloskey: Clin. Pharm. *6*, 839 (1987).
7 B. Holmes and E. M. Sorkin: Drugs *31*, 467 (1986).
8 S. F. Campbell and M. Davey: Drug Des. Deliv. *1*, 83 (1986).
9 K. S. Atwal, B. C. O'Reilly, E. P. Ruby, C. F. Turk, G. Aberg, M. M. Asaad, J. L. Bergey, S. Moreland and J. R. Powell: J. med. Chem. *30*, 627 (1987).
10 U.S. patents 3,930,022 and 4,076,843 assigned to E. R. Squibb and Sons, Inc.

11 CRC Handbook of Cardiovascular and Anti-inflammatory Agents, p. 207. Ed. M. Verderame, CRC Press, INC. Boca Raton, FL, 1986.
12 H. van de Waterbeemd, P.-A. Carrupt and B. Testa: J. med. Chem. 29, 600 (1986).
13 Drugs of the Future 11, 944 (1986) and references therein.
14 L. Anker, B. Testa, H. van de Waterbeemd, A. Bornand-Crausaz, A. Theodorou, P. Jenner and C. D. Marsden: Helv. chim. Acta 66, 542 (1983).
15 A. Forgione: Drugs of the Future 12, 944 (1987).
16 I. van Wijngaarden, C. G. Kruse, R. van Hes, J. A. M. van der Heyden and M. Th. Tulp: J. med. Chem. 30, 2099 (1987).
17 W. B. Wright, Jr., H. J. Brabender and R. A. Hardy, Jr.: J. Am. chem. Soc. 81, 1518 (1959).
18 J. Szmuszkovicz and P. F. Von Voigtlander: J. med. Chem. 25, 1125 (1982).
19 This compound was synthesized by J. Szmuszkovicz and coworkers at the Upjohn Company, Kalamazoo, Ml.
20 J. A. Clark and G. W. Pasternak: Neuropharmacology 27, 331 (1988).
21 S. Caccia, I. Conti, G. Vigano and S. Garattini: Pharmacology 33, 46 (1986).
22 S. Cortes, Z.-K. Liao, D. Watson and H. Kohn: J. med. Chem. 28, 601 (1985).
23 J. Conley, H. Kohn: J. med. Chem. 30, 567 (1987).
24 P. F. Von Voigtlander, E. D. Hall, M. Camacho Ochoa, R. A. Lewis and H. J. Triezenberg: J. Pharm. exp. Ther. 243, 542 (1987).
25 P. F. Von Voigtlander, R. A. Lewis and H. J. Triezenberg: Fedn Proc. 45, 677 (1986).
26 H. R. Munson, Jr., W. L. Smith, R. S. Alpin, R. F. Boswell, G. E. Jagdmann, Jr., Y. S. Yo, B. F. Kilpatrick, D. N. Johnson and W. N. Dannenburg: New Orleans ACS meeting, 1988.
27 A. L. Rovati: Scand. J. Gastroent. 11, 113 (1976).
28 J. Weiss, in: Proglumide and other Gastrin-Receptor Antagonists, p. 113. Ed. J. Weiss and S. E. Miederer. Excerpta Medica, Amsterdam 1979.
29 F. Makovec, R. Chiste, M. A. Bani, S. Pacini and L. A. Rovati: Arzneimittel-Forsch. 35, 1048 (1985).
30 I. Setnikar, M. Bani, R. Cereda, R. Chiste, F. Makovec, M. A. Pacini, L. Revel, L. C. Rovati and L. A. Rovati: Arzneimittel-Forsch. 37, 703 (1987).
31 H. J. Adams, R. A. Ronfeld and B. H. Takman: Local anesthetic agents, in: CRC Handbook of CNS Agents and Local anesthetics. Ed. M. Verderame. CRC Press, Inc., Boca Raton FL, 1986.
32 B. G. Covino: Rat. Drug Ther. 21, 8, 1 (1987).
33 Pharmacological Control of Hyperlipidaemia, p. 411. Ed. F. P. Bell. J. R. Prous Science Publishers, S.A. 1986.
34 B. Rosenberg, L. Van Camp, J. E. Trosko and V. H. Mansour: Nature (London) 222, 385 (1969).
35 A. Pasini and F. Zunino: Angew. Chem. Int. Ed. Engl. 26, 615 (1987).
36 S. E. Sherman and S. J. Lippard: Chem. Rev. 87, 1153 (1987).
37 D. T. Witiak, D. P. Rotella, J. A. Filppi and J. Gallucci: J. med. Chem. 30, 1327 (1987).
38 T. T. Yen: Life Sci. 41, 2349 (1987).
39 J. Szmuszkovicz, P. F. Von Voigtlander and M. P. Kane: J. med. Chem. 24, 1230 (1981).
40 H. P. Rang and M. M. Dale: Pharmacology, p. 119. Churchill Livingstone, 1987.
41 J. T. Pento: Drugs of the Future 12, 950 (1987).
42 S. Archer: A. Rev. Pharmac. Toxic. 25, 485 (1985).
43 S. Iversen: Chemistry in Britain 378 (1988).

Adrenergic receptor research: Recent developments

By George Kunos
Laboratory of Physiologic and Pharmacologic Studies, National Institute on Alcohol Abuse and Alcoholism, National Institutes of Health, Bethesda, MD 20892, USA

1	Introduction	151
2	Adrenergic receptors	151
3	Receptor classification	153
4	Transduction mechanisms	155
5	Receptor structure	157
6	Receptor regulation	158
6.1	Homologous regulation	159
6.2	Heterologous regulation	159
6.2.1	Inverse regulation of hepatic α_1- and β-receptors	159
6.2.2	Regulation of pulmonary β-receptors by interleukin-1	164
	References	166

1   **Introduction**

Adrenaline was discovered in 1895 as a component of adrenal medullary extracts possessing a vasopressor effect [1]. Soon after its discovery, it was recognized that adrenaline can produce a variety of biological effects, both stimulatory and inhibitory [2]. The discovery in 1906 by Dale of the selective antagonism of pressor responses to adrenaline by ergot alkaloids [3] led to the concept of dual, stimulatory and inhibitory, catecholamine receptors. This concept was further developed by Ahlquist [4] who was the first to classify adrenergic receptors based on pharmacological specificity rather than response type. Ahlquist's classification of α- and β-adrenergic receptors has stood the test of time and has proved to be a remarkably fertile concept. From a practical point of view, it has laid the ground for the development of new classes of therapeutic agents: the use of β-adrenergic antagonists in hypertension, coronary disease, and many other disorders makes these compounds the most widely used therapeutic agents, as recognized by the Nobel prize awarded to Sir James Black, the discoverer of propranolol [5]. Interestingly, the discovery of β-blocking agents was also the trigger for major theoretical advances, as it rekindled interest in Ahlquist's dormant concept. The ensuing research has led not only to further subclassification of adrenergic receptors, but also to major advances in our knowledge of receptor function and second messenger systems, and ultimately to the unraveling of the molecular structure of adrenergic receptors. Rather than representing the end of the road, these discoveries have opened new vistas for drug development and for better understanding how adrenergic receptors function. A comprehensive overview of all these developments is clearly beyond the scope of this review. Instead, selective examples of the regulation of adrenergic receptors will be discussed following a brief review of recent developments in our knowledge of the structure and function of adrenergic receptors.

2   **Adrenergic receptors**

Major progress in our understanding of adrenergic receptor mechanisms has been made through studies in four broad categories. Classical pharmacological studies have provided evidence for further subclassification or heterogeneity of adrenergic receptors, which often resulted in the development of novel therapeutic agents with increased specificity. Further refinement in such studies has become possible by the introduction of radioligand binding techniques in the mid seventies. Biochemical studies have provided new insight into mechanisms of signal transduction and the role of G proteins in the coupling of receptors to post-receptor pathways. The application of molecular biological techniques to adrenergic receptor research has resulted in the cloning and elucidation of the primary structure of the major adrenergic receptor subtypes in the last 3 years. This has opened new possibilities for clarifying how the structure of a receptor is related to its function at the molecular level. Finally, further progress has been made in our understanding of mechanisms of regulation of adrenergic receptors.

3   Receptor classification

Heterogeneity of the major classes of adrenergic receptors first became evident for the β-adrenergic receptor, which was subdivided into β_1 and β_2 by Lands et al. in 1967 [6]. This classification, which was based on the different relative potencies of adrenaline and noradrenaline in cardiac and smooth muscles, has stood the test of time. It has served as the basis for the development of subtype selective agonists and antagonists, of which β_1-receptor antagonists and β_2-receptor agonists have important therapeutic applications. Differential hormonal regulation also supports the distinct nature of β_1 and β_2-receptors: while thyroxine was found to selectively increase β_1-receptor responsiveness in vascular smooth muscle [7], glucocorticoids selectively increased β_2 while they decreased β_1-receptor density in 3T3-L1 preadipocytes [8]. The recent elucidation of the molecular structure of the β_2 [9] and β_1-adrenergic receptor [10] has indicated that these receptors are the products of distinct genes with different tissue distribution, and show limited (54 %) homology. From the point of view of receptor classification, the avian β-receptor extensively studied in the turkey erythrocyte has posed an interesting problem. While this receptor shows pharmacological similarity with mammalian β_1-receptors, it also displays some distinct differences which make it difficult to classify [11]. Interestingly, the molecular structure of the human β_1-adrenergic receptor [10] shows a greater, 69%, sequence homology with the turkey erythrocyte β-receptor [12] than with the human β_2-adrenergic receptor [13]. However, one should be cautious about overinterpreting levels of amino acid sequence homologies without additional information on the functional role of different domains of the receptor molecule. Further progress in understanding the structural basis of pharmacological differences between receptor subtypes can be expected from studies of mutant receptors generated by site-directed point mutations or deletions [14] or of 'chimeric' receptors with substitutions involving sequences from other receptor types [15]. However, it is not unlikely that subtle or not so subtle differences in the pharmacological 'properties' of a receptor may be induced by changes in its membrane environment rather than the structure of the receptor protein itself. α-adrenergic receptors have also been subclassified. In the early seventies, α-adrenergic inhibition of neurotransmitter release from sympathetic nerve terminals was proposed to involve a novel type of 'presynaptic' receptor called α_2, as opposed to the classical postsynaptic receptor which was called α_1 [16]. However, subsequent evidence for postsynaptic α_2 as well as presynaptic α_1-receptors was better accommodated by the redefinition of α_1- and α_2-receptors based on pharmacological specificity rather than anatomical location [17]. More recently, the possibility of further subgroups has been entertained both for α_1- [18] and α_2-receptors [19], in both cases based on differential sensitivities to selected receptor-active agents. The subclassification of α_1-receptors by Han et al. [20] is particularly interesting because, in addition to differential drug sensitivities, the receptor subtypes also differ in the biochemical pathways they activate: the α_1-subtype is proposed to be independent of inositol phospholipid hydrolysis and to control the opening of dihydropyridine-sensitive calcium channels to allow entry of extracellular calcium, while the α_{1b}-subtype appears to mediate IP_3-dependent release of intracellular calcium. It would be interesting to test whether coupling of α_1-receptors to different G proteins [21] may also be related to α_1-receptor subtypes. It should be kept in mind, however, that examples of overlapping pharmacological specificity make the subclassification of α_1-receptors uncertain at the present time [18].

The existence of α_2-receptor subtypes has been proposed by several investigators [19], based on differential sensitivities to prazosin as well as to some other ligands. What created renewed interest in α_2-receptor subtypes was the recent cloning of at least two different α_2-receptors with different chromosomal locations: the human platelet α_2-receptor encoded by a gene on chromosome 10 [22] and the human kidney α_2-receptor encoded by a gene on chromosome 4 [23]. While the ligand affinity profile of the former receptor expressed in a cultured cell line was found to be very similar to the proposed α_{2A}-subtype with low affinity to prazosin, the properties of the receptor from kidney were different from those of either the proposed α_{2A}- or the α_{2B}-subtype, the latter having high affinity for prazosin [19]. Clearly, more information is required before the structural basis of the *in vivo* heterogeneity of adrenergic receptor subtypes can be fully understood. Another development related to α_2-receptor subtypes has important therapeutic implications: Ruffolo and coworkers have reported that a novel compound, SK & F-104078, has a high degree of selectivity for postjunctional, as opposed to release modulating, α_2-receptors [24]. This activity profile would be particularly promising in the treatment of pulmonary hypertension, as the responsiveness of postjunctional α_2-receptors is increased under conditions of elevated pulmonary vascular tone [25]. Pharmacological differences between pre- and postjunctional α_2-receptors further complicate classification, as they do not parallel the differences between the proposed α_{2A}- and α_{2B}-receptors. Indeed, these latter subtypes appear to exist not only at post-junctional sites (see above) but also on nerve terminals: recent evidence indicates prazosin-sensitive (α_{2B}-?) receptors on noradrenergic terminals and prazosin-insensitive (α_{2A}-?) receptors on cholinergic terminals [26]. Pharmacological differences between presynaptic α_2-receptors located on noradrenergic vs 5-hydroxytryptaminergic nerve terminals have also been noted [27], but it is not clear whether or not they represent α_{2A}- and α_{2B}-receptors, respectively.

4 Transduction mechanisms

The biological effects of catecholamines are generally attributed to a change in the cellular concentrations of calcium of cyclic AMP (cAMP), which triggers a cascade of events leading to the ultimate response, i. e. a change in contractile, secretory, or neuronal activity. Activation of either subtype of β-receptors triggers a transient rise in cAMP, due to activation of adenylate cyclase, while activation of α_2-receptors results in inhibition of adenylate cyclase. α_2-Receptors generally function independently of cAMP, by triggering a transient rise in the cytosolic concentration of Ca^{2+} due to a phospholipase C-mediated breakdown of membrane polyphosphoinositides. The coupling of receptors to adenylate cyclase or to phospholipase C requires a third membrane component, one of a family of guanyl nucleotide binding or G proteins. G proteins are heterotrimers composed of α-, β- and γ-subunits. Upon activation of the receptor by an agonist ligand, tightly bound GDP is released from and GTP is bound to the α-subunit, which leads to its dissociation from the $\beta\gamma$-complex. It is in the dissociated form that the α-subunit is active in regulating various effector proteins such as adenylate cyclase, phospholipase C, phospholipase A_2, cyclic GMP phosphodiesterase, and certain ion channels. The α-subunit has GTPase activity, which converts the bound GTP to tightly bound GDP, and this inactivates the α-subunit leading to its reassociation with the $\beta\gamma$-dimer. A growing number of G proteins have been identified that differ in their α-subunits. Stimulation of adenylate cyclase by β-adrenergic receptors is via G_S, while inhibition by α_2-receptors is through G_i. The coupling of α_1-receptors to phospholipase C is via a less well defined G protein, although some effects triggered by α_1-receptors, such as activation of phospholipase A_2, may be mediated by a pertussis-toxin sensitive G protein, probably G_i [21]. This general scheme as well as the function and structure of G proteins have been the subject of numerous excellent reviews [28-30], and will not be discussed here.

Recent findings indicate that one type of adrenergic receptor may be coupled to more than one G protein [21]. As mentioned above, receptor subtypes may account for some of the examples for coupling to more than one G protein. However, the finding that expression of the cDNA for a single muscarinic receptor subtype in cells lacking endogenous receptor resulted in coupling to both adenylate cyclase and phosphoinositide turnover [31] indicates that the same receptor molecule may be involved in coupling to multiple transduction pathways. Other findings suggest that some effects mediated by β-adrenergic receptors, such as the inhibition of magnesium influx into S49 lymphoma cells [32], are not mediated by cyclic AMP. Whether or not a G protein is involved in such responses is not known. Similarly, the inhibition of adenylate cyclase by α_2-receptors appears to be neither sufficient nor necessary for many of the biological effects mediated by these receptors, such as platelet aggregation or inhibi-

tion of glucose-stimulated insulin release. There has been evidence to suggest that stimulation of α_2-receptors, as well as other receptors that inhibit adenylate cyclase, alkalinizes the interior of platelets and some cultured neuronal cells by accelerating Na^+/H^+ exchange [33, 34]. This effect, which appears to be closely linked to platelet secretion and aggregation, is not known to require a G protein. However, methodological problems may complicate the interpretation of some of these findings [35].
Activation of α_1-adrenergic and other calcium-linked receptors triggers a phospholipase C-mediated breakdown of polyphosphoinositides that yields two second messenger molecules: inositol 1,4,5-trisphosphate (IP_3) which releases calcium from intracellular stores [30], and diacylglycerol which activates protein kinase C [36]. Diacylglycerol lipase can then release fatty acids, predominantly arachidonic acid, which can be further metabolized through the cyclooxygenase, lipoxygenase and epoxygenase pathways. Recent evidence indicates, however, that receptor-mediated release of arachidonic acid can occur not only indirectly through phospholipase C, but also via a direct, G protein dependent activation of phospholipase A_2 [37], as demonstrated for α_1-receptors in a cultured thyroid cell line [21]. The potential role in neurotransmitter and hormone action of the arachidonate metabolites generated through this pathway is just beginning to emerge.

5   Receptor structure

The application of molecular biological techniques to adrenergic receptor research has resulted in an explosion of new knowledge on the structure of these receptors in the last three years. The sequence of one or more forms of each of the 4 major adrenergic receptor subtypes has been elucidated. Much of this information has been succinctly reviewed by Lefkowitz and Caron [38], and only a few salient points will be discussed here. All four adrenergic receptors are part of a gene family of G protein coupled receptors that also contains the muscarinic receptors, the serotonin 1A receptor and the light-sensitive protein, rhodopsin. These receptors show a high degree of sequence homology as well as a similar topographical organization in the cell membrane. All of these proteins appear to have seven hydrophobic, membrane-spanning α_1-helical domains, which are highly conserved and show the highest degree of sequence homology among the different receptors. Facing the extracellular space are the N-terminus segment and 3 loops connecting hydrophobic domains II-III, IV-V and VI-VII, while the C-terminal tail and 3 additional loops between domains I-II, III-IV and V-VI are on the cytoplasmic side of the membrane. Manipulation of the receptor genes has been used to dissect the functional importance of various regions or even single amino acids of the receptor molecules. These manipulations include point and deletion mutations [14, 38, 39] as well as the construction of chimeric receptors containing different parts of different receptors [15]. Limited proteolysis of the receptor protein has also been used for studying receptor structure/activity relationships [40, 41]. Overall, the results of such studies have allowed the following tentative conclusions to be made: 1) the hydrophobic α-helices may form a ligand binding pocket, with one [38] or more [41] potential binding sites identified with affinity probes. Studies with chimeric α_2/β_2-receptors suggest that different structural elements within the hydrophobic core are involved in agonist vs antagonist binding [15]; 2) within the hydrophobic domains, specific amino acid residues [aspartate 79, 113, 130 and 318] appear to be important in differentially influencing agonist and antagonist affinities [14, 39, 42]; 3) the cytoplasmic loops and carboxyl-terminal region are proposed to be involved in the coupling of the adrenergic receptors to various G proteins, although conflicting results have appeared in this regard [38]; 4) agonist-induced desensitization of β-adrenergic receptors is probably mediated by phosphorylation of serine and threonine residues in the carboxyl-terminus and possibly in the third cytoplasmic loop; 5) additional phosphorylation sites for protein kinases A and C are probably located in the three cytoplasmic loops, although the sites have not yet been localized.

6 Receptor regulation

Adrenergic receptors are under multiple regulatory influences, which constantly affect their physiological responsiveness as well as the level of their cellular expression. There are two major forms of regulation: homologous regulation is exerted by ligands of the receptor in question, and is expressed as agonist-induced desensitization or down-regulation, or antagonist- or denervation-induced supersensitivity or up-regulation. In contrast, heterologous regulation means that hormonal or other regulatory factors unrelated to a specific receptor can regulate the expression of that receptor.

6.1 Homologous regulation

Substantial evidence links the biochemical event of receptor phosphorylation to the loss of physiological responsiveness of the various adrenergic receptors, with most of the available evidence related to the β-adrenergic receptor, as reviewed recently [43]. The agonist-specific form of desensitization of the β-receptor appears to be related to phosphorylation by a novel type of enzyme, the β-receptor kinase, identified by Lefkowitz and coworkers [43]. Subsequent observations indicate that the agonist-occupied forms of other adrenergic receptors are also substrate for this enzyme. Site-directed mutagenesis studies with the β_2-adrenergic receptor strongly suggest that the sites of phosphorylation are serine and threonine residues in the carboxyl-terminus of the receptor [38]. As mentioned above, there appear to be additional sites of phosphorylation on both the α_1- and β_2-adrenergic receptors for protein kinases A and C, which might be involved in receptor desensitization that is less dependent on agonist occupancy [38]. Although the exact mechanism of how phosphorylation by the various kinases leads to desensitization is not yet clear, uncoupling of receptors from G proteins and/or sequestration of receptors are possibilities [38].

6.2 Heterologous regulation

A large variety of hormonal and metabolic factors have been shown to influence the physiological responsiveness as well as the cellular density of adrenergic receptors. Two examples of such regulation, in which our laboratory has been particularly interested, will be briefly reviewed here.

6.2.1 Inverse regulation of hepatic α_1- and β_2-receptors

Numerous physiological and pathological conditions have been associated with inverse, reciprocal changes in calcium-linked α_1- and cAMP-linked β-adrenergic receptors [44, 45]. This had led to the suggestion several years ago that α- and β-adrenergic receptors may be different forms of a common unit [46]. This explanation is no longer tenable in light of firm evidence that even the different subtypes of adrenergic receptors are distinct proteins and products of different genes. Instead, the reciprocal nature of these changes under most conditions could suggest that a common regulatory signal or factor is involved in the inverse changes in α- and β-receptors, which may represent a distinct form of receptor regulation with a common underlying mechanism. The system most attractive for studies of the 'conversion' phenomenon is the rat liver, in which tissue the adrenergic receptor mechanisms display unusual plasticity [45, 47]. In the liver of adult male rats the adrenergic activation of glycogenolysis is mediated by a typical calcium-linked, cAMP-independent, α_1- adrenergic mechanism, which is easily quantified in hepatocytes by measuring the activation of the rate limiting enzyme, glycogen phosphorylase [29]. However, the same response appears to convert to a typical β-receptor-mediated event under a number of conditions, including thyroid or glucocorticoid deficiency, malignant transformation, fetal state, etc. [47]. In addition to these *in vivo* conditions, a rapid (< 8 h) conversion from α_1- to β-adrenergic glycogenolysis has been observed *in vitro* in isolated hepatocytes kept in primary culture [48] or maintained in suspension in a serum-free buffer [49]. A common denominator among these *in vivo* and *in vitro* systems is that the conditions associated with the β-type response represent a lower level of differentiation, suggesting that the conversion of the adrenergic receptor response is linked to the process of cellular differentiation [47]. The *in vitro* system mentioned above has been adopted by several laboratories for further studies, because the inverse changes in receptor responses develop rapidly and the cells can be readily manipulated to explore the underlying mechanisms. For more details and references the reader is referred to a recent review [45].

In rat hepatocytes suspended in a serum-free medium, the time-dependent conversion from α- to β-adrenergic activation of phosphorylase is reflected by a decreased response to phenylephrine, the emergence of a response to isoproterenol, and a change from α- to β-receptor blockers being more effective in inhibiting the response to the mixed α-β agonist, adrenaline [49, 50]. Events closer to receptor activation are similarly affected: the phenylephrine-induced production of IP_3 is decreased and the isoproterenol-stimulated accumulation of cAMP is increased with time of incubation (E. J. N. Ishac and G. Kunos, manuscript in preparation). The selectivity of the 'conversion' is indicated by the lack of parallel changes in the phosphorylase, IP_3, and cAMP responses to other glycogenolytic hormones, such as vasopressin and glucagon. Activation of phosphorylase by the calcium ionophore, A23187, or by dibutyryl cAMP also remain unaffected, suggesting that the locus of the time-dependent change is at the level of the adrenergic receptors or in their coupling to post-receptor pathways, rather than in the more distal components of the transduction systems. Ligand binding studies using hepatocytes suspended in a serum-free buffer detected no significant changes in the number of α_1- and β-receptors at a time when the conversion of the receptor response was complete [49, 51], suggesting that the latter might be due to altered coupling. In contrast, in cultured hepatocytes increased β- [52–55] and decreased α-receptor densities [55] have been reported. Since the time course of the changes in receptor densities was found to be much slower [55] than the time course of the conversion of the receptor response under similar conditions [48], the primary event may be a change in receptor coupling: uncoupling of α_1-receptors from phospholipase C and increased coupling of β-receptors to adenylate cyclase precedes and may actually trigger the down-regulation of α_1- and upregulation of β-receptors, respectively. A recent study by Refsnes et al. [57] indicates, however, that the time course of the culture-induced increase in adenylate cyclase activity and β-receptor density are strictly parallel, with significant changes in both parameters detectable within 2 h. In that study, cells were plated and cultured in serum-containing media, and omitting the serum reduced but did not abolish the effects on receptor density and responsiveness [57]. Others report that a culture-induced increase in β-receptor density in plated hepatocytes has an absolute requirement for serum [51], while we found that in rat liver cells kept in suspension, the density of β-receptors remains unchanged with time whether or not serum is present. Thus, it appears that the increase in β-receptor density requires both serum and plating of the cells, while the emergence of a β-adrenergic activation of phosphorylase requires neither, which would dissociate the two events. Further work is required to explore the role of the above factors in the regulation of the expression of cell surface receptors.

Although the evidence discussed above suggest a time-dependent change in the coupling of adrenergic receptors in isolated hepatocytes, it is not yet clear what the mechanism of the altered coupling is. Ui and coworkers reported that primary culturing of rat liver cells led to a loss of the membrane substrate for pertussis-toxin induced ADP ribosylation (probably G_i), and that inhibition of ADP ribosyltransferases by nicotinamide prevented this loss as well as the parallel conversion of the adrenergic receptor response from α- to β-type [58]. In the basis of these finding, they proposed that culturing activates an endogenous ADP ribosyltransferase which would inactivate G_i, and this may be the cause of the receptor conversion by removing a G_i-mediated inhibition of β-adrenergic receptors. They also propose that the loss of α_1-receptor function is secondary to the emergence of a β-receptor response, and thus also depends on the inactivation of G_i [58]. Some observations are, however, at odds with this hypothesis. It has been reported that complete inactivation of hepatic G_i by *in vivo* treatment of rats with pertussis toxin has no effect on α_1-receptor induced phosphorylase activation or inositol phospholipid breakdown [59, 60]. Others have found that culturing of rat liver cells in the presence of pertussis toxin causes increased activation of adenylate cyclase by both isoproterenol and glucagon [57], which is different from the selective increase of the isoproterenol response after primary culturing. Furthermore, the effect of pertussis toxin was as or more evident at 24 h as at 7 h, which suggests that G_i activity is not lost after prolonged culturing of the hepatocytes [57]. Our own findings provide more direct evidence for the latter point. We have observed that *in vivo* treatment of rats with pertussis toxin resulted in a complete loss of the membrane substrate for pertussis-toxin catalyzed ADP-ribosylation and a partial conversion from α_1- to β-adrenergic glycogenolysis in hepatocytes obtained from these animals, which is in agreement with the findings of Itoh et al. [58]. However, when hepatocytes from control rats were incubated *in vitro* for 4 h, the conversion of the adrenergic receptor response from α_1- to β-type was more complete than that caused by *in vivo* treatment with pertussis toxin, yet there was no decrease in the ability of pertussis toxin to ADP-ribosylate the same, 41 kD, substrate in membranes obtained from the same cells (J. Kapocsi and G. Kunos, manuscript in preparation). It appears therefore that, at least in hepatocytes suspended in a serum-free buffer, prolonged incubation does not lead to inactivation of G_i, which thus can not account for the α_1-β conversion.

Alternative mechanisms for the altered coupling of adrenergic receptors have been proposed, and involve a time-dependent increase in membrane phospholipid breakdown and arachidonic acid release and a parallel increase in protein kinase C activity [45, 49, 50]. The involvement of altered phospholipid metabolism in the conversion phenomenon was suggested by the fact that glucocorticoid deficiency causes a similar change in receptor response, reversible by glucocorticoid treatment [61]. Glucocorticoids produce many of their biological effects through inhibition of phospholipase A_2-induced release of arachidonic by endogenous proteins called lipocortins [62]. It was suspected that loss of a similar mechanism in primary liver cells may be involved in the altered adrenergic receptor response. The following findings lend support to this hypothesis: a) partially purified lipocortin was able to reverse the time-dependent change from α_1- to β-adrenergic glycogenolysis [49]; b) the conversion could be also prevented if hepatocytes were incubated in the presence of fatty acid-free bovine serum albumine, which binds free fatty acids released by the ells [50]; c) the conversion could be triggered by exposure of freshly isolated hepatocytes to melittin, an activator of phospholipase A_2 [49] or to exogenous arachidonic acid [50]. Further experiments implicated a cyclooxygenase product of arachidonic acid metabolism: inhibitors of cyclooxygenase (indomethacin of ibuprofen) but not of lipoxygenase prevented the conversion of the receptor response caused by arachidonic acid or by prolonged *in vitro* incubation of the cells [50]. A role for phospholipase A_2 in inverse changes in α_1- and β-adrenergic receptor function has also been indicated by the results of recent experiments with rat pinealocytes: increased β- and decreased α_1-receptor responses triggered by exposure of rats to constant light could be mimicked by *in vitro* treatment of these cells by a phospholipase A_2 inhibitor [63].

Other findings suggest a role for protein kinase C in the culture-induced changes in adrenergic receptor function. Phorbol 12-myristate, 13-acetate (PMA), an activator of protein kinase C, caused translocation of the enzyme from cytosol to the plasma membrane of hepatocytes and also reduced the α_1- and increased the β-adrenergic component in phosphorylase activation without affecting phosphorylase activation by glucagon or vasopressin. Furthermore, prolonged incubation of the cells or their treatment with arachidonic acid caused a cytosol to membrane translocation of protein kinase C similar to that caused by PMA [45] (Ishac and Kunos, manuscript in preparation). These findings not only suggest that activation of protein kinase C is involved in the conversion response, but also suggest that the effect of arachidonic acid may be also mediated, at least in part, through protein kinase C. Conversely, activation of protein kinase C can contribute to the activation of phospholipase A_2, through the proposed phosphorylation and inactivation of lipocortin [64]. Therefore, these two pathways may synergistically interact in bringing about the inverse changes in α_1- and β-adrenergic receptor function. Observations from several laboratories indicate that the conversion of the hepatic adrenoceptor response requires the synthesis of new protein [45, 48, 50]. Since not only the increase in the β but also the decrease in the α-receptor response can be prevented by cycloheximide or actinomycin D, it is likely that the protein involved is not the receptor protein itself, but a regulatory factor that affects both receptors. One can further speculate that such a regulatory factor or factors may be responsible for the parallel activation of phospholipase A_2 and protein kinase C. In this regard it is interesting to note that interleukin 1 (IL-1), a monokine produced by macrophages, has been reported to affect the conversion of the adrenoceptor response observed in the rat pituitary gland. Upon isolation of rat anterior pituitaries, the adrenergic activation of ACTH release was shown to rapidly convert from a β- to an α- receptor mediated event [65], and the loss of the β but not the emergence of the α-response could be prevented by IL-1 [65]. It remains to be determined whether IL-1, which is also produced by Kupfer cells in the liver, may be involved along with some other protein factors in the culture-induced conversion of the hepatic adrenergic receptor response.

6.2.2   Regulation of pulmonary β-receptors by interleukin-1

Another example of the heterologous regulation of adrenergic receptors is the recently described effect of IL-1 and related monokines on the expression of pulmonary β-adrenergic receptors. It has been reported that a cultured, virus-transformed human lymphocyte cell line (IM-9 cells) as well as freshly isolated, normal lymphocytes produce and release protein factors that upregulate β-adrenergic receptors on A549 cultured human lung tumor cells [66]. One of the factors has been identified as IL-1 by immunoneutralization experiments, and purified as well as recombinant IL-1 have been shown to have a similar effect [66]. Interleukins 2 and 6 as well as interferon-α were without effect, while tumor necrosis factor-α (TNFα) had an effect similar to but weaker than the effect of IL-1. Glucocorticoids also upregulate pulmonary β-receptors, and were found to markedly potentiate the effect of IL-1 but not that of TNFα. Although the mechanism of this potentiation is not clear, it may be related to the ability of glucocorticoids to upregulate IL-1 receptors [67]. The effects of both IL-1 and glucocorticoids could be blocked by cycloheximide, suggesting that they involve the synthesis of new protein, probably the β-receptor protein [66]. Additional, unpublished observations from our laboratory indicate that IL-1 inhibits thymidine uptake and proliferation of A549 cells and increases their adhesion to the plastic tissue culture flask, which suggest increased differentiation. Furthermore, a series of antitumor agents including vincristine, vinblastine, colchicine and methotrexate, share the ability to upregulate β-adrenergic receptors in A549 cells. Together, these observations could suggest that regulation of β-adrenergic receptors is linked to the process of cellular differentiation and that upregulation of β-adrenergic receptors by IL-1 is part of its growth inhibitory, differentiation promoting effect in A549 cells.

It has long been known that stress can affect immune functions, and the regulatory effects described above represent an example of interactions between the sympatho-adrenal and the immune systems. This interaction may have implications for human pathology. Glucocorticoids are important in the treatment or prevention of certain respiratory diseases, such as bronchial asthma and neonatal respiratory distress syndrome. The therapeutic effectiveness of glucocorticoids in such disorders has been attributed, at least in part, to their ability to upregulate pulmonary β-adrenergic receptors and thus correct the β-adrenergic hypofunction characteristic of some of these disorders. IL-1 is present in lung tissue, where it is produced by pulmonary macrophages and may act as a paracrine regulator of β-receptors. This and the synergistic interaction of between IL-1 and glucocorticoids make one wonder whether a defect in the production, release, and/or cellular action of IL-1-like monokines in lung tissue may have a role in the pathomechanism of certain respiratory disorders.

References

1 G. Oliver and E. A. Schafer: J. Physiol. (London) 18, 230 (1895).
2 T. R. Elliott: J. Physiol. (London) 32, 401 (1905).
3 H. H. Dale: J. Physiol. (London) 34, 163 (1906).
4 R. P. Ahlquist: Am. J. Physiol. 153, 586 (1948).
5 J. W. Black, A. F. Crowther, R. G. Shanks, L. H. Smith and A. G. Dornhorst: Lancet I, 1080 (1964).
6 A. M. Lands, A. Arnold, J. P. McAuliff, F. P. Luduena and T. G. Brown: Nature 214, 597 (1967).
7 S. R. O'Donnell and J. C. Wanstall: Br. J. Pharmac. 88, 41 (1986).
8 M. T. Nakada, J. M. Stadel, K. S. Poksay and S. T. Crooke: J. Pharmac. exp. Ther. 31, 377 (1987).
9 R. A. F. Dixon, B. K. Kobilka, D. J. Strader, J. L. Benovic, H. G. Dohlman, T. Frielle, M. A. Bolanowski, C. D. Bennett, E. Rands, R. E. Diehl, R. A. Mumford, E. E. Slater, E. S. Sigal, M. G. Caron, R. J. Lefkowitz and C. D. Strader: Nature 321, 75 (1986).
10 T. Frielle, S. Collins, K. W. Daniel, M. G. Caron, R. J. Lefkowitz and B. K. Kobilka: Proc. natl Acad. Sci. USA 84, 7920 (1988).
11 K. P. Minneman, G. A. Weiland and P. B. Molinoff: Molec. Pharmac. 17, 1 (1980).
12 Y. Yarden, H. Rodriguez, S. K.-F. Wong, D. R. Brandt, D. C. May, J. Burnier, R. N. Harkins, E. Y. Chen, J. Ramachandran, A. Ullrich and E. M. Ross: Proc. nat. Acad. Sci. USA 84, 46 (1986).
13 B. K. Kobilka, R. A. F. Dixon, T. Frielle, H. G. Dohlman, M. A. Bolanowski, I. S. Sigal, T. L. Yang-Feng, U. Francke, M. G. Caron and R. J. Lefkowitz: Proc. natl Acad. Sci. USA 84, 46 (1987).
14 F. Z. Chung, C.-D. Wang, P. C. Potter, J. C. Venter and C. M. Fraser: J. biol. Chem. 263, 4052 (1988).
15 B. K. Kobilka, T. S. Kobilka, K. Daniel, J. W. Regan, M. G. Caron and R. J. Lefkowitz: Science 240, 1310 (1988).
16 S. Z. Langer: Biochem. Pharmac. 23, 1793 (1974).
17 S. Berthelsen and W. A. Pettinger: Life Sci. 21, 595 (1977).
18 J. McGrath and V. Wilson: Trends pharmac. Sci. 9, 162 (1988).
19 D. B. Bylund: Trends pharmac. Sci 9, 356 (1988).
20 C. Han, P. W. Abel and K. P. Minneman: Nature 329, 333 (1987).
21 R. M. Burch, A. Luini and J. Axelrod: Proc. natl Acad. Sci. USA 83, 7201 (1986).
22 B. K. Kobilka, H. Matsui, T. S. Kobilka, T. L. Yang-Feng, U. Francke, M. G. Caron, R. J. Lefkowitz and J. W. Regan: Science 238, 650 (1987).
23 J. W. Regan, T. S. Kobilka, T. L. Yang-Feng, M. G. Caron, R. J. Lefkowitz and B. K. Kobilka: Proc. natl Acad. Sci. USA 85, 6301 (1988).
24 R. R. Ruffolo, A. C. Sulpizio, A. J. Nichols, R. M. DeMarinis and J. P. Hieble: Naunyn Schmiedeberg's Arch. Pharmak. 336, 415 (1987).
25 R. J. Shebuski, E. H. Ohlstein, J. M. Smith Jr. and R. R. Ruffolo Jr.: J. Pharmac. exp. Ther. 242, 158 (1987).
26 J. Kapocsi, G. T. Somogyi, N. Ludvig, P. Serfozo, L. G. Harsing Jr., R. J. Woods and E. S. Vizi: Neurochem. Res. 12, 141 (1987).
27 G. Maura, A. Geminiani and M. Raiteri: Eur. J. Pharmacol. 116, 335 (1985).
28 A. G. Gilman: Cell 36, 577 (1984).
29 J. H. Exton: Am. J. Physiol. 248, E633 (1985).
30 M. J. Berridge and R. Irvine: Nature 312, 315 (1984).
31 A. Ashkenazi, J. W. Winslow, E. G. Peralta, G. L. Peterson, M. I. Schimerlik, D. J. Capon and J. Ramachandran: Science 238, 672 (1987).
32 M. E. Maguire and J. J. Erdos: J. biol. Chem. 255, 1030 (1980).

33 H. S. Banga, E. R. Simons, L. F. Brass and S. E. Rittenhouse: Proc. natl Acad. Sci. USA 83, 9197 (1986).
34 L. L. Isom, E. J. Cragoe Jr. and L. E. Limbird: J. biol. Chem. 262, 17504 (1987).
35 L. L. Isom, E. J. Cragoe Jr. and L. E. Limbird: J. biol. Chem. 263, 16513 (1988).
36 Y. Nishizuka: Nature 334, 661 (1988).
37 J. Axelrod, R. M. Burch and C. L. Jelsema: Trends Neurosci. 11, 117 (1988).
38 R. J. Lefkowitz and M. G. Caron: J. biol. Chem. 263, 4993 (1988).
39 C. D. Strader, I. S. Sigal, R. B. Register, M. R. Candelore, E. Rands and R. A. F. Dixon: Proc. natl Acad. Sci. USA 84, 4384 (1987).
40 H. G. Dohlman, M. Bouvier, J. L. Benovic, M. G. Caron and R. J. Lefkowitz: J. biol. Chem. 262, 14282 (1987).
41 S. K. F. Wong, C. Slaughter, A. E. Ruoho and E. M. Ross: J. biol. Chem. 263, 7925 (1988).
42 C. M. Fraser, F.-Z. Chung, C.-D. Wang and J. C. Venter: Proc. natl Acad. Sci. USA 85, 5478 (1988).
43 D. R. Sibley and R. J. Lefkowitz: Nature 317, 124 (1985).
44 G. Kunos: A. Rev. Physiol. Pharmac. 18, 291 (1978).
45 G. Kunos and E. J. N. Ishac: Biochem. Pharmac. 36, 1085 (1987).
46 G. Kunos and M. Szentivanyi: Nature 217, 1077 (1968).
47 G. Kunos: Trends pharmac. Sci. 5, 380 (1984).
48 F. Okajima and M. Ui: Archs biochem. Biophys. 213, 658 (1982).
49 G. Kunos, F. Hirata, E. J. N. Ishac and L. Tchakarov: Proc. natl Acad. Sci. USA 81, 6178 (1984).
50 E. J. N. Ishac and G. Kunos: Proc. natl Acad. Sci. USA 83, 53 (1986).
51 A. Tsujimoto, G. Tsujimoto, S. Azhar and B. B. Hoffman: Biochem. Pharmac. 35, 1400 (1986).
52 T. Nakamura, A. Tomomura, C. Noda, M. Shimoji and A. Ichihara: J. biol. Chem. 258, 9283 (1983).
53 M. Refsnes, D. Sandnes, O. Melien, T. E. Sand, S. Jacobsen and T. Christoffersen: FEBS Lett. 164, 291 (1983).
54 K. R. Schwarz, S. M. Lanier, E. A. Carter, C. J. Homcy and R. M. Graham: Molec. Pharmac. 27, 200 (1985).
56 T. Nakamura, A. Tomomura, S. Kato, C. Noda and A. Ichihara: J. Biochem. 96, 127 (1984).
57 M. Refsnes, D. Sandnes and T. Christoffersen: Eur. J. Biochem. 163, 457 (1987).
58 H. Itoh, F. Okajima and M. Ui: J. biol. Chem. 259, 15464 (1984).
59 C. J. Lynch, V. Prpic, P. F. Blackmore and J. H. Exton: Molec. Pharmac. 29, 196 (1986).
60 R. J. Uhing, V. Prpic, H. Jiang and J. H. Exton: J. biol. Chem. 261, 2140 (1986).
61 T. M. Chan, P. F. Blackmore, K. E. Steiner and J. H. Exton: J. biol. Chem. 254, 2428 (1978).
62 R. J. Flower, J. N. Wood and L. Parente: Adv. Inflammation Res. 7, 61 (1984).
63 J. Vanecek, D. Sugden, J. L. Weller and D. C. Klein: J. Neurochem. 47, 678 (1986).
64 F. Hirata: J. biol. Chem. 256, 7730 (1984).
65 M. Boyle, G. Yamamoto, M. Chen, J. Rivier and W. Vale: Proc. natl Acad. Sci. USA 85, 5556 (1988).
66 L. Stern and G. Kunos: J. biol. Chem. 263, 15876 (1988).
67 T. Akahoshi, J. J. Oppenheim and K. Matsushima: J. exp. Med. 167, 924 (1988).

From podophyllotoxin glucoside to etoposide

By H. Stähelin and A. von Wartburg
Preclinical Research, Sandoz Ltd., Basel, Switzerland

1	Introduction	171
2	History	172
3	Podophyllum aglucone lignans	178
4	Genuine glycosides	183
5	4'-O-Alkyl homologues of podophyllum glucosides	186
6	Acyl derivatives of podophyllum glucosides	187
7	Condensation products of podophyllum glucosides with aldehydes	187
8	SP–G and SP–I (Proresid®)	191
9	Podophyllinic acid hydrazides and related compounds	196
9.1	Unsubstituted hydrazides of podophyllum lignans	197
9.2	Carbonyl condensation products of acid hydrazides	198
9.3	Alkyl and aryl hydrazides of podophyllinic and picropodophyllinic acid	200
9.4	Acyl derivatives of podophyllinic acid hydrazide	202
10	Neopodophyllotoxin and podophyllinic acid	203
11	Demethylenepodophyllotoxin, sikkimotoxin and related compounds	206
12	4'-Demethylepipodophyllotoxin aglucones	208
13	4'-Demethylepipodophyllotoxin glucoside and derivatives	210
13.1	Synthesis	210
13.2	Condensation products of 4'-demethylepipodophyllotoxin glucoside with carbonyl compounds	214
14	Miscellaneous lignan compounds	217
14.1	Podophyllotoxin carbamates	217
14.2	Desoxypodophyllotoxin and derivatives	219
14.3	Desoxypodophyllinic acid derivatives	221
14.4	Apopicropodophyllotoxins	222
14.5	Podophyllol and congeners	223
15	Teniposide and etoposide	224
15.1	Cytostatic and oncostatic effects in preclinical systems	225
15.2	Immunosuppression	228
15.3	Activity against microorganisms	229
15.4	Cytotoxicity, reversibility of cellular effects	229
15.5	Toxicology	231
15.6	Delayed toxicity	232
15.7	Resistance	233
15.8	Preclinical pharmacokinetics, galenical problems	235
15.9	Human pharmacokinetics	237
15.10	Clinical use	238
16	Mechanism of action	241
16.1	Spindle poison activity vs inhibition of entry into mitosis	241
16.2	At which point of the cell cycle do VM and VP act?	243

16.3	Attempts to explain the mechanism of action of VM and VP at the molecular level	244
16.4	Interaction with topoisomerase, and other possible mechanisms . .	246
17	Structure activity relationships	250
18	Conclusions	255
	References	258

Abbreviations

VM	teniposide
VP	etoposide
P	podophyllotoxin
D	4'-demethyl
E	epi
G	glucoside
B	benzylidene
n	number of experiments
i. v.	intravenous(ly)
i. p.	intraperitoneal(ly)
s. c.	subcutaneous(ly)
p. o.	per os (oral)
n. t.	not tested

1   **Introduction**

In the field of cancer research, there is not only a flood of original articles, but also an abundance of reviews and books on a variety of topics, including chemotherapy. Several reviews have appeared in the past few years dealing with drugs of the podophyllum type, particularly etoposide [1]. So why publish yet another one?

First of all, etoposide and – less so – its predecessor teniposide (together often referred to as „the epipodophyllotoxins") have not only another mechanism of action than the previously known podophyllum compounds, but have become major anticancer drugs, and the way they were found and developed has been reported only in fragments and sometimes inaccurately. Indeed, a tortuous path led chemists and biologists from the long-known podophyllotoxin to its glucoside and from there to etoposide.

During the course of the work devoted to the search for better cancer drugs, many hundreds of compounds related to podophyllotoxin have been synthesized or extracted from the plant and tested in our pharmaceutical laboratories. While a considerable part of the chemistry involved has been published, only a minor part of biological results has found its way into print. Therefore, results of testing for inhibition of tumor cell proliferation *in vitro* and *in vivo* are presented here for a number of compounds belonging to the more important groups of podophyllum lignans on which we have been working for more than 20 years. Since certain aspects of the relationship between structure and mechanism of action have been rather neglected until very recently, a special section is devoted to this topic.

It is not our intention to give a comprehensive and balanced review on all parts of podophyllum chemistry and biology. Clinical applications of teniposide and etoposide are treated only very superficially since this is the main topic of several previous reviews. The present study is predominantly an account of the work done at Sandoz Ltd and should enable the interested reader to complement his knowledge and, perhaps, to draw his own conclusions regarding structure activity relationships in this class of chemical compounds and to form an opinion regarding possibilities of further developments in the drugs of the podophyllum family.

2   History

Podophyllum emodi Wall., which grows in the Himalayan region, and the American *Podophyllum peltatum* L. (may apple, mandrake) are old medicinal plants. Both species are herbaceous perennials belonging to the family of the Berberidaceae. The dried roots (named podophyllum) were used by the natives of both continents as cathartics and anthelmintics. A resinous alcoholic, water-insoluble extract of the rhizomes, called podophyllin (or resina podophylli) has long been included in the pharmacopoeia of the US and several European countries. The literature on the biological effects and the constituents of both plants was reviewed extensively and admirably by Kelly and Hartwell in 1954 [2]. The present section will therefore concentrate on an overview of the developments in podophyllum drugs since 1953.

The renewed interest in the podophyllum plant in the 1940's was generated by Kaplan [3] who demonstrated the curative effect of podophyllin in tumorous growths (condylomata acuminata), and subsequently by King and Sullivan [4] who found the antiproliferative effect of podophyllin to be similar to that of colchicine at the cellular level.

Podophyllotoxin, the main constituent of podophyllin, was first described by Podwyssotzki in 1880 [5]. The correct structure was proposed by Hartwell and Schrecker [6] and later confirmed by Gensler et al. [7] by means of a first total synthesis. Beginning in 1947, Hartwell and collaborators started an extensive chemical investigation of podophyllin which resulted in the isolation of (besides podophyllotoxin) α- and β-peltatin from American podophyllin and 4'-demethylpodophyllotoxin from the Indian plant [8]. In addition, desoxypodophyllotoxin and the inactive dehydropodophyllotoxin were found by Kofod and Jørgensen [9]. A third podophyllum species, described by Chatterjee and Mukerjee [10] as *Podophyllum sikkimensis*, yielded sikkimotoxin [11]. All these compounds are structurally closely related and belong to the class of natural substances called lignans, plant products containing the 2,3-dibenzylbutane skeleton. An excellent review on the chemistry of podophyllum was published by Hartwell and Schrecker in 1958 [12].

In the early 1950's, chemists in the pharmaceutical research department of Sandoz Ltd set out to test the hypothesis that some podophyllum lignans are present in the plant as glycosides which may be lost in the course of the then used extraction procedures. That glycosides may be more useful as therapeutics than their aglycones is well known for digitalis and strophanthus preparations, an area where Sandoz chemists and pharmacologists had years of experience. It was expected that podophyllum glycosides would be more water-soluble and less toxic than aglycones. While our studies were going on, the β-D-glucoside of picropodophyllotoxin was reported [13]. However, this compound is biologically nearly inactive, poorly water-soluble and, in addition, probably an artifact which may not occur as such in the plant.

Starting from fresh or carefully dried rhizomes of *P. emodi*, we were indeed able to isolate two glycosidic lignans which proved to be podophyllotoxin β-D-glucopyranoside [14, 15] and 4'-demethylpodophyllotoxin β-D-glucopyranoside [16]. Both glucosides, less toxic and hydrophobic than the aglucones, were also found in the American *P. peltatum*, which in addition contained the glucosides of α- and β-peltatin [17–19].

Thus the hypothesis proved to be correct and the expectations – better water solubility and lower toxicity – were fulfilled. However, the glucosides proved not only less toxic, but also considerably less potent than their aglucones.

In our attempts to improve on the therapeutic index, large series of derivatives were prepared both from glucosides and aglucones. Among the chemical modifications, condensation products of podophyllum glucosides with aldehydes and derivatives of podophyllinic acid hydrazide aroused particular interest based on their chemical and biological properties. Two preparations were selected as potential anticancer agents for extensive testing *in vitro* in animals and in man: namely SP–G, the condensation product of the podophyllum glucoside fraction with benzaldehyde, and SP–I, podophyllinic acid ethyl hydrazide. Based on favorable clinical results, both preparations were commercialized in 1963 under the experimental designations SP–G (later to be called Proresid oral®) and SP–I (Proresid intravenous®).

After these two preparations had been developed for clinical use, the search for still better compounds in the podophyllum series continued; altogether more than 550 derivatives were prepared and tested. Some of them will be discussed below.

In early 1962, analysis of the cytostatic potency of SP–G by means of a cell culture assay [20] revealed that the activity of the known components could not account for the potency of SP–G. It was therefore concluded that small quantities of unknown, highly active by-products must be present. It had also been observed that SP–G produced a considerable increase in the life span of mice inoculated with leukemia L–1210, an effect not exerted by the hitherto isolated components. These exciting pharmacological results [21] stimulated us to search for new cytostatic principles in SP–G as well as in the crude glucoside fraction of *Podophyllum* species. Careful fractionation of SP–G lots by systematic and extensive chromatographic procedures [22] – guided by biological testing – first yielded the anomer of podophyllotoxin benzylidene glucoside in which the phenyl group of the aldehyde residue occupies the axial position. Next came two further benzylidene derivatives, that of podorhizol glucoside and of 4'-demethyldesoxypodophyllotoxin glucoside. Both parent lignan glucosides have also been encountered as genuine components of *P. emodi* (named first lignan F and H). However, the activity of these 3 benzylidene derivatives proved too low to explain the potency of SP–G. We then came across a fourth component which was identified as the benzylidene derivative of desoxypodophyllinic acid glucopyranosylester which was indeed highly potent in cell cultures but lacked activity against leukemia L–1210. Its parent-free glucoside (lignan J) had also been isolated from *P. emodi* [23].

In late 1964, after two years of endeavors to trace down the component responsible for the activity of SP–G in leukemia L–1210, another compound could be isolated (named benzylidene lignan P) which turned out to not only greatly inhibit cell proliferation *in vitro* but also to produce a remarkable increase of survival time in leukemic mice at low doses [22]: it was obviously the long-sought antileukemic factor. The rather unusual structure of this compound was found to be 4'-demethylepipodophyllotoxin benzylidene-1-β-D-glucopyranoside. As expected the yield was very low, SP–G contained less than 0.25 % (at that time, this scantiness posed more problems than it would nowadays). Likewise, the free glucoside could be detected in *P. emodi* extracts only in very small amounts.

When the effect of benzylidene lignan P was investigated in cultures of chick embryo fibroblasts [24] in early 1965, it immediately became clear by microscopical methods that we were dealing with a different mechanism of action: while podophyllotoxin and its then known derivatives are spindle poisons and cause an increase in mitotic figures, i. e. they inhibit proliferation by arresting the cells during mitosis (in metaphase, like colchicine), benzylidene lignan P prevents cells from entering mitosis and reduces the mitotic index to almost zero within less than an hour. This surprising discovery [21] was, together with the antitumor effect, reason enough to further investigate this class of compounds in a joint (chemical and biological) effort. The discovery of the effect of etoposide on DNA, about 10 years later [25], was a first step in understanding the new mechanism of action at the molecular level. A further step was the demonstration that the effects of teniposide and etoposide an DNA are mediated by topoisomerase [26, 27]

The poor availability of 4'-demethylepipodophyllotoxin glucoside from natural sources prevented more extensive pharmacological studies and chemical modifications. We therefore turned our attention to a synthetic approach. Starting from the easily accessible podophyllotoxin, a number of synthetic steps were elaborated (see section 13) among which a newly developed stereoselective glycosidation reaction and the prevention of isomerization in the course of removal of the protective groups were particularly intriguing [28]. In the course of these investigations, the first total syntheses of podophyllotoxin-β-D-glucopyranoside, epipodophyllotoxin-β-D-glucopyranoside and 4'-demethylepipodophyllotoxin-β-D-glucopyranoside were achieved.

With synthetic lignan P at our disposal it was possible to prepare numerous derivatives; first of all a large series of condensation products with carbonyl compounds. Most of the obtained cyclic acetals and ketals were active in mouse leukemia L–1210. Two of them, the thenylidene derivative VM 26 (teniposide, VM–26, original Sandoz code no. 15–426, NSC 122819) and the ethylidene derivative VP 16–213 (etoposide, VP–16, original Sandoz code no. 16–213, NSC 141540) were selected for in depth pharmacological evaluation and clinical testing, mainly on the basis of their antitumor effects in animals.

VM 26 was synthesized in late 1965, VP–16 about nine months later. Teniposide was first tested in man in 1967, etoposide in 1971. Results with these two compounds in the clinical studies with cancer patients arranged by Sandoz were rather encouraging, but by the mid-seven-

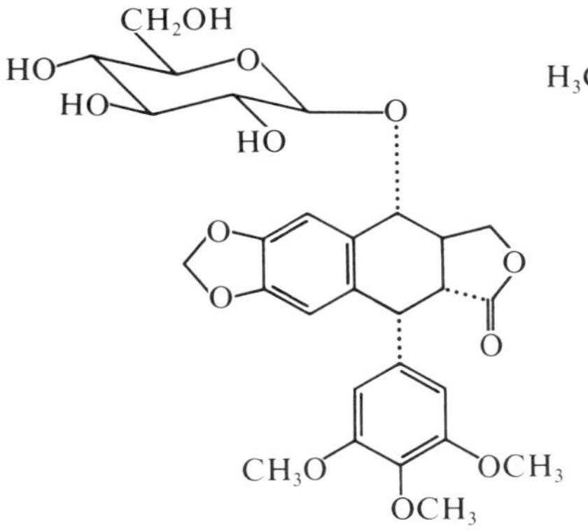

Podophyllotoxin glucoside

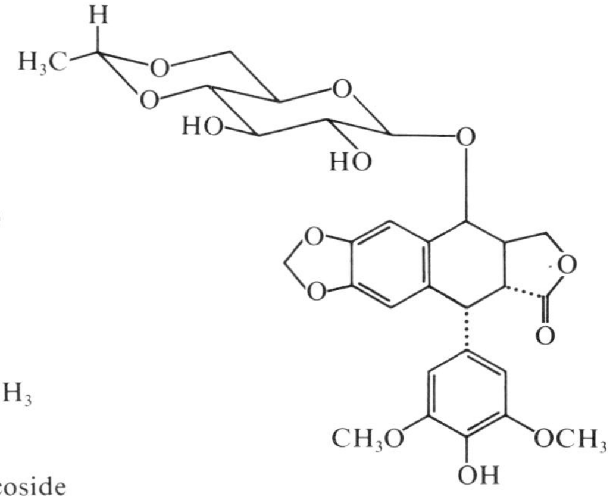

Etoposide

ties cancer research was no longer among the priorities in the pharmaceutical division of the company. VM and VP were licensed out to the US company Bristol-Myers in 1978 despite the fact that Sandoz had already commercialized VM in some countries in 1976 under the designations Vumon® and Vehem®. The American company, with its extensive involvement in the cancer field, successfully continued the clinical development of the two epipodophyllotoxin derivatives and introduced etoposide as VePesid® in the US market in 1983 (incidentally, etoposide was approved in November 1983 for testicular cancer by the US Food and Drug Administration (FDA) on the same day as the immunosuppressant cyclosporin A (Sandimmune®), a compound developed on the chemical and biological side by groups directed by the authors of this article [29, 30] and whose specific immunosuppressive activity was discovered by one of us [31]).

Short chronology of podophyllum drugs

1820   Podophyllin is included in the US pharmacopoeia
1861   Bentley mentions local antitumor effects of podophyllin
1880   Podwyssotzki isolates podophyllotoxin
1942   Kaplan describes effects of podophyllin in benign tumors, condylomata acuminata; the drug disappears from the US pharmacopoeia
1946   King and Sullivan report mechanism of action of podophyllin: stop of cell division in metaphase of mitosis

1951	Hartwell and Schrecker determine the correct structure of podophyllotoxin; begin of clinical trials with systemic administration of some podophyllum compounds
1954	Discovery of glucosides of podophyllotoxin and peltatins in the podophyllum plant by the Sandoz chemists Renz, von Wartburg and coworkers
1956–1959	Condensation of lignan glucosides with aldehydes by von Wartburg and coworkers; preparation of SP–G by Angliker and coworkers, synthesis of SP–I by Rutschmann and Renz; pharmacological testing by Stähelin and Emmenegger
1962	Biological analysis by Stähelin suggests the presence of very small amounts of till then unknown, highly active compounds in SP–G with *in vitro* and *in vivo* antitumor effects
1963	First commercialization of podophyllum drugs for systemic cancer treatment (SP–G and SP–I)
1963–1965	Isolation of the „anti-leukemia" factor in SP–G by close chemical/biological collaboration (Keller, Kuhn, von Wartburg, Stähelin) and characterization as demethylepipodophyllotoxin-benzylidene-glucoside (DEPBG)
1965	Stähelin establishes a mechanism for DEPBG which is new for podophyllum compounds: inhibition of entry of cells into mitosis; synthesis and first biological testing of teniposide
1967	Kuhn and von Wartburg elaborate a stereoselective synthesis of demethylepipodophyllotoxin-glucoside, suitable for large-scale production; synthesis and first biological testing of etoposide
1967	Start of clinical trials of teniposide
1971	Start of clinical trials of etoposide
1974	Loike et al. report DNA fragmentation by teniposide and etoposide
1976	Commercialization of teniposide as Vumon® in some countries
1978	Sandoz hands over further development of teniposide and etoposide to Bristol-Myers
1982	Long et al. find interaction of etoposide with the enzyme topoisomerase II
1983	Approval by the FDA of etoposide for testicular cancer

3 Podophyllum aglucone lignans

In the early 1950's, when work in this field began at Sandoz [14], the following natural lignans were known: podophyllotoxin (P) (**1**), picropodophyllotoxin (= picropodophyllin) (**2**), 4'-demethylpodophyllotoxin (DP) (**3**), desoxypodophyllotoxin (desoxy-P, silicicolin) (**4**), α-peltatin (**6**) and β-peltatin (**5**). They were isolated either from the dried roots (= podophyllum) or the resinous extract (= podophyllin) of *P. emodi* and *P. peltatum*.

P (**1**), isolated as early as 1880 by Podwyssotzki [5], represents the main compound of both *Podophyllum* species; its structure was definitively established by Hartwell and Schrecker [6] and later confirmed by Gensler [7] by means of an elegant total synthesis. More recently, several additional syntheses of P and its 1-epimer (epipodophyllotoxin, EP) have been published (see [32, 33]). The absolute configuration, deduced by Schrecker and Hartwell [34], could later be corroborated in our laboratories by X-ray analysis of 2'-bromo-

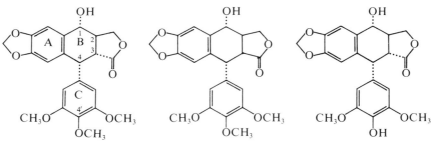

1 Podophyllotoxin   *2* Picropodophyllotoxin (Picropodophyllin)   *3* 4'-Demethylpodophyllotoxin

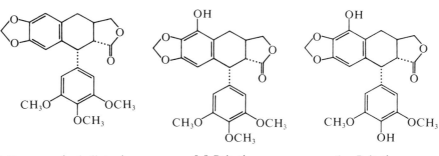

4 Desoxypodophyllotoxin   *5* β-Peltatin   *6* α-Peltatin

podophyllotoxin [35]. Picropodophyllotoxin (**2**) is easily formed by base-catalyzed rearrangement of P and is probably an artifact generated in the course of the isolation procedure. Desoxypodophyllotoxin (**4**) was first isolated from other plants, only later from podophyllum (see [12]).

All these natural substances from podophyllum species are structurally closely related and belong to the class of lignans, which are plant products containing the 2,3-dibenzylbutane skeleton. Early work on the chemistry and pharmacology of podophyllin and its main constitutents has been well presented and reviewed by Kelly and Hartwell [2] and by Hartwell and Schrecker [12].

Podophyllin is still in use as a first line local treatment of condylomata acuminata, a venereal, proliferative skin disease caused by a papilloma virus; solutions of P are also recommended for that purpose [36]. A number of attempts were made to treat malignant tumors in humans with podophyllotoxin or peltatins, either by local application in the case of cancers of accessible sites, or systemically (parenteral or oral administration) in malignancies of other organs. Except for very superficial skin cancers and condylomata acuminata, the overall results were disappointing and certainly inferior to those obtained with the anticancer drugs developed in the late forties and early fifties, such as mechlorethamine, aminopterin, amethopterin and others. The clinical use of P and the peltatins for treating systemically malignant tumors ceased in the 1950's.

The mechanism of the arrest of mitosis in metaphase by P and derivatives (including peltatins) is the same as that of colchicine, namely an inhibition of the assembly of tubulin into microtubules. The latter constitute the mitotic spindle which, during mitosis, brings about the arrangement of the chromosomes in the metaphase plate and their separation in anaphase. The inability of cells to form microtubules prevents them from dividing: the cells remain arrested in metaphase with clumped chromosomes and eventually die. This effect is brought about by binding of the respective agents to tubulin. The binding sites for colchicine and P or its congeners, but not for the vinca alcaloids, do overlap. Binding of P is more easily reversible than that of colchicine [37, 38]. P has no effect in gout [2] where colchicine is used extensively, while, vice versa, colchicine is hardly used in condylomata acuminata.

Table 1
Cytostatic and toxic effects of "early" podophyllum lignans

compound No.		ID-50 P-815 μg/ml	sarcoma % tumor inhibit.	L-1210 % incr. life span	LD-50 mouse mg/kg
1	podophyllotoxin (P)	0.005	29	35	35
2	picropodophyllotoxin	0.2	n.t.	n.t.	n.t.
3	4'-demethyl-P	0.007	58	10	>120
4	desoxy-P	0.002	var.*	27	52
5	β-peltatin	0.001	28	23	27
6	α-peltatin	0.001	57	2	13

* results highly variable.

ID-50: concentration inhibiting proliferation of mouse tumor cells (P-815 mastocytoma) *in vitro* by 50 %; the respective values are the average of several experiments and can therefore usually be regarded as well established (however, in order not to give the impression of unwarranted accuracy, values are usually presented with only one significant digit). Sarcoma: mouse sarcomas 37 and 180 (the latter tumor was used less frequently) were inoculated s.c. and evaluated by tumor weight; for each experiment, the result with the optimal dose is recorded, optimal dose being the highest non-toxic dose (i.e. a dose producing at most 17 % lethality or 10 % body weight loss). With L-1210, optimal dose is the one producing the highest median increase in life span over untreated controls. For sarcomas and L-1210 (the latter inoculated s.c. into the hind leg), i.p. treatment started 1 day after tumor inoculation and was continued daily for 8 days. The majority of the results for tumor inhibition are average figures of 2 or more experiments. LD-50: dose killing 50 % of the animals upon single administration; application was i.v. or i.p., no major difference being observed between these two routes in cases where both were tested.

Some of our results obtained with the early-known podophyllum aglucones are summarized in Table 1. The methodology used in our own experiments can be summarized briefly as follows.

Cytostatic potency of compounds and some qualitative aspects of inhibition of cell proliferation can best be assessed in cell culture assays. For quantitative evaluation we used cultures of P-815 mouse mastocytoma cells [20]. This system had several advantages over other cell culture assays in use at that time: cells do not adhere to the substrate and can therefore be easily counted; they are fast growing and results have a high reproducibility. A better qualitative evaluation of the cytological *in vitro* effects of such drugs is possible in cultures of proliferating cells which do not detach from the substrate even when they are damaged. Such a system, developed by us [21] made use of primary cultures of chick embryo fibroblasts [24, 39, 40] and was applied extensively in the evaluation of the podophyllotoxin derivatives synthesized in our company and proved particularly valuable in distinguishing between spindle poisons and cytostatic drugs with other mechanisms of action. These fibroblast cultures, originating from vessel walls of 15-day-old chick embryos, permit, besides determining the lowest antimitotic concentration, the rating of non-specific cell damage elicited by the compounds. Particularly for counting mitoses and establishing the mitotic index the system proved very useful since the cells in mitosis – which easily detach and get lost in ordinary monolayer cultures – remain in place due to the use of a plasma clot. For the evaluation of antitumor activity in animals, transplantable tumors of the mouse (sarcoma 37, sarcoma 180, leukemia L-1210) and the rat (Walker carcinosarcoma, Yoshida sarcoma) were used, applying the usual methodology (see [24] and [41]).

P, about as active as colchicine and vincaleukoblastine as an inhibitor of cell proliferation *in vitro*, shows moderate effects in mouse sarcoma while increasing the life span of mice inoculated with leukemia L-1210 to a certain degree, but significantly. DP is about as active *in vitro* as P, apparently less toxic (result of a single experiment, therefore of limited reliability), of moderate activity in mouse sarcomas and exhibits little effect on the survival time in L-1210. The peltatins are about 5x more potent than P *in vitro;* in L-1210, α-peltatin is inactive, while β-peltatin moderately increases survival time (in this model, increases of median life span of 25 % or more are usually considered significant). It thus seems that 4'-demethylation in this type of molecule rather decreases activity in leukemia L-1210. Picropodophyllotoxin has long been known to be much less active than its stereoisomer P [42]. Desoxy-P is slightly more potent than P as a mitotic poison and of about the same activity in animal tumors (Table 1).

Hartwell and Schrecker [12] have listed the ratios between maximum tolerated and minimum effective doses in mice bearing sarcoma 37 for these and other compounds; the ratios of P and the peltatins were more favorable than that of DP, and picropodophyllotoxin did not produce any tumor inhibition at all.

The mechanism of action of the compounds listed in Table 1 is known. They act as spindle poisons in our fibroblast cultures; they produce the typical figures of arrested metaphases ("c-mitosis", also called "ball-metaphase" because of the clumped chromosomes), and an increase of the mitotic index. The mechanism of action, in relation to the chemical structure, of some of these and other P deriva-

Podophyllotoxin series   Iso-compounds

tives will be discussed in more detail later. All compounds mentioned in the following sections exhibit, if not mentioned otherwise, the spindle poison type of cytostatic activity.

Prominent structural features of P are the rigid 2:3-transoriented, 5-membered lactone ring and the four contiguous asymmetric centers; 16 different optically active forms are possible, grouped in 8 pairs of enantiomorphic modifications (or DL-mixtures). Their biological activities are listed in Table 2.

Table 2
Stereochemistry at ring B of podophyllotoxin as related to cytostatic potency

	configuration			ID-50 P-815
	$H_{C-1}:H_{C-2}$	$H_{C-2}:H_{C-3}$	$H_{C-3}:H_{C-4}$	μg/ml
podophyllotoxin	trans	trans	cis	0.005
epipodophyllotoxin	cis	trans	cis	0.03
picropodophyllotoxin	trans	cis	trans	0.2
epipicropodophyllotoxin	cis	cis	trans	1
isopodophyllotoxin	trans	trans	trans	1
epiisopodophyllotoxin	cis	trans	trans	0.02
isopicropodophyllotoxin	trans	cis	cis	0.5
epiisopicropodophyllotoxin	cis	cis	cis	0.6

The racemate of picropodophyllotoxin is about 30x less active than P (the "natural" form is about 16x less active, its optical antipode about 2000x). Since these compounds (with the exception of P), as far as they have been tested, do not show any interesting *in vivo* activity [12], the structure-activity relationships will not be discussed here. We will come back to epipodophyllotoxin and its glucoside later. Several ethers of EP were synthesized; all turned out to be less active than EP itself. The only derivative of this series with a cytostatic potency similar to EP was DL-epiiso-P-methyl ether.

4 Genuine glycosides

As outlined in the history section, our investigations began with the search for glycosidic compounds because the then known aglucones had not shown useful therapeutic effects in man. The procedures were therefore adjusted to preserve glycosides which may be lost or degraded in the course of isolation procedures in use at that time. Based on experience with cardiac glycosides (digitalis, strophanthus, squill etc.), fresh or carefully dried rhizomes of podophyllum species were extracted, using procedures inhibiting enzyme activity. The obtained mixture of glycosides was separated by partition chromatography and yielded four pure main components: podophyllotoxin-β-

7 Podophyllotoxin-β-D-glucopyranoside

8 4'-Demethylpodophyllotoxin-β-D-glucopyranoside

9 β-Peltatin-β-D-glucopyranoside

10 α-Peltatin-β-D-glucopyranoside

D-glucopyranoside (**7**) and 4'-demethylpodophyllotoxin-β-D-glucopyranoside (**8**) from the Indian *P. emodi* [14, 15, 16] (they occur also in the American *P. peltatum*), as well as the β-D-glucopyranosides of β- and α-peltatin (**9, 10**) from *P. peltatum* [17, 18, 19]. These lignan glycosides were readily soluble in water and, as anticipated, easily split into aglucone and D-glucose by emulsin and other enzyme preparations containing β-glucosidases.

Table 3
Cytostatic and toxic effects of genuine glycosides

compound No.		ID-50 P-815 μg/ml	sarcoma % tumor inhibit.	L-1210 % incr. life span	LD-50 mouse mg/kg
7	P glucoside (PG)	6	40	7	297
8	4'-demethyl-PG (DPG)	2	n.t.	0	n.t.
9	β-peltatin glucoside	5	n.t.	n.t.	>200
10	α-peltatin glucoside	>10	n.t.	n.t.	>200
11	podorhizol glucoside	1	n.t.	n.t.	n.t.
12	demethyldesoxy-PG	2	n.t.	0	>400
13	desoxypodophyllinic acid glucopyranosylester	0.005	30	23	150

The first animal experiments with the β-D-glucosides of P and DP in non-tumorous rats and guinea pigs showed a considerably reduced toxicity in comparison with the respective aglucones. However, enthusiasm decreased somewhat when it was found that the glucosides' therapeutic potency was also diminished substantially: the concentrations required to arrest cell division in metaphase of mitosis in chick embryo fibroblast cultures were higher by 2–4 orders of magnitude for the glucosides than for the aglucones (Table 3) [39, 40]. A comparison of Tables 1 and 3 shows the decrease in cytostatic potency brought about by glucosidation which is particularly striking with α-peltatin (a factor of more than 10 000). While PG shows some tumor inhibiting activity in sarcoma 37, it is inactive in leukemia L-1210, and so is DPG.

The decrease in absolute potency, as observed *in vitro* with the glucosides, is, by itself, not an indication of inferiority as to therapeutic utility. But the results with the glucosides in mouse tumors were not very encouraging either (see Table 3 and [39, 40]). In addition, the duration of action of the glucosides *in vivo* seemed to be quite short.

The rather disappointing biological results with genuine podophyllum glycosides and the possibility that, after oral administration, unpredictable cleavage to more toxic aglucones in the intestinal tract might occur, were reason enough to embark on a more extensive program for chemical modifications, comprising glycosides as well as aglucones. Parallel to the synthetic approach (see below), efforts were made to find new genuine plant constituents, also with the objective to broaden the knowledge on structure-activity relationships. Early work, especially using thin layer chromatography to follow or guide separation procedures, revealed the presence of several additional sugar-containing lignans in the plant. Systematic preparative chromatography of enriched fractions finally led to three pure minor components which were named lignans F, H and J. Their structure could be elucidated by degradation and correlation reactions as well as by spectroscopic evidence.

Lignan F, podorhizol glucoside (**11**), represents one of the rare glycosides which are cleaved by acids as well as by bases, and has an un-

11 Podorhizol-β-D-glucopyranoside

12 4′-Demethyldesoxy-podophyllotoxin-β D-glucopyranoside

13 Desoxypodophyllinic acid-1β-D-glucopyranosyl ester

usual structure [43]. Lignan H is the glucoside of 4'-demethyldesoxypodophyllotoxin (**12**) [44], and lignan J is not a glycoside but an ester derivative, the β-D-glucopyranosylester of desoxypodophyllinic acid (**13**) [23].

The striking structural features of these new compounds suggested that they might act as precursors or by-products of podophyllum lignan biosynthesis. All three display cytostatic activity (Table 3); this is especially remarkable for podorhizol glucoside (**11**) which does not contain the usual tetrahydronaphthalene ring but is a derivative of a dibenzylbutyrolactone. Lignans of this type have not been encountered before in *Podophyllum* species. The ester bond of desoxypodophyllinic acid glucopyranosylester is, even *in vitro* (e. g. in cell cultures), easily split, leading, after closure of the lactone ring, to desoxypodophyllotoxin; this may explain the high cytostatic effect of lignan J in tissue culture (see also history section)

5   4'-O-Alkyl homologues of podophyllum glucosides

In one of our first attempts to gain more information on structure-activity relationships, various 4'-O-alkyl homologues (**14**) of podophyllum glucosides were prepared. Alkylation was performed with DPG (**8**) and α-peltatin glucoside (**10**), both containing a free phenolic OH-group suitable for reaction with diazoalkanes. Thus, reaction of

Podophyllotoxin series   Peltatin series

14 4'-O-Alkyl ethers   *15* O-Acyl derivatives
R = H   R = Acyl residue
R' = Alkyl residue   R' = CH₃ or acyl residue

DPG with carbon-14 labeled diazomethane yielded radioactive PG [45] which proved useful for pharmacokinetic studies. With higher diazoalkanes, the 4'-ethyl-, -propyl-, -butyl- and -isoamyl-homologues of PG were obtained. Their proliferation inhibiting effect in fibroblast cultures was of the same order of magnitude as that of PG or DPG, and if injected i. p. into mice bearing Ehrlich ascites tumors, mitotic arrest in the tumor cells lasted about as long as with DPG or somewhat longer. Analysis of side effects of these compounds (diarrhea, emesis, microscopic changes in the intestines, bone marrow depression etc.) revealed no therapeutic advantages over PG or its 4'-demethyl derivative [39, 40].

6 Acyl derivatives of podophyllum glucosides

For reasons outlined above, a number of acyl derivatives of podophyllum glucosides were prepared [40]. Esterification of the sugar hydroxyl groups with fatty acids of different chain lengths (C2, C3, C4, C16 and C18) led to the corresponding tetraacyl derivatives of PG and β-peltatin glucoside (**15**). DPG and α-peltatin glucoside, which in addition contain a phenolic OH-group in an aromatic ring, yielded the corresponding pentaacyl compounds. The acylated derivatives were only slightly soluble in water and resistant – at least *in vitro* – to β-glucosidases. Their cytostatic effects turned out to be inferior to those of the free glucosides in fibroblast cultures and in the Ehrlich ascites tumor of mice. Their potency decreased with increasing chain length of the fatty acid residue [39, 40].

7 Condensation products of podophyllum glucosides with aldehydes

Noticing that blocking all of the sugar hydroxyl groups did not lead to therapeutically improved derivatives, a more selective substitution in the glucose moiety was attempted by condensation reactions with aldehydes [39, 40]. In a first experiment PG was converted with benzaldehyde in the presence of Lewis acids to P-benzylidene-β-D-glucopyranoside (**16 a**). The assumed structure of this cyclic acetal was confirmed by chemical and spectroscopic evidence. The aldehyde residue occupies the energetically favorable equatorial position, as can be deduced from NMR studies (the axial anomer is also formed to a very small extent).

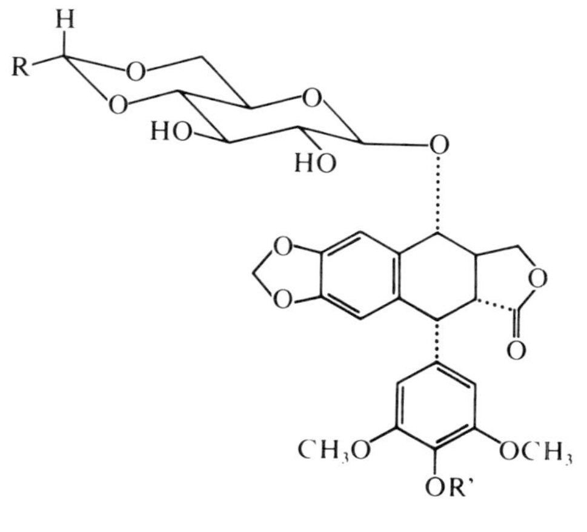

Cyclic acetals

16 (a–d) Podophyllotoxin series

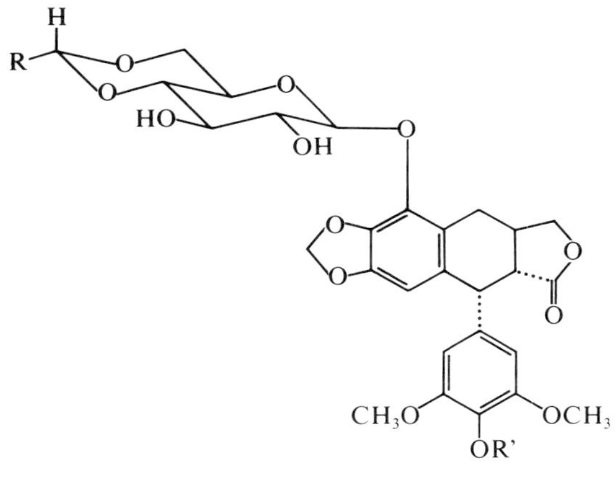

17 (a–c) Peltatin series

R = Aldehyde residue
R' = H or CH_3

The acid-catalyzed condensation reaction turned out to be a versatile method to produce cyclic acetals of lignan glucosides (**16, 17**) both in the podophyllotoxin and the peltatin series. Using various aldehydes as carbonyl reagents a number of these derivatives were prepared, among them the anisylidene and thenylidene glucosides. Cyclic acetals of podophyllum glucosides are only slightly soluble in water and *in vitro* resistant to the action of β-glucosidases. They are easily cleaved by weak acids into the corresponding aldehyde and the intact glucoside.

Table 4
Effects of aldehyde condensation products of podophyllum glucosides

compound No.	R'	R	ID-50 P-815 μg/ml	sarcoma % tumor inhibit.	L-1210 % incr. life span	LD-50 mouse mg/kg
16a	CH₃	phenyl	3	n.s.	5	240
16b	CH₃	p-methoxyphenyl	2	26	16	>300
16c	H	phenyl	0.8	16	29	160
16d	H	p-methoxyphenyl	0.3	28	27	330
17a	CH₃	phenyl	3	n.t.	6	>360
17b	H	phenyl	1	n.t.	2	>350
17c	H	p-methoxyphenyl	3	43	3	230

Results of biological testing of 2 of the aldehyde condensation products of PG are given in Table 4. As to cytostatic potency, benzylidene- (**16 a**) and anisylidene-glucosides (**16 b**) of P are about as active as PG. Chick embryo fibroblast cultures treated with these compounds clearly demonstrate spindle poison activity with a large increase of the mitotic index. The tumor-inhibiting effects in mouse sarcomas and leukemia L-1210 are nil or marginal at best; the toxic potential is quite low. The animals die within less than 24 h (sometimes within 10 min) which indicates that the cause of death is not arrest of cell division; the latter kills animals usually not before 48 h after administration of the drug. The LD-50 of the benzylidene glucoside is higher by a factor of about 1.5 when the drug is given i. p. instead of i. v.

Quite high amounts (close to the toxic dose) of solvents, e. g. polysorbate 80, had to be used in order to prepare clear aqueous solutions of these hydrophobic compounds; therefore, though animals injected with only the solvent were used as controls, it cannot be excluded that toxicity of the solvent may, in some cases, have had a marginal influence on lethality.

Condensation of benzaldehyde to the glucose changes biological activity in a similar way whether glucose is attached to the podophyllotoxin nucleus in position 1 (PG) or in position 8 (peltatin glucosides) (Table 4).

Carbon–14 labeled PG and its benzylidene derivative [45] were used to measure absorption, excretion and distribution of these compounds. While absorption of the glucoside after oral application to rats is negligible, more than 35 % of orally administered PBG appears in the bile and urine within 33 h. When distribution of radioactivity after p. o. application of 25 mg/kg of PBG and after i. v. administration of 25 mg/kg of PG was compared, relatively high levels were found in bone marrow, liver and kidney with both compounds; the lowest drug concentration was found in the brain. Blood levels are not very different, half life in the blood being slightly longer for the benzylidene derivative than for the free glucoside [39, 40].

Aldehyde condensation products were also prepared with the glucosides of 4'-demethylpodophyllotoxin and of α-peltatin (which is also demethylated in position 4') [40]. The biological results (Table 4) show, on the one hand, that 4'-OH derivatives of PG aldehyde condensation products have a higher cytostatic potency (by a factor of 4 to 7) than the corresponding 4'-methoxy compounds, and on the other hand, that condensation with aldehyde is more effective with DPG than with PG (comp. Table 3); the latter holds true not only for *in vitro* activity, but also for tumor inhibition in mice. The antitumor effects of DP benzylidene and anisylidene glucoside are, however, still borderline or insignificant. What is more interesting is the fact that these two compounds (**16 c, 16 d**) produce a much lower increase of the mitotic index in fibroblast cultures than the respective free glucoside or the aldehyde condensation products of PG (see also [46]; that must be interpreted to indicate that they exhibit an additional mechanism of action (besides spindle poison activity) which leads to an inhibition of entry into mitosis. At high concentrations, no mitoses at all could be found in cultures treated with these compounds, the cells were prevented completely from going into the mitotic phase of the cell cycle. We will come back to this aspect of 4'-demethylation and aldehyde condensation to the glucose in section 16. Another aspect of the difference between the aldehyde condensation products of PG and of DPG is the duration of action in Ehrlich ascites; the demethylated derivatives produce a much longer lasting mitotic arrest in tumor celles than the methylated ones [40].

Aldehyde condensation products of α-peltatin glucoside (**17 b, 17 c**) have cytostatic activity *in vitro* and *in vivo* similar to β-peltatin benzylidene glucoside (**17 a**) and considerably higher than the free glucoside of α-peltatin; there is some inhibition of the growth of sarcoma 37 and 180, but no antitumor effect in leukemia L-1210 (Table 4).

8   SP–G and SP–I (Proresid)

The benzylidene glucoside of P (**16 a**) (sometimes also called SPG 400) exhibited some properties – chemical stability, good absorption in the gastrointestinal tract, positive effects in a few cancer patients [24, 47] – which made it worth investigating more closely. In the course of these studies, benzylidene condensation was performed not only with pure PG, but also with a podophyllum extract which still contained other podophyllum glucosides, and compounds of a different structure. Production of this non-refined preparation is simpler and cheaper than that of the pure benzylidated glucoside; it consists of about 80 % PBG, about 10 % DPBG, small amounts of the respective (free) glucosides, traces of aglucones and some unidentified products.

The code name for this preparation was SPG 827 (or just SP–G). Some years later, the commercial preparation of SP–G as well as that of SP–I (see below) was given the designation Proresid®, in some countries Proreside® or Proresipar®. Solubility of SP–G in aqueous media is poor, powerful solvents such as dimethyl sulfoxide and Tween® *(polysorbate)* 80 have to be used in order to obtain clear solutions with optimal and reproducible activity *in vitro* and *in vivo* (including oral administration).

Table 5
Cytostatic and toxic effects of SP-G and SP-I

	ID-50 P-815 µg/ml	sarcoma % tumor inhibit.	L-1210 % incr. life span	LD-50 mouse mg/kg
SP-G	0.5	47 (n = 13)	65 (n = 21)	214 i.v. (n = 10)
SP-I	0.5	46 (n = 17)	17 (n = 13)	283 i.v. (n = 14)

Some pharmacological results with SP–G are listed in Table 5. When they are compared with the cytostatic effects of the main constituent (PBG, Table 4), it becomes evident that inhibition of cell multiplication *in vitro* and of tumor growth *in vivo* is superior with the „crude" preparation (it is therefore important to distinguish between the pure PBG and SP–G, a distinction which unfortunately has been neglected in many publications). With SP–G, growth of sarcoma 37 is inhibited significantly (though not extremely well), while prolongation of survival time in leukemic mice is considerable. In the latter case (L-1210), there is no difference between experiments with i. v. and those with i. p. administration of the drug. When the L-1210 tumor cells are inoculated i. p. (instead of s. c. into the hind leg, our standard procedure), i. p. treatment becomes much more effective, and increases in life span of up to 190 % are obtained. At the US Cancer Chemotherapy National Service Center (CCNSC), SP–G produced similar results, e. g. 41–158 % increase in life span in L-1210 murine leukemia which were inoculated i. p. (NSC No. of SP–G is 42076 D). Oral administration of SP–G prolongs the life span of mice inoculated with L-1210 only to a limited degree, probably because doses of the drug which would produce high enough blood and tissue levels are associated with severe gastrointestinal side effects. Some antitumor effect of SP–G is also observed when L-1210 is inoculated intracerebrally (murine lymphocytic leukemia L-1210 was tested more extensively since this tumor model was for years considered to have a particularly high predictability value for human malignancies [48]).

Survival of mice inoculated i. p. with Ehrlich ascites tumor is also prolonged considerably by i. p. (but not by oral) SP–G treatment. Transplantable rat tumors were used as well to evaluate the antitumor activity of SP–G: results with the Walker carcinosarcoma (inoculated s. c.) were highly variable and often poorly dose-dependent; percentage of inhibition with active i. p. doses ranged from 0 to 100 %; growth of the Yoshida sarcoma was also reduced to a significant but variable degree.

In fibroblast cultures SP–G clearly exhibits the characteristics of a spindle poison with an early onset of action; depending on the drug concentration, ana- and telophases disappears completely 10–45 min after addition of the drug, accompanied by an increasing number of arrested metaphases. At very high concentrations, entry of cells into mitosis is inhibited as well. These and other features of the effect of SP–G in fibroblast cultures have been described in detail [24]. SP–G affects neither respiration or glycolysis of normal or tumor cells *in vitro* nor the growth of bacteria or yeast. Repeated treatment of mouse tumor cells *in vitro* with partially cytostatic concentrations leads to an increase of the ID–50 of cell proliferation by a factor of 2 [24], thus to a low degree of resistance.

Chronic toxicity studies in rats, cats and dogs that had been given SP–G orally did not reveal an organ-specific toxicity except for lipid depletion in the adrenals; however mitotic arrest was observed in organs with proliferating cells, including testes; the principal side effects were diarrhea and emesis in dogs treated orally. Except for a slight leukopenia in dogs given sublethal doses, no signs of bone marrow toxicity were recorded [49]. In the course of studies with toxohormone it was found that repeated administration of SP–G to rats leads to a considerable reduction of the level of serum iron and a concomitant increase of the iron content of the liver [50].

Thymidine incorporation into mouse tumor cells proliferating *in vitro* is not inhibited by SP–G; DNA synthesis and chromosome reduplication continue even in the presence of active concentrations of the drug, thus producing polyploidy [51]. This is in accordance with the assumption that the primary mechanism of action of SP–G is inhibition of the mitotic spindle, similar to colchicine. SP–G shows, like other podophyllum compounds which have been investigated [52], a considerable binding to human serum proteins, less to serum proteins of other species.

In rats [53] and mice [54] SP–G is able to suppress the immune response to foreign erythrocytes quite substantially, in equitoxic doses to about the same degree as 6-mercaptopurine. The preparation also suppresses the symptoms of Freund adjuvant arthritis in rats [55], a model of an autoimmune disease related to rheumatoid arthritis in man (for clinical use of Proresid® in rheumatoid arthritis see below).

SP–G had the disadvantage of not being a single chemical entity and could hardly be given parenterally; it was also presumed that its main side effect (diarrhea) was somehow related to the route of administration. Therefore, a podophyllum compound with better water solubility was sought which could be injected intravenously. A derivative with the desired properties was chosen among the compounds with an open lactone ring (this structural class had been derivatized extensively). It is the 2-ethyl hydrazide of podophyllinic acid (**20 b**) (see section 9). This drug can be brought into aqueous solution with the help of ethanol, thus allowing the preparation of ampoules whose content can be diluted for i. v. injection or infusion. The compound was designated SPI 77 or just SP–I (generic name: *mitopodozide*).

Some of the biological test results of SP–I are given in Table 5. As to cytostatic potency *in vitro*, SP–I is equal to SP–G. Inhibition of the growth of mouse sarcoma 37 by daily i. p. or i. v. treatment is very similar to that of SP–G as well and is obtained with doses of the same order, namely about 70–80 mg/kg/day parenterally. SP–I is inactive when given orally. Similar results (60% inhibition) in the treatment of sarcoma 37 by SP–I were obtained by Benz [56]. On the other hand, in mouse leukemia L-1210, SP–I produces only 17% increase in life span, a figure statistically ($p < 0.01$) but not biologically significantly different from zero. Growth of the Walker carcinosarcoma in rats is inhibited by 55% (mean of 4 experiments) by SP–I given i. v. or i. p., variability of results being much less than with SP–G in this tumor model.

In cultures of chick embryo fibroblasts, the effects of SP–I are clearly those of a spindle poison: stop in metaphase, clumping of chromosomes, accumulation of arrested mitoses. Upon removal of the drug after 1 h of treatment, the fibroblasts soon resume normal mitotic activity, i. e. ana- and telophases reappear within less than an hour and reach normal values 3–4 h after placing the cultures in control medium [24] (this is in contrast to other spindle poisons, e. g. vincaleukoblastine, where, after a 1-h exposure, late mitotic phases do not reappear for more than 9 h). However, several hours later, without adding the drug again, a second nadir of mitoses occurs and the number of ana- and telophases is zero 24 h after removal of SP–I (but this time without increase of metaphases). The biochemical basis for this delayed effect (which occurs after approximately one cell generation) is not clear. Thymidine incorporation into P-815 mastocytoma cells is not inhibited by SP–I, rather slightly stimulated [51]. No effect of this drug was observed when respiration and glycolysis of tumor and normal (kidney) cells were measured and the drug added *in vitro;* sarcoma 37 slices (but not kidney slices) of mice which had been treated with SP–I, however, exhibited a strongly reduced respiratory and glycolytic rate, perhaps due to a drug effect on the blood supply of the tumor. Similarly to SP–G, SP–I treatment in rats leads to a decrease in serum iron. When SP–I is added to a balanced salt solution used to perfuse anesthetized rats, the flow decreases considerably, indicating a vasoconstrictor effect of the drug; SP–G is less active in this respect [21].

Chronic toxicity studies in dogs and monkeys (treatment once or twice daily i. v. for 3 months) revealed no specific organ toxicity of SP–I. Higher doses produced emesis and diarrhea. Mitotic arrest and its consequences were seen in the intestinal epithelium, in lymphatic organs and particularly in the testis, but not in the bone marrow.

Treatment with medium to high doses led to leukopenia from which the animals recovered without interruption of drug application [57]. In accordance with this, the distribution of tritium-labeled SP–I in dogs showed a low drug concentration in the bone marrow (but a high one in liver and kidneys) [58].

Of some interest is the immunosuppressive effect of SP–I in experimental systems in view of the clinical use of this drug in diseases – besides cancer – with an autoimmune component such as rheumatoid arthritis (see below). In contrast to SP–G, SP–I showed no immunosuppressive activity in rats [21] and produced a weak effect in mice immunized with heterogeneic red cells [54]. In another model, however, in runt disease elicited by injection of allogeneic spleen cells in newborn mice, SP–I reduced the pathological symptoms considerably [59].

Clinical testing of SP–G in cancer with an oral preparation (soft gelatin capsules containing a solution of SP–G in Tween® 80) started in 1959/60; somewhat later SP–I was added to the clinical program, and many patients were treated with both preparations. By the end of 1962 about 1000 treated cases had been recorded; 12% of them had shown improvement (reduction of tumor mass or reestablishment of the ability to perform daily/professional duties). Solid tumors of many different localizations and types were investigated. The main side effect was diarrhea; nausea/vomiting and, in a few cases, alopecia were also observed, while leukopenia was rare and usually not serious. Due to the fact that SP–G can be taken orally, home treatment was possible, and patients could titrate their individual dosis themselves because the first signal of overdosing is a side effect which can be perceived by the patient and is comparatively harmless (diarrhea, in contrast to leukopenia). Leukemias were soon found to respond to SP–G and SP–I treatment only very exceptionally in a clinically relevant way.

The first clinical publications appeared in 1961 [60] and were followed by numerous others (about 300 were registered by 1970), e. g. Ravina [61], Falkson et al. [62], Stamm [63], Aquilini [64], Gosalvez [65], Larsen et al. [66], Vaitkevicius and Reed [67], Beumer and Porton [68], Krishnamurthi et al. [69], Gebhardt and Becher [70], Schmidt [71], Chakravorty et al. [72], Estévez et al. [73], and many others. Later a short review was given by Trüb and Pohlschmidt [74]. Special types of administration of SP–G and SP–I were also investigated, e. g. local application to the tumor [75] (for which special galenical forms of SP–G had been developed) and intra-arterial infusion of SP–I [76, 77, 78]. With the help of Tween® 80, an injectable intravenous form of SP–G was developed, but its clinical use posed too many problems and was soon abandoned. One of the earliest attempts to assess chemosensitivity of human tumors by culturing the malignant cells and exposing them to different anticancer drugs *in vitro* also included SP–I [79]. As in experimental animal systems [24], it was also possible in human tumors to assess the cytostatic effects of SP–G and SP–I *in situ* by analyzing the different mitotic phases in tumor biopsies during treatment [80]. In certain cases of psoriasis, a positive effect of (oral) SP–G was observed [81]. Because of the low toxicity of SP–G and SP–I in bone marrow, these drugs were frequently combined with X-ray treatment, and sensitization of human tumors for X-rays by SP–G has also been reported [82].

Although objective, positive effects were found in a significant percentage of cancer patients, and in many cases a remarkable improvement of the general condition of the treated subjects was observed, the long-term results with Proresid® did not achieve the level obtained with some of the newer and more aggressive anticancer drugs which became available at the time. Therefore the clinical use of SP–G and SP–I as anticancer agents decreased gradually despite their good tolerability.

Other clinical indications which were investigated with these two drugs are some diseases with an (at least partial) autoimmune basis, particularly rheumatoid arthritis. Chlud et al. [83, 84] observed positive clinical effects of SP–I in about two thirds of rheumatoid arthritis patients, while SP–G was hardly effective. Clinical improvements were also seen in patients with colitis ulcerosa [85] and multiple sclerosis [86, 87]; however, SP–I did not show any effect in the experimental allergic encephalomyelitis of the rat, an animal model for multiple sclerosis [88]. SP–G is still being used in some clinical centers in Scandinavia in the treatment of rheumatoid arthritis (see e. g. [89]).

9 Podophyllinic acid hydrazides and related compounds

A podophyllotoxin derivative with an open lactone ring, called podophyllic acid (**22**), a 2,3-cis(picro)-derivative, has long been known (see [2]). All early attempts at opening the lactone ring involved the use of alkaline media and led therefore to the 2,3-cis (picro-)-configuration. We will adopt the nomenclature of Rutschmann and Renz [90]: podophyllinic (2,3-trans) and picropodophyllinic acid.

In the course of experiments conducted to synthesize amines with the ring system of P, Rutschmann and Renz [90] achieved a stereoselective cleavage of the lactone ring, using hydrazine in buffered solution. The resulting main product was identified as podophyllinic acid hydrazide (**18 a**); the substituents at C–2 and C–3 displayed the intact trans-configuration. Hydrazine cleavage turned out to be a general method which could be applied to other lignans and their glucosides [40] of the podophyllotoxin type. In the picro series, hydrazides of picropodophyllinic acid with 2,3-cis configuration (e. g. **18 c**) were obtained.

Altogether more than 220 podophyllum compounds with an open lactone ring were prepared. Only a selection of them will be mentioned here because many were inactive or biologically not interesting. The activity of a few derivatives with an open lactone ring is reported by Hartwell and Schrecker [12] and more recently several compounds with substitution of the lactone configuration by other ring structures have been prepared by Gensler and others and tested (for a short review see [91]). With the exception of the podophyllinic acid ethyl hydrazide (**20 b**) (SP–I, see above), and a small number of the 1-O-glucosides of podophyllinic and picropodophyllinic acid [40], no biological results of the compounds with an altered lactone ring synthesized in our company have been published so far; those given below were obtained in the laboratory of one of the authors [21].

9.1 Unsubstituted hydrazides of podophyllum lignans

The primary products of the reaction between podophyllum lignans and hydrazine are all of low activity biologically, and the differences between the 2,3-trans and the 2,3-cis (picro) form are again minor in most cases (Table 6).

With the exception of α- and β-peltatinic acid hydrazides, these compounds were tested only in fibroblast cultures. Cytostatic activity of podophyllinic acid hydrazide (**18 a**) is slightly higher than that of podophyllinic acid (see below). Here too, glucosidation (**18 b**) results

18 (a–l) Unsubstituted acyl hydrazides

Table 6
Cytostatic potency of podophyllum lignan hydrazides as assayed in fibroblast cultures

compound No.	R	R'	R"	config. at C-2	C-3		ID-50 µg/ml
18a	αOH	CH$_3$	H	β	α	podophyllinic acid hydrazide	1
b	αG	CH$_3$	H	β	α	1-glucosido-podophyllinic acid hydrazide	>100
c	αOH	CH$_3$	H	β	β	picropodophyllinic acid hydrazide	2
d	αG	CH$_3$	H	β	β	1-glucosido-picropodophyllinic acid hydrazide	>100
e	H	CH$_3$	OH	β	α	β-peltatinic acid hydrazide	0.2
f	H	H	OH	β	α	α-peltatinic acid hydrazide	0.2
g	H	H	OH	β	β	α-picropeltatinic acid hydrazide	3
h	βOH	CH$_3$	H	β	α	epipodophyllinic acid hydrazide	5
i	βOH	CH$_3$	H	β	β	epipicropodophyllinic acid hydrazide	5
k	H	CH$_3$	H	β	β	desoxypicropodophyllinic acid hydrazide	2
l	H	CH$_3$	H	α	β	isodesoxypodophyllinic acid hydrazide	2

G = O-β-D-glucopyranosyl.

in a considerable decrease of potency. The hydrazides of the peltatinic acids (**18 e, 18 f**) are about 5x more active than those of the podophyllinic acid. β-Picropeltatinic acid hydrazide could not be tested reliably because it was not possible to bring it into a clear aqueous solution.

9.2   Carbonyl condensation products of acid hydrazides

Condensation of hydrazides of podophyllinic and picropodophyllinic acid with aldehydes or ketones resulted in a number of acyl hydrazones [90]. With few exceptions they were tested for biological activity in fibroblast cultures only; some of their glucosides were also tested for effects on mitoses in Ehrlich ascites tumors of mice (see [40]).

From podophyllotoxin glucoside to etoposide   199

19 (a–w) Podophyllinic and picropodophyllinic acid hydrazones

Table 7
Inhibition of mitoses in fibroblast cultures by acyl hydrazones of podophyllinic and picropodophyllinic acid

compound No.	config. at C-3	R	R'	ID-50 µg/ml
19a	α	H	isopropylidene	3
b	α	G	isopropylidene	>100
c	α	H	hexylidene	1
d	α	H	dodecylidene	1
e	α	H	octadecylidene	0.3
f	α	H	cyclohexylidene	0.5
g	α	H	benzylidene	1
h	α	G	benzylidene	>100
i	α	H	anisylidene	1
k	α	H	o-chlorobenzylidene	3
l	α	H	m-sulfobenzylidene (Na salt)	>10
m	α	H	naphthylidene	0.3
n	α	H	phenylethylidene	0.4
o	α	H	thenylidene	0.3
p	α	H	1-methylpiperidylidene	1
q	β	H	ethylidene	1
r	β	H	isopropylidene	10
s	β	G	isopropylidene	>100
t	β	H	hexylidene	1
u	β	H	benzylidene	0.6
v	β	H	thenylidene	3
w	β	H	1-methylpiperidylidene	3

G = β-D-glucopyranosyl.

In the series of podophyllinic acid hydrazones which are N-substituted with aliphatic residues (**19**), cytostatic potency increases somewhat with increasing length of the aliphatic chain (Table 7), but is of the same order of magnitude as that of the podophyllinic acid hydrazide itself. Glucosidation in position 1 again greatly reduces activity.

The dodecylidene derivative (**19 d**) was also tested in a preliminary way *in vivo* and found to be inactive in inhibiting the growth of sarcoma 37. Condensation with substituted aromatic and heterocyclic aldehydes does not improve activity (Table 7). Phenylethylidene podophyllinic acid hydrazide (**19 n**) was given to the Cancer Chemotherapy National Service Center in Bethesda, Maryland, and assayed for tumor inhibition in mice (sarcoma 180 and leukemia L–1210); no significant anticancer activity was found.

The respective picro-derivatives of many of the above-mentioned hydrazones and a few additional ones were also prepared and tested in tissue culture. The results of some of them (**19 q–w**) are given in Table 7. In most cases, the difference between the 2,3-trans and the 2,3-cis form is small. All these derivatives exhibited the spindle poison type of activity. The ID–50 values of Table 7 must, however, be considered as approximate only, for the following reasons: many compounds were tested only once; for a large part of them the concentration steps used in testing were rather large (10-fold); and the assay in fibroblast cultures (counting mitotic phases) is quantitatively less well reproducible than the P–815 mastocytoma assay (cell count).

Some analogous derivatives of the peltatins were also prepared and tested. Their activity in fibroblast cultures is of the same order of magnitude as that of the respective podophyllinic acid hydrazones. Thus, α-peltatinic acid ethylidene hydrazide has an ID–50 of 1 μg/ml; β-peltatinic acid ethylidene hydrazide is somewhat more active with an ID–50 of 0.2 μg/ml.

9.3 Alkyl and aryl hydrazides of podophyllinic and picropodophyllinic acid

Hydrogenation of the above-mentioned hydrazones yields N-alkyl and -aryl hydrazides (**20 a–o**) [90] in which the N = C double bond is saturated. One of them is podophyllinic acid ethyl hydrazide (SP–I) (**20 b**), discussed in section 8. Some of these substituted hydrazides were also obtained by direct alkylation of podophyllinic acid hydrazide with alkyl halogenides, yielding mono- or dialkyl derivatives, depending on the reaction conditions. A considerable number of them were synthesized and tested for cytostatic activity.

20 (a–o) Alkyl and aryl-alkyl hydrazides of podophyllinic and picropodophyllinic acid

Table 8
Cytostatic activity of alkyl and aryl hydrazides of podophyllinic and picropodophyllinic acid

compound No.	config. at C-3	R	ID-50 fibrobl. µg/ml	ID-50 P-815 µg/ml
20a	α	methyl	>3	n.t.
b	α	ethyl (SP-I)	0.6	0.5
c	α	n-propyl	0.3	n.t.
d	α	isopropyl	2	n.t.
e	α	hexyl	0.2	0.2
f	α	dodecyl	0.06	0.05
g	α	tetradecyl	0.05	n.t.
h	α	cyclohexyl	2	n.t.
i	α	monobenzyl	0.1	n.t.
k	α	phenylethyl	0.6	n.t.
l	β	methyl	3	n.t.
m	β	ethyl	n.t.	10
n	β	hexyl	>10	n.t.
o	β	benzyl	>3	n.t.

The hydrogenated compounds with the 2,3-trans configuration are usually more active than the corresponding hydrazone derivatives, while in the picro series (2,3-cis configuration) the opposite is the case. Conversion to the 2,3-cis configuration reduces cytostatic potency considerably among the saturated derivatives. Furthermore, there is a remarkable increase in activity with increasing alkyl chain length (Table 8).

The dodecyl derivative (**20 f**) was investigated more extensively, because of its comparatively high potency regarding inhibition of cell proliferation *in vitro*. Among other things, its effects on cell division in the bone marrow of normal and adrenalectomized rats were studied and related to general toxicity. Adrenalectomy increased susceptibility of proliferation of bone marrow cells to podophyllinic acid dodecyl hydrazide, but increased general toxicity (i. e. lethality) even more; the latter could be antagonized by administration of corticosteroids. With this podophyllum compound, the relationship of cytostatic potency to bone marrow toxicity was much more favorable than that of a colchicine derivative in clinical use at the time [92] (for a while we assumed that the comparatively low bone marrow toxicity may be related to the high serum protein binding of the compound, which was particularly strong in human serum). However, oncostatic effects in mice (Ehrlich ascites tumor, sarcoma 37, sarcoma 180 and mastocytoma P–815) were not convincing and there were considerable galenical problems due to low water solubility. Therefore, this drug was not further developed.

Though about 10x less potent *in vitro* than the dodecyl hydrazide of podophyllinic acid, the ethyl hydrazide was more active in solid mouse tumors and less problematic concerning galenical and stability aspects; it was chosen for development as a parenteral form of a podophyllum compound and as a complement to SP–G; it was commercialized in 1963 (see section 8).

The 4'-demethyl derivatives of these two podophyllinic acid hydrazides (ethyl and dodecyl) show about the same potency in fibroblast cultures as the 4'-methoxy compounds.

Alkyl hydrazides of peltatinic acid [90] were more active than the corresponding podophyllinic acid derivatives. The ethyl hydrazides of α- and β-peltatinic acid both had an ID–50 of 0.02 µg/ml (i. e. about 25 × lower than the P derivative), while the dodecyl derivatives of the peltatinic series were only slightly more potent than those of the P series.

9.4 Acyl derivatives of podophyllinic acid hydrazide

Finally, a number of N-monoacyl derivatives of podophyllinic acid hydrazide were prepared (**21 a–h**). They exhibit cytostatic activities of a lower degree than the corresponding compounds with N-alkyl residues (Table 9).

Surprising is the much higher potency of the phenyl-thioureido derivative compared to the phenyl-ureido compound. Though the lauroyl is more active than the monoacetyl derivative, there is no general trend in the sense that more lipophilic compounds are more active.

21 (a–h) N-Acyl derivatives of podophyllinic acid hydrazide

Table 9
Cytostatic activity of N-acyl derivatives of podophyllinic acid hydrazides in fibroblast cultures

compound No.	R	ID-50 µg/ml
21a	acetyl	30
b	lauroyl	2
c	maleinyl	1
d	succinoyl	7
e	phenylureido	10
f	phenylthioureido	0.3
g	phthaloyl	2
h	p-aminobenzenesulfonyl	2

10   Neopodophyllotoxin and podophyllinic acid

Pharmacological results with podophyllinic acid hydrazide derivatives had indicated that the 2,3-trans lactone group was not an indispensable structural element for antimitotic effects of podophyllum compounds. Most of the active agents of this series represent derivatives of podophyllinic acid (**27**). This at that time unknown hydroxy acid displayed the trans-configuration of the substituents at C–2 and C–3 as P (**1**), and seemed highly interesting from the chemical as well as the pharmacological point of view; ways to its synthesis were sought.

All attempts to prepare podophyllinic acid (**27**) by alcaline cleavage of the lactone ring failed; epimerization and formation of (2,3-cis)picropodophyllinic acid (**22**) (originally named podophyllic acid; see [90]) proved to be kinetically the predominant reaction. For the same reason, conversion of podophyllinic acid hydrazide (**18 a**)

27 Podophyllinic acid ⇌̸ 1 Podophyllotoxin ⟶ 2 Picropodophyllotoxin ⟶ 22 R = H: Picropodophyllinic acid

23 R = CH$_3$: Methylester

or podophyllinic acid methyl ester (25) into the parent hydroxy acid turned out to be unsuccessful.

Our new approach to synthesize the desired podophyllinic acid was based on earlier findings concerning transesterification reactions. It could be shown that desoxy-P (4) undergoes transesterification through treatment with methanol and perchloric acid to yield a mixture of unchanged lactone and desoxypodophyllinic acid methylester (24). The acid catalyzed methanolysis proceeded in an equilibrium reaction with preservation of the vital configuration at C–3 [23]. Application of this procedure to P, using methanol and zinc chloride, produced a mixture of unchanged P, podophyllinic acid methylester (25) and a new isomeric 1,3-lactone, named neopodophyllotoxin (26). It was assumed that base-catalyzed epimerization with the 1,3-lactone could not occur due to steric hindrance, thus enabling a stereoselective cleavage of the lactone ring to the desired 2,3-trans hydroxy acid (27). It was indeed possible to convert neopodophyllotoxin into podophyllinic acid [93, 94]. The structure of the quite stable hydroxy acid could easily be proven by chemical correlation reactions, NMR spectra and comparison of molecular rotations.

Podophyllinic acid (27) turned out to have a quite low cytostatic potency. Its methyl ester (25) is almost 100× more active (Table 10). The latter compound, though of low toxicity and high *in vitro* activity, does not increase the survival time of mice inoculated with leukemia L–1210 and exhibits a weak and variable inhibitory effect on the growth of mouse sarcoma 37 but a pronounced activity in rats inoculated with the Walker carcinosarcoma. Another feature of this

From podophyllotoxin glucoside to etoposide

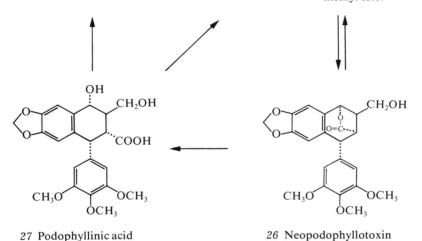

4 R = H Desoxypodophyllotoxin

1 R = OH Podophyllotoxin

24 R = H Desoxypodophyllinic acid methyl ester

25 R = OH Podophyllinic acid methyl ester

27 Podophyllinic acid

26 Neopodophyllotoxin

Table 10
Podophyllinic acid and related compounds

compound No.		ID-50 P-815 μg/ml
22	picropodophyllinic acid	10
23	picropodophyllinic acid methyl ester	0.1
24	desoxypodophyllinic acid methyl ester	0.1
25	podophyllinic acid methyl ester	0.04
26	neopodophyllotoxin	12
27	podophyllinic acid	3
	l-glucosido-podophyllinic acid methyl ester	>10
	4,6-benzylidene glucosido-podophyllinic acid methyl ester	>10
	isopodophyllinic acid (DL)	11
	0,0-benzylidene podophyllinic acid methyl ester	6
	β-peltatinic acid methyl ester	0.04
	α-peltatinic acid methyl ester	0.05

compound is its very high binding to human serum proteins. Glucosidation in position 1 of podophyllinic acid methyl ester reduces its potency more than 100-fold, and the aldehyde condensation product of this glucoside is also inactive. The intermediate product in the synthesis of podophyllinic acid, neopodophyllotoxin (**26**), is somewhat less potent than the acid. Another aldehyde condensation product, involving the primary and secondary OH-groups of (**25**), O,O-benzylidene podophyllinic acid methyl ester, still exhibits some activity (Table 10).

Picropodophyllinic acid, formerly called podophyllic acid (**22**), is only about 3x less potent than podophyllinic acid, and a similar ratio holds for picropodophyllinic acid methylester (**23**) as compared to the 2,3-trans ester (**25**).

11 Demethylenepodophyllotoxin, sikkimotoxin and related compounds

In 1950, Chatterjee and Datta [11] isolated a new lignan from rhizomes of *Podophyllum sikkimensis* R. Chatterjee and Muckerjee which they named sikkimotoxin. According to the proposed structure (**28 a**), sikkimotoxin is closely related to P, differing only in the replacement of the methylenedioxy group by two methoxyls [95]. A resinous fraction of *P. sikkimensis* was found to be damaging to sarcoma 37 (see footnote 14 of ref. [2]). Since neither sikkomotoxin nor the plant could be obtained for pharmacological evaluation, picrosikkimotoxin (**28 b**) and sikkimotoxin (6,7-O-demethylene-6,7-O-dimethyl-P) (**28 a**) were synthesized in our laboratories by the late E. Schreier [96, 97]. The structures of both synthetic compounds could be unequivocally proven by stereochemical correlation with picropodophyllotoxin. However, the physical data of synthetic sikkimotoxin and its picro derivative did not agree in all respects with the published values. Therefore, some doubts arose as to the correctness of the proposed structure or the purity of the compounds from natural sources. The availability of synthetic sikkimotoxin allowed the preparation of derivatives with an open methylene dioxy group (unpublished results) which will be discussed below.

28 (a–h) Sikkimotoxin and derivatives

Table 11
Cytostatic potency of sikkimotoxin and derivatives

compound No.	R	R'	$H_{C-1}:H_{C-2}$	$H_{C-2}:H_{C-3}$	$H_{C-3}:H_{C-4}$		ID-50 P-815 µg/ml
28a	OH	CH₃	trans	trans	cis	sikkimotoxin	0.1
b	OH	CH₃	trans	cis	trans	picrosikkimotoxin	>10
c	OH	CH₃	cis	trans	cis	episikkimotoxin	6
d	OH	CH₃	cis	cis	trans	epipicrosikkimotoxin	>10
e	OH	CH₃	cis	trans	trans	epiisosikkimotoxin	>10
f	H	CH₃	–	trans	cis	desoxysikkimotoxin	0.2
g	H	CH₃	–	cis	trans	desoxypicrosikkimotoxin	>10
h	H	H	–	trans	cis	6,7-O-demethyldesoxysikkimotoxin	0.3

Synthetic sikkimotoxin (**28 a**) inhibits proliferation of P–815 mastocytoma cells *in vitro* with an ID–50 of 0.1 µg/ml and is thus about 20x less potent than P (Table 11); in fibroblast cultures, it shows a spindle poison type of activity [21].

The range of chemical reactions characteristic for lignans of the P type is practically the same in the sikkimotoxin series, e. g. isomerization to 1-epi- and picro-compounds. The stereoisomers of sikkimotoxin which have been tested are mostly inactive in the mastocytoma test at a concentration of 10 µg/ml: picrosikkimotoxin, DL-picrosikkimotoxin, DL-epiisosikkimotoxin (**28 e**), DL-epipicrosikkimotoxin (**28 d**). Only episikkimotoxin (**28 c**) weakly reduces proliferation of these mouse tumor cells with an ID–50 of 6 µg/ml and exhibits a spindle poison type effect in fibroblast cultures (Table 11).

A number of other analogues of sikkimotoxin were also prepared and tested [21]. Transesterification by zinc chloride-catalyzed methanolysis, which in the case of P had led to podophyllinic acid methyl ester (**25**) and neopodophyllotoxin (**26**) [93], proceeded in the same way with sikkimotoxin, yielding the 6,7-O-demethylene-6,7-O-dimethyl analogues of podophyllinic acid methyl ester and neopodophyllotoxin [97].

Selective cleavage of the methylene dioxy ring was, as with P, also performed with desoxy-P. The main compound obtained, (**28 h**), was methylated to yield desoxysikkimotoxin (**28 f**) which in turn was epimerized with bases to the picro compound (**28 g**). Desoxypicrosikkimotoxin has also been obtained by total synthesis [96]. For results as to cytostatic potency see Table 11.

Demethylation of sikkimotoxin (**28 a**) in positions 6 and 7 and/or 4' results in a reduction of cytostatic potency, 6,7,4'-trisdemethylsikkimotoxin being about 30x less active *in vitro* than sikkimotoxin. In fibroblast cultures, they arrest cells in metaphase, like P, though the metaphase chromosomes seem to be somewhat less clumped than under the influence of P.

In addition to 6,7-demethylene compounds related to P, selective cleavage of the methylene dioxy ring was also obtained in the peltatin series. The free phenolic groups of such lignans could be alkylated to the corresponding ether derivatives. Their cytostatic potency is comparatively low. Surprisingly, 6,7-demethylene-α-peltatin was completely inactive (ID–50 > 10 μg/ml *in vitro*).

Other compounds of the sikkimotoxin type, those with an altered lactone ring, exhibited little or no cytostatic activity as well. Among them were neosikkimotoxin (the analogue of neopodophyllotoxin) and the corresponding analogues of podophyllinic acid methyl ester and podophyllinic acid ethyl hydrazide.

All sikkimotoxin derivatives tested in fibroblast cultures exhibited the spindle poison type. The more active among them were assayed *in vivo* in mouse leukemia L–1210, but none of them increased the survival time of the tumor-bearing animals [21].

12   4'-Demethylepipodophyllotoxin aglucones

While the 4'-demethyl derivative of P (**3**) belongs to the group of the early-known podophyllum compounds (see section 3) and the epi-isomer of P was described in 1951 (see [12]), a P derivative with both structural changes was not known before the „minor" podophyllum lignans were isolated. DEP (**29**) can be obtained by epimerization of DP or from P by selective cleavage of the methoxy group at C–4' [98] (see next section). The cytostatic activity of DEP and some of its derivatives (**30 a–m**) is given in Table 12.

4'-Demethylepipodophyllotoxin (DEP) is about 10x less potent than 4'-demethylpodophyllotoxin and only slightly less active than epipodophyllotoxin [46]. It is a typical spindle poison (large increase of the mitotic index over controls) and inactive in mouse leukemia L–1210. DEP exhibits a low toxicity and reduces the growth of sarco-

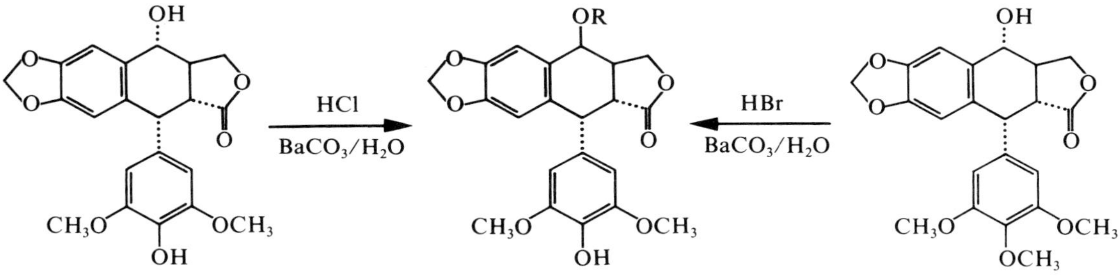

3 4'-Demethylpodo-phyllotoxin

29 R = H 4'-Demethylepi-podophyllotoxin

30 (a–m) R = Acyl

1 Podophyllotoxin

Table 12
Cytostatic activity of 1-O-acyl derivatives of 4'-demethylepipodophyllotoxin (DEP) aglucone

compound No.	R	ID-50 P-815 µg/ml	rMI	L-1210 % incr. life span
29	H	0.06	10	7
30a	pivaloyl	>3	n.t.	n.t.
b	nonanoyl	>10	n.t.	n.t.
c	benzoyl	0.5	10	n.t.
d	α-naphthoyl	>10	n.t.	n.t.
e	furoyl	0.04	10	0
f	thenoyl	0.4	18	n.t.
g	α-glyceryl	0.2	1	n.t.
h	carbamoyl	0.2	17	9
i	methylcarbamoyl	0.3	5	11
k	phenylcarbamoyl	0.08	3	68
l	p-chlorophenylcarbamoyl	0.4	0.2	74
m	β-pyridylcarbamoyl	0.04	2	62

rMI = relative mitotic index, the number of mitotic figures per 1000 cells in (for 6 h) treated fibroblast cultures divided by the same number in control cultures.

mas 37 and 180 to some degree. It was also tested for immunosuppressive effects in mice (inhibition of hemagglutinin formation) and found to be inactive.

With the exception of the α-glyceryl ether of DEP (**30 g**), the derivatives not belonging to the carbamoyl series all exhibit spindle poison activity when tested in fibroblast cultures. On the other hand, the carbamate derivatives, with the exception of carbamoyl DEP, have relative mitotic indices below 6, indicating that they are not spindle poisons with pure antitubulin activity (acting during mitosis), but also inhibitors of the entry of the cells into mitosis. Besides this structure activity relationship, there are other correlations of interest. As mentioned earlier, the lowest drug concentration producing a significant inhibition of cell proliferation in chick embryo fibroblast cultures (i. e. reducing the number of late mitotic phases – ana- and telophase – by 50–70%) differs from the ID–50 in P–815 mastocytoma cell cultures by a factor of usually less than 2 for spindle poisons. For the derivatives producing a low or no increase of the mitotic index (i. e. those inhibiting the entry into mitosis), this ratio is usually around 5, the ID–50 in P–815 cultures being lower than the ID–50 in fibroblasts. A second, more important correlation becomes evident on inspection of Table 12: a significant therapeutic effect in mouse leukemia L-1210 is found only in those derivatives which produce a low relative mitotic index. Though there are exceptions to this rule (podophyllotoxin, for instance, produces a large increase in mitotic index and prolongs the life span of L-1210-inoculated mice to a moderate but statistically significant degree), it seems that in the podophyllum series a biologically interesting antitumor effect in this animal model is found almost exclusively with derivatives which show some degree of inhibition of entry into mitosis. This aspect will be discussed again below.

13   4'-Demethylepipodophyllotoxin glucoside and derivatives
13.1   Synthesis

As mentioned in the section on history, progressive narrowing-down by systematic chromatography of enriched fractions of SP–G led to the compound which is responsible for the activity of that drug in mouse leukemia L-1210: 4-demethylepipodophyllotoxin benzylidene glucoside (benzylidene lignan P, DEPBG) (**39 i**). This compound, which produces a considerable increase in the life span of leukemic mice at low doses, is the prototype of the class of aldehyde condensation products of DEPG to which teniposide and etoposide belong. However, DEPG occurs only in minimal amounts in the plant and more extensive pharmacological evaluation of this compound and its derivatives was only possible when we succeeded in the development of a stereoselective synthesis of lignan P, suitable for production at a larger scale [28].

The synthesis of 4'-demethylepipodophyllotoxin glucoside (38) presented several problems, due to stereochemical peculiarities and the sensitivity of the aglucone to acids and bases. The final synthetic approach, emerging from numerous preliminary experiments, passes through 3 main phases: 1) preparation of DEP (29); 2) glucosidation reaction and 3) removal of the protecting groups.

The hitherto unknown DEP could have easily been obtained from the natural DP by epimerization via the 1-chloride [98]. However, because of the scarcity of DP in the plant extracts, we turned our attention to a chemical modification of P, the easily available main lignan of Podophyllum. Using HBr, the methoxy group at 4' could be cleaved selectively. The intermediate 1-bromoderivative was then hydrolyzed to 4'-demethylepipodophyllotoxin (29) [98].

For the glycosidation step, classical methods (using tetra-O-acetyl-α-D-glucopyranosyl bromide and a suitable catalyst) which proved successful in the synthesis of PG (7) [99] could not be applied satisfactorily for stereochemical reasons and the high reactivity of the 1-epi hydroxyl group. A new general glycosidation procedure for epi-derivatives was then elaborated. It could be shown that P and EP (32) reacted smoothly with 2,3,4,6-tetra-O-acetyl-β-D-glucopyranose at low temperature in the presence of BF_3-etherate to yield identical products, recognized as the tetra-acetate of EPG (33) [100]. In an analogous way, DP and DEP (protected as benzyloxycarbonyl derivatives [35 and 36]) furnished the same compound, namely the protected tetra-acetate (37) of DEPG [28]. This new glycosidation reaction leads exclusively to 1-epi derivatives and is also highly specific with respect to the glycosidic linkage. Moreover, the method is not restricted to glucose; other hexoses, e. g. β-D-galactose, are also suitable sugar compounds.

The last important step consisted in the development of an expedient method to prepare glycosides from their corresponding acetyl deriva-

Synthesis of podophyllotoxin-β-D-glucopyranoside and epipodophyllotoxin-β-D-glucopyranoside

1 Podophyllotoxin

tetra-O-acetyl-α-D-glucopyranosyl bromide
CH₃CN/Hg(CN)₂
→ **31**

ZnCl₂-MeOH
→ **7** Podophyllotoxin-β-D-glucoside

32 Epipodophyllotoxin

2,3,4,6-tetra-O-acetyl-β-D-glucopyranose
BF₃-etherate
→ **33**

Zn(OAc)₂
EtOH
→ **34** Epipodophyllotoxin-β-D-glucoside

2,3,4,6-tetra-O-acetyl-β-D-glucopyranose
BF₃-etherate (from **1** to **33**)

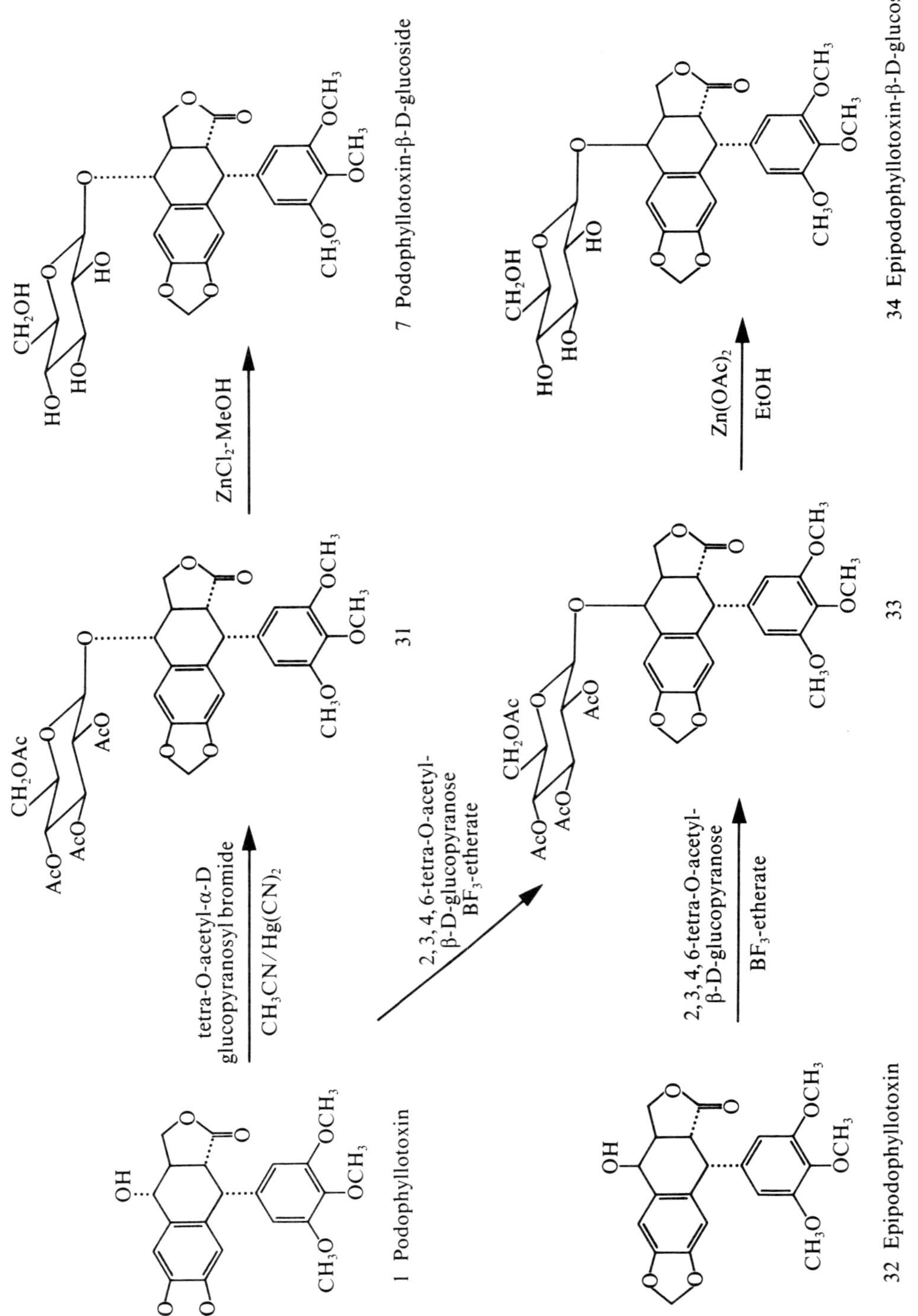

Synthesis of 4'-demethylepipodophyllotoxin-β-D-glucopyranoside

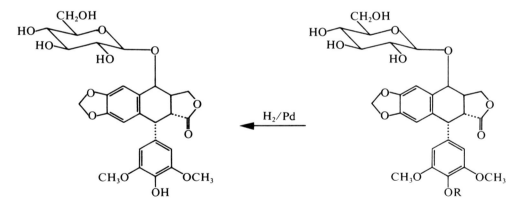

35 R = COOCH$_2$C$_6$H$_5$, C-1 = αOH
36 R = COOCH$_2$C$_6$H$_5$, C-1 = βOH

38 4'-Demethylepipodophyllotoxin-
    β-D-glucopyranoside

tives, respecting their acid and base sensitivity. This could be accomplished by methanolysis of tetra-O-acetyl-PG (**31**) in the presence of ZnCl$_2$ to yield the free glucoside (**7**), leading to the first synthesis of podophyllotoxin-β-D-glucopyranoside [99]. In an analogous way,

EPG (**34**) was obtained by transesterification from the parent acetate (**33**) [100]. In the case of the acetyl derivative (**37**), the protective groups were successively removed by zinc acetate-catalyzed methanolysis, followed by hydrogenolysis, to furnish the required 4'-demethyl-epipodophyllotoxin-β-D-glucopyranoside (**38**) [28].

13.2 Condensation products of 4'-demethylepipodophyllotoxin glucoside with carbonyl compounds

The availability of 4'-demethylepipodophyllotoxin glucoside (**38**) (see previous section) permitted the preparation of numerous derivatives. Based on previous experience (see history section), condensation reactions with aldehydes were of special interest. Condensation of DEPG with carbonyl compounds was generally achieved by reaction of the glucoside with carbonyl compounds in the presence of acids or Lewis acids [101]. Among the two stereoisomers formed, differing in the configuration at the newly introduced asymmetric carbon atom, the one with an equatorial bond of the aldehyde residue predominates almost exclusively.

39 (a–r)

Condensation products of 4'-demethylepipodophyllotoxin-β-D-glucopyranoside with aldehydes

Table 13
Cytostatic activity of 4'-demethylepipodophyllotoxin glucoside (DEPG) and some of its aldehyde condensation products

compound No.		ID-50 P-815 μg/ml	L-1210 % incr. life span
38	DEPG	4	34
	R		
39a	H	0.06	56
b	CH_3 (etoposide)	0.05	167
c	C_2H_5	0.009	97
d	C_5H_{11}	0.009	65
e	CH_3COCH_2	0.1	109
f	$(CH_3)_2CHCOCH_2$	0.07	127
g	$C_2H_5SCH_2$	0.02	104
h	$(CH_3)NCH_2CH_2$	2	317
i	C_6H_5	0.007	97
k	$oMeC_6H_4$	0.009	61
l	$oMeOC_6H_4$	0.01	46
m	$oClC_6H_4$	0.004	24
n	$mO_2NC_6H_4$	0.04	52
o	1-naphthyl	0.01	95
p	2-tiophene (teniposide)	0.005	121
q	O=	0.6	165
r	H_3C—N	2	261

More than 60 condensation products of DEPG with aldehydes were prepared and tested for cytostatic activity *in vitro* and *in vivo* [101]. Some of them (**39 a–r**) are listed in Table 13.

DEPG itself is, analogous to other P glucosides, *in vitro* of low cytostatic activity and produces only a moderate increase in life span in leukemic mice. As to its mechanism of action, it is a pure or almost pure spindle poison, and does not inhibit entry of cells into mitosis, since in fibroblast cultures it produces a high relative mitotic index [46].

Condensation of aldehydes to DEPT glucoside causes a large increase of cytostatic potency, larger than in the P, the EP or the DPT glucoside series when comparing the benzylidene derivatives: condensation of DEPG with benzaldehyde makes the former 600x more potent as an inhibitor of cell proliferation in P-815 mastocytoma cultures. At the same time, it completely changes the mechanism of action and greatly enhances its oncostatic effect in mouse leukemia

L-1210. At about the time when these compounds were under evaluation, Goldin et al. [48] compared different anticancer screening procedures as to their predictability for the treatment of human malignancies, and found leukemia L-1210 to be the most valuable of the models available at that time. Our selection of P derivatives for more detailed evaluation therefore relied heavily on results with this murine tumor.

The benzylidene derivative (**39 i**), having been the first of the DEPG derivatives obtained and tested (see section 2), was investigated fairly extensively for its cytostatic and other effects. Parallel to the assessment of cytostatic potency in mastocytoma cultures (see Table 13), compounds were tested in chick embryo fibroblast cultures in the first stage of screening in order to evaluate their type of effect on dividing as well as on resting cells. In this system, after 6 h of incubation with the drug, there was, instead of the usual increase of the number of cells in mitosis, a complete disappearance of all mitotic phases. This made it evident from the beginning that we were dealing with another mechanism of action than that of spindle poisons. Another difference to the spindle poisons became apparent when DEPT benzylidene glucoside was tested for acute toxicity in mice; while after i. v. or i. p. injection of spindle poisons of the podophyllotoxin type the animals usually succumb within hours, with the new compound they do so only after 3–4 days or later. This also suggested a new mechanism and an increased persistence of some effect(s). Since the biological activity of DEPBG is, in most respects, very similar to that of the thenylidene derivative (teniposide) (**39 p**), discussed extensively below, its pharmacological effects will not be dealt with here in more detail. In contrast to teniposide, the benzylidene derivative shows some activity also upon oral administration in animals. The results given in Table 13 (and Table 19) for etoposide's and teniposide's anti-L-1210 activity are mean values of early results with daily parenteral application of the drug. It was soon found that other treatment schedules are more effective [41].

Some structure activity relationships become evident on inspection of Table 13. They will not be discussed in detail here. Table 13 also shows that, in this series, there is hardly any correlation between cytostatic potency *in vitro* and antitumor effect in L-1210. Of special interest are compounds (**39 h**) and (**39 r**): they are water soluble, have a low activity *in vitro* and prolong the survival time of leukemic mice dramatically. In our hands, the methylpiperidylidene derivative (**39 r**) cured a large percentage of mice inoculated s. c. with one million L-1210 cells when optimal i. v. treatment schedules were used. This compound was also tested by the US Cancer Chemotherapy National Service Center (accession No. B 116351) and proved active against B 16 melanoma in mice. Kreis and Soricelli [102] compared it to etoposide in a number of tumor models and found VP-16, in general, somewhat superior. A major drawback of this hydrophilic derivative is the fact that it is not active upon oral administration. Recently, other etoposide analogues with increased water solubility and good activity in animal tumors have been described [103].

Results with some aldehyde condensation products of DEPG not listed in Table 13 have been published [101]. One of them is the furfurylidene derivative which has recently been described as differing from VM regarding susceptibility of resistant cells and regarding DNA damage [104]. Keller et al. [101] have also synthesized and tested condensation products of DEPG with ketones and aldehyde condensation products of DEPT-β-D-galactoside. The latter free glycoside has a low *in vitro* activity (ID-50 in P-815 cultures is 4μg/ml) and shows no significant effect in leukemia L-1210. Its aldehyde condensation products are more active *in vitro* but of very modest anti-L-1210 effect [101]. Obviously, the stereochemistry of glucoside derivatives is more favorable for this anti-tumor activity than that of galactoside analogues.

14 Miscellaneous lignan compounds
14.1 Podophyllotoxin carbamates

Among the esters of P with carbamic acid, the unsubstituted carbamate (**40 a**) is one of the most active *in vitro* (Table 14) and was, because of its favorable ratio between cytostatic potency *in vitro* and toxicity in animals, and for other reasons, investigated more thoroughly.

Similarly to SP–I, another nitrogen-containing derivative (see section 8), it shows good tumor inhibition in the mouse sarcomas 37 and 180 (superior to that of SP–I), and no significant effect in mouse leukemia L-1210. Inhibition of tumor growth is also considerable in the two solid rat tumors tested, Walker carcinosarcoma and Yoshida sarcoma (68 and 67 %, resp.), and proliferation of tumor cells is strongly

40 (a–f) Podophyllotoxin-carbamate and related compounds

Table 14
Cytostatic activity of podophyllotoxin carbamate and related compounds

compound No.	R	R'		ID-50 P-815 µg/ml
40a	H	CH_3	P-carbamate	0.02
b	CH_3	CH_3	P-methylcarbamate	0.06
c	C_6H_5	CH_3	P-phenylcarbamate	0.1
d	$CONH_2$	CH_3	P-allophanate	0.02
e	H	H	DP-carbamate	0.06
f	C_2H_5	$CONHC_2H_5$	DP-1,4'-bis-N-ethylcarbamate	0.5

reduced in the Brown-Pearce carcinoma, implanted in the testis of rabbits. Immunosuppressive activity in mice was found to be weak [21]. Another interesting feature of PT carbamate is its high binding to human serum proteins; at "therapeutic" concentrations, >98 % of the drug present in human (or rabbit) serum is bound to albumin, as was ascertained by dialysis experiments. Assays in fibroblast cultures with and without serum showed a lesser affinity for this drug in the serum proteins of other species (mouse, rat, cow, horse) compared to that of man. With this highly protein-bound compound it could be shown that the amount of albumin taken up by tumor cells by way of pinocytosis may bring enough protein-bound drug into a cell ("piggyback pinocytosis") to inhibit proliferation [105]. The question whether protein binding of anticancer drugs is a desirable characteristic has not been resolved since the differential uptake of these proteins by tumor cells vs. normal cells would have to be known. P carbamate was found to be superior to podophyllinic acid hydrazide (SP-I) in experimental tumors and did not show any unexpected toxic effects in rats or monkeys, and was, in

addition, chemically more stable than the latter compound; therefore, in 1967, it was tested in man under the clinical code VB-74. The results in a limited number of patients were, however, not encouraging enough and clinical tests were not continued; furthermore, another anticancer drug of the podophyllum series (VM-26, see below) was being developed at that time and it seemed more promising.
Of the esters of carbamic acid of P, some are listed in Table 14. A few of them, e. g. methyl carbamate (**40 b**), allophanate (**40 d**) and phenyl carbamate (**40 c**) as well as the DP carbamate (**40 e**), were assayed *in vivo;* they inhibit the growth of mouse sarcomas less than the P carbamate, do not prolong survival time of leukemic mice and were considered less promising than P carbamate.
Similar derivatives were also synthesized in the peltatin series. Their *in vitro* activity as to inhibition of proliferation of P-815 mastocytoma cells was quite high (β-peltatin carbamate, α- and β-peltatin methyl carbamates showed ID-50's of 0.002, 0.002 and 0.001 μg/ml, resp.), but none of those tested was significantly active *in vivo* (slight or no activity in sarcoma 37, no activity in L-1210). An example of the large influence of "minor" structural modifications is the finding that β-peltatin ethyl carbamate is 70x more active than the respective diethyl carbamate (ID-50's are 0.001 and 0.07, resp.). Results of the peltatin carbamates are not given in a table.

14.2   Desoxypodophyllotoxin and derivatives

The medical use of plants – wild chervil, *Anthriscus* – containing 1-desoxy-P (**4**) dates back to at least the 10th century (see [106]). This compound was, however, isolated only in 1940 and later from different plants, among them *Podophyllum peltatum;* Hartwell et al. then discovered a compound in *Juniperus silicicola* which they named silicicolin and found to be identical to desoxypodophyllotoxin (see [12]). These authors report activity of the compound against sarcoma 37.
Desoxy-P (silicicolin) (**4**) and some derivatives were also prepared in our company and tested for cytostatic activity. Potency of desoxy-P *in vitro* is slightly higher than that of P itself, and its acute toxicity and oncostatic activity in mice are also of the same order of magnitude as those of P (see Table 1). Silicicolin exhibits a strong binding to human serum proteins.

4 Desoxypodophyllotoxin

41 (a–h) Isomers and derivatives

Table 15
Cytostatic activity of desoxypodophyllotoxin (desoxy-P, silicicolin) and derivatives

compound No.	R	R'	R'	configuration $H_{C-2}:H_{C-3}$	$H_{C-3}:H_{C-4}$		ID-50 P-815 µg/ml
4	OCH_3	OCH_3	OCH_3	trans	cis	desoxy-P	0.002
41a	OCH_3	OCH_3	OCH_3	cis	trans	desoxypicro-P	0.03
b	OCH_3	OCH_3	OCH_3	trans	trans	(+)-isodesoxy-P	0.02
c	OCH_3	OCH_3	OCH_3	trans	trans	(−)-isodesoxy-P	>10
d	OCH_3	OCH_3	OCH_3	cis	cis	isodesoxy-picro-P	0.05
e	OCH_3	OH	OCH_3	trans	cis	4'-demethyl-desoxy-P	0.005
f	OCH_3	OHC_3	H	trans	trans	5'-desmethoxy-isodesoxy-P (DL)	0.4
g	H	OCH_3	H	trans	trans	3',5'-desmethoxy-isodesoxy-P (DL)	3
h	H	H	H	trans	trans	3',4',5'-desmethoxy-isodesoxy-P (DL)	0.3

The cytostatic activity *in vitro* of some stereoisomers of desoxy-P are given in Table 15. The picro form (**41 a**) of desoxy-P is roughly 10x less potent than desoxy-P, and so is (+)-isodesoxy-P (**41 b**), while its (−)-isomer (**41 c**) is inactive (and extremely hydrophobic); inactivity of (−)-isodesoxy-P was confirmed by Zavala et al. [107]. Hartwell and Schrecker [12] found no oncostatic effect of these compounds.

Some variations regarding substitution on ring C of desoxy- and isodesoxy-P were investigated as well. 4'-demethyl-desoxy-P (**41 e**) is almost as potent as desoxy-P, while the removal of 1, 2 or all 3 methyl groups from isodesoxy-P (**41 e–h**) results in a considerable decrease of activity (Table 15).

Connecting two desoxy-P molecules in position 1 results in bis(desoxy-PT) which retains, rather surprisingly, a considerable activity, higher than that of, e. g., PG.

When the hydroxy group in position 1 of P is oxidized to the ketone, the resulting podophyllotoxone has a low cytostatic potency *in vitro* (ID–50) around 1 µg/ml), and the corresponding picro analogue, picropodophyllotoxone, is practically inactive (results not shown).

14.3   Desoxypodophyllinic acid derivatives

1-Desoxy-P (silicicolin) (4) derivatives with an open lactone ring were prepared mainly by substitution in the alcoholic group in position 2 of the silicicolinic acid methyl ester because the free acid is not accessible.

This ester itself (24) has a considerable *in vitro* potency regarding inhibition of cell proliferation (though lower than that of the corresponding podophyllinic acid derivative, see Table 10) and a low toxicity, but *in vivo* it does not inhibit tumor growth; on the contrary, the weight of the tumors (sarcoma 37) of treated mice was, in several experiments, higher than that of controls [21]. Another remarkable ob-

24 Desoxypodophyllinic acid methylester

42 (a–e) Derivatives

Table 16
Cytostatic activity of desoxypodophyllinic acid derivates in P-815 mastocytoma cell cultures

compound No.	R	R'	R"	ID-50 P-815 µg/ml
24	CH_3	H	CH_3	0.1
42a	CH_3	CH_3CO	CH_3	0.6
b	CH_3	H_2NCO	CH_3	1.5
c	CH_3	C_6H_5NHCO	CH_3	0.6
d	C_2H_5	H	CH_3	1
e	CH_3	H	H	0.2

servation with this compound was that mice treated i. v. with high but non-lethal doses remained in a paralyzed state resembling general anesthesia for about half an hour. This and some other derivatives (**42 a–e**) are listed in Table 16. Prolongation of the alkyl side chain in position 3 does, in contrast to the corresponding hydrazide derivatives, not increase potency, as a comparison between silicicolinic acid methyl and ethyl esters shows.

14.4 Apopicropodophyllotoxins

Unsaturated podophyllum compounds, without oxygen in position 1 and with a double bond in ring B have been known since 1933 (see [12]). Some derivatives of this "apo" type were prepared by J. Rutschmann and E. Schreier. The cytostatic potency of some of them is given in Table 17.

43 α-apopikropodophyllotoxin
(α-apopicropodophyllin)

44 β-apopicropodophyllotoxin
(β-apopicropodophyllin)

Table 17
Cytostatic activity of apopicropodophyllotoxins

compound No.		ID-50 P-815 µg/ml
43	(−)-α-apopicropodophyllotoxin	0.005
44	(+)-β-apopicropodophyllotoxin	0.002
	(−)-β-apopicropodophyllotoxin	2
	(+)-4'-demethyl-β-apopicropodophyllotoxin	0.005
	(+)-3',4'-demethyl-β-apopicropodophyllotoxin	0.5
	5'-desmethoxy-β-apopicropodophyllotoxin (DL)	0.1
	3',5'-desmethoxy-β-apopicropodophyllotoxin (DL)	2
	3',4',5'-desmethoxy-β-apopicropodophyllotoxin (DL)	0.04

(−)-α-Apopicropodophyllotoxin (**43**) is about as potent as P and (+)-β-apopicropodophyllotoxin (**44**) as active as silicicolin, while the antipode, the (−)-isomer, has almost no effect. Others also found β-apopicro-P to be very potent *in vivo*, about equal to P [108], or even more effective than P *in vitro* [109]. The 4'-demethyl derivative of the (+)-isomer was also tested *in vivo* and turned out to inhibit growth of sarcomas 37 and 180 by 30–40 % but to produce no significant increase in life span of leukemic mice. A second demethylation at ring C results in a still more hydrophilic, but considerably less active compound. It is interesting to note that removal of all 3 methoxy groups of ring C reduces the potency of β-apopicropodophyllotoxin less than the loss of 1 methoxy group and still less than removal of the 2 methoxy groups at C-3' and C-5', and that this order is not the same in desmethoxy derivatives of isodesoxypodophyllotoxin (compare Table 15). The completely desmethoxylated derivative produced, in a single experiment in sarcoma 37, a very good inhibition of tumor growth but was not investigated further because of its extreme hydrophobicity. (The respective desmethoxy derivatives of picropodophyllotoxin and of isopodophyllotoxin, synthesized by E. Schreier, were still less active, but in both cases the 3', 4', 5'–desmethoxy compound was more potent than the mono-desmethoxy and bis-desmethoxy derivative; results not given).

14.5   Podophyllol and congeners

Another series of podophyllotoxin derivatives with an open or otherwise modified lactone ring are podophyllol (**46**) and its congeners.

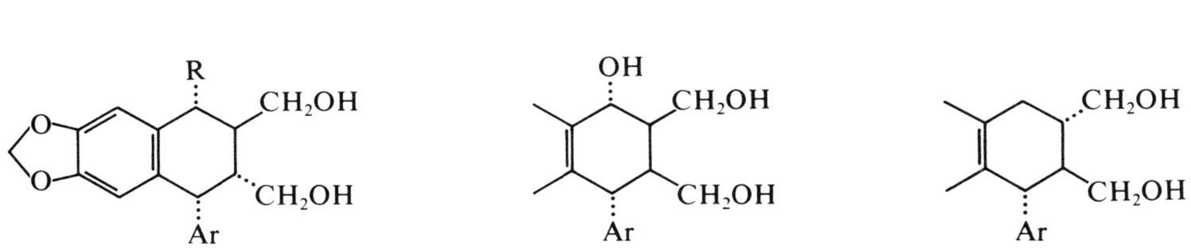

45 R = H: 1-Desoxypodo-
       phyllol

46 R = OH: Podophyllol

47 Picropodophyllol

48 Desoxyiso-
   podophyllol

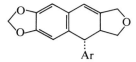

49 Anhydropodophyllol
50 Anhydrodesoxy-podophyllol
51 Bisanhydropicro-podophyllol

Ar = 3, 4, 5-trimethoxyphenyl

Table 18
Cytostatic potency of podophyllol and congeners in fibroblast (Fbb) and P-815 mastocytoma cultures

compound No.		Fbb ID-50 µg/ml	ID-50 P-815 µg/ml
45	1-desoxypodophyllol	0.3	0.4
46	podophyllol	0.2	0.3
47	picropodophyllol	6	10
48	desoxyisopodophyllol	0.4	0.4
49	anhydropodophyllol	10	>10
50	anhydro-1-desoxypodophyllol	0.01	0.02
51	bisanhydropicropodophyllol	0.7	0.2

Some of them are quite active regarding cytostatic effect *in vitro*, particularly the anhydro-1-desoxypodophyllol (**50**) (Table 18). This latter compound and 1-desoxypodophyllol (**45**) were also tested in transplantable mouse tumors (sarcoma 37, leukemia L-1210), but found to be nearly inactive in tolerated doses. In this group of derivatives, the difference in potency between the 2,3-trans and the 2,3-cis compounds is considerable.

Analogous derivatives with alcoholic groups were prepared in the peltatin series as well; cytostatic potency *in vitro* of α- and β-peltatinol is of the same order of magnitude as that of podophyllol.

15   Teniposide and etoposide

Among the aldehyde condensation products of 4'-demethylepipodophyllotoxin glucoside, the thenylidene derivative was the first to be selected for in depth evaluation in view of a possible clinical testing.

This decision was based, among other things, on its high cytostatic potency *in vitro* and its antitumor effect in mous leukemia L-1210 (the importance which was attributed at that time to this tumor model is also reflected in the fact that anti-L-1210 activity was specifically mentioned in the title of the first publication on this compound [110]). Main criteria for the selection, in 1969, of the other derivative for clinical testing, DEP ethylidene glucoside, were its superior antitumor effect, particularly in leukemia L-1210, its chemical stability and (probably connected to the latter property) good activity upon oral administration. The ethylidene derivative was chosen for clinical development at a moment when the thenylidene derivative had already been in clinical testing for some time and the desirability of an oral preparation had risen. Since both derivatives are qualitatively quite similar, they will be dealt with together in the following sub-sections.

15.1   Cytostatic and oncostatic effects in preclinical systems

As already mentioned earlier, the most prominent feature of the aldehyde condensation products of DEP glucoside, besides good anti-L-1210 activity, is that they inhibit entry of cells into mitosis and therefore reduce the mitotic index of a proliferating cell population, while all previously known podophyllum compounds arrest cells in metaphase and lead to an increased number of cells in (abnormal) mitosis. This characteristic has been emphasized by using the expression "new mechanism" in the title if the first full paper on 4'-demethyl-1-0-[4,6-0-(2-thenylidene)-β-D-glucopyranosyl] epipodophyllotoxin (VM 26, teniposide) [41]. The inhibitory effect of VM and VP, 4'-demethyl-1-0-[4,6-0-(ethylidene)-β-D-glucopyranosyl] epipodophyllotoxin (etoposide, VP-16), on the beginning of cell division – mitotic – events becomes microscopically visible in chick embryo fibroblast cultures within 1–2 h after addition of the drug, with high concentrations even somewhat earlier [21]. From this it could be concluded that one of the main points of attack of VM and VP has its position in the cell cycle shortly before entry into mitosis, i. e. in G2 phase (see also section on mechanism of action). At sufficiently high drug concentrations, cells leaving the G0/G1 phase of the cell cycle begin to disintegrate [41]. Quantitatively, VM is roughly 10x more potent than VP regarding inhibition of proliferation of P-815 mouse mastocytoma cells, ID-50 being 0.005 μg/ml (approx. 8 nM) for VM and 0.05 μg/ml (approx. 80 nM) for VP (see Table 19 and ref. [41,111]).

VP and, particularly, VM are quite hydrophobic; at 37 °C, about 4 μg VM and 120 μg VP can, without solvent, be dissolved in 1 ml distilled water [21]. One problem in animal tests was, therefore, the preparation of clear aqueous solutions for i. v. and i. p. injection (results of experiments in which drugs are injected i. p. or s. c. in non-aqueous solutions or in aqueous suspensions can, of course, not be interpreted as to effective dose without blood level determinations). This problem could be satisfactorily managed by dissolving the drugs first in ethanol or, preferably, dimethylsulfoxide (DMSO), then adding Tween® 80 (*polysorbate 80)* or Cremophor EL® *(polyethoxylated castor oil)* and then water or a balanced salt solution [41,111].

The difference in survival time of animals injected with a single lethal dose, between spindle poisons and DEPT glucoside acetals, including VM and VP, has already been mentioned (see section 16). The mechanism of lethal toxicity is, therefore, also different. It can be assumed that acute lethal effects of podophyllotoxin and related spindle poisons are due to an action on the central nervous system (which is rich in microtubules); in humans too, central effects have been described after podophyllin poisoning [112]. With VM it was found that a lethal dose produces agranulocytosis and that the blood of moribund mice contains gram-negative bacteria, in one case tentatively identified as *Pseudomonas* [21]. In subacute toxicity tests in mice, LD-50 increases with increasing intervals between drug administration when the compound is given three times i. v., with VM from about 5 mg/kg (per injection) upon daily administration to about 25 mg/kg when administered at 10-day intervals; further spacing does not additionally elevate LD-50 [21]. This may be a parameter for estimating the duration of effects and their reversibility in the whole animal.

Our own results with VM and VP in experimental tumor systems can be summarized as follows (see [41,111]). Both compounds show significant oncostatic (i. e. *in vivo* tumor inhibiting) activity in Ehrlich ascites, sarcoma 37, sarcoma 180, leukemias L-1210 and P-1534, mastocytoma P-815 (mouse tumors) and in the Walker carcinosarcoma of the rat. VP was also found active in the intratesticularly inoculated Brown-Pearce carcinoma of the rabbit and in inhibiting the growth of rat mammary tumors elicited by dimethylbenzanthracene treatment [21]. In most models, results with VP are superior to those with VM. Although VM inhibits tumor growth when given orally, the results are variable and the doses required for significant activity 10–50 times higher than when the drug is injected i. v. [21].

In leukemia L-1210, which was investigated quite extensively, increase in life span is high – higher than with almost all other cancer chemotherapy drugs – and, depending on the experimental conditions, some animals are cured. Cures attain nearly 100 % when 10^5 tumor cells or less are inoculated intraperitoneally and drug administered (starting one day later) by the same route; in this way the drug reaches most of the tumor cells in a higher concentration. This latter experimental setup was mostly used by the Cancer Chemotherapy National Service Center of the US National Cancer Institute which, therefore, obtained even better results in L-1210 than we ourselves; we had predominantly tested with subcutaneous (in the hind leg) tumor inoculation, the drug being given i. p. or – closer to the clinical situation – i. v. While the difference in antitumor effect was minor whether drug was given i. v. or i. p., there was a substantial difference when the schedule of drug administration was changed from daily to twice weekly, the latter producing longer survival times and more cures. With VM, the optimal schedule as to cures in L-1210 was drug application every fourth day when i. v. treatment was started one day after s. c. inoculation of one million tumor cells [21]. A strong schedule dependency of the effects of VM in L-1210 has also been found by others [113].

While VM has been investigated less extensively in experimental systems by others, VP was tested by many authors and in a large number of animal tumors. Results in leukemia L-1210 were similar to those obtained in our laboratories, and additional tumors were found to respond to treatment with VP, e. g. the B 16 melanoma, P-388 leukemia, Lewis lung carcinoma and others. Achterrath et al. have summarized preclinical results with VP, including inhibition of experimental tumors [114, 115]. Rose and Bradner [116] have performed extensive studies with VP and also give references to earlier work. These authors investigated some combinations as well and obtained especially good results when leukemia P-388 was treated with a combination of VP and mitomycin C. We studied combinations of VM and VP with cyclophosphamide and other anticancer drugs in leukemia L-1210. When non-toxic, equiactive single doses of VM and cyclophosphamide were combined, increase on life span was highest (and superadditive) when both drugs were given i. v. simultaneously; injecting both drugs at intervals of up to 4 days was less beneficial [21]. Synergism with alkylating agents has also been found by others, e. g. Dombernowsky and Nissen [117]. (It should be mentioned here that a really complete evaluation of a possible, optimal synergism is almost not feasible since it would necessitate the testing of so many variables as to (priming and subsequent) doses, schedules and time intervals that many hundreds of animals would have to be used for only two drugs in a single experimental tumor). In a minority of cases, VM showed better antitumor activity than VP, e. g. in Lewis lung carcinoma of the mouse [118] and in intracerebrally inoculated L-1210 [21,41]. Most of the earlier reports on preclinical activity of VM and VP have been reviewed in a comparative analysis of these two drugs [119].

15.2 Immunosuppression

Since our interests were not confined to cancer chemotherapy, many of our compounds, including podophyllotoxin derivatives, were also examined for immunosuppressive activity [54, 120, 121]. Therefore, VM and VP were investigated regarding inhibition of the immune response at an early stage.

VM treatment reduces the hemagglutinin titer of mice immunized with sheep red cells significantly. VP has a somewhat higher efficacy than VM in this respect. Both derivatives decrease the hemagglutinin titer by a factor of up to >100 in mice, depending on the treatment schedule, and less so in monkeys; they are able to prevent the paralytic symptoms of experimental allergic encephalomyelitis in rats (an autoimmunity model for multiple sclerosis) and VP treatment increases the survival time of homologous skin transplants in mice [122]. Morita et al. [123] also found a depressing effect of VP on hemolytic plaque-forming cells in the spleens of immunized mice. Both compounds also inhibit the symptoms of Freund's adjuvant arthritis in rats, a model for rheumatoid arthritis and susceptible to immunosuppressive effects; VP is very effective in this respect, a treatment of 3 mg/kg/day s. c., given during the latency period, produces a 97% inhibition of the secondary joint swellings [124]. VP has also been used for suppression of the host vs. graft reaction in allogeneic bone marrow transplantation in man [125].

15.3 Activity against microorganisms

Both epipodophyllotoxin derivatives exhibit some antimicrobial effects, though only at concentrations which are much higher than those inhibiting proliferation of mammalian cells. No activity was found against gram-negative bacteria, while some gram-positive germs were affected; *Sarcina lutea, Micrococcus pyogenes, Streptococcus haemolyticus* and *Bacillus subtilis* were inhibited. Minimal inhibitory concentrations (MIC) of VM 26 were 3–12 μg/ml, staphylococci *(M. pyogenes)* being the most sensitive [126]; slightly higher concentrations of VP prevented the growth of streptococci (MIC was 8 μg/ml for *Str. aronson*) and staphylococci (MIC 16 μg/ml) [122]. Quite recently, Calame et al. obtained similar results [127]; they established a bactericidal effect of VP and concluded that this compound should have some effect on gram-positive bacterial infections in patients.

In experiments of a rather preliminary nature, we tested the antiviral effects of VM and VP. The results indicate that both compounds inhibit to some extent plaque formation by vaccinia and herpes simplex virus in cultured cells. Nishiyama et al. [128] found a definite inhibitory effect of VP on plaque formation by herpes simplex virus and attribute this effect to impairment of the function of host cell topoisomerase II. *In vivo,* antiviral activity (survival time increase of mice infected with vaccinia or influenza virus) was absent or negligible [21]. Interestingly, Markkanen et al. [129] found a considerable anti-herpes simplex virus effect *in vitro* of P and some congeners, while VP was active only at very high concentrations.

15.4 Cytotoxicity, reversibility of cellular effects

In fibroblast cultures it has been established that cells which are arrested in metaphase by a podophyllum preparation of the spindle poison type (SP-G) cannot resume progression in the cell cycle when the drug is washed out; the arrested cells finally die; but cells which did not enter mitosis while the drug was present can go through a normal mitosis after removal of the drug [24]. Similar results were obtained with podophyllinic acid ethyl hydrazide (SP-I) [130].

After treatment of chick embryo fibroblasts in tissue culture with the DEPT glucoside derivatives VM and VP, the course of events is different. At higher drug concentrations, after a very short period of mitotic inhibition in metaphase, some cells (presumably those in the process of entering mitosis) begin to disintegrate and no cell divisions can be seen any more. The number of decomposing cells increases steadily, leaving behind droplets consisting, among other things, of chromatin. Cells in interphase look microscopically unaffected (see Fig. 5 in ref. [46]). At lower but still effective drug concentrations one can see a disappearance of mitotic figures in proliferating fibroblast cultures, with no obvious microscopic alterations of the cells. When the cultures are washed free of drug after one or several hours, there is, in contrast to the spindle poisons, no reappearance of normal mitoses; the course of events is the same as without washing [41] (it has to be assumed, however, that there is a certain degree of reversibility within a narrow concentration range at the lower limit of activity).

The persistence (or, vice versa, the reversibility) of inhibition of cell proliferation upon drug removal after treatment with a cytostatic compound for a certain period is of considerable importance for understanding the activity of the compound at the level of the whole organism, and also for the treatment schedule. This aspect was therefore investigated more in detail with VM and VP, using P-815 mouse mastocytoma and L-1210 leukemia cell cultures. Both compounds and both cell strains gave similar results, equiactive concentrations of VP being, of course, higher than those of VM. Proliferating cells were treated for periods of between 20 min and 24 h with different drug concentrations, then washed free of the drug and the proportion of surviving cells (i.e. those capable of further multiplication) determined either by colony formation in agar or by a dilution assay (the latter based on the finding that about 63 % of untreated cultures which, in the average, contain only one healthy P-815 cell, give rise to macroscopically visible growth, as is to be expected according to the Poisson distribution if all single cells can proliferate).

These assays showed, e.g., that, after treatment of non-synchronized P-815 cells with VP for one hour, the surviving fraction was 0.72 at 1 μg/ml, 0.05 at 3.2 μg/ml and about 0.006 at 10 μg/ml. With an exposure time of only 20 min, the surviving fraction is about 0.03 when VP is used at a concentration of 10 μg/ml. If such a short exposure time is able to kill more than 90 % of the cells, it must be concluded that either the drug acts in all phases of the cell cycle or it cannot be removed completely from a sensitive structure in the cell by washing and dilution in fresh medium. The relationship between surviving fraction, drug concentration and time of exposure can be expressed in a formula which is, however, adequate for only part of the time-concentration range [131,132].

The time course of death of the treated cells has not been followed in detail, but it has been found that 8 h after a 1-h treatment of a P-815 culture with a VP concentration which prevents > 99 % of the cells from further multiplication, the dye exclusion test is positive for less than 10 % of the cells [131]. This indicates that this test – which is often used to evaluate the viability of cells – is not able to identify those cells which cannot proliferate any more and are destined to disintegrate soon.

Cells of a murine fibroblast line (Earle's L cells), when exposed to VP for 1 h and then transferred to fresh, drug-free medium, become very large and multinucleated within the next 10 days, but they do not multiply any more [21].

Killing of tumor cells by VP was also assayed *in vivo*. Mice were inoculated with L-1210 cells and 6 days later treated once i.v. with 50 mg/kg of VP; one day later, the number of living L-1210 cells was determined in different organs by an *in vivo* bioassay [133]. In comparison to inoculated but untreated animals, the surviving fraction of tumor cells in the treated mice was approximately 0.1 in the lung, 0.06 in the liver, 0.1 in the brain and 0.002 in the spleen [21]. The high degree of elimination of tumor cells in the spleen is apparently not due to a pharmacokinetic feature of the drug since VP concentrations in the spleen are, at least in rats, lower than e.g. in the liver. Ovalicin [120], a drug known to decrease the mitotic count in spleens of immunized animals but inactive in leukemia L-1210, was used as control compound; it did not reduce the number of L-1210 cells significantly, not even in the spleen and despite repeated high doses [21].

15.5   Toxicology

While in cell cultures VP is about 10x less potent than VM regarding cytostatic activity in mouse tumor cells, in acute toxicity tests (single application) in mice the thenylidene derivative is only about 3–4 times more toxic than the ethylidene derivative [111].

Equitoxic total doses of VP for a 2-week period are smaller when the drug is given daily than when it is given to mice i.v. twice weekly [111], in accordance with the higher therapeutic effect of the latter treatment schedule.

Subchronic (4 weeks) and chronic (26 weeks) toxicity tests were performed in our toxicology department (J. Grauwiler, R. Griffith, G. Rüttimann and H. Weidmann) using rats, dogs and monkeys. The results were similar for VM and VP and have been summarized for VP by Achterrath et al. [114]. They are unremarkable in so far as the alterations found were to be expected with cytostatic drugs: leukopenia, anemia, thrombocytopenia (not always present), bone marrow depression, thymus involution, colitis, hair loss etc. No clear-cut organ toxicity, unrelated to the cytostatic effect, was found, except for a severe fibrous peritonitis and some liver changes which develop during the "recovery" period in rats after a 4-week intraperitoneal treatment with VM [126,134]. The real significance of this finding was realized only later (see section on delayed toxicity).

Given to pregnant rats, VP elicits teratogenic and other embryotoxic effects (see [114]). When the compound is administered to rats or mice at a suitable period of the pregnancy, it is possible to kill all offspring with a dose which does not seriously harm the dams [21, 135]; in mice, the same can be achieved with podophyllotoxin [21, 136]. Mutagenicity was not found in microbial systems (Ames test) [137, 138], but Sieber et al. [139] and others [138] observed mutagenic effects of VM and VP in mammalian systems. DeMarini et al. [140] showed VM to be mutagenic in mouse lymphoma cells at extremely low concentrations and related this activity to effects on topoisomerase (see section 16.4).

15.6   Delayed toxicity

Avery et al. [141], when observing mice which had been cured from leukemia L5178Y by VM treatment, found that, depending on the dose, these mice began to die from day 50 on and did so up to about 9 months after treatment. These deaths were not due to leukemic relapse and occurred also in VM-treated animals not inoculated with tumor cells. This finding was only made in animals treated by the i.p., not by the s.c. and i.m., route. Chemical peritonitis and ascites formation were found in the affected mice. In a subsequent paper, more extended studies with i.p. administration of VM were reported by the same group [142]. These authors found liver changes in the animals dying of this delayed toxicity and suggested a specific liver damaging effect of VM, despite the fact that they had not observed such changes in animals treated by the s.c. and i.m. route. We extended these studies by including the i.v., the oral and the intrapleural administration, by using rats besides mice and by treating the animals with VP as well [134].

We could confirm the results of the group in Memphis with VP and found a fairly clear time separation of "early" deaths – due to cytostatic effects, mainly on the blood-forming tissue – and "late" deaths, due to chronic peritonitis. VP, upon repeated i.p. injection, also elicits delayed toxicity in mice and rats, and does so also after (single or repeated) intrapleural administration in rats. No delayed toxicity was found after i.v. injection of VM and VP nor after oral application of VP. Thus, it had to be concluded that delayed toxicity is due to a local effect of these drugs, causing chronic inflammation, and that the hepatotoxicity found by Hacker and Roberts was secondary to this inflammation and not a direct drug effect. However, the question arose whether local application (i.p., intrapleural, intrathecal, into joints etc.) of VM or VP in man would also lead to similar consequences. To our knowledge, no such clinical observations have been reported.

We ourselves did not observe the delayed toxicity in our animals cured of leukemia by i.p. treatment with VM or VP because we usually reinoculated s.c. the survivors with leukemic cells in order to evaluate a possible immune response to the tumor cells. Most or all of them died of leukemia with tumor formation at the site of inoculation; a few may have died from delayed toxicity, but this was not recognized, and death was attributed to leukemia.

15.7 Resistance

Tissue culture cell lines often become less susceptible to the effects of drugs when they are treated with the respective compound at nonlethal concentrations for a prolonged period of time. This was tested for VM with murine P-815 mastocytoma cells. Cells were treated continuously with sublethal drug concentrations and their sensitivity to the cytostatic effect of VM was assessed periodically by determining the 50% inhibitory concentration (ID-50). The resistance factor (i.e. the ratio of ID-50 at time t to ID-50 at time zero) reached 2 after 4 weeks of VM treatment and about 6 after 4 months; there it stayed despite continuing drug exposure to the culture for another 6 weeks (in the same experiment, podophyllotoxin carbamate was tested and the P-815 cells were found to develop a resistance of only marginal degree) [143]. Staphylococci, on the other hand, develop a higher resistance to VM and do so more quickly; their resistance factor, based on MIC determinations, was about 160 after 7–8 subcultures [122]. Lee et al. [144] obtained much higher resistance factors (up to 2000) in leukemia L-1210 cells cultured *in vitro* and treated with VM; they could associate the resistance with changes in the flux of the drug across the cell membrane, particularly increased efflux [145].

The mechanism of this resistance was not investigated in our laboratories, nor was the question studied whether the resistant P-815 cells are also less susceptible to other anticancer drugs. Different mechanisms of resistance to single drugs are known [146]. A special type is a multidrug resistance which is at least partially due to an enhanced drug transport out of the cells, brought about by increased production of a glycoprotein, termed P170, which facilitates efflux (other mechanisms are also possible, see e.g. [146,147]. Cells showing this type of resistance are usually less susceptible to a group of chemically unrelated anticancer drugs: etoposide/teniposide, anthracyclines, vinca alcaloids and others. Multidrug resistance can be antagonized by calcium channel blockers [148], some neuroleptics (calmodulin inhibitors), including thioridazine [149], and by cyclosporin A [150,151]. The relationship between X-ray-induced resistance to VP on the one hand and chemically induced resistance and cross-resistance to other drugs on the other hand has been discussed by Lock and Hill [152].

Other aspects of the clinically important topic of resistance of cells to VM/VP will be mentioned in the section dealing with the mechanism of action.

15.8 Preclinical pharmacokinetics, galenical problems

Pharmacokinetics of VM and VP in animals is rather unremarkable. In rats, after i.v. administration of tritium-labeled VM 26, there is a certain initial drug accumulation in the liver, kidneys and adrenals; no significant retention is found after 24 h; the liver, kidneys and large intestine are the organs with the highest concentration; drug content in a transplanted tumor (Walker carcinosarcoma) is about equal to that of the blood. 67 % of the injected radioactivity appears in the bile within 24 h (bile fistula rats) and more than 50 % is excreted in the feces. In the dog, excretion of VM in the urine is higher than in the feces [126].

Results with labeled VP given to rats i.v. or p.o. are similar to those obtained with VM, except that after oral administration the majority is excreted in the feces [153]. Pharmacokinetics of tritium-labeled VP was also studied in monkeys; here, roughly two thirds of i.v. injected radioactivity is excreted by the kidneys [114]. VP, at high local concentrations (intracarotid infusion), has been found to reversibly disrupt the blood-brain barrier in rats [154], which may enhance its own penetration into the brain and that of other compounds. Due to the comparatively low reversibility of the cellular cytostatic effect of VP (see section 15.4), blood levels are not a good parameter for the duration of the cytostatic effect. We found that the resumption of mitotic activity in the duodenal epithelium of rats after a single i.v. dose of VP occurs much (about 48 h) later than would be expected from blood levels; the expected value was calculated based on the level of freely diffusible compound (see paragraph on protein binding, below) [155]. Therefore, blood levels are not given here for the experimental animals.

In the attempt to find an optimal galenical form for oral administration of VP, a large number of experiments were performed in different animal species. Determining the optimal dose for increase of life span in leukemic mice showed that this dose is lowest when the drug is given in a galenical form in which the compound is completely dissolved and does not precipitate upon contact with water [21]; for this purpose, the surfactants polysorbate 80 or polyethoxylated castor oil are very helpful (this experience was also crucial in the development of the immunosuppressive drug Sandimmune® where severe absorption problems had to be solved [156]). Based on excretion of radioactivity in the bile of rats, it can be calculated, in comparison to animals to which the drug was given i.v., that about 40 % of labeled VP given orally is absorbed from the gastrointestinal tract [153]. In dogs, it was shown that bioavailability of oral VP is about 10x higher when it is given as a clear aqueous solution (containing Tween® 80) than when a suspension is administered in capsules [157]; and in prolonged oral toxicity tests VP given for 4 weeks as a suspension elicited some leukopenia at a dose of 30 mg/kg/day, while the dose of 10 mg/kg/day administered as a Tween®-containing solution induced agranulocytosis and was lethal [158]. In rhesus monkeys VP is absorbed quickly (peak of blood level after 45 min) and to about 40 % when administered orally as a clear aqueous solution [159]; in 4-week toxicity tests, VP suspension at an oral dose of 100 mg/kg/day did not produce leukopenia in monkeys, but administered as a Tween®-containing solution it elicited leukopenia at 30 mg/kg/day and agranulocytosis at 100 mg/kg/day [158].

Tumor cells cultured *in vitro* accumulate VM and VP. The content of P-815 mastocytoma cells after 1-h incubation in growth medium (containing 10 % horse serum) of tritium-labeled VM is about 14x higher than the concentration in the medium; for VP this factor is about 10. The accumulation factor does not depend (within limits) on the absolute drug concentration. Washing once (centrifugation of the cells and resuspension in fresh, drug-free medium) removes about 68 % of VM from the cells, while about 97 % of VP leaves the cell upon washing. A second washing removes again about 50 % of the remaining VP [160]. When labeled VP is added to rabbit blood, the drug concentration in the erythrocytes is about 70 % of that in the plasma [21]. Allen [161] studied the accumulation of VM and VP in L-1210 cells more in detail and found the steady-state accumulation of labeled VM to be about 10x greater than that of VP. There are high and low affinity intracellular binding sites. A significant amount of labeled drug is irreversibly retained within the nuclei and does not exchange when the cells are incubated in medium containing unlabeled drug. That this quantity is of the order of 10 % may also be deduced from the finding that the kinetics of death of P-815 cells is the same whether they are incubated in a solution containing 1 μg/ml VP or in a drug-free solution after treatment for 1 h with 10 μg/ml VP [21].

Another pharmacokinetic parameter, drug binding to plasma proteins, has also been investigated, as was done with the podophyllum preparations of the spindle poison type (see above). The difference in activity of VM and VP in fibroblast cultures when incubated in medium with and without serum already indicated a considerable binding of the drugs, particularly VM, to serum proteins. Protein binding of VM in rat serum was about equal to that in human serum. The percentage of bound VP (i.e. not ultrafiltrable) was not significantly concentration-dependent in the range 1–10 μg/ml. In mouse serum about 78 % of VP was bound, and in human serum 88 % (at a total concentration of 1 μg/ml). Ultrafiltration in an albumin solution showed that most or all of this binding is due to albumin [21]. When experimenting *in vitro* where these drugs often act in serum-containing media, consideration of protein binding (or of the freely diffusible concentration of the drug) is an important issue (see also [162]). Binding of VM and VP to human plasma proteins will be discussed again in the section on human pharmacokinetics.

A further component of pharmacokinetics is the metabolism of VM and VP in the organism. Our knowledge of this particular element of the effects of the compounds is far from complete and still somewhat controversial. It has been well reviewed by Clark and Slevin [106] and will not be discussed here in detail (except for that aspect which affects the biochemical mechanism af action): the metabolites identified so far *in vivo* are less active than the parent compounds [163]. In addition, the rapid onset of action of VM and VP in cell culture systems rather speaks against a metabolic alteration before the molecules become active. We conclude that, at present, metabolism does not seem to contribute significantly to our understanding of the drugs' effects.

15.9   Human pharmacokinetics

Pharmacokinetics of VM and VP have been studied in man quite extensively. Allen and Creaven have investigated both compounds comparatively and in detail by using labeled drug [164]. They could fit the plasma decay kinetics after i.v. infusion to a three-compartment open model for VM and a two-compartment open model for VP. Excretion rate of VP, particularly by the kidney, is higher than that of VM; this difference may contribute to the better tolerability

of the former. On the other hand, the fraction metabolized is higher for VM. By determining protein binding to human serum albumin by means of dialysis these authors could calculate that, at typical plasma concentrations, 99.4% of VM and 94% of VP are protein-bound; thus the part of VP able to diffuse freely out of the vascular bed is roughly 10x higher than that of VM (this in turn may increase tolerability for VM which is 10x more potent in vitro, and is more slowly eliminated than VP, from which one would expect a more than 3–4 fold better tolerability for VP as compared to VM). They also deduced from their data a great propensity for tissue storage. A more recent review of the clinical pharmacokinetics of teniposide and etoposide and their significance for human therapy has been given by Clark and Slevin [106]. The terminal half-life of VP is less than that of VM, and the plasma and renal clearances of VP are greater. Metabolism of both drugs and the identified and possible metabolites are discussed extensively by these authors; they conclude that our knowledge regarding metabolism is still insufficient. Other summaries of the clinical pharmacokinetics of VP have been given by Raettig et al. [115], Creaven [165], Evans et al. [166], D'Incalci and Garattini [167,168] and others. Pflüger et al. [169] analyzed the pharmacokinetics of this compound using mass spectrometry and found that the β-phase elimination half-life becomes shorter with increasing specificity of the detection method. Drug accumulation was not detected and intrapatient variability was found to be small. Cisplatin seems to change the human body's handling of VP. Oral bioavailability of VP is about 50%, with a considerable interpatient variation [106]; D'Incalci et al. found a higher absorption from the gastrointestinal tract, particularly with the drinking ampoules [170].
Metabolic degradation of VM and VP in man (and animals) have been investigated in a number of studies, VP more extensively than VM. Results are still not clear and somewhat controversial (see [106,171]) and will not be discussed here.

15.10 Clinical use

As mentioned in the history section, clinical testing of VM started in 1967. The first dosing experiments were done in Switzerland, mainly by H. R. Marti in Aarau. Soon, however, the drug was also investigated in other countries, e.g. USA (here primarily by O. Selawry and co-workers who had already clinically studied the earlier podophyllum drugs, SP-G and SP-I), France and Denmark. The first preliminary clinical reports appeared from US hospitals in the same year as the first full preclinical paper on teniposide [41], indicating that the Swiss (preclinical and clinical) investigators were somewhat slow in publishing their results. The first reports [172,173] were concerned primarily with tolerated dose and toxicity. They established leukopenia as the main side effect and already indicated some therapeutic activity in an astrocytoma and in lymphomas. A careful dose-finding study then provided a first baseline for a starting dose with a weekly schedule and again suggested effectiveness in brain tumors [174]. Another early phase I study in France, involving a large number of patients, found activity mainly in lymphomas [175]. The first clinical results of the European Organization for Research on the Treatment of Cancer (EORTC) with VM, a drug with a new structure and a new mechanism, led G. Mathé to designate VM 26 as "la drogue de 1971" [176]. VM was also included in an early study assaying *in vitro* sensitivity of human tumor cells in primary culture with the goal of selecting the most promising drug [177]. Further clinical studies produced new findings and gave hints to other types of tumors which may be sensitive to VM [178,179 and many others]. Sonntag et al. [180] compared different treatment schedules and found a comparatively high responder rate with VM monotherapy in previously treated lymphoma patients. Goldsmith and Carter [181], reviewing the early data on VM in 1973, concluded that VM 26 could be a valuable addition to the therapeutic armamentarium. In 1977, Rozencweig et al. [182] published a comparative analysis of preclinical and clinical results obtained with VM and VP, in which most of the early studies with VM are reported, and in which they emphasize the presumable activity of this compound in brain tumors. While effectiveness in brain tumors has often tentatively been correlated to lipophilicity, it must be kept in mind that the blood-brain barrier is not the same in cerebral tumors as in normal brain tissue; higher levels in brain tumors and their surroundings as compared to healthy central nervous system tissue have been documented for compounds less lipophilic than VM (see, e.g. [24]). A more recent overview on – particularly clinical – experience with VM has been given by O'Dwyer et al. [183]. Grem et al. [184] have extensively reviewed the use of teniposide in the treatment of leukemia – which was mainly inaugurated by Rivera et al. in Memphis (Tenn.) –, both as a single agent and in combination with other drugs, particularly cytosine arabinoside. VM was also used in ablation regimens before (autologous or allogeneic) bone marrow transplantation (see [184]).

Etoposide was first given to human patients in 1971; early clinical studies with the intravenous form were performed in Switzerland [185], Denmark, France and the US. In the first published study, mainly lymphomas responded, but also a few solid tumors [186]; this latter fact and the possibility of giving VP (in contrast to VM) orally made the ethylidene derivative from the beginning somewhat more interesting for clinical researchers than the thenylidene compound. Other reports soon followed [187,188] documenting activity in several different human malignancies, e.g. acute monocytoid leukemia [189]. Early Swiss studies with VP had already noted activity in small cell carcinomas of the lung [190] and compared i.v. administration with two oral forms [190,191]. Effectiveness in lung cancer was also shown in early US studies [192]. Other investigators of that period used only the oral form and reported relatively good effects in patients with ovarian cancer [193]. The results with VP up to 1976 were summarized by Rozenczweig et al. [182]. Etoposide soon found more widespread use than teniposide, and results with this derivative were reviewed at conferences in 1981 [194] and 1983/84 [1,195]. Sinkule [196] and O'Dwyer et al. [197] reviewed the preclinical data as well as the clinical results with VP.

Beneficial effects of VP are documented primarily in testicular and small-cell lung cancers, lymphomas, leukemias and Kaposi's sarcoma. Some activity has also been reported in other tumors, e.g. non-small-cell lung cancer, ovarian carcinoma, gestational choriocarcinoma, breast cancer, neuroblastoma and hepatocellular carcinoma. In a number of other types of malignancies, response rates were lower or sufficient data are not available (see also [198]). Of course, except for the phase 1 studies, most patients were treated with combinations of VP with other drugs, e.g. cisplatin. A quite successful regimen for lung cancer was developed by the Swiss Group for Clinical Cancer Research (SAKK); it included adriamycin, etoposide and cisplatin (AVP) (see [199]). Particularly gratifying are the positive results in children, e.g. with recurrent malignant solid tumors [200].

A more recent use of VP is based on the fact that its clinically dangerous side effects are mostly limited to leuko- and thrombocytopenia and that this toxicity is usually manageable. It consists in the treatment of leukemias or disseminated cancers with very high doses (sometimes in combination with other drugs) which is then followed by transplantation of (purged) autologous bone marrow to rescue the almost completely eradicated hematopoietic tissue [201, 202, 203, 204]. Along a similar line, VP has also been employed, together with total body irradiation or other drugs, in the preparation for allogeneic bone marrow transplantation in advanced hematological malignancies [205,206]; for prevention of graft-versus-host disease, some of these patients also received cyclosporin A, developed by the same groups as etoposide. Another use of VP is in preconditioning for marrow transplantation before ablative treatment with cyclophosphamide and total body irradiation in acute lymphocytic leukemia [207,208]. Etoposide has also been tested experimentally in models for *in vitro* purging of bone marrow from tumor cells before autologous transplantation in leukemias and lymphomas [209,210]. Wood and Jacobs [211] took advantage of the high affinity of etoposide for blood platelets: they loaded thrombocytes of a patient with refractory immune thrombocytopenic purpura with that drug and reinfused them; they so obtained a durable remission of the thrombocytopenia, presumably by delivering the compound in this way selectively to macrophages.

16   Mechanism of action
16.1   Spindle poison activity vs. inhibition of entry into mitosis

King and Sullivan [4] have shown that the mechanism of the cytostatic action at the cellular level of podophyllin is the same as that of colchicine, i.e. an inhibition of the formation of the mitotic spindle, resulting in an arrest of the cell division process in metaphase and a clumping of the chromosomes (c-mitosis). Later it was shown that, at the subcellular level, this is due to binding of podophyllotoxin (or colchicine) to tubulin, preventing these macromolecules to form the microtubules which constitute the fibers of the mitotic spindle (see [212]). C-mitoses are readily recognized microscopically in stained preparations of cell or tissue cultures; the diagnosis "spindle poison activity" is therefore an easy matter, made even more simple by the fact that the number of (c-)mitoses increases with time under the influence of a spindle poison (see e.g. Fig. 2 in ref. [46]).

All constituents of podophyllin and all compounds of the podophyllum series known up to 1965, including those synthesized in our laboratories, have — as far as they have been tested — an activity of the spindle poison type. Evidence that the anti-microtubular activity of at least some of these compounds is the predominant or exclusive mechanism of cytostatic action derives from a comparison of tubulin binding affinity with cell proliferation inhibitory potency and also with antitumor effects *in vivo* [213]. A comparison of inhibition of microtubule assembly by P and analogues as reported by Loike et al. [214] with our ID-50 values in P-815 mastocytoma cultures points in the same direction. Therefore, since our results regarding cytostatic potency show a very good correlation with the ability of the drugs to induce c-mitoses, it can be assumed that podophyllum compounds which produce these arrested metaphases bind to tubulin and that the appearance of c-mitotic figures in cell or tissue cultures reflects inhibition of microtubule assembly.

Only a minority of the podophyllum drugs do not increase the relative mitotic index (for definition see Table 12) beyond 6 in fibroblast cultures with an incubation time of 6 h, among them 4'-demethylpodophyllotoxin benzylidene glucoside [46] and some 4'-demethylepipodophyllotoxin aglucones (see Table 12). It can be hypothesized that compounds producing c-mitoses but only a low increase of the mitotic index are not pure spindle poisons but have in addition the capacity to inhibit to some degree the entry of cells into mitosis at those concentrations which also prevent the formation of the mitotic spindle.

Structural requirements in the podophyllum series for this second mechanism (inhibition of entry into mitosis) do not become entirely clear from our data obtained in fibroblast cultures. It is obvious that demethylation of the P molecule in position 4' is required, but this by itself is not sufficient. Attachment of glucose (condensed with an aldehyde) in epi-position at C-1 (see [46]) or some specific substituents with epiconfiguration in position 1 (see Table 19) seem necessary. If this "specific substituent" is glucose with an aldehyde condensed to it (as in VM and VP), the second mechanism becomes much more potent and virtually supersedes spindle poison activity. One could assume that P derivatives have two mechanisms at the cellular level: spindle poison activity and inhibition of entry into mitosis. The balance between these two effects depends on chemical substitution, particularly in position 1 and 4' of the P molecule. At one end of the range of the ratio of the two effects is P and many of its derivatives which can be considered to be pure spindle poisons. At the other end of the range are, among the derivatives discussed here, the aldehyde condensation products of DEP glucoside. Their cytostatic potency depends on their activity to prevent cells from entering the mitotic phase of the cell cycle (i.e. producing a stop in G2). On the other hand, at higher concentrations most or all of these derivatives still have the capacity to inhibit formation of the mitotic spindle, like VM [41]. It has been pointed out earlier [215] that the spindle poison activity of VM becomes evident at about the same concentration as that of P glucoside, P benzylidene and thenylidene glucosides. This suggests that the effect of glucoside derivatives on the mitotic spindle is little affected by those substitutions which bestow upon the molecule the second activity, that which arrests cells in the G2 phase.

A more detailed analysis of the relationships between structure and mechanism of action is not within the scope of this review; the reader may derive some hints from Table 2 of ref. [46] and from Table 19 of the present publication. Unclear is, e.g., the contribution of the epi-configuration of the substituents in position 1 of the P molecule to the therapeutic effect in the whole organism. A few aspects of structure-mechanism relationships will be discussed below when dealing with more recent work of others in this field.

It has already been mentioned that there is an inverse correlation between "therapeutic" effect of P derivatives in mouse leukemia L-1210 and the increase of the mitotic index which they produce in fibroblast cultures. Thus, in this and other animal models and apparently also in cancer patients, the second mechanism of the P derivatives is the more efficacious one.

16.2   At which point of the cell cycle do VM and VP act?

By observing fibroblast or mastocytoma cell cultures at different times after drug addition it became clear that VM and VP prevent cells from entering the first part of the mitotic phase (prophase) within a few hours or less than an hour, depending on the concentrations [41, 215]; this effect was therefore named "preprophase activity" to indicate the difference in point of attack in the cell cycle in comparison to the pure spindle poisons which have a "metaphase activity" [41]. At higher drug concentrations the interval between beginning of drug treatment and disappearance of prophases is shorter than the duration of the G2 phase, indicating that cell cycle progression is arrested within that phase [216]; at lower concentrations the delay may be as long as or longer than the G2 phase which would suggest a stop late in the S phase. Measuring DNA content of VP-treated cells by cytophotometry after Feulgen staining or by cytofluorometry after acridine orange staining [217] showed an accumulation of cells with a 4n DNA content, also indicating – together with a lack of mitotic figures – a stop in G2.

A stop in G2 phase was then also found by other investigators by cytofluorometry and other methods after treatment of different human and other cell lines with VM and VP [218, 219, 220]. Drewinko and Barlogie also showed the G2 block to be most pronounced when treatment commenced in the S phase, and a delay of S phase transit at higher concentrations of VP [221,222].

16.3   Attempts to explain the mechanism of action of VM and
        VP at the molecular level

While the action of podophyllotoxin and other spindle poisons had been studied and elucidated at the biochemical or molecular level (see, e.g., [223]), only few hints as to the molecular basis of "preprophase activity" of VM and VP were available in the midseventies. We will discuss here first some findings at the biochemical level which do not seem to be able to explain inhibition of cell proliferation by VM and VP.

We had found that, in contrast to the spindle poison SP-I, VM [224] and VP [216, 217] inhibit incorporation of thymidine into DNA of P-815 mouse mastocytoma cells. However, the concentrations required for a significant effect are higher than those reducing cell multiplication. In addition, DNA synthesis continues and is not inhibited to a similar degree as thymidine incorporation, and DNA content per cell increases [217, 225], as does cell size; 12 h after a 1-h treatment of P-815 cells with VP, cell volume has roughly doubled [21]. This suggested that thymidine incorporation is not necessarily a reliable parameter for DNA synthesis and that the effects of these drugs which lead to a stop of cell proliferation may not primarily be directed to DNA synthesis (in accordance with the point of attack of these compounds in the cell cycle, see section 16.2). Synthesis of RNA and proteins is not inhibited by either VM or VP: the content per cell of these macromolecules increases [217, 225]. Changes in the nucleotide pools in P-815 cells treated with VP have also been observed but were considered not to be able to explain the basic mechanism of action of this type of compounds [226, 227]. These data suggested that VM and VP act by (a) mechanism(s) different from alkylating agents, antimetabolites, spindle poisons and other drugs; they showed, on the other hand, some similarity between the effects of the epipodophyllotoxin derivatives and X-rays regarding cell proliferation, mitosis, precursor incorporation and macromolecular synthesis [225]. In connection with this similarity it is interesting to note that VP has been reported to produce "radiation recall" in patients [228, 229]. Gupta [230] also concluded that the genetic lesions produced by VM and VP are probably very similar to those produced by X-rays, and others have generated VP-resistant cell lines by fractionated X-irradiation [231]. Misra and Roberts [218] also came to the conclusion that the primary site of VM action does not appear to be DNA, RNA or protein synthesis.

Loike et al. [25, 232] reported that VP and VM at high concentrations inhibit transport of thymidine and uridine into HeLa cells. However, no such inhibition was found with the concentrations used by us [217] nor by others [233] using mouse leukemia cells. In view of the high concentrations necessary for this effect and the lack of a close correlation between thymidine incorporation and DNA synthesis, nucleoside transport inhibition does not seem to be a plausible mechanism to explain the reduction of cell multiplication.

Gosalvez et al. have attempted to correlate cytostatic effects of VM with inhibition of cellular or mitochondrial respiration [234–236]. The effects of VM on oxygen consumption of cells or mitochondria *in vitro* were obtained at 12.5 μM of VM; it is highly doubtful whether results with concentrations which are roughly 1000x higher than those inhibiting cell proliferation can contribute to the understanding of the mechanism of cytostatic action. Besides that, using the Warburg method, we found no significant inhibition of respiration when VM was added to tumor cells *in vitro* up to concentrations of 15 µM; and the oxygen consumption of tumor, kidney or liver slices taken from animals which had been treated with high doses of the drug was not different from that of slices obtained from control animals [126]. In addition, interference with respiratory energy production would probably damage tissues with a high oxygen need (brain, kidney) before tumor cells.

16.4 Interaction with topoisomerase, and other possible mechanisms

Inhibition of thymidine incorporation by VM (and, later, VP) had given a first hint as to an effect of this type of drug on DNA, but further analysis could not substantiate a relevant direct action on DNA synthesis (see above). A finding suggesting that VP somehow affects chromatin was reported by Huang et al. [237] who observed chromosomal aberrations in VP-treated cells. Rao and Rao [220] also found microscopical chromosome alterations in VM-treated cells and suggested that chromosome damage may be a contributing factor for the halt of cells in the G2 phase of the cell cycle, similar to the surveillance mechanism in G2 which eliminates cells with altered DNA, as proposed by Tobey [238].
Loike et al. first reported in 1974 that VM and VP convert high molecular DNA in HeLa cells into smaller (still acid-insoluble) fragments [25,239]. They interpreted their results as suggesting single-strand chromosome breaks. After removal of the drug the cells are able to repair the DNA fragmentation. The lowest concentration (1 µM) of VP producing a slight fragmentation is, albeit higher than, still of an order of magnitude similar to the minimal inhibitory concentration for HeLa cell proliferation of about 0.2 µM [111]. Loike and Horwitz observed no alteration in the sedimentation properties of purified DNA of HeLa cells or adenovirus following incubation with VP, indicating that strand breakage was not a direct effect of drug on DNA. These authors also noted temperature dependence of DNA fragmentation. Single-strand breaks and fragmentation of DNA were also found to occur within one minute *in vivo* when mice bearing ascitic L-1210 tumor cells were treated i.p. with VM [240]; Roberts et al. related the alteration of DNA by VM to its oncolytic effect. Pointing in the same direction is the observation that agents which inhibit DNA repair increase the therapeutic effect of VP in an-

imals [241]. These findings were extended with a more sensitive method by Wozniak and Ross [242] who reported single- and double-strand breaks and DNA-protein cross-links upon treatment of L-1210 cells or isolated nuclei with VP; they also could correlate these effects with inhibition of L-1210 cell colony formation by VP. Kalwinsky et al. [243] studied the concentration dependence of DNA breakage by VP and found a good correlation between arrest in the G2 phase and induction of DNA strand breaks. They concluded that DNA strand scission is the initial event in the sequence leading to growth inhibition and death of VP-treated cells.

The reports mentioned above (and others), led by that of Loike and Horwitz [239], clearly indicated an effect of VM and VP on the integrity of DNA in treated cells. However, the biochemical basis of it was unclear, particularly in view of the fact that purified DNA is not affected [226,239] and that VP has no effect on the polymerization step of DNA synthesis in a cell-free system nor does it form a complex with purified DNA [226]. At a meeting in Syracuse, N.Y., in spring 1982, B. Long mentioned to one of us (H. St.) that he had found an interaction of VM and VP with the enzyme topoisomerase. He and his co-workers had studied a number of podophyllotoxin derivatives concerning the ability to elicit DNA breakage and to inhibit proliferation in a human lung carcinoma cell line and observed a parallel between inhibition of catenation activity of eukaryotic type II topoisomerase and DNA breakage or cytostatic activity [244–246]. Thus, type II topoisomerase was proposed as the primary target of VM and VP. This enzyme had already been found to be involved in DNA cleavage by intercalative antitumor drugs (e.g. anthracyclines) which act by stabilizing a cleavable complex between the enzyme and DNA, as do VM and VP which are, however, not intercalative [247]. Chen et al. found VM to be 5-10-fold more potent than VP in the stimulation of topoisomerase II-mediated DNA cleavage [247]; the same factor separates cytostatic potency of these drugs in cell culture, thus supporting the hypothesis that interference with the breakage-reunion reaction of topoisomerase may be the main mechanism of action of VM and VP. Similar observations were reported by Ross et al. [248]. Resistance to VP was suggested to be correlated to decreased sensitivity of topoisomerase II to VP and, at the same time, to intercalating drugs [249].

A number of further studies corroborate the role of topoisomerase II in the activity of VM and VP. Long and Brattain [246] found a good correlation between DNA breakage and inhibition of catenation activity of topoisomerase among a number of P derivatives including VM and VP. How the effects of the drugs on the enzyme bring about cytotoxicity is not so obvious. Rowe et al. [250] and others [251] favor the hypothesis that formation of a cleavable complex between topoisomerase and DNA – stimulating the enzyme-mediated DNA cleavage – is the crucial event and that inhibition of the catalytic activity of the enzyme is less likely to be responsible for cytotoxicity. Long [163] proposed a model for topoisomerase inhibition that may explain the relationship between double-strand DNA breakage and cytotoxicity. Kerrigan et al. [252] reached similar conclusions. This assumption was supported by further data and it was also suggested that one type of multidrug resistance (involving, besides VM and VP, anthra-cyclines and other drugs) is due to alterations of topoisomerase structure [253] (see also [254,255]). In any case, it seems that resistance of tumor cells often involves not only several drugs (multidrug resistance), but in many cases, including VP, several mechanisms (multifactorial resistance) [256]. Single- and double-strand DNA breaks, induced by VM and VP, are rapidly repaired after drug removal in HeLa cells [239] and in another human carcinoma cell line [257]. This occurs with drug concentrations and exposure times which do, in other cells (e.g. P-815 mastocytoma or L-1210 leukemia [131]), suppress further proliferation of, or kill, almost all cells. Perhaps some chromatin damage, undetected by the methods used to assay DNA cleavage, remains. Besides that, Long et al. [258] concluded that cytotoxicity of VM and VP is more closely related to double- than to single-strand breaks. The minor groove of the DNA double helix seems to be the place where VM interacts with topoisomerase since compounds which bind there inhibit the drug's topoisomerase-mediated effects [259,260].

When comparing the effects of VM or VP on different cell types, it has to be born in mind that usually rapidly proliferating cells are more susceptible to these compounds [258] (in our hands too, HeLa cells were less sensitive to VP than P-815 cells [111], the latter having a much shorter doubling time); and, within the same cell line, DNA cleaving is more pronounced when the cells are in a proliferative as compared to a quiescent state [261]. In general, topoisomerase II levels are found to be correlated with cell or tissue growth rate [262,263], and within the cell cycle drug sensitivity is highest during the S and particularly the G2 phase, correlated with topoisomerase content [264]. In the same line, it was found that nascent, replicating DNA is the preferential target for drugs interacting with topoisomerase II [259]. Some observations, however, cast some doubt on the assumption that topoisomerase-mediated DNA breaks are the only factor responsible for the cytotoxicity of these drugs [264,265].

The findings concerning the relationship between proliferative state, topoisomerase content and susceptibility of cells to VM and VP may explain a number of observations related to differences in sensitivity of various animal and human tumors to drugs acting via topoisomerase. Of some practical importance may be the observation that cytosine arabinoside enhances the VP-induced DNA cleavage [266]. However, other data cast doubt on the assumption that the topoisomerase concentration in the cells is a sufficient determinant for the cytostatic activity of VM [267].

Other data again suggest that the effects of VM and VP have some similarity to those of X-rays. Cells from patients suffering from ataxia-teleangiectasia, known to be highly susceptible to the effect of X-rays, are also more sensitive to VP than cells from healthy subjects [268], and it was hypothesized that cells from ataxia-teleangiectasia patients may have abnormal topoisomerase II activity [269].

Several groups have, based on experimental data with VP, postulated free radical formation and activation by 3'-demethylation with aldehyde formation to a reactive catechol, semi-quinone and ortho-quinone [270–276]. In the same line, Teicher et al. [277] showed that higher oxygen tensions increase the cytostatic effect of etoposide for tumor cells *in vitro* and *in vivo* (as is the case with X-rays), but not for bone marrow cells *in vitro*. It has also been reported that the catechol derivative of VP (orthodihydroxy conformation in ring C) exhibits a slightly higher cytotoxicity than the parent compound in chinese hamster ovary cells, but not in another cell line [273]. Long [163] found 3'-demethyl-VP and the corresponding orthoquinone to produce only low levels of DNA double-strand breaks and to be clearly less cytotoxic than VP itself. Another oxidative metabolite of VP has been identified by Broggini et al. [278], namely a 1,2,3,4-tetrahydro-VP, a molecule with an additional aromatic ring. To what extent these and possibly other metabolites are generated *in vivo*, particularly in human tumors, and to which degree they contribute to the therapeutic effect – or are even the chief actors in it – remains to be seen. In any case, these findings lead to the conclusion that, due to free radical formation, a direct effect of VM/VP on DNA, not involving topoisomerase, cannot be excluded with absolute certainty. Recently, a low frequency but saturable binding of VM and VP to DNA has been described, and two functional domains of the molecule have been postulated [279].

17   Structure activity relationships

It is not the purpose of this section to present a detailed analysis of structure activity relationships in the large group of podophyllotoxin derivatives. Many of these emerge from the present text or other reviews, e.g. that of Jardine [280]. We will discuss here only some aspects of the relationship between structure and mechanism of action, namely the question which structural elements of the molecule are responsible for the "preprophase" activity, i.e. the inhibition of entry into mitosis, as opposed to spindle poison activity.

When dealing with molecules which act by two or more mechanisms, the pharmacologist and clinician (sometimes in contrast to the biochemist) is primarily interested in that mechanism which becomes operative at the lowest concentration. Mechanisms entering the scene only at significantly higher drug concentrations are of practical relevance only when the respective concentrations are reached at the site of action *in vivo* and when these mechanisms elicit long-lasting effects and are not soon superseded by those mechanisms acting at lower concentrations. For example, VM has, according to our results, spindle poison activity, but only at a concentration almost 1000x higher than preprophase activity; the former effect is therefore irrelevant for practical purposes. Or, vice versa, 4'-demethylpodophyllotoxin exhibits a minimal DNA breaking effect at a concentration of 10 μM [281] but inhibits cell proliferation down to about 0.02 μM and increases the mitotic index considerably at this concentration [46]; thus it has to be considered as a spindle poison. In the following discussion, therefore, substantial weight will be given to drug concentrations. Since it seems that, at the lowest concentration, VM and VP affect primarily cells in the G2 phase of the cell cycle, we will refer to this effect as G2-activity ("interphase activity" would also be appropriate since the exact position of the relevant biochemical event in the cell cycle is not known).

Our own conclusions regarding the question whether a drug acts by an anti-microtubular (spindle poison) mechanism or by inhibiting the entry into mitosis (G2-activity, presumably by inducing DNA breaks and involving topoisomerase II) are primarily based on the number and aspect of mitoses in chick embryo fibroblast and P-815 mastocytoma cell cultures. Other parameters in which drugs which act by these two types of mechanisms differ, have already been mentioned: inhibition of thymidine incorporation into DNA, reversibility of the cytostatic effect, difference between fibroblasts and P-815 cells as to 50 % inhibitory concentrations, time of death of animals treated with a single lethal dose (which is related to the lesion leading to death, see above) and increase in survival time of L-1210 inoculated mice.

In 1972, we listed the 4 chemical alterations in the podophyllotoxin molecule which bring about the change from a "pure" spindle poison to an (almost) pure "G2 poison": demethylation in position 4', epimerization in position 1, presence of glucose in position 1 and aldehyde condensation to the glucose [46] (it may be of interest to note that only one of these alterations – the aldehyde condensation – is "artificial", since molecules with the other changes occur in nature). The contributions of each of the 4 structural changes to the shift in the balance between the two mechanisms is not straightforward but depends on the presence or absence of the other 3 alterations. Compounds with all possible combinations of these alterations are listed, together with some relevant experimental data, in Table 19.

Almost all authors agree that 4'-demethylation is an important prerequisit for G2-activity; it creates a reactive hydroxy group which has been implicated in the mechanism of the cytostatic effect of VM and VP (see above). Gupta et al. [282], however, came to the conclusion, based on cross-resistance patterns, that 4'demethylation is not an absolute requirement but only enhancing the "VM-like" cytostatic effect, and Long [163] could induce DNA scission with some compounds with a blocked 4'-OH group. By itself, 4'-demethylation is certainly not sufficient, demethylpodophyllotoxin is still a spindle poison (see above); if 4'-demethylation alone bestowed upon the molecule a reasonable G2-activity, α-peltatin would have to be ex-

Table 19
Structure-activity relationships regarding 4'-demethylation (D), l-epimerization (E), glucosidation (G) and benzaldehyde condensation (B) of podophyllotoxin (P)

compound	ID-50 P-815 μM a)	potency P-815 P = 1	relat. mitotic index b)	tubulin assembly ID-50, μM c)	colch. binding Ki, μM d)	DNA breaks μM e)	L-1210 % incr. life span a)
P	0.012	1	12	0.6	0.51	>100	35
DP	0.018	0.67	14	0.5	0.65	10	10
EP	0.082	0.15	9.0	5	12	>100	11
DEP	0.14	0.086	9.5	2	–	1	7
PG	10	0.0012	10	–	180	–	0
DPG	3.6	0.0033	7.9	–	–	100	0
EPG	>35	<0.0003	–	–	–	–	7
DEPG	7.8	0.0015	26	–	–	100	34
PBG	5.7	0.0021	14	–	39	–	5
DPBG	1.2	0.010	3.7	–	–	1	29
EPBG	0.50	0.024	10	–	–	–	60
DEPBG	0.01	1.2	0.10	–	–	0.1	91
VM 26	0.0076	1.6	0.07	>100	–	0.1	121
VP 16	0.078	0.15	0	>100	–	1	167

a) values taken from Stähelin ([41, 46, 111]); b) explanation of relative mitotic index see table 12; values taken from Stähelin ([46] and unpublished results); c) values (taken from Loike et al. [214]) represent ID-50's für microtubule assembly; d) values (taken from Kelleher [213]) represent Ki values for inhibition of colchicine binding to mouse brain tubulin; e) values (adapted from data of Long et al. [271]) represent the lowest concentrations increasing single-strand DNA break frequency to an evaluable level.

pected to show more DNA breaking effect than it does [281]. 4'-Demethylation together with isomerization (to "epi") or with glycosylation in position 1 still does not abolish the spindle poison character (see relative mitotic index in Table 19). Long et al. [281] report a certain increase of single-strand break frequency in a lung adenocarcinoma cell line treated with 1 μM of DEP; we found this compound to inhibit P-815 cell proliferation down to a concentration of about 0.1 μM and to increase, in this concentration range, the mitotic index considerably; in leukemia L-1210 it is inactive. These data show that DEP is an antitubular compound at the lowest effective concentration but, at somewhat higher levels, also shows the other (G2-) effect. On the other hand, DPBG produces only a slight increase in mitotic index and has modest, but significant anti-L-1210 activity [46]; since this compound and DEP are roughly equally potent regarding DNA scission [281], we may assume that the antitubular potency of the latter is higher and therefore predominant as to cytostatic mechanism (this is in agreement with the finding that DEP is about 14x more active as a cytostatic agent *in vitro*). If DNA breaking activity is mainly responsible for the anti-L-1210 effect of podophyllum compounds, should then two derivatives with about equal DNA scission potency not produce similar increases in survival time in this tumor model? To explain this apparent discrepancy we can assume that of two drugs with equal effects on DNA that compound will be less useful in L-1210 which has the higher antitubular potency, since the latter – which contributes but little to the anti-L-1210 activity – reduces the highest tolerated dose.

Glucosidation in position 1 always reduces the cytostatic potency considerably (this also holds true for induction of single-strand DNA breaks [281]), as well in the podophyllotoxin as in the demethylpodophyllotoxin series, and does not change the mechanism of action. In our hands, all free glucosides were spindle poisons; glucose reduces, but does not abolish antitubular activity of podophyllotoxin and derivatives. Why VP does not show any inhibition of the assembly of brain microtubules in a cell-free system even at a concentration of 100 μM [232] is not clear. The system (fibroblasts) in which we observed inhibition of the formation of the mitotic spindle – which has to be interpreted as inhibition of microtubule assembly (see, e.g., [38]) – seems to be much more sensitive to this drug effect than measurement of microtubule assembly by following turbidity changes: while the latter method requires about 5 μM of podophyllotoxin for a significant effect [232], concentrations of the order of 0.01 μM are sufficient to exhibit spindle poison activity in fibroblast cultures (see above). Even the podophyllotoxin concentration found by Kelleher [213] to inhibit colchicine binding to tubulin (0.51 μM) is still 50x higher than that which leads to c-mitoses in fibroblasts in tissue culture. This factor of 50 or more also holds for other spindle poisons [213]. We therefore assume that the discrepancy between the results of Loike and Horwitz [232] and ours [41,215] regarding inhibition of assembly of tubulin into microtubules by VM and VP can be explained by different sensitivities of the methods used and that our observation of an initial spindle poison activity of VM and VP in chick embryo fibroblast cultures really indicates an inhibition of the assembly of microtubules. A contributing factor, already mentioned, is the intracellular accumulation of these drugs as compared to the concentration outside the cells. Enhancement of vincristine neurotoxicity by VM and VP in patients [283] may also be an indication of a certain anti-microtubular effect intrinsic to the epipodophyllotoxins.

Gupta [230,284] has investigated the relationship between spindle poisons of the podophyllum series and VM/VP by means of mutant mammalian cell lines resistant to either podophyllotoxin or VM; he found that the podophyllotoxin-resistant lines are usually not resistant to VM and vice versa. VM-resistant lines exhibit increased resistance to other drugs, including doxorubicin. On the other hand, Gupta and Singh [285], again using resistant chinese hamster ovary cell lines, found cross-resistance between VM and DEPG; in addition, in their hands the latter compound did not increase the mitotic index and produced DNA strand breaks at high concentrations from which they concluded that DEPG has a mechanism of action similar to VM and VP. Long et al. [281] also found this drug to produce single-strand DNA breaks, albeit only at high concentrations (100 μM). From our results we conclude that this free glucoside (without aldehyde condensed to the glucose) is primarily a spindle poison which increases the mitotic index in chick embryo fibroblasts considerably [46] (the result regarding effect of DEPG on mitotic index given in that 1972 paper was confirmed later); that this compound has moderate but significant anti-L-1210 activity [46] is in agreement with the (low) DNA breaking potency. The lack of an increase in the mitotic index of chinese hamster cells when treated with DEPG may perhaps be due to a slightly reduced sensitivity of this species to spindle poison activity (Syrian (golden) hamsters are known to be more resistant to colchicine [38] or podophyllotoxin [21] than, e.g., mice and rats) and, therefore, the balance between preprophase and spindle poison activity may be different in hamster cells. It must also be remembered that DNA-breaking potency depends on the rate of proliferation of the cells as well (see above).

That, in the podophyllotoxin molecule, glucose in position 1 is not absolutely necessary for G2- (and anti-L-1210) activity has been mentioned (see Table 12). Thurston et al. [286] synthesized a number of new derivatives and found 4'-demethylepipodophyllotoxin-1-ethylether to have an inhibitory activity on topoisomerase II comparable to that of VP.

18   Conclusions

The introduction of etoposide and teniposide into cancer chemotherapy is one example for the way, starting from old folk remedies, new single chemical entities of therapeutic value are developed. Some of the more important steps of this development have been mentioned in the section on history. From the heuristic point of view it may, however, be interesting to mention a few motives, rationales, strokes of luck etc. which contributed to the final result.

After the reports of Bentley in the 19th century and that of Kaplan in 1942 on local effects of podophyllin in the treatment of superficial tumors, further chemical and biological work was performed mainly in the United States and led to the isolation of – besides the long-known podophyllotoxin – several other aglucones which, however, were not successful in clinical tests when using systemic administration.

The discovery by Sandoz chemists in Basel that the drugs are present in the plant as glucosides opened up a new area of chemical and biological research. Some of the goals of the work done in our company were the obvious ones: increase of the therapeutic index and of duration of action, enhancement of water solubility to facilitate parenteral administration, prevention of metabolic degradation of the compounds, the discovery of a form which is absorbed from the gut, etc.

One of the rather early steps in our endeavors with podophyllum glycosides was the condensation of aldehydes to the glucose moiety; the rationale behind this was to change the pharmacokinetic behavior (increasing duration of action) and to prevent the splitting off of the glucose by glucosidases in the body. While this chemical alteration of a genuine glucoside led only to a partial pharmacological success (SP-G) in the case of podophyllotoxin, it later turned out to be a crucial element in arriving at etoposide. Thus, we have here a case of serendipity: aldehyde condensation was performed in order to stabilize podophyllotoxin glucoside and ameliorate its pharmacokinetic behavior (which it did), but its unexpected, much more useful effect was that it converted a related molecule (demethylepipodophyllotoxin glucoside, present in the plant in very small amounts and not even known at that time) into compounds with a completely different mechanism of action, higher potency and considerably improved therapeutic properties. Two of these compounds are etoposide and teniposide.

Several other "rational" approaches were not succesful, e.g. acylation of the glucose moiety, opening of the methylene-dioxy ring, and particularly opening of the lactone ring in order to arrive at more water-soluble products. Though much work was invested in this latter attempt (see section 9) and a compound (SP-I) with a certain clinical activity in cancer and rheumatoid arthritis resulted from it, a major success was not achieved. Peltatins are sometimes more potent than the corresponding podophyllotoxin compounds, but the peltatin derivatives prepared in our company did not show, as far as this has been tested, any therapeutic advantages; this may be related to the fact that the 1-epi configuration is not possible in the peltatin series. A number of other unsuccessful approaches emerge from this review; a few have not even been mentioned.

SP-G and SP-I, the first podophyllum drugs commercialized for systemic treatment of malignancies, have often been plainly qualified as being "too toxic". This is misleading, since their therapeutic index was not much worse than that of the then available anticancer drugs, but the maximal achievable therapeutic effect finally turned out to be inferior to the then emerging other chemotherapies. Their toxicity (diarrhea in the case of SP-G) was, however, more conspicuous than that (leukopenia) of most other antitumor compounds. Had we only a few not "too toxic" anticancer drugs (in cancer-eradicating doses), the cancer problem would, of course, largely be solved.

After the finding, first at the cellular, then at the biochemical level, that the action of podophyllin and podophyllotoxin on cell division is essentially the same as that of colchicine (see sections 2 and 3) and after a considerable amount of work on the chemistry and biology of podophyllum lignans, the discovery in 1965, of a new mechanism of action of a certain type of podophyllotoxin derivatives and the simultaneous demonstration of their improved therapeutic properties in animal models opened the way to a great step forward in the use of "offspring" of this old medicinal plant in cancer chemotherapy. It cannot be overemphasized that this development was only possible because of a close cooperation between chemists and pharmacologists/cell biologists over many years.

Elucidation of the biochemical basis of the new mode of action (prevention of entry of cells into mitosis as opposed to stop during mitosis) also took some time and is not yet complete. Interference with the activity of topoisomerase II by the podophyllum compounds (aldehyde condensation products of 4'-demethylepipodophyllotoxin glucoside) acting in G2-phase of the cell cycle is probably the principal mechanism by which etoposide and teniposide inhibit cell proliferation. It is interesting to note that the quinolones, a group of recent antibacterial agents of clinical importance, also operate via a closely related enzyme (usually called gyrase in procaryotes). In a similar way as the effect of antimetabolites (acting on DNA replication) and spindle poisons is connected to eukaryotic cell proliferation, the action of drugs interfering with topoisomerase II is – luckily – largely restricted to those cells which prepare themselves for cell division. While topoisomerase may also affect gene expression, one of its main roles is in DNA replication and cell division; therefore, it is a suitable target for agents aiming specifically at inhibition of these events. Since such drugs usually affect only a small minority of cells in the body, namely those preparing for or in the process of division, they should rather be called cytostatic then cytotoxic, the latter meaning (non-specifically) poisonous for (all) cells.

As has been documented in this review, the therapeutic characteristics of podophyllotoxin derivatives are related to the mechanism of action, those acting predominantly in the G2-phase being the more useful ones as compared to the ones which chiefly affect the mitotic spindle. It is difficult to decide (and irrelevant for practical purposes) whether all active compounds of the podophyllum type exhibit both mechanisms, but for certain categories of derivatives this is certainly true. Essential is the balance between the two, the action on DNA/topoisomerase and the spindle poison activity, i.e. the ratio of the minimal drug concentrations at which these two effects become significantly operative in the cell. In a molecule with widely differing minimal concentrations for the two mechanisms, the one which needs much higher doses can, for practical purposes, be neglected because it will not operate in the whole organism.

The structural characteristics which determine the balance between spindle poison and "G2"-activity habe been discussed in section 16. We still maintain that the four changes of the podophyllotoxin molecule, elaborated in 1972, namely 4'-demethylation, 1-epimerization, 1-glucosidation and condensation of an aldehyde to the glucose, are essential for the transformation of a (pure?) spindle poison into an almost pure inhibitor of entry into mitosis. A possible exception may be certain non-glycosylated derivatives of demethylepipodophyllotoxin, some of which were discussed in section 12; however, their preclinical characteristics, including toxicology, have not been evaluated sufficiently to form an opinion on the question whether there is a chance that they could be superior to etoposide as therapeutic agents.

Further investigations of the mechanism of action of the epipodophyllotoxins are certainly warranted. Some of the unsettled questions are: Is interference with topoisomerase-activity really the principal or even the only mechanism operating under *in vivo* clinical conditions? How does this lead to cell death? What is the role of drug metabolism and the resulting metabolites, again under clinical conditions? Is topoisomerase involved in some or most clinical cases of resistance to the epipodophyllotoxins tenisposide and etoposide? Should the mentioned similarities between cellular effects of VM/VP and X-rays be studied more thoroughly, should these drugs be classified as radiomimetics?

What can we expect from the future? Clinical evaluation of etoposide and teniposide will certainly make further progress, e.g. in the direction of new indications, optimization of dosage schedules and combination with other drugs etc. Another problem concerns the possibility of finding podophyllum compounds of higher clinical utility. It would be futile to try to give a definite answer to this question now. It is obvious that just finding a derivative with a higher cytostatic potency is not sufficient; what matters is the therapeutic index or specificity, i.e. the difference between effects on the tumor and those on normal (proliferating) tissues. Here is certainly room for improvement, and one of the aims of the present review is to put the reader in a better position to decide where to look – and where not – for such improvement.

References

1 B.F. Issell, F.M. Muggia and S.K. Carter (eds): Etoposide (VP-16), current status and new developments. Academic Press, Orlando 1984.
2 M.G. Kelly and J.L. Hartwell: J. natl Cancer Inst. *14*,967 (1954).
3 I.W. Kaplan: New Orleans Med. Surg. J. *94*,388 (1942).
4 L.S. King and M. Sullivan: Science *104*,244 (1946).
5 V. Podwyssotzki: Arch. exp. Path. Pharmac. *13*,29 (1880).
6 J.L. Hartwell and A.W. Schrecker: J. Am. chem. Soc. *73*,2909 (1951).

7   W.J. Gensler and C.D. Gatsonis: J. Am. chem. Soc. *84*,1748 (1962); J. org. Chem. *31*,4004 (1966).
8   J.L. Hartwell and W.E. Detty: J. Am. chem. Soc. *70*,2833 (1948); *72*,246 (1950).
9   H. Kofod and C. Jørgensen: Acta chem. scand. *8*,1296 (1954); *9*,346 (1955).
10 R. Chatterjee and S.K. Mukerjee: Indian J. Physiol. all. Sci. *4*,61 (1950).
11 R. Chatterjee and D.K. Datta: Indian J. Physiol. all. Sci. *4*, 7 (1950).
12 J.L. Hartwell and A.W. Schrecker: Fortschr. Chem. org. NatStoffe *15*,83 (1958).
13 M.V. Nadkarni, J.L. Hartwell, P.B. Maury and J. Leiter: J. Am. chem. Soc. *75*,1308 (1953).
14 A. Stoll, J. Renz and A. von Wartburg: J. Am. chem. Soc. *76*,3103 (1954).
15 A. Stoll, J. Renz and A. von Wartburg: Helv. chim. Acta *37*,1747 (1954).
16 A. Stoll, A. von Wartburg, E. Angliker and J. Renz: J. Am. chem. Soc. *76*,5004 (1954).
17 A. Stoll, A. von Wartburg, E. Angliker and J. Renz: J. Am. chem. Soc. *76*,6413 (1954).
18 A. Stoll, A. von Wartburg and J. Renz: J. Am. chem. Soc. *77*,1710 (1955).
19 A. von Wartburg, E. Angliker and J. Renz: Helv. chim. Acta *40*,1331 (1957).
20 H. Stähelin: Med. exp. *7*,92 (1962).
21 H. Stähelin: unpublished results.
22 C. Keller, M. Kuhn, A. von Wartburg and H. Stähelin: unpublished results (1963/64).
23 M. Kuhn and A. von Wartburg: Helv. chim. Acta *46*,2127 (1963).
24 H. Stähelin and A. Cerletti: Schweiz. med. Wschr. *94*,1490 (1964).
25 J.D. Loike, S.B. Horwitz and A.P. Grollmann: The Pharmacologist *16*,209 (1974).
26 B.H. Long and A. Minocha: Proc. Am. Ass. Cancer Res. *24*,1271 (1983).
27 W. Ross, T. Rowe, B. Glisson, J. Yalowich and L. Liu: Cancer Res. *44*,5857 (1984).
28 M. Kuhn and A. von Wartburg: Helv. chim. Acta *51*,1631 (1968); Helv. chim. Acta *52*,948 (1969).
29 A. Rüegger, M. Kuhn, H. Lichti, H. R. Loosli, R. Huguenin, C. Quiquerez and A. von Wartburg: Helv. chim. Acta *59*,1075 (1976).
30 J. F. Borel, A. Rüegger and H. Stähelin: Experientia *32*,777 (1976).
31 H. Stähelin: Internal Report of 31 Jan. 1972.
32 T. Kaneko and H. Wong: Tetrahedron Lett. *28*,517 (1987) and earlier work quoted here.
33 D.I. Macdonald and T. Durst: J. org. Chem. *53*,3663 (1988).
34 A.W. Schrecker and J.L. Hartwell: J. org. Chem. *21*,381 (1956).
35 T.J. Petcher, H.P. Weber, M. Kuhn and A. von Wartburg: J. chem. Soc., Perkin Trans. *II*,288 (1973).
36 K.R. Beutner: Semin. Dermat. *6*,10 (1987).
37 L. Wilson: Ann. N. Y. Acad. Sci. *253*,213 (1975).
38 P. Dustin: Microtubules. Springer-Verlag, Berlin 1978.
39 A. Cerletti, H. Emmenegger and H. Stähelin: Actual. pharmac. (Paris) *12*,103 (1959).
40 H. Emmenegger, H. Stähelin, J. Rutschmann, J. Renz and A. von Wartburg: Arzneimittel-Forsch. (Drug Res.) *11*,327,459 (1961).
41 H. Stähelin: Eur. J. Cancer *6*,303 (1970).
42 J. Leiter, V. Downing, J.L. Hartwell and M.J. Shear: J. natl Cancer Inst. *10*,1273 (1950).
43 M. Kuhn and A. von Wartburg: Helv. chim. Acta *50*,1546 (1967).
44 A. von Wartburg, M. Kuhn and H. Lichti: Helv. chim. Acta *47*,1203 (1964).

45 A. Stoll, J. Rutschmann, A. von Wartburg and J. Renz: Helv. chim. Acta *39*,993 (1956).
46 H. Stähelin: Planta med. *22*,336 (1972).
47 A. Weder: Schweiz. med. Wschr. *88*,625 (1958).
48 A. Goldin, A.A. Serpick and N. Mantel: Cancer Chemother. Rep. *50*,173 (1966).
49 Sandoz Ltd.: Exposé über die Cytostatica SP-I und SP-G Sandoz (1963).
50 H. Stähelin, K. Zehnder and F. Kalberer: unpublished observations (1962).
51 K. Batz, F. Kalberer and H. Stähelin: Experientia *20*,524 (1964).
52 H. Stähelin: Experientia *16*,306 (1960).
53 H. Stähelin: unpublished observations (1964).
54 S. Lazary and H. Stähelin: In: 5th Internat. Congr. Chemother. Vol. III, p. 317. Ed. K.H. Spitzy. Verlag Wiener med. Akademie, Vienna 1967.
55 D. Wiesinger: unpublished observations (1967).
56 A. Benz: Dermatologica *136*,368 (1968).
57 W.R. Schalch, W. Meier-Ruge and J. Grauwiler: unpublished results (1962–1965).
58 W. Meier-Ruge, F. Kalberer and J. Grauwiler: Klin. Wschr. *42*,1024 (1964).
59 E. M. Lemmel and K. Nouza: Fol. biol. *12*,253 (1966).
60 O. Hubacher: Krebsarzt *16*,369 (1961); Schweiz. med. Wschr. *91*,1316 (1961).
61 A. Ravina: Presse méd. *71*,807 (1963).
62 G. Falkson, A.G. Sandison and J. van Zyl: S. Afr. J. Radiol. *2*,1 (1964).
63 O. Stamm: Münch. med. Wschr. *106*,41 (1964).
64 E. Aquilini: Minerva ginec. *16*,902 (1964).
65 M. Gosalvez: Rev. esp. Onc. *11*,357 (1964).
66 V. Larsen, J. Brockner and O. Storm: Acta chir. scand. Suppl. *343*,203 (1965).
67 V.K. Vaitkevicius and M.L. Reed: Cancer Chemother. Rep. *50*, 565 (1966).
68 H.M. Beumer and W.M. Porton: Ned. T. Geneesk. *110*,1877 (1966).
69 S. Krishnamurthi, V. Shanta and M.K. Nair: Cancer *20*,822 (1967).
70 K.H. Gebhardt and R. Becher: Med. Welt *18*,1705 (1967).
71 C.G. Schmidt: In: Krebsforschung und Krebsbekämpfung, Vol. VI, p. 309. Ed. E. Bock. Urban und Schwarzenberg, München 1967.
72 R.C. Chakravorty, S.K. Sarkar, S. Sen and B. Mukerji: Br. J. Cancer *21*,33 (1967).
73 R.A. Estévez, A.A. Carugati, O.T. Estévez and M. Speroni: Rev. Lat.-Amer. Quimioter. Antineoplas. *2*,101 (1968).
74 C.L.P. Trüb and L. Pohlschmidt: Therapie Gegenw. *117*,358 (1978).
75 M. Neiger: Practica Oto-rhino-laryng. *26*,268 (1964).
76 K.A. Newton, G. Westbury and W.F. White: Cancer Chemother. Rep. *43*,33 (1964).
77 K.A. Newton: Br. J. Radiol. *40*,823 (1967).
78 J. Bilder: Zentbl. Chir. *93*,337 (1968).
79 H. Limburg and M. Krahe: Dtsch. med. Wschr. *89*,1938 (1964).
80 O. Stamm and H. Stähelin: Cancer *18*,1096 (1965).
81 J. Gomez Orbaneja, L. Iglesias Diez and J.L. Sanchez Lozano: Dermatologica *145*,348 (1972).
82 A. Wiskemann: Strahlentherapie *143*,338 (1972).
83 K. Chlud, E. Prohaska and Ch. Pfaller: Z. Rheumaforsch. *26*,250 (1967).
84 K. Chlud, E. Prohaska, J. Zeitlhofer, Ch. Pfaller and B. Friza: Münch. med. Wschr. *110*,88 (1968).
85 K. Chlud: Münch. med. Wschr. *111*,2503 (1969).

86   E. Neumayer: Wien. med. Wschr. *119*,281 (1969).
87   W. Danielczyk: Wien. Klin. Wschr. *82*,697 (1970).
88   Sandoz Ltd.: Document on immunosuppressive effects of SP-I (1968).
89   B. Norberg, K. Berglund, U. Edström. L. Rydgren and G. Sturfeldt: Scand. J. Rheumat. *14*,271 (1985).
90   J. Rutschmann and J. Renz: Helv. chim. Acta *42*,890 (1959).
91   T.W. Doyle: In: Issell et al. (Eds) ref. [1], p. 15 (1984).
92   H. Stähelin: Presented at 10th Int. Congr. Cell Biology, Paris (1960).
93   M. Kuhn und A. von Wartburg: Experientia *19*,391 (1963).
94   J. Renz, M. Kuhn und A. von Wartburg: Justus Liebigs Annln Chem. *681*,207 (1965).
95   R. Chatterjee and S.C. Chakravarti: J. Am. Pharmac. Ass. *41*,415 (1952).
96   E. Schreier: Helv. chim. Acta *46*,75 (1963).
97   E. Schreier: Helv. chim. Acta *47*,1529 (1964).
98   M. Kuhn, C. Keller-Juslén and A. von Wartburg: Helv. chim. Acta *52*,944 (1969).
99   M. Kuhn and A. von Wartburg: Helv. chim. Acta *51*,163 (1968).
100 M. Kuhn and A. von Wartburg: Helv. chim. Acta *51*,1631 (1968).
101 C. Keller-Juslén, M. Kuhn, A. von Wartburg and H. Stähelin: J. med. Chem. *14*,936 (1971).
102 W. Kreis and A. Soricelli: J. Cancer Res. clin. Onc. *95*,233 (1979).
103 W.C. Rose, G.A. Basler, S.W. Mamber, M. Saulnier, A.M. Casazza and S.K. Carter: Proc. Am. Ass. Cancer Res. *29*,353 (1988).
104 C.K. Mirabelli, J.O. Bartus, S.M. Mong, G.A. Hofmann, H.F. Bartus, M.R. Mattern, R. Gupta and R.K. Johnson: Proc. Am. Ass. Cancer Res. *29*,328 (1988).
105 H. Stähelin: 9th Int. Cancer Congr., Tokyo, Abstracts, vol. I, p. 342, 1966.
106 P.I. Clark and M.L. Slevin: Clin. Pharmacokinet. *12*,223 (1987).
107 F. Zavala, D. Guenard, J.P. Robin and E. Brown: J. med. Chem. *23*,546 (1980).
108 M. Maturova, J. Malinsky and F. Santavy: J. natl Cancer Inst. *22*,297 (1959).
109 O. Buchardt, R.B. Jensen, H.F. Hansen, P.E. Nielsen, D. Andersen and I. Chinoin: J. pharm. Sci. *75*,1076 (1986).
110 H. Stähelin: Proc. Am. Ass. Cancer Res. *10*,86 (1969).
111 H. Stähelin: Eur. J. Cancer *9*,215 (1973).
112 A. Stoudemire, N. Baker and T.L. Thompson: Am. J. Psychiat. *138*,1505 (1981).
113 J. Venditti: Cancer Chemother. Rep. *55*,35 (1971).
114 W. Achterrath, N. Niederle and R. Raettig: In: Etoposid, derzeitiger Stand und neue Entwicklungen in der Chemotherapie maligner Neoplasien, p. 1. Eds S. Seeber, G.A. Nagel, W. Achterrath, C.G. Schmidt and R. Raettig. W. Zuckschwerdt, München 1981.
115 R. Raettig, N. Niederle and W. Achterrath: In: Etoposid (VP 16-213) in der Therapie maligner Erkrankungen, p. 3. Eds J. Schwarzmeier, E. Deutsch and K. Karrer. Springer-Verlag, Wien/New York 1984.
116 W.C. Rose and W.T. Bradner: In: B.F. Issell et al., ref. [1], p. 33 (1984).
117 P. Dombernowsky and N.I. Nissen: Eur. J. Cancer *12*,181 (1976).
118 T. Colombo, M. Broggini, M. Vaghi, G. Amato, E. Erba and M. D'Incalci: Eur. J. Cancer clin. Onc. *22*,173 (1986).
119 M. Rozencweig, D. VonHoff, J.E. Henney and F.M. Muggia: Cancer *40*,334 (1977).
120 S. Lazary and H. Stähelin: Experientia *24*,1171 (1968).
121 J.F. Borel, C. Feurer, H.U. Gubler and H. Stähelin: Agents Actions *6*,468 (1976).
122 S. Lazary: unpublished results.

123 M. Morita, A. Haji, A. Goto, N. Hattori and Y. Hasegawa: Fol. pharmac. jap. *87*,77 (1986).
124 D. Wiesinger: unpublished results.
125 W. Gassmann, H.U. Wottge, H. Loeffler and W. Müllerruchholtz: Bone Marrow Transplant. *3* (Suppl. 1),206 (1988).
126 Sandoz Pharma: Preclinical Brochure "VM 26, Cytostatic" (1967).
127 W. Calame, R. van der Waals, N. Douwes-Idema, H. Mattie and R. van Furth: Antimicrob. Agents Chemother. *32*,1456 (1988).
128 Y. Nishiyama, H. Fujioka, T. Tsurumi, N. Yamamoto, K. Maeno, S. Yoshida and K. Shimokata: J. gen. Virol. *68*,913 (1987).
129 T. Markkanen, M.L. Mäkinen, E. Maunuksela and P. Himanen: Drugs. exp. clin. Res. *7*,711 (1981).
130 H. Stähelin: In: Experimentelle und klinische Erfahrungen mit Podphyllin-derivaten in der Tumortherapie, p. 19 Eds H. Lettré and S. Witte. Schattauer, Stuttgart 1967.
131 H. Stähelin: In: G.K. Daikos, Progress in Chemotherapy, Vol. III, p. 819. Proc. 8th Internat. Congr. Chemother., Athens 1973, Hellenic Soc. Chemotherapy, Athens 1974.
132 H. Stähelin: In: Workshop on clinical usefulness of cell kinetic information for tumour chemotherapy, p. 21 Ed. M. van Putten. REP-TNO, Rijswijk 1974.
133 H.E. Skipper, F.M. Schabel, M.W. Trader and J.R. Thomson: Cancer Res. *21*,1154 (1961).
134 H. Stähelin: Eur. J. Cancer *12*,925 (1976).
135 H. Schön: unpublished results.
136 B.P. Wiesner, J. Yudkin: Nature *176*,249 (1955).
137 U. Friederich: SAKK (Schweizerische Arbeitsgruppe für klinische Krebsforschung) Bulletin *3*(3),3 (1983).
138 R.S. Gupta, A. Bromke, D.W. Bryant, R. Gupa, B. Singh and D.R. McCalla: Mutagenesis *2*,179 (1987).
139 S.M. Sieber, J. Whang-Peng, C. Botkin and T. Knutsen: Teratology *18*,31 (1978).
140 D.M. DeMarini, K.H. Brock, C.L. Doerr and M.M. Moore: Mutat. Res. *187*,141 (1987).
141 T.L. Avery, D. Roberts and R.A. Rice: Cancer Chemothr. Rep., Part 1, *57*,165 (1973).
142 M. Hacker and D. Roberts: Cancer Res. *35*,1756 (1975).
143 M. Kalberer-Rüsch: unpublished results.
144 T. Lee and D. Roberts: Cancer Res. *44*,2981 (1984).
145 T. Lee and D. Roberts: Cancer Res. *44*,2986 (1984).
146 J.A. Moscow and K.H. Cowan: J. natl Cancer Inst. *80*,14 (1988).
147 R.A. Kramer, J. Hakher and G. Kim: Science *241*,694 (1988).
148 L.M. Slater, S.L. Murray, M.W. Wetzel, P. Sweet and M. Stupecky: Cancer Chemother. Pharmac. *16*,50 (1986).
149 S. Akiyama, N. Shiraishi, Y. Kuratomi, M. Nakagawa and M. Kuwano: J. natl Cancer Inst. *76*,839 (1986).
150 R. Osieka, S. Seeber, R. Pannenbäcker, D. Soll, P. Glatte and C.G. Schmidt: Cancer Chemother. Pharmac. *18*,198 (1986).
151 P.R. Twentyman, N.E. Fox and D.J.G. White: Br. J. Cancer *56*,55 (1987).
152 R.B. Lock and B.T. Hill: Intern. J. Cancer *42*,373 (1988).
153 Sandoz Ltd.: VP 16-213, Preclinical Brochure (1970).
154 M.K. Spigelman, R.A. Zappulla, J.A. Strauchen, E.J. Feuer, J. Johnson, S.J. Goldsmith, L.I. Malis and J.F. Holland: Cancer, Res. *46*,1453 (1986).
155 H. Stähelin: Krebsinformation (Swiss Cancer League) *8* (3),79 (1973).
156 H. Stähelin: Progr. Allergy *38*,19 (1986).
157 J. Franz: unpublished results.

158 G. Rüttimann: unpublished results.
159 O. Wagner: unpublished results.
160 A. Grieder: unpublished results (1971).
161 L.M. Allen: Cancer Res. *38*,2549 (1978); Drug Metab. Rev. *8*,119 (1978).
162 R. Bailey-Wood, C.M. Dallimore, T.J. Littlewood and D.P. Bentley: Br. J. Cancer *52*,613 (1985).
163 B.H. Long: Natl Cancer Inst. Monogr. *4*,123 (1987).
164 L.M. Allen and P.J. Creaven: Eur. J. Cancer *11*,697 (1975).
165 P.J. Creaven: In: B.F. Issell et al., ref. [1] p. 103, (1984).
166 W.E. Evans, J.A. Sinkule, P.R. Hutson, F.A. Hayes and G. Rivera: In: B.F. Issell et al., ref. [1] p. 117, (1984).
167 M. D'Incalci and S. Garattini: In: Cancer Chemotherapy and Biological Response Modifiers, Annual 9, p. 67. Eds H.M. Pinedo, D.L. Longo and B.A. Chabner. Elsevier 1987.
168 M. D'Incalci and S. Garattini: In: Cancer Chemotherapy and Biological Response Modifiers, Annual 10, p. 57. Eds H.M. Pinedo, D.L. Longo and B.A. Chabner. Elsevier 1988.
169 K.H. Pflüger, L. Schmidt, M. Merkel, H. Jungclas and K. Havemann: Cancer Chemother. Pharmac. *20*,59 (1987).
170 M. D'Incalci, P. Farina, C. Sessa, C. Mangioni, V. Conter, G. Masera, M. Rocchetti, M. Brambilla Pisoni, E. Piazza, M. Beer and F. Cavalli: Cancer Chemother. Pharmac. *7*,141 (1982).
171 K. Hande, L. Anthony, R. Hamilton, R. Bennett, B. Sweetman and R. Branch: Cancer Res. *48*,1829 (1988).
172 F.M. Muggia: Proc. Am. Ass. Cancer Res. *11*,58 (1970).
173 G. Trempe, M. Sykes, C. Young and I. Krakoff: Proc. Am. Ass. Cancer Res. *11*,79 (1970).
174 F.M. Muggia, O.S. Selawry and H.H. Hansen: Cancer Chemother. Rep., part 1, *55*,575 (1971).
175 EORTC Clinical Screening Group: 7th Int. Congr. Chemother., Prague, Abstract B-8/11 (1971).
176 G. Mathé: Paris Match, 13 March (1971).
177 H. Limburg: Arch. Gynäk. *211*,384 (1971).
178 P. Dombernowsky, N.I. Nissen and V. Larsen: Cancer Chemother. Rep., part 1, *56*,71 (1972).
179 EORTC Co-operative Group: Br. med. J. *2*,744 (1972).
180 R.W. Sonntag, H.J. Senn, G. Nagel, K. Giger and P. Alberto: Eur. J. Cancer *10*,93 (1974).
181 M.A. Goldsmith and S.K. Carter: Eur. J. Cancer *9*,477 (1973).
182 M. Rozencweig, D.D. Von Hoff, J.E. Henney and F.M. Muggia: Cancer *40*,334 (1977).
183 P.J. O'Dwyer, M.T. Alonso, B. Leyland-Jones and S. Marsoni: Cancer Treatment Rep. *68*,1455 (1984).
184 J.L. Grem, D.F. Hoth, B. Leyland-Jones, S.A. King, R.S. Ungerleider and R.E. Wittes: J. clin. Oncol. *6*,351 (1988).
185 K.W. Brunner: Internal Report (1971).
186 N.I. Nissen, V. Larsen, H. Pedersen and K. Thomsen: Cancer Chemother. Rep., part 1, *56*,769 (1972).
187 EORTC, Clinical Screening Group: Br. med. J. *3*,199 (1973).
188 J.J. Wang and D.S. Chervinsky: Proc. Am. Ass. Cancer Res. *14*,110 (1973).
189 J.-L. Amiel, L. Schwarzenberg, M. Musset, F. de Vassal, M. Hayat and G. Mathé: Bull. Cancer *61*,419 (1974).
190 W.F. Jungi, H.J. Senn, C. Beckmann and P. Radielovic: Krebsinformation (Swiss Cancer League) *9* (2),50 (1974).

191 K.W. Brunner, R.W. Sonntag, F. Cavalli and H.J. Ryssel: Krebsinformation (Swiss Cancer League) 9 (2),49 (1974).
192 P.J. Creaven, S.J. Newman, O.S. Selawry, M.H. Cohen and A. Primack: Cancer Chemother. Rep., part 1, 58,901 (1974).
193 G. Falkson and H. Falkson: Proc. Am. Ass. Cancer Res. 15,160 (1974).
194 S. Seeber, G.A. Nagel, W. Achterrath, C.G. Schmidt and R. Raettig, (Eds): Etoposid, In: Aktuelle Onkologie, vol. 4. W. Zuckschwerdt Verlag, München 1981.
195 J. Schwarzmeier, E. Deutsch and K. Karrer, (Eds): Etoposid (VP 16-213) in der Therapie maligner Erkrankungen. Springer-Verlag, Wien 1984.
196 J.A. Sinkule: Pharmacotherapy 4,61 (1984).
197 P.J. O'Dwyer, B. Leyland-Jones, M.T. Alonso, S. Marsoni and R.E. Wittes: New Engl. J. Med. 312,692 (1985).
198 R. Marsoni, D. Hoth, R. Simon, B. Leyland-Jones, M. De Rosa and R.E. Wittes: Cancer Treatment Rep. 71,71 (1987).
199 F. Cavalli, A. Goldhirsch and R. Joss: In: B.F. Issell et al., ref. [1] (1984), p. 163.
200 F. Kung, F.A. Hayes, J. Krischer, D. Mahoney jr., B. Leventhal, G. Brodeur, D.H. Berry, R. Dubowy and S. Toledano: Invest. New Drugs 6,31 (1988).
201 L. Vellekoop, G. Spitzer, M. Keating, K. McCredie and E. Hersh: Proc. Am. Soc. clin. Oncol. 21,444 (1980).
202 S.N. Wolff and M.F. Fer: Proc. Am. Ass. Cancer Res. 23,134 (1982).
203 N.H. Mulder, A.F. Meinesz, D.T. Sleijfer, P.E. Postmus, E.G.E. de Vries, C.T. Smit Sibinga and R. Vriesendorp: In: Autologous Bone Marrow Transplantation and Solid Tumors, Eds J.G. McVie, O. Dalesio and I.E. Smith. Raven Press, New York 1984.
204 R.A. Stahel, R.W. Takvorian, A.T. Skarin and G.P. Canellos: Eur. J. Cancer clin. Oncol. 20,1233 (1984).
205 K.G. Blume, S.J. Forman, M.R. O'Donnell et al.: Blood 66 (suppl. 1), 883A (1985); Blood 69,1015 (1987).
206 A.R. Zander, S. Culbert, S. Jagannath et al.: Cancer 59,1083 (1987).
207 A. Gratwohl, A. Osterwalder, A. Lori, C. Nissen and B. Speck: Exp. Hemat. 13 (suppl. 17),11 (1985).
208 B. Häfliger, A. Gratwohl, A. Tichelli, A. Würsch and B. Speck: Sem. Oncol. 14 (Suppl. 1),134 (1987).
209 N. Ciobanu, E. Paietta, M. Andreeff, P. Papenhausen and P.H. Wiernik: Exp. Hemat. 14,626 (1986).
210 A. De Fabritiis, A. Pulsoni, A. Sandrelli, F. Simone, S. Amadori, G. Meloni and F. Mandelli: Bone Marrow Transplant. 2,287 (1987).
211 L. Wood and P. Jacobs: Am. J. Hemat. 27,63 (1988).
212 J.A. Snyder and R.C. Mc Intosh: A. Rev. Biochem. 45,699 (1976).
213 J.K. Kelleher: Molec. Pharmac. 13,232 (1977); Cancer Treatm. Rep. 62,1443 (1978).
214 J.D. Loike, C.F. Brewer, H. Sternlicht, W.J. Gensler and S.B. Horwitz: Cancer Res. 38,2688 (1978).
215 H. Stähelin and G. Poschmann: Oncology 35,217 (1978).
216 A. Grieder, R. Maurer and H. Stähelin: Experientia 29,772 (1973).
217 A. Grieder, R. Maurer and H. Stähelin: Cancer Res. 34,1788 (1974).
218 N.C. Misra and D. Roberts: Cancer Res. 35,99 (1975).
219 A. Krishan, K. Paika and E. Frei: J. Cell. Biol. 66,521 (1975).
220 A.P. Rao and P.N. Rao: J. Cell Biol. 67,351a (1975); J. natl Cancer Inst. 57,1139 (1976).
221 B. Barlogie and B. Drewinko: In: Pulse Cytophotometry, p. 226. Eds E. Göhde, J. Schumann and Th. Büchner. European Press, Ghent 1976.
222 B. Drewinko and B. Barlogie: Cancer Treatm. Rep. 60,1295 (1976).

223   L. Wilson, J.R. Bamburg, S.B. Mizel, L.M. Grisham and K.M. Creswell: Fedn Proc. *33*,158 (1974).
224   P. Gradwohl and H. Stähelin: 10th Int. Cancer Congr., Houston (1970), abstract No. 644.
225   A. Grieder, R. Maurer and H. Stähelin: Cancer Res. *37*,2998 (1977).
226   J. Ostrowski: unpublished results (1970/71).
227   A. Grieder and R. Maurer: Experientia *30*,703 (1974).
228   J.A. Fontana: Cancer Treatm. Rep. *63*,224 (1979).
229   R.J. Giever, R.S. Heusinkveld, M.R. Manning and G.T. Bowden: Int. J. Radiat Oncol. Biol. Phys. *8*,921 (1981).
230   R.S. Gupta: Cancer Res. *43*,1568 (1983).
231   B.T. Hill and A.S. Bellamy: Int. J. Cancer *33*,599 (1984).
232   J.D. Loike and S.B. Horwitz: Biochemistry *15*,5435 (1976).
233   J.J. Wang and D.S. Chervinsky: Proc. Am. Ass. Cancer Res. *14*,110 (1973).
234   M. Gosalvez, J. Perez-Garcia and M. Lopez: Eur. J. Cancer *8*,471 (1972).
235   M. Gosalvez, M. Blanco, J. Hunter, M. Miko and B. Chance: Eur. J. Cancer *10*,567 (1974).
236   M. Gosalvez, R. Garcia-Cañero and H. Reinhold: Eur. J. Cancer *11*,709 (1975).
237   C.C. Huang, Y. Hou and J.J. Wang: Cancer Res. *33*,3123 (1973).
238   R.A. Tobey: Nature *254*,245 (1975).
239   J.D. Loike and S.B. Horwitz: Biochemistry *15*,5443 (1976).
240   D. Roberts, S. Hilliard and C. Peck: Cancer Res. *40*,4225 (1980).
241   A.M. Arnold and J.M.A. Whitehouse: Br. J. Cancer *43*,733 (1981).
242   A.J. Wozniak and W.R. Ross: Proc. Am. Ass. Cancer Res. *23*,197 (1982); Cancer Res. *43*,120 (1983).
243   D.K. Kalwinsky, A.T. Look, J. Ducore and A. Fridland: Cancer Res. *43*,1592 (1983).
244   B.H. Long and A. Minocha: Proc. Am. Ass. Cancer Res. *24*,321 (1983).
245   A. Minocha and B.H. Long: Biochem. biophys. Res. Commun. *122*,165 (1984).
246   B.H. Long and M.G. Brattain: In: B.F. Issell et al., ref. [1] p. 63, (1984).
247   G.L. Chen, L Yang, T.C. Rowe, B.D. Halligan, K.M. Tewey and L.F. Liu: J. biol. Chem. *259*,13560 (1984).
248   W. Ross, T. Rowe, B. Glisson, J. Yalowich and L. Liu: Cancer Res. *44*,5857 (1984).
249   B. Glisson, R. Gupta, S. Smallwood, E. Cantu and W. Ross: Proc. Am. Ass. Cancer Res. *26*,340 (1985).
250   T. Rowe, G. Kupfer and W. Ross: Biochem. Pharmac. *34*,2483 (1985).
251   L.F. Liu: Proc. Am. Ass. Cancer Res. *29*,524 (1988).
252   D. Kerrigan, Y. Pommier and K.W. Kohn: Natl Cancer Inst. Monogr. *4*,117 (1987).
253   B. Glisson, R. Gupta, S. Smallwood and W. Ross: Cancer Res. *46*,1934 (1986).
254   M. Potmesil and W.E. Ross (eds): Natl. Cancer Inst. Monogr. *4*, several articles (1987).
255   J.A. Moscow and K.H. Cowan: J. natl Cancer Inst. *80*,14 (1988).
256   B.K. Sinha, N. Haim, L. Dusre, D. Kerrigan and Y. Pommier: Cancer Res. *48*,5096 (1988).
257   B.H. Long, S.T. Musial and M.G. Brattain: Cancer Res. *45*,3106 (1985).
258   B.H. Long, S.T. Musial and M.G. Brattain: Cancer Res. *46*,3809 (1986).
259   J.M. Woynarowski, R.D. Sigmund and T.A. Beerman: Biochim. biophys. Acta *950*,21 (1988).
260   J.M. Woynarowski, R.D. Sigmund and T.A. Beerman: Proc. Am. Ass. Cancer Res. *29*,274 (1988).

261 J. Markovits, Y. Pommier, D. Kerrigan, J.M. Covey, E.J. Tilchen and K.W. Kohn: Cancer Res. *47*,2050 (1987).
262 W.G. Nelson, K.R. Cho, Y. Hsiang, L.F. Liu and D.S. Coffey: Cancer Res. *47*,3246 (1987).
263 D.M. Sullivan, M.D. Lathan and W.E. Ross: Cancer Res. *47*,3973 (1987).
264 K. Chow and W.E. Ross: Molec. cell. Biol. *7*,3119 (1987).
265 K. Chow, C.K. King and W.E. Ross: Biochem. Pharmac. *37*,1117 (1988).
266 M. Bakic, D. Chan, B.S. Andersson, M. Beran, L. Silberman, E. Estey, L. Ricketts and L.A. Zwelling: Biochem. Pharmac. *36*,4067 (1987).
267 J. Hwang, J. Whang, S. Shyy, L.F. Liu: Proc. Am. Ass. Cancer Res. *29*,324 (1988).
268 W.D. Henner and M.E. Blazka: J. natl Cancer Inst. *76*,1007 (1986).
269 P.J. Smith, C.O. Anderson and J.V. Watson: Cancer Res. *46*,5641 (1986).
270 B.K. Sinha, M.A. Trush and B. Kalyanaraman: Biochem. Pharmac. *32*,3495 (1983).
271 J.M.S. van Maanen, C. de Ruiter, P.R. Kootstra, M.V. Lafleur, J. De Vries, J. Retèl and H.M. Pinedo: Eur. J. Cancer clin. Oncol. *21*,1215 (1985).
272 B.K. Sinha, M.A. Trush and B. Kalyanaraman: Biochem. Pharmac. *34*,2036 (1985).
273 J.M.S. van Maanen, J. De Vries, D. Pappie, E. van den Akker, M.V.M. Lafleur, J. Retèl, J. van der Greef and H.M. Pinedo: Cancer Res. *47*,4658 (1987).
274 N. Haim, J. Nemec, J. Roman and B.K. Sinha: Cancer Res. *47*,5835 (1987).
275 J.M.S. van Maanen, U.H. Verkerk, J. Broersen, M.V.M. Lafleur, J. De Vries, J. Retèl and H.M. Pinedo: Free Rad. Res. Commun. *4*,371 (1988).
276 H.H.J.L. Ploegmakers and W.J. van Oort: J. autom. Chem. *10*,135 (1988).
277 B.A. Teicher, S.A. Holden and C.M. Ross: J. natl Cancer Inst. *75*,1129 (1985).
278 M. Broggini, C. Rossi, E. Benfenati, M. D'Incalci, R. Fanelli and P. Gariboldi: Chem.-Biol. Interactions *55*,215 (1985).
279 W.E. Ross: Proc. Am. Ass. Cancer Res. *29*,535 (1988).
280 I. Jardine: In: Anticancer Agents Based on Natural Product Models, p. 319. Eds J. Cassady and J. Douros. Academic Press, Orlando 1980.
281 B.H. Long, S.T. Musial and M.G. Brattain: Biochemistry *23*,1183 (1984).
282 R.S. Gupta, P.C. Chenchaiah and R. Gupta: Anti-Cancer Drug Design *2*,1 (1987).
283 J.D. Griffiths, R.J. Stark, J.C. Ding and I.A. Cooper: Cancer Treatm. Rep. *70*,519 (1986).
284 R.S. Gupta: Cancer Res. *43*,505 (1983).
285 R.S. Gupta and B. Singh: J. natl Cancer Inst. *73*,241 (1984).
286 L.S. Thurston, H. Irie, S. Tani, F. Han, Z. Liu, Y. Cheng and K. Lee: J. med. Chem. *29*,1547 (1986).

Emerging concepts towards the development of contraceptive agents

By Ranjan P. Srivastava and A. P. Bhaduri*
Division of Medicinal Chemistry, Central Drug Research Institute, Lucknow 226 001, India

1	Introduction	268
2	Present knowledge of steroid-receptor interaction	271
2.1	Entry of steroid into the cell	273
2.2	Nature of the cytoplasmic receptor	273
2.2.1	Nature of the estrogen receptor	273
2.2.2	Nature of the progesterone receptor	274
2.3	Steroid-induced changes in cytoplasmic receptors	275
2.3.1	Estrogen receptor	275
2.3.2	Progesterone receptor	275
2.3.3	Mode of steroid-receptor interaction	275
2.4	Translocation of receptors from cytoplasm to nucleus	276
2.5	Interaction of the steroid-receptor complex with nuclear components	276
2.6	Dissociation of steroids from receptor sites	279
2.7	Tissue response evoked by steroids	279
3	Present status of estrogens, antiestrogens and antiprogestins in the context of steroid-receptor interaction	280
4	Current status of LHRH agonists and antagonists	288
4.1	LHRH agonists	288
4.2	LHRH antagonists	289
5	Present status of menses inducers	291
6	Approaches towards male contraceptives	300
7	Immunological approach to contraception: contraceptive vaccines	303
8	Conclusions	305
	Acknowledgment	306
	References	306

* CDRI Communication No. 4488.
To whom all correspondence should be made.

1 Introduction

In the present scenario of an ever-increasing global population the necessity of limiting births can not be denied. It seems that even nature is conscious of this fact since the number of pregnancies which are spontaneously interrupted has been found to be much higher than the number carried to term.

Historically, the development of chemical contraceptives started with the attempt to inhibit ovulation which was followed by interference with post-ovulatory events. The ability of certain hormones (chemicals) to cause reversible infertility in females by inhibiting pregnancy, when given orally, was first reported by Sturgis and Albright [1]. The first report of the clinical use of a chemical contraceptive was by Pincus [2] in the late fifties. A couple of decades later a major research program was launched by WHO for developing safe, acceptable and effective methods of fertility regulation [3].

Based on the observation of Pincus and Rock [4] that progesterone-estrogen combination administered orally gives good cycle control and is well accepted, many different combinations have been tested in the schedule day 5 to day 22 of the cycle [5]. Because of certain undesirable side effects of estrogen, clinicians became conscious of the need for avoiding this hormone or for using it in drastically reduced doses and this led to the use of gestogens. Various gestogens used to inhibit pregnancy are listed in Figure 1.

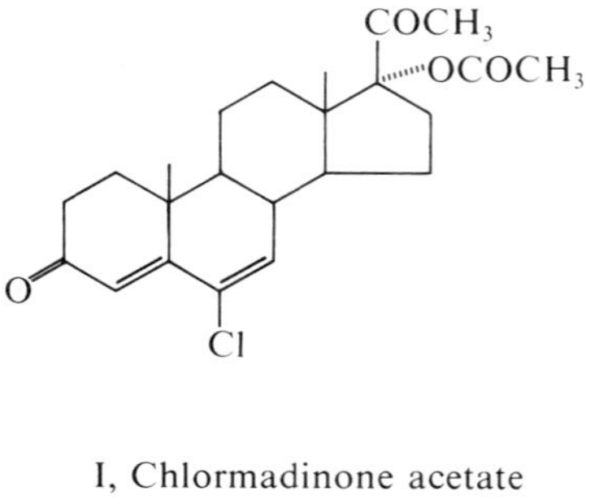

I, Chlormadinone acetate

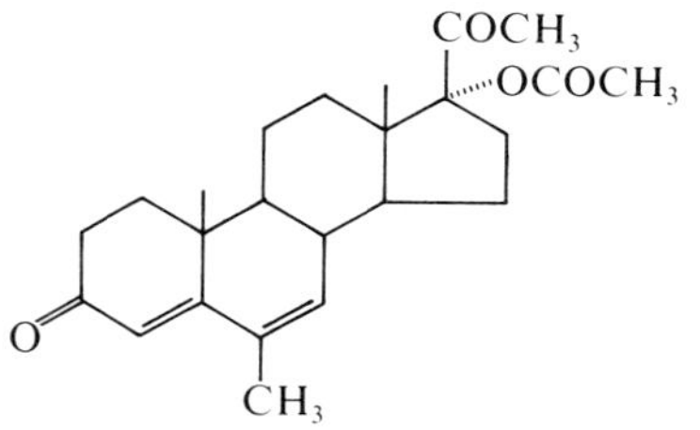

II, Megestrol acetate

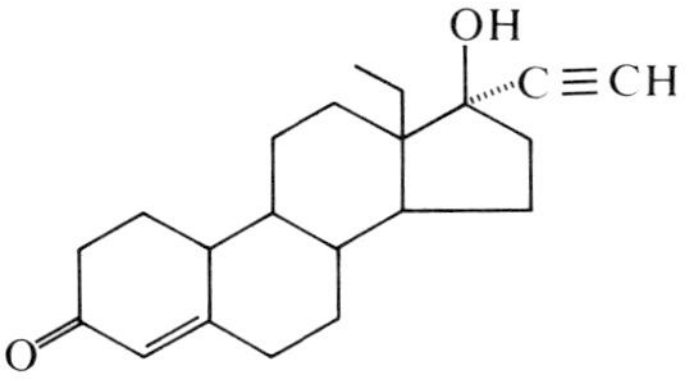

III, Norgestrel

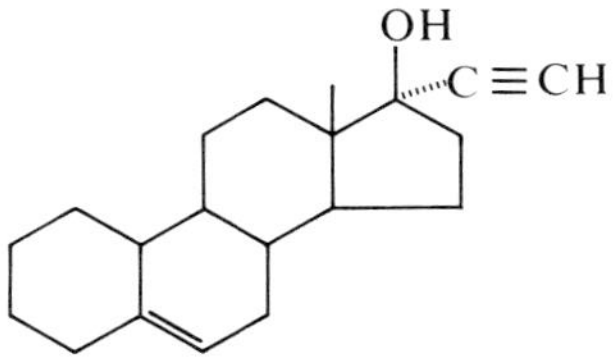

IV, Cingestrol

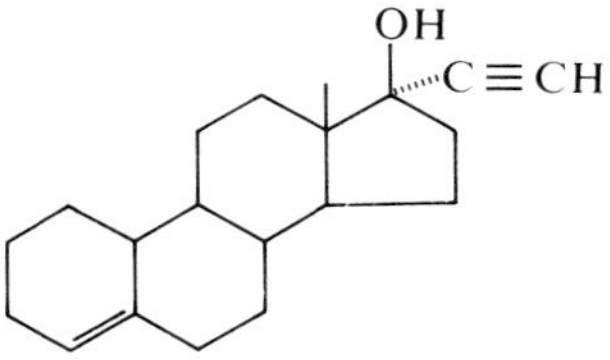

V, Lynestrenol

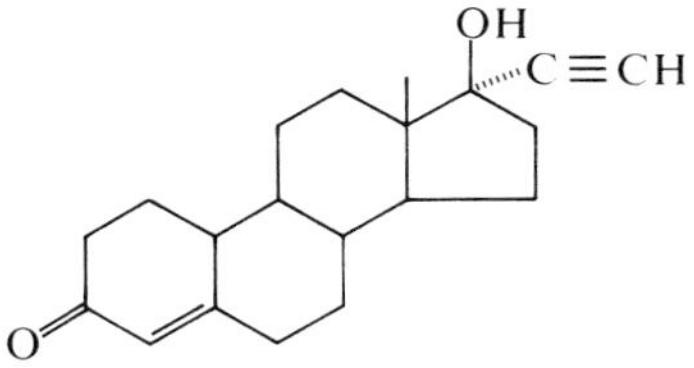

VI, Norethisterone

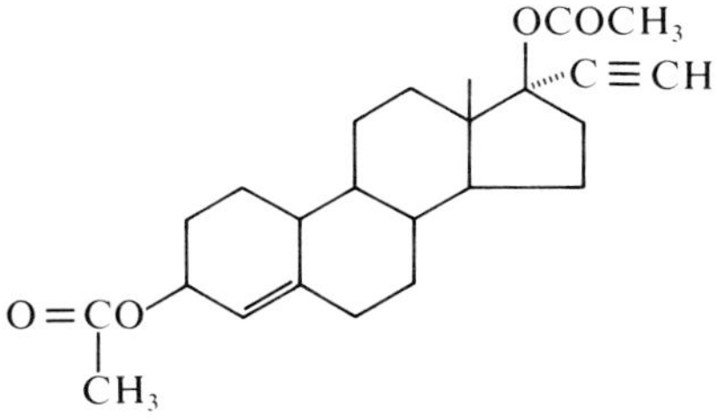

VII, Ethynodiol diacetate

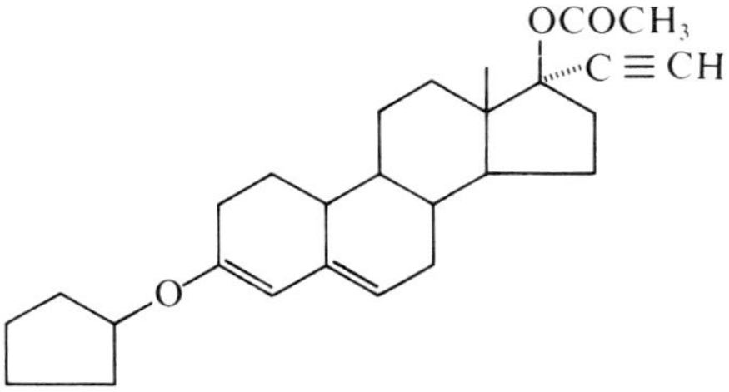

VIII, Quingestanol acetate

Figure 1
Gestogens used for contraception (pregnancy inhibition)

Unfortunately these gestogens did not prove clinically useful because of their side effects. They not only disturbed the menstrual pattern but also failed to prevent pregnancy in heavier women [6]. Various steroid hormones, singly and in combination, have been clinically evaluated [7] as long-acting injectable contraceptives (Table 1).

Table 1
Injectable contraceptives

Compound No.	Name	Total dose (mg)
	Monthly injectable hormonal preparations	
Progesterone alone		
IX	Norethisterone oenanthate	25
X	Medroxyprogesterone acetate	50
XI	20β-hydroxy-19-norprogesterone phenylpropionate	50
XII	Lynestrenol phenylpropionate	50
XIII	17α-hydroxy-19-norprogesterone caproate	200
XIV	Dihydroxyprogesterone acetophenide	200
XV	17α-hydroxyprogesterone caproate	500
Estrogen alone		
XVI	Estradiol undecylate	30
Estrogen-progestin combination		
XVII	dl-Norgestrel + estradiol hexahydro-benzoate	25 + 5
XVIII	Medroxyprogesterone acetate + estradiol cypionate	25 + 5 & 50 + 10
XIX	Norethisterone oenanthate + estradiol undecylate	30 + 50
XX	Medroxyprogesterone acetate + estradiol valerate	100 + 20
XXI	Dihydroxyprogesterone acetophenide + estradiol oenanthate	150 + 10
XXII	Dihydroxyprogesterone acetophenide + estradiol-3-benzoate, 17β-butyrate	150 + 10
XXIII	17α-hydroxyprogesterone caproate + estradiol valerate	500 + 10
Three-monthly injectable hormonal preparations		
XXIV	depot-Medroxyprogesterone acetate	150
XXV	Norethisterone oenanthate	200
I	Chlormadinone acetate	250
Six-monthly injectable hormonal preparation		
XXIV	depot-Medroxyprogesterone acetate	at various doses

Serious cycle irregularities were observed with these preparations and this was attributed to insufficient levels of the steroids in the circulation. Successful pregnancy requires adequate progestational support of the endometrium, both for the preparation for nidation and for the maintenance of the placenta [8], and any compound which depresses the progesterone level in the luteal phase is likely to cause endometrial breakdown and subsequent menstrual discharge. On the basis of this concept a number of antiprogestational agents have been evaluated for their pregnancy-inhibiting activity. Recent emphasis in the area of chemical contraception is on the development of an agent which if taken post-coitum for a short period, possibly 5–6 days, before the expected date of menses, would lead to the induction of menses even in cases of early pregnancy. In principle, the induction or regulation of menses can be achieved by interfering with the preparation of the endometrium for implantation, either by blocking the progesterone receptors or by the direct action of the agent on the endometrium or the endometrial-blastocyst junction. Menstrual regulation is also possible by inducing luteolysis which in turn leads to a decrease in the progesterone level and an interruption of pregnancy.

Immunological control of fertility is a fascinating approach and various attempts have been made in this direction. Since gonadotropins are known to play a role in fertility regulation, immunization of human females with hCG conjugates has been tried [9]. Another possible immunological approach is based on the observation that proteins specific to decidual cells are elaborated following transformation of stromal cells and hence an antiserum specific to decidual tissue might exhibit abortifacient activity [10]. Raising antibodies against spermatozoa seems to offer yet another approach to contraception. Interference with sperm motility and capacitation and use of medicated IUDs for the prevention of pregnancy have also been reported.

The present review deals with various approaches to fertility regulation in females and males, but does not include clinical experiments with prostaglandins for terminating pregnancy.

2 Present knowledge of steroid-receptor interaction

In view of the tremendous interest in the development of hormone antagonists as contraceptive agents, a proper understanding of the present status of hormone-receptor interaction becomes essential. Knowledge of the molecular basis of reproduction is an essential prerequisite for an understanding of fertility impairment in females. The mechanism of steroid-receptor interaction [11–13] (Figure 2) as proposed in 1960, was primarily concerned with estrogen-receptor interaction. The mechanism was assumed to be as follows: (i) the ste-

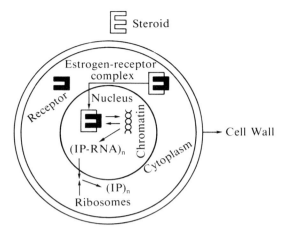

$(IP)_n$: specific proteins ($n \geq 1$)

Figure 2
Mechanism of steroid-receptor interaction

roid enters the cells of target tissues, (ii) it binds to specific receptors (proteins) located in the cytoplasm, (iii) the steroid-receptor complex that is formed translocates to the nucleus, and (iv) translocation results in the alteration of the pattern of gene expression.

This model of steroid-receptor interaction has been accepted by other workers [14–43] and refinements have been made in it. These refinements are mainly concerned with the entry of steroid hormone into the cell, the nature of the cytoplasmic receptor, the steroid-induced changes in cytoplasmic receptors, the translocation of receptors from the cytoplasm to the nucleus, the interaction of the steroid-receptor complex with the nuclear components, dissociation of the steroid from receptor sites and the tissue response evoked by the steroid. It appears that the basic nature of all steroid-receptor interactions is the same and that the sequence of events in estrogen-receptor interaction can serve as a model for explaining the molecular basis of all steroid-receptor interactions.

2.1 Entry of steroid into the cell

Experiments of Jenson & Jacobson [44] have shown that estradiol and estrone readily diffuse into the target cells because of their lipophilic property. In view of the large number of extracellular binding sites in the intact tissues which can influence uptake of steroid, Baulieu [45] has determined the equilibrium binding constants of whole tissue uptake of estradiol and cell-free binding and has reported a difference of the order of 5×10^{-9} in the binding constants. Milgram [46] has proposed that entry of estrogen into the cell is protein-mediated.

2.2 Nature of the cytoplasmic receptor

The cytoplasmic receptor is basically a protein whose specific affinity for the steroid is high ($K_d = 2 \times 10^{-10}$ to 1×10^{-9} M) [47–50]. Steroid receptors have been characterized on the basis of their sucrose density gradients.

2.2.1 Nature of the estrogen receptor

This receptor protein exists as a 8S molecule in a hypotonic (low-salt) medium [51, 52] and as a 4S molecule in a hypertonic (0.4 M KCl) medium [52–54]. According to Notides and Nielson [55] the cytoplasmic 4S estrogen receptor is an asymmetric protein with a molecular weight of 76,000. This confirms the earlier data of Yamamoto and Alberts [32 a]. The estrogen receptor has not been purified sufficiently to undertake studies on the preferential binding sites of estrogen. However, the sequence of amino acid residues involved in nuclear binding has been suggested [56]. It has been proposed that the human estrogen receptor possesses three hydrophobic domains. The first has 120 amino acids, while the second containing between 120 and 300 amino acids is rich in cysteine, lysine and arginine but poor in leucine and proline. The third hydrophobic region having a number of positively charged amino acids is assumed to be located on the surface of the receptor and consequently it can bind to nucleic acids. The sequence of amino acids present near the N-terminus of the estrogen receptor [57] is reported to be Arg-Pro-Gln-Leu-Lys.

2.2.2 Nature of the progesterone receptor

After purification by a combination of ammonium sulphate precipitation and affinity chromatography, the human uterine progesterone receptor has a sedimentation constant of 4S in a high-salt (0.4 M KCl) medium and of 8S at low ionic strength. It is an asymmetric protein with a molecular weight of 45,000 [58, 59]. Its asymmetric nature may be responsible for facilitating the entry of the hormone-receptor complex into the nucleus. The progesterone receptor, unlike other steroid hormone receptors, is found in relatively high concentrations in the target cells after estrogen exposure and in early pregnancy [24, 60]. The preferential binding sites of progesterone and the sequence of amino acids present at the active sites of the receptor have been elucidated. Teutsch [61] has suggested that steroid binding site in the progesterone receptor is in the form of two hydrophobic clefts. The smaller one which holds the steroid backbone is approximately 11 A° in length and has at either end a hydrogen binding zone with 0–3 and 0–17 or 0–20 oxygens at positions 3, 17 and 20 of the steroid. The second one is about 20 A° in length and extends above and below the plane of the steroid. The planes of the two grooves are roughly perpendicular to each other, and together form a cross-shaped (+) hydrophobic site which can tightly lock the steroid in place. This locking may be responsible for the antihormonal activity of 11β-substituted 19-norsteroids [61]. The amino acid residues involved in the progesterone-receptor interaction have been identified as histidine and methionine by Holmes et al. [62]. There is evidence [63] to suggest that progesterone binds to the receptor by at least two hydrogen bonds through oxygen atoms at positions 3 and 20 of the steroid molecule. According to Holmes [62], the methionine or histidine residue, adjacent to the D ring, may act as a hydrogen donor to the oxygen atom at the position 20 of the steroid molecule (0–20). Examination of steric models on the basis of affinity labelling with 11α- and 16α-(bromo[$2'-^3$H]-acetoxy)progesterones (BAP), has revealed that 11α-BAP alkylates position 1 of a histidine residue, while 16α-BAP alkylates position 3 of histidine and methionine residues. This affinity labelling does not occur in the presence of an excess of progesterone.

2.3 Steroid induced changes in cytoplasmic receptor

2.3.1 Estrogen receptor

Jensen et al. [35] have proposed that a cytoplasmic 4S receptor first binds to estradiol to form a 5S estrogen – receptor complex which translocates to the nucleus. They observed a slight difference in the sedimentation of cytoplasmic 5S and nuclear 5S complexes [64]. According to Yamamoto [65] the conversion of a 4S receptor to a 5S complex occurs more rapidly in the presence of DNA. It has been reported that phosphorylation of tyrosine of the calf uterine estrogen receptor by cytoplasmic kinase induces the binding of estrogen with the receptor [66]. The selection of phosphorylation of tyrosine by cytoplasmic kinase may require the presence of basic (lysine or arginine) and acidic (glutamate or aspartate) amino acids on the N-terminal site of the phosphorylated tyrosine [67].

2.3.2 Progesterone receptor

Two progesterone binding proteins have been found in the cytosol of chick oviduct [68], one of which aggregates to form a complex greater than 8S whereas the other does not aggregate and remains in 4S form.

2.3.3 Mode of steroid-receptor interaction

Two possible modes of steroid-receptor interaction have been proposed. The first envisages an equilibrium between two forms of receptors, one possessing a high affinity for the steroid and the other a lower affinity. Binding of the steroid with the high affinity form removes some of the latter from the existing equilibrium and possibly induces the low affinity receptors to change to the higher affinity form. This equilibrium model has been proposed by Manod et al. [69]. The second mode of interaction envisages only one form of receptor in the absence of steroid but in the presence of a steroid a change in the conformation of the receptor occurs. This is the basis of the 'induced fit' model proposed by Koshland [70]. Both models assume a steroid-induced transformation of receptors into another form. Presence of DNA has also been suggested as a cause of transformation of the monomer 4S progesterone receptor to its 8S tetramer [43].

2.4   Translocation of receptors from cytoplasm to nucleus

It has been observed that the steroid, after its entry into the target cell, is found in the nucleus of the cell [13, 71, 72] and that this process of translocation does not involve any energy-utilizing system [49]. The rate of translocation is directly proportional to the concentration of the steroid-receptor complex in the cytoplasm [73]. Translocation is not linked to any other cellular function and appears to be an inherent property of the receptor [49]. Perhaps the high binding affinity of steroid-receptor complexes for nuclear sites and the asymmetric nature of the receptor protein are the basic factors for translocation. On the basis of studies carried out in a cell-free system [74], it has been proposed that the driving force for translocation is the concentration gradient of the steroid-receptor complex in the two compartments which in turn is due to the binding of the receptor to some nuclear component (acceptor).

2.5   Interaction of the steroid-receptor complex with nuclear components

The specific interaction of the steroid-receptor complex with nuclear components can account for both the accumulation of receptors in nucleus and nuclear response to the hormone. A number of studies have been carried out to identify the specific nuclear binding components. The term 'acceptor' has been used to designate the nuclear sites that specifically bind to the steroid-receptor complex. Maurer and Chalkley [75], Teng and Hamilton [76] and King et al. [77] have shown that the estrogen-receptor complex interacts with nuclear chromatin. Spelsberg et al. [78] have shown that the progesterone-receptor complex binds to the chromatin extracted from the oviduct (the target tissue) and the specificity of the binding is due to the acidic proteins of chromatin. Fractionation of these proteins has furnished a number of non-histone proteins, designated as AP_3, which confer the binding ability on chromatin. These results indicate the existence of an acceptor site on the chromatin to which a hormone-receptor complex can bind. Unlike the progesterone-receptor complex, the estrogen-receptor complex does not specifically bind to a single chromatin component such as AP_3 acidic proteins. However, further studies [79–81] have revealed that steroid-receptor complexes interact directly with DNA. Liao et al. [82, 83] have shown that the estrogen-receptor complex binds to ribonucleoprotein particles in its target tissue.

The studies of Muller et al. [84, 85] have helped to explain the relationship between steroid action and gene expression. Tsai et al. [86] found that estrogen increases the number of new initiation sites for RNA synthesis on oviduct chromatin (supposedly equal to the number of transcribed genes), even though it is known that there are only one or two ovalbumin genes in chick DNA [87, 88]. It appears that a large part of the genome is influenced by the steroid-receptor complex and that many gene loci are involved. According to Yamamoto et al. [89], the estradiol-receptor complex produces its effect by binding to a small number of high affinity sites on the genome, but possesses low affinity for non-specific DNA sequences. These non-specific loci, because of their large number, completely mask the presence of the high affinity sites. It has been suggested that up to 10^3 specific sites, with affinities in the range of 10^{-8} to 10^{-10}M can exist without being detected by bulk binding assays currently in use. The steroid 17β-estradiol stimulates the binding of its receptor proteins to DNA. This interaction has been found to be of low affinity (1×10^{-4}M) [65, 90] and non-specific with respect to the DNA base sequence. Binding of receptor to double-stranded RNA has not been observed.

Based on the observation that a large number of receptor molecules translocate to the nucleus but only a relatively small number of genes are stimulated, Yamamoto and Alberts [89, 91] proposed a two-site model for DNA binding of the steroid-receptor complex. Interaction of a small number of high affinity binding sites and a large number of low affinity sites have been proposed. The binding of the steroid-receptor complex to high affinity binding sites has been reported to cause the hypomethylation of MSPI (nuclease, restriction enzyme) site situated at 611 base pair (bp) in the upstream region of vitellogenin II gene [92–94]. Similarly, an interaction of the steroid-receptor complex with DNA in this region of the gene can explain the presence of deoxyribonuclease I (DNase I) hypersensitivity at -700 bp upstream from the gene [95]. This DNase I hypersensitivity and the hypomethylation of MSPI site require the presence of estrogen. On the basis of these studies, Jost et al. [96] suggested that the steroid-receptor complex binds to the 5' upstream region of the gene where both hypomethylation of MSPI and DNase I hypersensitivity are observed. They also demonstrated that the nuclear estrogen-receptor complex binds preferentially to DNA fragment situated at the 5' end of the gene containing the MSPI hypomethylation site. The most probable binding site [96] for the estrogen-receptor complex is the region G–C–C–T–t–A–C–C–G–G–A–G–C–T–G–A–A–A–G–A–A–C–A–C . . . . of vitellogenin II gene. The 5' flanking sequence of the human estrogen-receptor gene starts from -128 to the capsite of nucleotide $+1$ [97]. The estrogen-receptor mRNA sequence begins at nucleotide $+1$ and ends at the start of the poly 'A' tail in nucleotide $+6322$ [56]. Three putative polydenylation signal sites (AATAAA), two potential cAMP-dependent kinase phosphorylation sites, potential nuclear transfer signal (Arg–Pro–Glu–Leu–Lys), 5'-ATG codon and the TGA termination codon have also been observed in the sequence [56] of the estrogen-receptor gene [56]. No potential N-linked glycosylation site is present.

A study of the preferential binding site for progesterone-receptor complex has been carried out by Mulvihill et al. [97]. They used a DNA cellulose competitive binding assay method to measure the extent of displacement of the chick DNA-cellulose by a purified cloned fragment of genomic DNA. Based on a comparison of various sequences of homology, a sequence that may result in a preferential binding of chick oviduct progesterone-receptor complex with hormone responsive genes [97] was proposed as

$$\text{ATC} \genfrac{}{}{0pt}{}{\text{CC}}{\text{TT}} \text{ATT} \genfrac{}{}{0pt}{}{\text{A}}{\text{T}} \text{TCTG} \genfrac{}{}{0pt}{}{\text{G}}{\text{T}} \text{TTGTA}$$

It has been suggested that specific double-stranded DNA sequences are recognized *in vitro* by the oviduct progesterone-receptor complex [97]. O'Malley et al. [98] have studied the binding of progesterone-receptor subunit 'A' to several chick gene DNAs. Sequence preference has revealed a marked retention of certain DNA fragments. Restriction endonuclease mapping suggests the presence of multiple receptor interaction sites flanking the 5' terminus of the ovalbumin gene. One of the preferential binding sites has been reported to be present between -135 and -247 bp upstream from the start of transcription. This region contains a $18/A+T$ rich base pair sequence desirable for the binding site itself.

2.6 Dissociation of steroids from receptor sites

Anderson et al. [99], with the help of exchange assay techniques, have demonstrated that estrogen receptors can lose one ligand and then rebind to a fresh ligand. This was confirmed by Spelsberg et al. [100] who have suggested that steroid is probably metabolized an secreted from the cell and the free receptor is recycled back into the receptor-acceptor site pathway.

2.7 Tissue response evoked by steroids

A model to explain tissue response to estrogen has been proposed by Szego et al. [101]. According to this model lysosomes act as carriers for steroids which in turn cause the lysosome to carry certain proteins to the nucleus. The evidence in support of this model is the in-

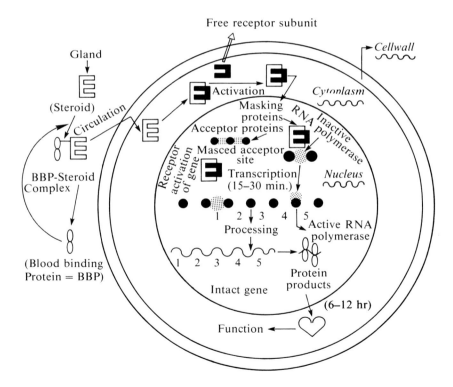

Figure 3
Basic pathway of the steroid-receptor interaction in target cells

crease in the levels of lysosomal enzymes in the nucleus at 2 and 10 min after estradiol injection. Estrogen is also reported to be bound by lysosome-rich cytoplasmic fractions. On the basis of the foregoing discussions, a graphic representation [100] of the 'two compartment steroid-receptor interactions' is shown in Figure 3. But the understanding of drug-receptor interaction would remain incomplete without taking into account the 'single compartment mode' of hormone-receptor interaction proposed by Gorski et al. [102]. They have suggested that receptors with or without bound estrogen, are present in the nuclear fraction. Binding of estrogen to the receptor changes the conformation of the macromolecule which either remains immobilized at its nuclear site or binds to some nuclear constituent. The nature of the latter is not clear but specific sequences of the DNA or the nuclear matrix or some complex of the two are attractive possibilities. Nuclear matrix is considered a nuclease-resistant nuclear structure but the concept of nuclear matrix is controversial [103]. Studies on the endocrinology of pregnancy and understanding of hormone-receptor interactions have significantly contributed to the development of hormone antagonists. The salient features of various developments in this area are discussed below.

3   Present status of estrogens, antiestrogens and antiprogestins in the context of steroid-receptor interaction

Impairment of fertility by antiestrogens or antiprogestins is an attractive approach to the design of antifertility agents. Studies on the molecular substructures of estrogen which are responsible for its estrogenic activity have helped immensely in the design of antiestrogens and the findings are reviewed here. On the basis of substructure analysis, estrogen agonists and antagonists can be classified into the following categories: (i) steroids, (ii) doisynolic acids, (iii) stilbenes, (iv) gem-diarylethylenes, (v) triarylethylenes and (vi) triarylpropiones. However, this classification does not mean that the estrogen agonist and antagonist activity of other molecules are less important [104–107].

Two substructures of steroidal estrogens responsible for their agonist activity are the C_3-phenolic hydroxyl and C_{17} β-hydroxy groups. The distance between two hydroxy functions has a profound influence on the binding affinitiy of the steroid. The importance of this distance in stilbene has been highlighted [108–110]. Besides these two hydroxy groups, the importance of other subunits such as the ethyl residue is evident from the study of the binding affinity (RBA) of various analogues of stilbene to estrogen receptor [105, 111–117]. The great importance of the geometry of stilbene is evident from the results of RBA studies with compounds in which the tetrasubstituted ethylenic double bond has been hydrogenated. It is interesting to note that only the meso isomer of the hydrogenated product has high affinity while the *d* and *l* isomers are poor ligands [118]. (−) *Trans*-doisynolic acid and allenoic acid are potent estrogen agonists [119] and possibly possess the binding sites of estradiol. Diarylethylenes were possibly the first category of compounds to provide a lead for the design of potential antiestrogens. Studies on it and on triarylethylenes such as tamoxifen, nafoxidine, clomiphene and centchroman have revealed the very important role of the aminoalkylether sidechain in these compounds [120–124]. This sidechain is essential for antiestrogenic activity while its absence imparts agonist activity. Its disposition in space is also critical [125]. Antiestrogenic activity of triarylethylene is closely related to its molecular geometry and is species-dependent; *cis*-tamoxifen and *cis*-clomiphene are agonists while their *trans*-isomers have both agonist and antagonist properties [126, 127]. *Trans*-tamoxifen is almost a total agonist in mice, an antagonist in chicks and a mixed agonist and antagonist in rats [120, 128–130]. Assessment of the role of the subunits in non-steroidal estrogens/an-

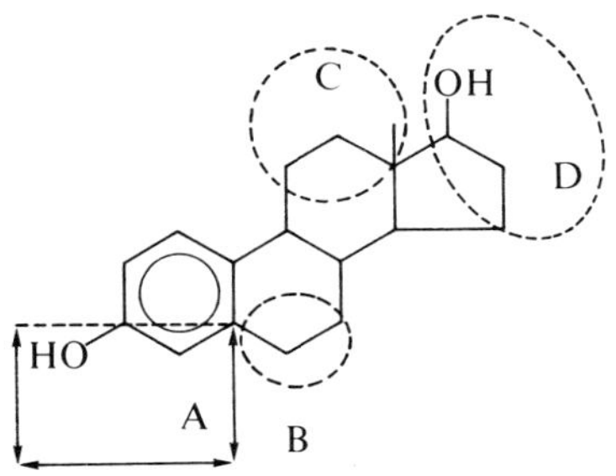

XXVI, Estradiol 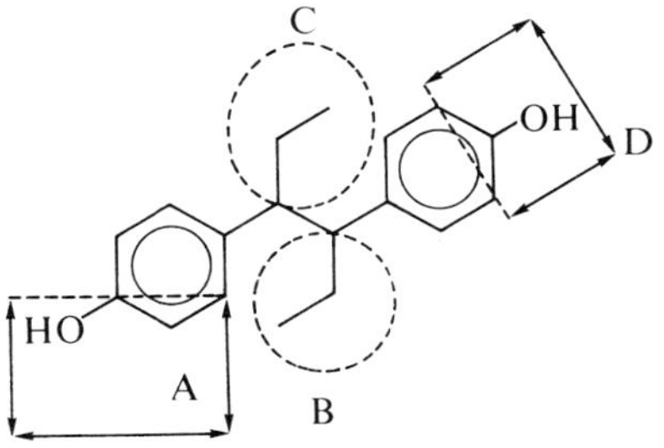   XXVII, Hexestrol

Figure 4
Substructural units important for the activity of various estrogen agonists/antagonists

tiestrogens would remain incomplete without taking into consideration their metabolites. For example hydroxy tamoxifen, a major metabolite of tamoxifen, is a better antiestrogen than the parent compound [131]. Substructural units of various estrogen agonists/antagonists are described in Figure 4.

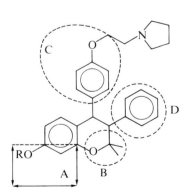

XXVIII, Allenolic acid

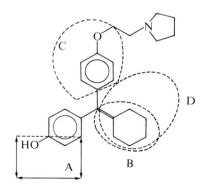

XXIX, Cyclohexane carboxylic acid

XXX, Centchroman
R = CH$_3$

XXXI, Diarylethylenes

Figure 4 (continued)

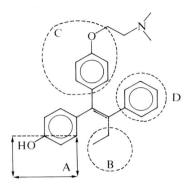

XXXII, Hydroxytamoxifen

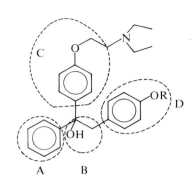

XXXIII, MER-25
R = CH₃

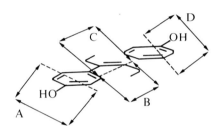

XXXIV, Diethylstilbestrol

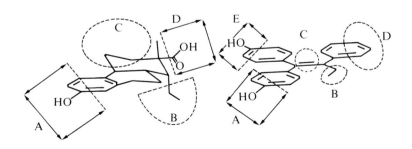

XXXV, Doisynolic Acid   XXXVI, Triaryl ethylenes

Figure 4 (continued)

Following the reports of antifertility activity of triaryl propenones [132, 133], new antiestrogens such as trioxyphen and LY-117018 were synthesized [134, 135]. Various studies carried out by changing the subunits of estrogen agonists and antagonists have shown that inter-

XXXVII, Trioxyphen   XXXVIII, LY-117018

action of these compounds with the estrogen receptor is responsible for their activity. It is likely that the antagonist action of a compound is due to its ability to inhibit the binding of estrogen to the estrogen receptor. Possibly the interactions of estrogen-estrogen receptor complexes and antiestrogen-estrogen receptor complexes to DNA are different and in the overall mechanism of estrogenic action, the alkylaminoalkyl sidechain plays a significant role.

The search for an antiprogestin as a contraceptive agent also represents a major area of research. Antiprogestins, if administered during very early pregnancy, may bind to endometrial progesterone receptors and thereby cause interference which may disturb the maintenance of the secretory endometrium essential for pregnancy. Design of antiprogestins, having selective antiprogestational activity therefore, requires a very careful analysis of substructures. Gestrinone (R 2323), the first antiprogestin to be synthesized, is characterized by a $C_{13}-$ ethyl group [136] and a $\Delta^{4, 9, 11}$-triene system [137]. The antiprogesterone activity of this compound has been demonstrated in several species of animals [138–140]. Effect of substituents at C_7 and C_{11} has been studied at Roussel Uclaf [141]. The presence of substituents at C_{11} as in Moxestrol (R 2858), RU 16117 and C_{11}-methoxy

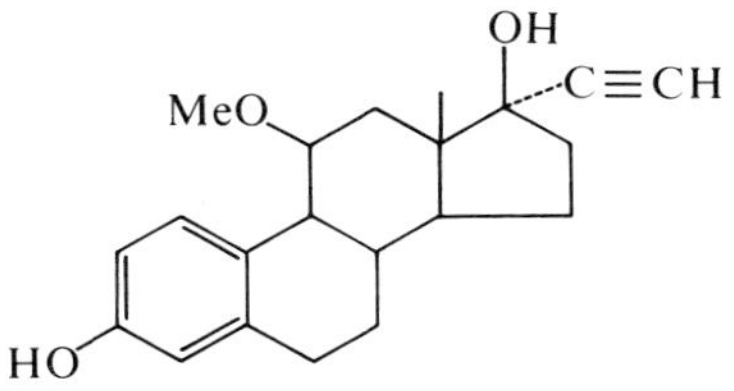

XXXIX, Moxestrol

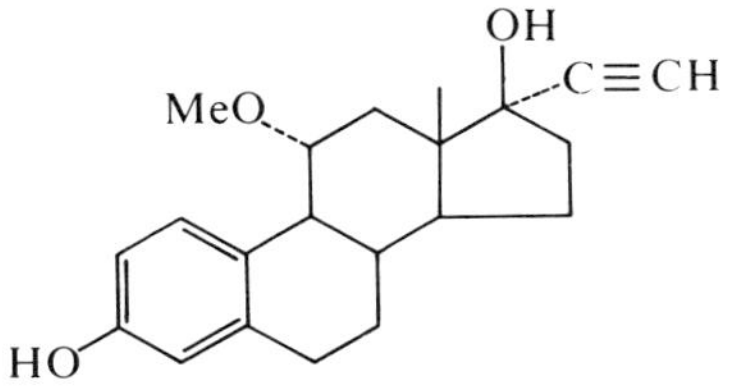

XL, RU 16117

dienes (Norethisterone series) results in better antiprogestational activity. Studies have also been carried out with substituents at C_{13} and C_{17} [142]. Introduction of an aromatic ring or a long sidechain at C_{11} has furnished interesting results. A long sidechain has been found to increase affinity to progesterone receptor while a 11β-phenyl group confers antiprogestational activity. Structures of important C_{11} substituted derivatives are given [143–145] in Figure 5.

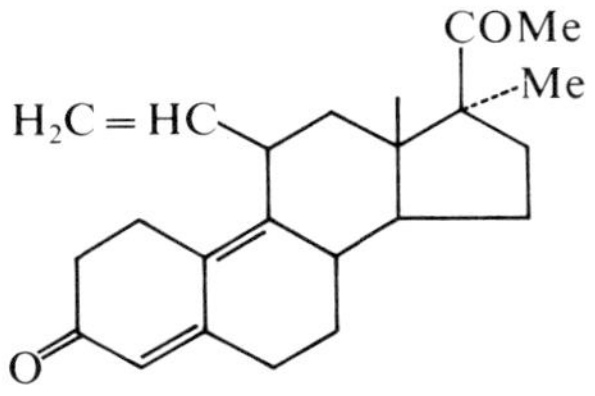

XLI, RU 25253

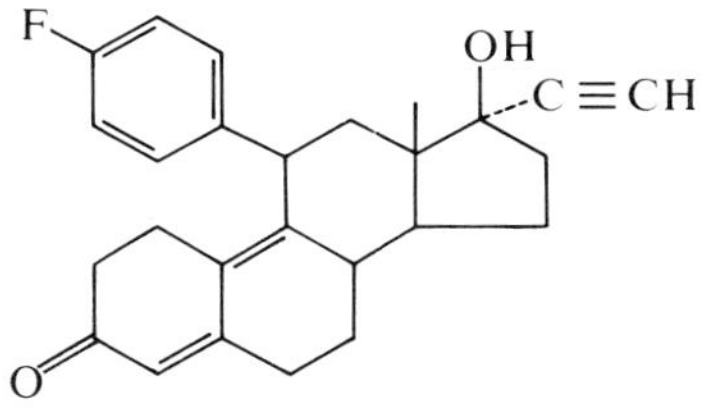

XLII, RU 25593

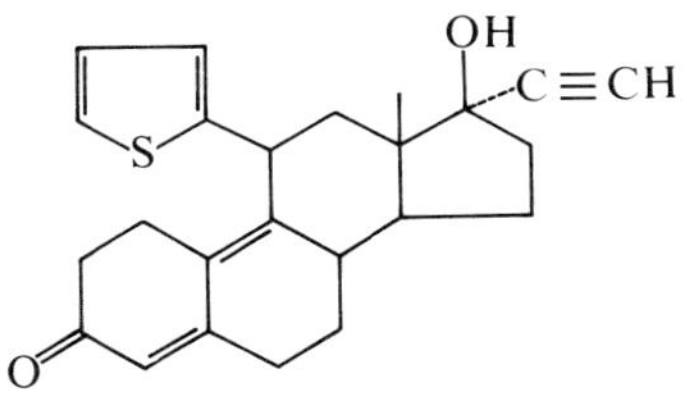

XLIII, RU 25055

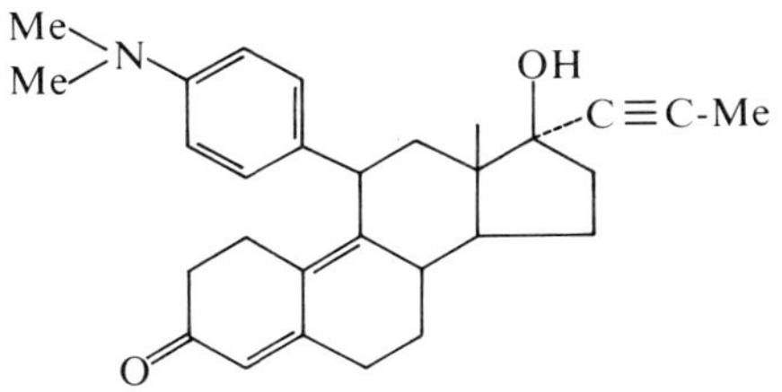

XLIV, RU 38486 or RU 486

Figure 5
Chemical structures of C_{11} derivatives

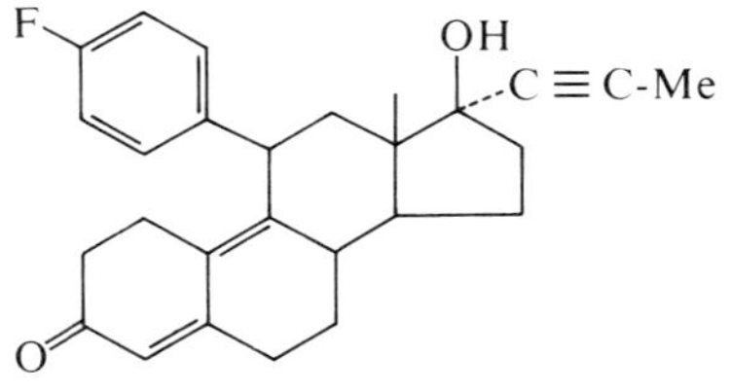

XLV, RU 38140

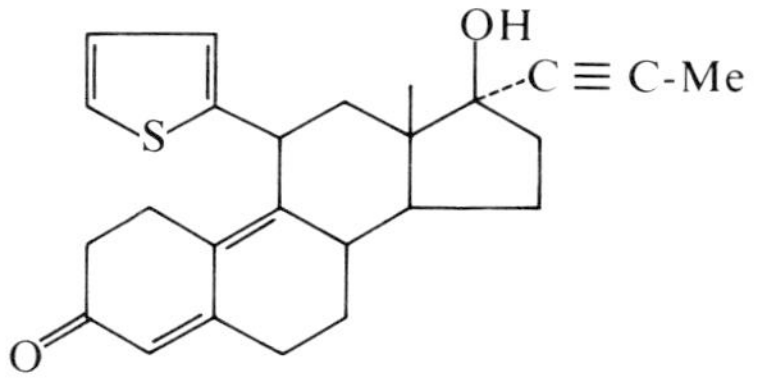

XLVI, RU 28289

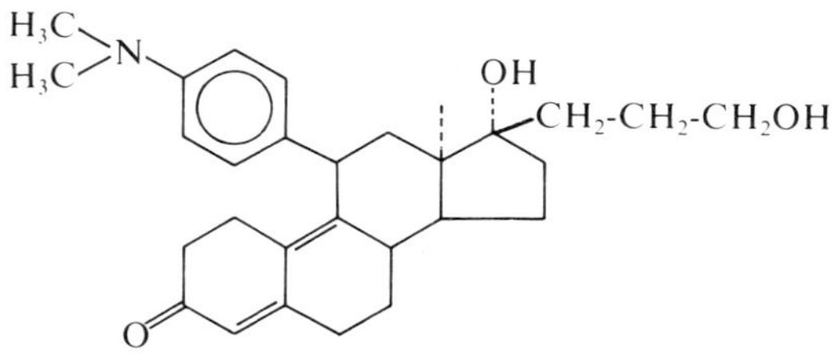

XLVII, ZK 98299

Biological evaluation of C_{11} substituted derivatives has shown that these compounds can interact with the glucocorticoid receptor. The synthetic antiprogestin RU 486 (17β-hydroxy-11β-[4-dimethylaminophenyl]-17α-[1-propynyl]estra-4,9-diene-3-one, Roussel-Uclaf, Paris, France), when given to normally cyclic women in midluteal phase, induces menstruation by local action on the endometrium, very likely at the progesterone-receptor level [146]. Proliferation of endometrium and withdrawal bleeding can be redeveloped after the exposure to this compound. It does not affect serum cortisol or pituitary gonadotropins [147, 148]. *In vitro* and *in vivo* experiments in various animal species have suggested that RU 486 has no agonist activity but is antagonist to both progesterone and synthetic glucocorticosteroid (dexamethasone). It reduces the inhibition of corticotropin (ACTH) secretion induced by corticosteroid in rat pituitary cells [149] and the action of dexamethasone on cultured L-929 fibroblasts [150]. It also antagonizes the effects of corticosteroid on liver glycogen and tryptophan pyrrolase, thymolysis and diuresis [151]. In nonpregnant women, RU 486 (= 1 mg/kg of b. wt per day) interrupts the luteal phase without affecting the pituitary adrenal axis, thereby indicating that it is rather an antiprogesterone than an antiglucocorticoid. There is a significant increase in plasma corticotropin, β-lipotropin and cortisol concentration in case of pregnancy interruption by RU 486 (= 4 mg/kg per day) [152]. RU 486 has been reported to antagonize the acute as well as chronic effects of dexamethasone (*in vivo* and *in vitro*) when administered to different target tissues in the animal and does not exhibit any agonist activity even at large doses [151]. When RU 486 is administered at 6 mg/kg along with 1 mg of dexamethasone it reduces the inhibitory effect of dexamethasone on the pituitary adrenal axis. This indicates that RU 486 is an antiglucocorticoid that disrupts the pituitary feedback of both cortisol and dexamethasone. The potent antagonist effect of RU 486 is directly related to its strong interaction with the cytosolic glucocorticoid receptor (GR), since its relative binding affinity for rat thymus GR is approximately 3 times that of dexamethasone at 0 °C. The affinity of activated RU 486-GR complex for DNA is much lower than that of dexamethasone. The *in vitro* nuclear uptake of [^{3}H] RU 486 and that of [^{3}H] dexamethasone were compared after incubation with minced thymus at 25 ° and 37 °C [153]. A low specific uptake of [^{3}H] RU 486 was observed irrespective of the concentration or the period of incubation. These observations lead to the conclusion that glucocorticoid agonists interact with the nonactivated receptor to form a more stable activated complex (low K^{-1}). RU 486 interacts strongly with the nonactivated receptor but the resulting complex is less able to translocate properly into the nucleus for evoking the glucocorticoid response. The antiglucocorticoid activity can be minimized along with optimization of the antiprogestational effect by just modifying the dose and the time of administration of RU 486. It has been found that RU 486 does not exhibit any glucocorticoid activity at a dose of 100 mg/day in normally cyclic women [152]. Recently it has been suggested that the antiglucocorticoid activity of the drug has no clinical relevance at the doses used for fertility control [154]. Some progesterone agonist activity and gonadotropin inhibitory activity of RU 486 have also been reported in literature [155], probably due to interference with events involved in the regulatory mechanism. RU 486 has also been utilized as a menses inducers in women (administration during the first week after the expected date of menses) at a total dose of 800 mg given over 4 days [156]. However, a large-scale trial ist necessary to confirm its safety and to determine its optimal mode of administration.

Structural modification of RU 486 has also been carried out with a view to dissociate antiprogestational and antiglucocorticoid activities. Of the various modified compounds the potent antiprogestin ZK 98299 (XLVII) has been found to possess lower antiglucocorticoid activity [157].

4 Current status of LHRH agonists and antagonists

The isolation, structure elucidation and synthesis of luteinizing hormone releasing hormone (LHRH), which controls the release of luteinizing hormone (LH) and follicle-stimulating hormone from the pituitary, have helped to identify it as yet another target for the control of fertility. The amino acid sequence in human LHRH is as follows: pGlu^1-His^2-Trp^3-Ser^4-Tyr^5-Gly^6-Leu^7-Arg^8-Pro^9-Gly^{10}-NH$_2$ [158, 159]. Earlier work on the chemistry and pharmacology of LHRH analogs has already been reviewed [158–163] and only the recent advancements are described here. Agonists and antagonists of LHRH have been used to impair fertility. The clinical relevance of these agonists and antagonists in contraception and the salient changes made in the structure of LHRH are the important aspects dealt with in the present review.

4.1 LHRH agonists

A potent LHRH agonist when used continuously decreases the stimulatory response and may lead to complete tachyphylaxis to the production of biologically active gonadotropins [164]. In principle the periodic use of LHRH agonists can also impair fertility. For example, its use in midluteal phase may decrease hyperestrogenicity [165] while its use during the period in which hyperprolactinemia of lactation occurs, may block the symptomatic stimulation. These sequences of events may be responsible for contraceptive response of LHRH agonists [166, 167]. The hyperstimulation and tachyphylaxis described earlier lead to continuous secretion of biologically inactive gonadotropin [168–171]. Continuous LHRH treatment of rats and mice elevates the α-LH subunit mRNA but depresses the β-subunit in pituitary [172] which causes hyper- and neoplastic changes in the pituitary. It has been proposed that a potent LHRH agonist possesses high hydrophobicity [173] due to which it competes with the membrane bound receptors [174] for the same binding sites on phospholipid membrane [163]. Based on this concept, molecular modifications have been made in LHRH agonists to reduce their rate of degradation in the biophase. Incorporation of a D-amino acid at position 6 and an additional substituent at position 10 of LHRH in order to impair the catabolic sites of the latter (positions 5, 6, 9 and 10) represent the salient structural modifications for obtaining the agonist [175, 176]. The potencies of various LHRH agonists have also been correlated with their *in vitro* receptor affinities [177]. Of various LHRH agonists studied, Nafarelin [(D-3-(2-naphthylalanine)^6] LHRH (XLVIII) by nasal route has been shown to be the most potent.

LHRH agonists at low doses have been reported as female contraceptive agents without ovarian suppression [178, 179] but they cause endometrial hyper- and neoplasia, probably due to the development of an unopposed estrogen state [180]. Combination of a LHRH agonist with a progestogen [181], or its use as a luteolysisinhibiting agent has not shown any advantages over its use as a routine contraceptive.

4.2 LHRH antagonists

Competitive LHRH antagonists, superior to endogenous LHRH, may successfully block the gonadotropin function even at a low dose. They do not have a stimulatory phase and also do not interfere with receptor down-regulation [182]. Gonadal suppression by LHRH antagonists increases the endogenous. LHRH secretion [163]. The lack of receptor down-regulation and increased LHRH synthesis are better induced by antagonist than agonists [166]. LHRH antagonists alone or in combination with prostaglandins, have been successfully used [183] as interceptives (luteolysis).

It has been found that replacement of His^2 residue in LHRH by D-amino acids leads to antagonist activity. Based on this observation various structural modifications have been carried out (Table 2) of which the salient ones are as follows.

Position 1': Hydrophobic N-blocked amino acid to block proteolysis of N-terminus and to increase hydrophobicity of the molecule.

Position 2': D-aromatic amino acid to control the antagonist character of the molecule.
Position 3': Aromatic D-amino acid which not only removes the residual agonist activity but also prevents proteolysis and stimulates hydrophobicity.
Position 6': Aromatic D-amino acid to prevent proteolysis and increase hydrophobicity.
Position 10': D-Ala or aza-Gly-NHEt to prevent proteolysis.

Table 2
Potent LHRH antagonists

Compound No.	Description of antagonist	Claimed advantages	Ref.
XLIX	[N-Ac-Pro^1, D-pF-Phe^2, D-Nal(2)^{3,6}] LHRH	Class of potent, extremely hydrophobic analogs	[158, 159], [184]
L	[N-Ac-D-pCl-Phe^1, D-pCl-Phe^2, D-Trp^3, D-Arg^6, D-Ala^{10}] LHRH	High potency, prolonged duration of action	[185]
LI	[N-Ac-D-Nal(2)^1, D-pF-Phe^2, D-Trp^3, D-Arg^6] LHRH	Improved potency (under clinical trial)	[186, 187]
LII	[N-Ac-D-Nal(2)^1, D-pCl-Phe^2, D-Trp^3, D-Arg^6, D-Ala^{10}] LHRH	– ditto –	[186, 187]
LIII	[N-Ac-D-Nal(2)^1, D-pCl-Phe^2, D-Pal(3)^3, D-Arg^6, Trp^7, D-Ala^{10}] LHRH	Improved potency	[188, 189]
LIV	[N-Ac-D-Nal(2)^1, D-pCl-Phe^2, D-Trp^3, D-Arg^6, Phe^7, D-Ala^{10}] LHRH	– ditto –	[188–190]
LV	[N-Ac-D-Nal(2)^1, D-pCl-Phe^2, D-Trp^3, D-hArg(Et$_2$)^6, D-Ala^{10}] LHRH	Improved duration of action	[191]
LVI	[N-Ac-D-Nal(2)^1, D-α-Me, pCl-Phe^2, D-Trp^3, D-Arg^5, D-Tyr^6, D-Ala^{10}] LHRH	Lower antiovulatory activity	[192]
LVII	[N-Ac-D-Nal(2)^1, D-pCl-Phe^2, D-Pal(3)^3, Arg^5, D-Glu(pMeOPh)^6, D-Ala^{10}] LHRH	Low MCD potency	[193]
LVIII	[N-Ac-D-Nal(2)^1, D-pCl-Phe^2, D-Pal(3)^{3,6}, Arg^5, D-Ala^{10}] LHRH	Improved antiovulatory activity	[194]
LIX	[N-Ac-D-Nal(2)^1, D-α-Me, pCl-Phe^2, D-Pal(3)^3, Arg^6, Lys(iPr)^8, D-Ala^{10}] LHRH	Improved gut wall transit with high potency and low MCD	[195]
LX	[N-Ac-D-Nal(2)^1, D-pCl-Phe^2, D-Trp^3, D-Lys(iPr)^6, Lys(iPr)^8, D-Ala^{10}] LHRH	– ditto –	[195]
LXI	[N-Ac-D-Nal(2)^1, D-α-Me, pCl-Phe^2, D-Pal(3)^3, D-Arg^6, hArg(Et$_2$)^8, D-Ala^{10}] LHRH	– ditto –	[195]

Com- pound No.	Description of antagonist	Claimed advantages	Ref.
LXII	[N-Ac-D-Nal(2)^1, D-pCl-Phe^2, D-Pal(3)^3, Lys(Nic)^5, D-Lys(Nic)^6, Lys(iPr)^8, D-Ala^{10}] LHRH	Very low MCD with decreased antiovulatory activity (under clinical trial)	[164]
LXIII	[N-Ac-D-Nal(2)^1, D-pCl-Phe^2, D-Pal(3)^3, D-hArg(Et$_2$)^6, hArg(Et$_2$)^8, D-Ala^{10}] LHRH	Reduced MCD with a high antiovulatory activity (under clinical trial)	[166]
LXIV	[N-Ac-D-Nal(2)^1, D-pCl-Phe^2, D-Pal(3)^{3,6,} hArg(Et$_2$)^8, D-Ala^{10}] LHRH	– ditto –	[166]
LXV	[N-Ac-D-Nal(2)^1, D-pCl-Phe^2, D-Pal(3)^{3,6,} Pal(3)^5, hArg(CH$_2$CF$_3$)$_2^8$, D-Ala^{10}] LHRH D-Ala^{10}] LHRH	Very low MCD potency	[166]

5 Present status of menses inducers

The concept of menses inducers owes its origin to the need of a simple and safe method of birth control and is concerned with the use of a chemical agent as an oral pill. The use of this chemical agent is recommended for a short span of time before the expected date of menses (even in cases where the probability of pregnancy exists). The overall pressure of this agent per annum would be substantially less compared to combination pills since a sexually active woman using only an effective menses inducer and no other method of birth control would theoretically need 12–13 exposures per annum. The objective behind the development of menses inducer is, therefore, to avoid unnecessary physiological complications.

A menses inducer elicits its biological response directly or indirectly either on the endometrium or on the endometrium-blastocyst junction and makes the bioenvironment conducive to pregnancy failure. Rational design of a chemical agent capable of interference at the endometrial-blastocyst junction, would requires an understanding of membrane behavior at the site of action. It has been suggested [196] that modification of the membrane occurs before the acquisition of fertility and that free radicals originating from ascorbate and/or estradiol contribute significantly towards the membrane changes preceding fertility. The exact nature of the qualitative and quantitative changes occurring at the blastocyst-endometrial junction have not been elucidated. However, interference with the free radical involved in regulating the adhesiveness of the blastocyst and implantation reaction is an attractive concept for design of antifertility drugs. Adhesion of the blastocyst to the endometrial tissue can be reversed if a site selective chemical agent interacts either with the free radicals or with the membrane protein or brings about changes in the lipid symmetry. If such a chemical agent does not involve the endocrine hormones it would be more acceptable as a contragestational agent. In the light of this requirement, the obvious choice for designing this type of contragestational agents falls on heterocycles. A number of heterocyclic compounds have been evaluated as pregnancy inhibiting agents [197–222] but the mechanism of action of many of these compounds has not been reported. Even where the mode of action of a contragestational heterocycle has been proposed, its use as menses inducer has either not been evaluated or not reported. In order to give a clear picture of this situation, in the present review the contragestational efficacy of heterocycles has been discussed under two heads: one deals with completely elucidated mechanisms of action while the other relates to incomplete information. The emphasis while reviewing the first category of compounds is on their potential as menses inducers.

The mode of action of the most effective heterocycle, namely 3-(2-ethylphenyl)-5-(3-methoxyphenyl)-H-1,2,4-triazole (LXVI; DL III-IT), has been studied in animals by Galliani et al. [223]. They have reported that it acts directly on the utero-placental complex

LXVI, DL III-IT

when administered during the early post-implantation period. The salient observations made by these workers are:
(i) Progesterone does not antagonize the contragestational effect of LXVI and the circulating level of progesterone does not change during the period of contraception. This rules out any luteolytic effect.
(ii) In a hypophysectomized and ovariectomized animal, whose preg-

nancy is maintained with appropriate hormonal treatment, LXVI induces pregnancy arrest and adrenalectomy does not interfere with its contraceptive action. This suggests that pituitary, ovaries and adrenals are not involved in the contraceptive effect.

(iii) Compound LXVI has been found to be devoid of hormonal or antihormonal activity in standard endocrinological tests [223]. Studies [223, 224] on its mode of action have revealed that it causes selective degeneration of decidual cells and increases the number of progesterone receptors which is followed by a marked decrease in the ODC (ornithine decarboxylase) activity in the product of conception. However, this compound does not directly inhibit ODC activity. It has, therefore, been suggested that LXVI interferes with the chain of events by which progesterone regulates the mitotic activity of decidual and trophoblastic cells necessary to the placenta for achieving morphological and functional maturity. While these observations did stimulate research on heterocycles as candidate menses inducers, the substructures (pharmacophores) in heterocycles responsible for eliciting pregnancy interceptive activity remained unidentified. Substructural analysis of heterocycle molecules requires the evaluation of new heterocycles as menses inducers. As a first step in this direction, positional isomers in triazoles (LXVII, LXVIII) and compounds in which one of the nitrogen atoms has been replaced by other heteroatoms (LXIX) have been evaluated as menses inducers [225].

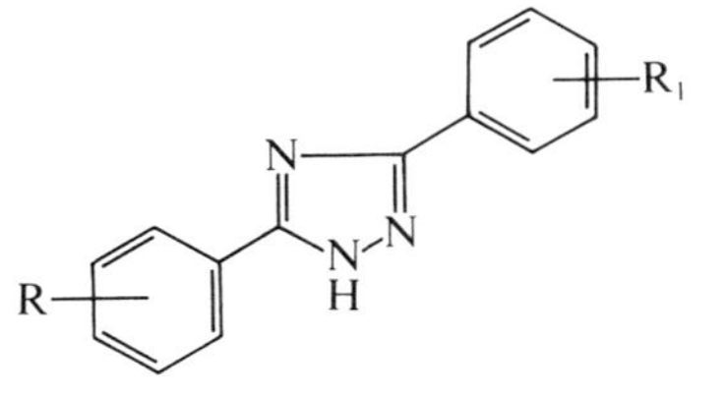

LXVII

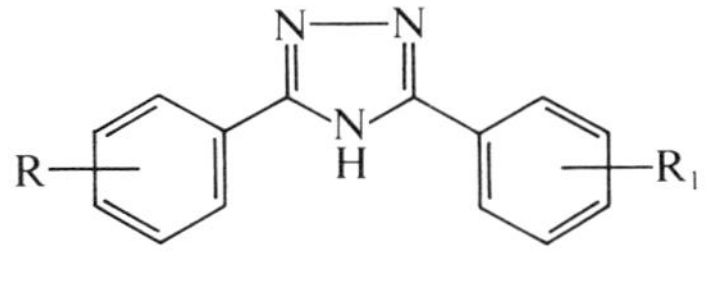

LXVIII

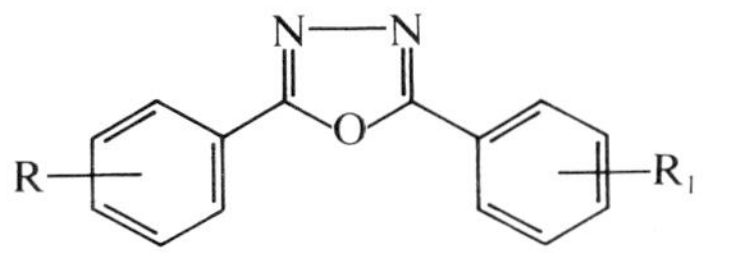

LXIX

It has been found that isomeric diaryl triazoles (LXVII, LXVIII) and oxadiazoles (LXIX) given i. p. exhibit 100 % fetal resorption in rats and hamsters but by oral route their activity is extremely poor [226, 227]. In an extension of this study, evaluation of pregnancy-inhibiting activity of simpler heterocycles possessing only one phenyl ring namely monoaryl furans and isooxazoles has been carried out [228]. However, none of them showed any noteworthy activity.
Oral administration of 2', 3', 5'-tri-O-acetyl-6-azurine (LXX; TA-AZUR), a representative of six-membered heterocycles, at a dose of

LXX, TA-AZUR

300 mg/kg on days 8, 9, 10, 11 and 12 of pregnancy has been reported to inhibit pregnancy [198] in rabbits. TA-AZUR is supposed to act as an abortifacient. Its effect is manifested even when administered in late pregnancy. No antiovulatory effect has been observed. It is not toxic at the administered dose. Another example of a six-membered heterocycle which has pregnancy-inhibiting activity is that of 4-aryl tetrasubstituted, 1,4-dihydropyridine [229]. The incorporation of various substructures in these compounds is aimed at providing an inbuilt mechanism for altering membrane permeability and interfering with redox reactions. It has been observed that 2,6-dimethyl-3,5-dimethoxycarbonyl-4-substituted-1,4-dihydropyridines (LXXI–LXXVI) exhibit 100 % pregnancy-inhibiting activity in hamsters and rats at a dose of 10 mg/kg [228, 229].

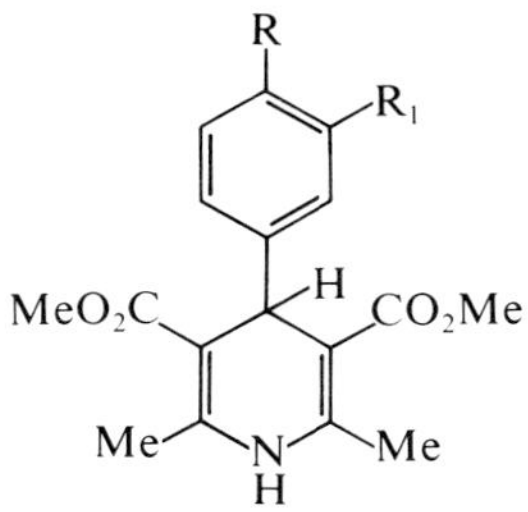

LXXI, R = COOMe, R_1 = H
LXXII, R = NHCOMe, R_1 = H

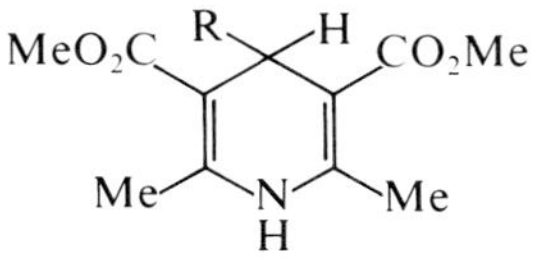

LXXIII, R = -CH(H)-CH(OH)-CH(OH)-CH(H)-CH(H)-CH(OH)-CH(OH)-CH(H)-CH$_2$OH

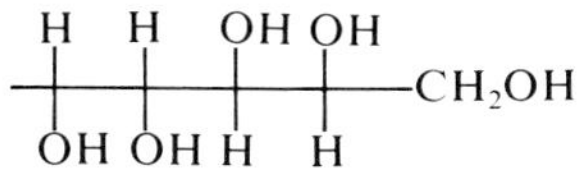

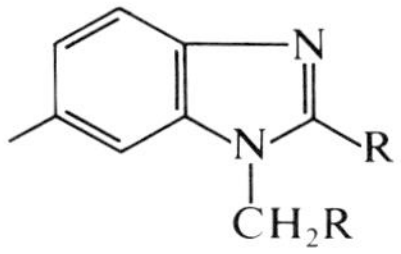

LXXIV, R =

LXXV, R =

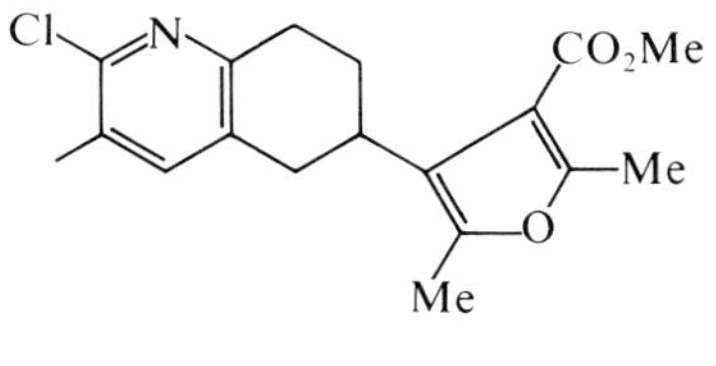

LXXVI, R =

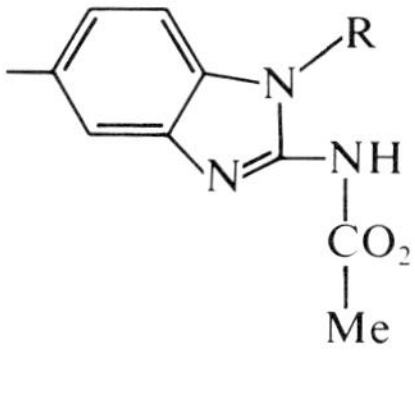

Compounds in which a quinoline ring is annealed to other heterocycles have been prepared as menses inducers on the assumption that they may specifically reach the uterus since the quinoline alkaloid quinine, at high doses, has abortifacient activity [230]. Based on this concept, various 2,3-disubstituted quinolines, a few of their dihydroanalogs and various heterocycles annealed to quinoline have been evaluated for pregnancy-inhibiting activity [231]. Of these, 2-oxo-3-cyano-1,4-dihydroquinoline (LXXVII; CDRI 84–182), 3-amino-6,7-dimethoxypyrazolo[3,4-d]quinoline (LXXVIII, CDRI 85–83) and

Tricyclic heterocycles

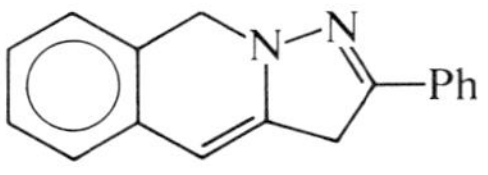

LXXXVI

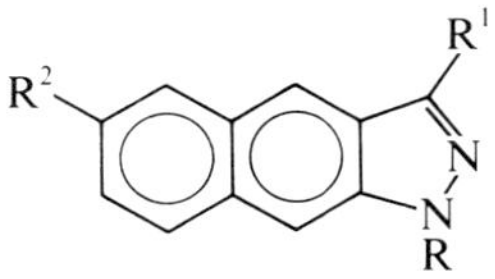

LXXXVII

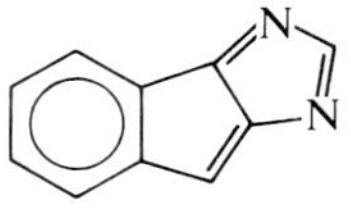

LXXXVIII

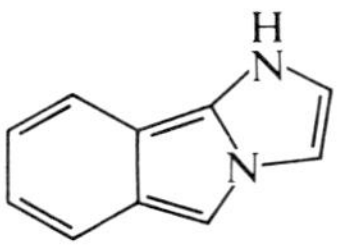

LXXXIX

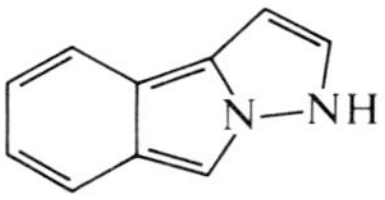

XC

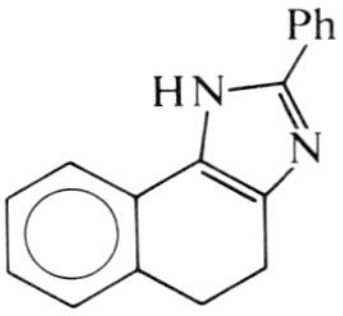

XCI

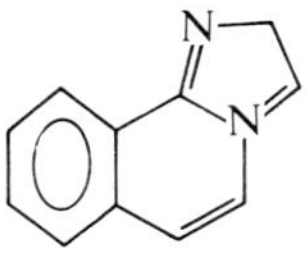

XCII

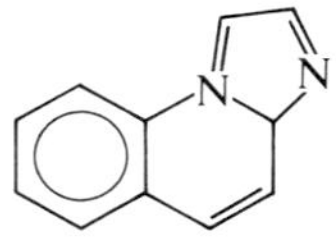

XCIII

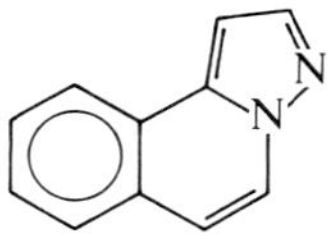

XCIV

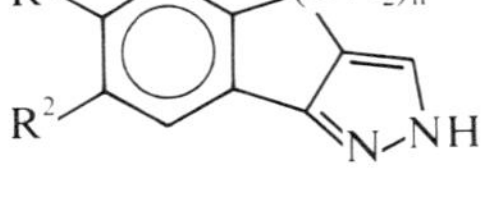

R = 4-pyridyl
R¹ = R² = H;
n = 2

XCV

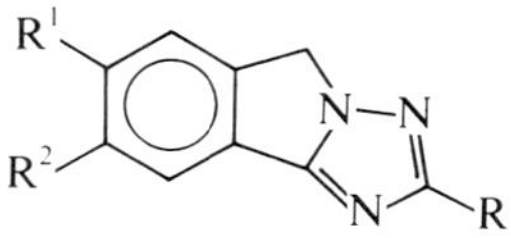

R = Substituted phenyl

XCVI

Figure 6 (continued)

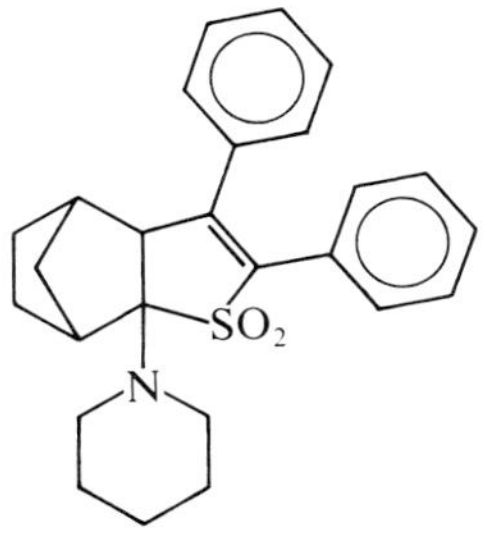

XCVII

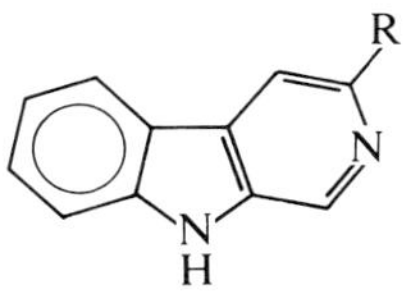

XCVIII

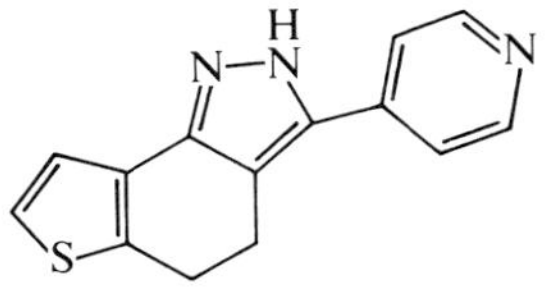

XCIX

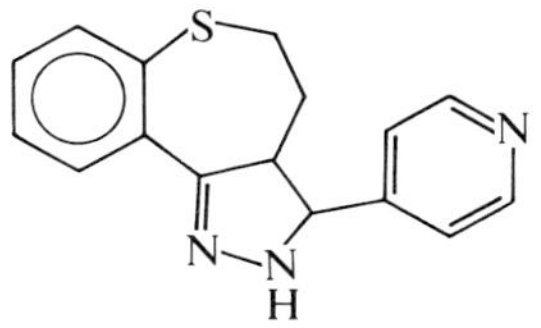

XCX

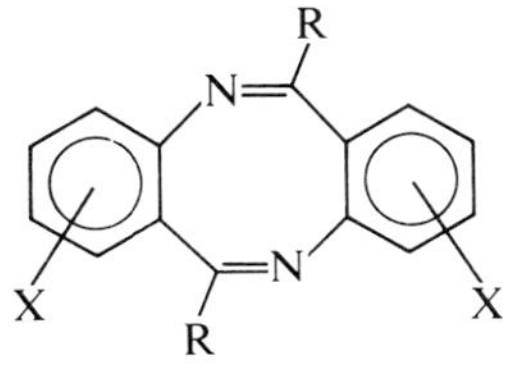

CI

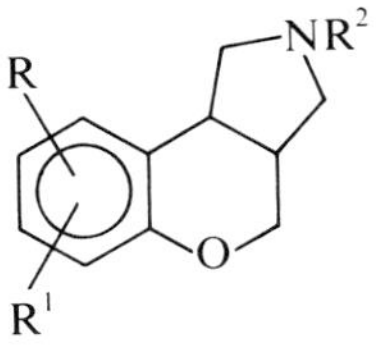

CII

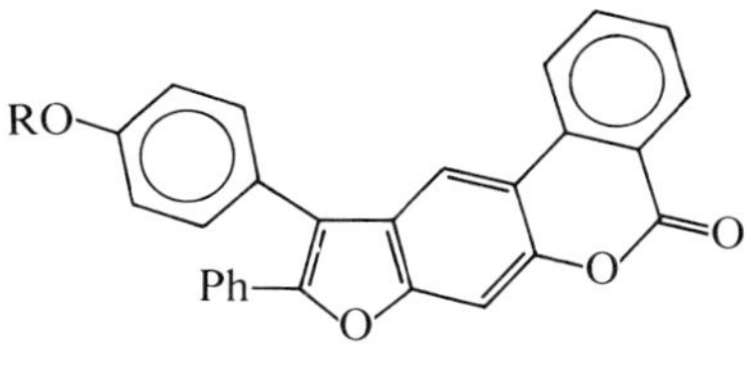

Polycyclic heterocycles

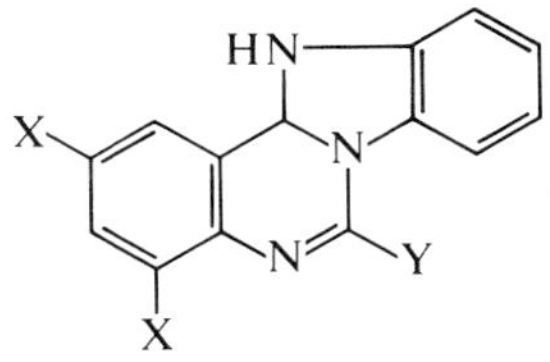

R = H, aminoalkyl

CIII

X = H, Br;
X = −CH=CH.C$_6$H$_4$(p-OCH$_3$), CH$_3$

CIV

Figure 6 (continued)

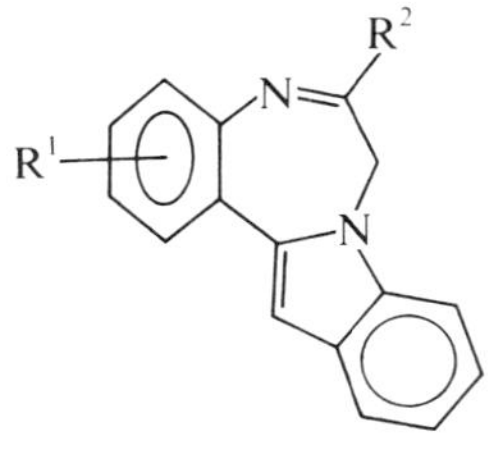

CV 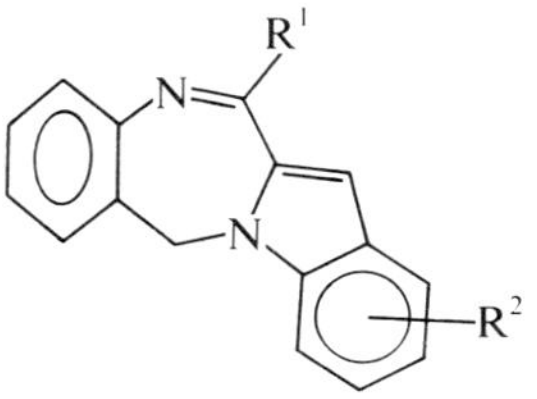 CVI

Figure 6 (continued)

6 Approaches towards male contraceptives

A safe, effective, reversible, orally administered contraceptive for males would be a more acceptable method of fertility regulation than periodic abstinence, coitus interruptus, condoms and vasectomy. Development of such a contraceptive would require a better understanding of the salient features of male reproductive physiology. The spermatogenic cycle in the human male is of 74 days and is regulated by pituitary gonadotropins which in turn are modulated by gonadal peptides [236]. Sperm maturation occurs in the epididymis but the sequence of events leading to it is a complex and still incompletely comprehended process. Transport of the mature sperm is associated with secretion of various products by the prostate and seminal vesicles [237]. In view of this, selective inhibition of spermatogenesis without any interference with testicular steroid secretion, libido and important metabolic functions of the secreted androgens may not be possible easily. Possibility of azo-spermia due to prolonged use of antispermatogenic agents also can not be ruled out. In order to avoid these problems, the epididymal maturation process and the ovum-penetrating ability of the mature sperm are considered as suitable targets for interference by male contraceptives. Clinical experience has shown that antiandrogens can inhibit spermatogenesis in a safe manner. Antiandrogens are weak antagonists which compete with dihydrotestosterone for the same binding sites on androgen receptor without affecting sperm maturation. Design of antiandrogens has been done on the basis of various structural modifications of 17α-acetoxy progesterone and their present status is discussed here. Cyp-

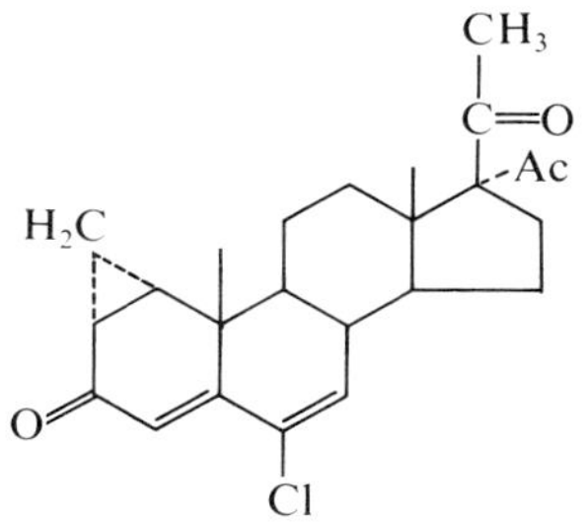

CVII, CA

roterone acetate (CVII; CA), an antiandrogenic and progestogenic steroid, in a daily dose of 5-20 mg was found to induce marked oligospermia along with reduction of sperm motility [238–241]. But as it also suppressed gonadotropin and plasma androgen levels [242, 243], combination of CA with an androgen was evaluated. Various combinations subjected to clinical trials were: DMPA + testosterone cypionate (TC) or testosterone enanthate (TE) (CVIII) [244–246], 17-hydroxyprogesterone caproate + TE (CIX) [247], medroxyprogesterone acetate (MPA) + TE (CX) [247], levonoregestrel + testosterone (CXI) [248, 249]. However, none of the combinations gave satisfactory results since consistent azospermia along with varying degrees of oligospermia were observed with all the combinations. A series of novel testosterone esters have been synthesized for evaluation as long-acting injectable androgens. On of them has been reported to maintain the physiological testosterone level in castrated male monkeys [250]. Various other antiandrogens designed as male contraceptive agents are shown in Figure 7 [251–253].

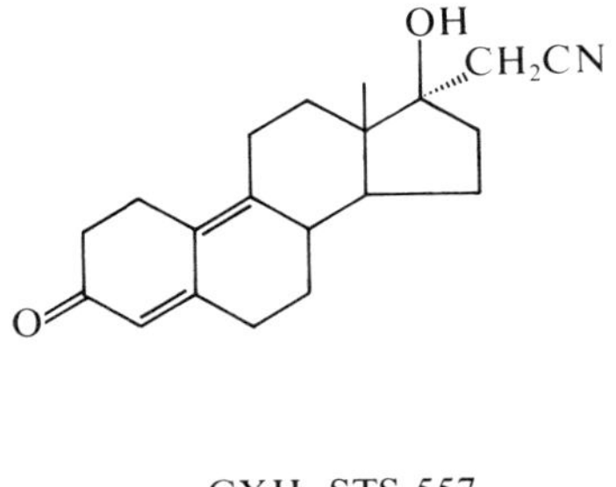

CXII, STS-557

Figure 7
Potent antiandrogens as male contraceptives

CXIII, TSAA-291   CXIV, SKF-7690

STS-557 has been found to be progestonic, antigonadotrophic and antispermatogenic but has no effect on sperm maturation. Its affinity to androgen receptor is due to the presence of 4-en-3-one and 17β-OH groups. SKF-7690, a β-norsteroid, partially inhibits fertility only at selected doses. It has no hormonal properties.
Clinical experience with gossypol (CXV) [254] as a male fertility regulating agent during the period 1978–80 has been unsatisfactory,

CXV

since it causes hypokalemia and the effect is irreversible [255, 256]. It is interesting to note that only (-)enantiomer of gossypol is active [257]. A glycoside fraction from the plant *Trypterigium wilfordii* has shown marked antifertility effect in male rats [258]. In view of the reversible antifertility activity shown by α-chlorohydrin (3-chloropropane-1,2-diol) [259], various sugars having chlorine at 6- und 6'-positions were synthesized. These have been found to exhibit reversible inhibition of certain enzymes of the glycolytic pathway of epididymal sperm, but are neurotoxic [260, 261].
LHRH agonists were also suggested [262, 263] for use as male contraceptives but have been discarded because of their interference with the establishment of azospermia. LHRH antagonists, however, have

a better prognosis. In low doses they block the pituitary function in men [264]. Studies in animals have shown that adequate supplementation with testosterone reduces the development of azospermia but this stage is reached in 2–3 spermatogenic cycles [265, 266].

7 Immunological approach to contraception: contraceptive vaccines

This approach has attracted attention because of certain possible advantages in the use of a vaccine. These advantages are: (i) long duration of action, (ii) absence of metabolic side effects, (iii) can be easily administered by paramedical personnel, and (iv) easy preparation of the synthetic components. The key to the immunological approach is the selection of a specific antigen which would selectively interfere with a critical event in the reproductive process and not cross-react with other tissue antigens. The first report of immunization of a human subject was published nearly half a century back [267]. Almost three decades ago a number of reports [268] appeared on the immunization of male and female guinea pigs, rabbits, rats, humans, sheep, chicken etc. In almost all the species, results were as expected, in regard to antibody formation leading to prevention of spermatogenesis, immobilization of sperms and impairment of fertility. Uteroglobin, a pregnancy-specific protein of rabbit, has been considered as a candidate for active immunization [269]. Tyler and Bishop have reviewed [268] the status of immunization with sperm antigens. According to Debrowoloski [270], specific antiplacental serum can also inhibit pregnancy in guinea pigs and rabbits. This has been confirmed by another report [271]. Immunization of baboons with human placental lactogen has been reported to cause abortion [272, 273]. Heterologous (antologous) placental globulin is an interesting candidate since it has been found to abort rhesus monkeys without any significant damage to the animals [273]. Probably, antiplacental globulin in the presence of complement reacts with trophoblastic tissue and destroys it and then abortion follows. In this experimental model, neither monkey placental lactogen (MPL) nor monkey chorionic gonadotropin (MCG) appeared to be the responsible antigenic element. An unidentified placental antigen(s) is presumed to be the causative factor [273]. Other approaches that have been attempted are the use of hCG, zona pellucida, anti-LHRH antibodies etc. Advances made in the understanding of the chemistry of chorionic and pituitary gonadotropins [274] have helped in developing an immunological approach to neutralize the action of chorionic gonadotropin. Immunization of the human female with hCG results in formation of antibody that possibly neutralizes the hormonal action in very early pregnancy. Antibodies raised against hCG react with hCG and human LH. It is known that the selectivity towards hCG is based on the fact that hCG is composed of α- and β-subunits and that the α-subunit is similar to that of pituitary gonadotropins, while the β-subunit is different. Antibodies raised against the β-subunit may neutralize endogenous hCG without interfering with endogenous LH. In fact it has been reported [9] that tetanus toxoid conjugated to the β-subunit of hCG causes antibody formation in rabbits, monkeys and goats and is capable of neutralizing the hormonal action of native hCG. Since hCG stimulates progesterone secretion by the corpus luteum, its absence may lead to insufficient production of progesterone necessary for successful implantation or maintenance of gestation. In this approach a synthetic peptide is incorporated at the C-terminal (109–145) amino acid of the β-subunit of hCG [275] which is then conjugated with diphtheria toxoid (DT) as carrier [276, 277]. This conjugate along with muramyldipeptide as adjuvant [278] shows > 95 % contraceptive efficacy in baboons without any adverse side effects [279]. Antibodies induced by this β-hCG vaccine cross-react only with baboon chorionic gonadotropin (bCG). Another immunological approach is based on the possibility of inducing abortion by an antiserum specific to decidual tissue, as it is known that proteins specific to decidual cells are elaborated following transformation of stromal cells [10]. Interference with sperm membrane or antibodies against spermatozoa is another possible approach to contraception. Yet another approach that has been attempted is the prevention of zona penetration by spermatozoa utilizing antibodies against zona material [280]. Pregnancy inhibition by anti-LHRH antibodies is an interesting possibility. In baboons these antibodies cause a perceptible drop in the chorionic gonadotropin level which leads to a decrease in progesterone level, followed by bleeding and termination of pregnancy [281]. However, the time of administration of antibodies is very critical since a delay of ten days failed to exhibit any effect on pregnancy. Their mode of action could be either luteolysis or interference with placental functions [282]. However, it is difficult to decide whether the drop in chorionic gonadotropin level is the primary action of the antibodies or is the result of the termination of pregnancy brought about by the antibodies by an alternative mechanism. Inhibition of chorionic gonadotropin production *in vitro* by antibodies has been reported [281]. Chorionic gonadotropin is essential for maintenance of early pregnancy and reduction of the hormone level at this stage by anti-LHRH antibodies possibly leads to abortion. Studies on the effect of anti-LHRH antibodies after targeted delivery to the uterus would be interesting since it may furnish a new approach to contraception.

Fertility regulation has also been attempted using sperm antigen. Immunization of female baboons with mouse lactate dehydrogenase (LDH-C_4) vaccine has provided promising initial results [283]. Although there is increasing interest in such a vaccine, its stage of development is still many years behind that of β-hCG vaccine.

8   Conclusions

With the growing understanding of the endocrinology of pregnancy, the target sites identified for interference by means of a pregnancy-inhibiting agent continue to change. For example, ovulation inhibitors have been replaced by hormone antagonists and the present emphasis appears to be on the use of specific antiprogestins and menses regulators. Immunological control of fertility is an attractive approach but a complete understanding of the effect of a vaccine on human subjects has not been achieved. The limited clinical experience with β-hCG vaccine would suggest that a vaccine consisting of a synthetic peptide conjugated with a carrier along with an adjuvant may give fruitful results. A heterodimer of β-hCG whose heterologous component could be the LH of some other species needs to be evaluated. In the near future the results of experiments with vaccines against hypothalamic LHRH and various sperm-specific antigens, besides β-hCG, are expected to become available. The sperm-specific antigens may include specific enzymes. Yet another approach to preventing pregnancy by immunological means is based on the concept of depriving the placenta of vital nutrients, such as vitamins, if carrier proteins can be identified and specific antibodies raised for the same. Synthesis of oligopeptides mimicking antigenic components of male or female antigens would require greater knowledge of the amino acid sequences of these components and the conformation needed for obtaining optimum biological response. It appears that the development of a satisfactory fertility control vaccine would require a long time and during this period the use of contragestational agents is bound to become more acceptable. It is expected that in the near future the search for new steroidal and non-steroidal menses-inducers would yield interesting results. It appears that a non-steroidal menses-inducer may require targeting to the site of action. The future of this class of menses-inducers rests on their being effective by the oral route of administration. While steroids have an advantage as an oral contraceptive, the development of completely non-hormonal steroid having specific action on the utero-blastocyst junction requires basic studies to identify the structural subunits responsible for such action. The advent of antiprogestins, though exciting, must await analysis of large-scale clinical data before deciding on their usefulness. In the various approaches to fertility control at present, the only common factor appears to be the awareness to avoid unnecessary physiological complications. The end of this century is bound to furnish a chemical agent for fertility control which could be taken for a lesser number of days in a cycle. Moreover, a specific enzyme as an antigen may be the vaccine of the future.

Acknowledgment

Thanks are due to Dr. S. Bhattacharya for rendering all possible help in the preparation of the manuscript.

References

1   S. H. Sturgis and R. Albright: Endocrinology 26, 68 (1940).
2   G. Pincus: The control of fertility. Academic Press, New York, London 1965.
3   Wld. Hlth. Org.: Expanded Programme of Research, Development and Research Training in Human Production, Programme, strategy and Implementation. HR/71. 9, Feb. 1972.
4   G. Pincus and J. Rock: Ann. N. Y. Acad. Sci. 71, 677 (1958).
5   R. I. Dorfman: Burgers Medicinal Chemistry, Part II, Vol. IV, p. 941. Ed. M. E. Wolff. Wiley-Interscience, New York, Chichester, Brisbane, Toronto 1979.
6   M. P. Vessey, E. Mears, L. Andolsek and M. Orgrino-Oven: Lancet 1, 915 (1972).
7   G. Benagiano: Regulation of Human Fertility, p. 323. Ed. E. Diczfalusy. Bogtrykkeriet Forum, Copenhagen 1977.

8 S. J. Segal: Frontiers in Reproduction and Fertility Control, Pt. II, p. 170. Ed. R. O. Greep and M. A. Koblinsky. MIT Press, Cambridge, London 1977.
9 S. J. Segal: Contraception 13, 125 (1976).
10 D. G. Porter and C. A. Finn: Frontiers in Reproduction and Fertility Control, Pt. II, p. 146, Ed. R. O. Greep and M. A. Koblinsky. MIT Press, Cambridge, London 1977.
11 J. Gorski, D. Toft, G. Shyamala, D. Smith and A. Notides: Rec. Prog. Horm. Res. 24, 45 (1968).
12 E. V. Jensen: Proc. natn. Acad. Sci. (USA) 59, 632 (1968).
13 G. Shyamala and J. Gorski: J. Cell Biol. 35, 125 (1967).
14 E. E. Baulieu, I. Jung, J. P. Blondeau and P. Robel. Adv. Biosci. 7, 179 (1971).
15 E. E. Baulieu and P. Robel: Some Aspects of the Actiology and Biochemistry of Prostatic Cancer, p. 74. Ed. K. Griffiths and C. G. Pierrespoint (1970).
16 J. D. Baxter and G. M. Tomkinds: Proc. natn. Acad. Sci. (USA) 68, 932 (1971).
17 M. Beato, W. Braudle, D. Biescurg and C. E. Scheris: Biochim. biophys. Acta 208, 125 (1970).
18 I. S. Edelman: J. Steroid Biochem. 6, 147 (1975).
19 S. Liao & S. Fang: Some Aspects of the Actiology and Biochemistry of Prostatic Cancer. Ed. K. Griffiths and C. G. Pierrespoint (1970).
20 S. Fang, K. M. Anderson and S. Liao: J. biol. Chem. 224, 6584 (1969).
21 W. I. P. Mainwaring: J. Endocr. 45, 531 (1969).
22 W. I. P. Mainwaring: Some aspects of Actiology and Biochemistry of Prostatic Cancer. Ed. K. Griffiths and C. G. Pierrespoint (1970).
23 D. Marver, D. Goodman and I. S. Edelman: Kidney Int. 1, 210 (1972).
24 E. Milgram, M. Atger and E. E. Baulieu: Steroids 16, 741 (1970).
25 A. Munck: J. Steroid Biochem. 3, 567 (1972).
26 B. W. O'Malley, D. O. Toft and M. R. Sherman: Proc. natn. Acad. Sci. (USA) 67, 501 (1970).
27 B. W. O'Malley, D. O. Toft and M. R. Sherman: J. biol. Chem. 246, 1117 (1971).
28 M. R. Sherman, P. L. Corvol and B. W. O'Malley: J. biol. Chem. 245, 6085 (1970).
29 G. E. Swaneek, L. L. H. Chu and I. S. Edelman: J. biol. Chem. 245, 5382 (1970).
30 O. Unhjem, K. J. Tveter and A. Aakraag: Acta endocr. 62, 153 (1969).
31 W. G. Wiest and B. R. Rao: Adv. Biosci. 7, 251 (1971).
32a K. R. Yamamoto and B. Alberts: Proc. natn. Acad. Sci. (USA) 69, 2105 (1972).
32 C. Wira and A. Munck: J. biol. Chem. 245, 3436 (1970).
33 E. E. Baulieu: Rec. Prog. Horm. Res. 27, 351 (1971).
34 E. V. Jensen and F. R. DeSombre: A. Rev. Biochem. 41, 203 (1972).
35 E. V. Jensen, S. Mohla, T. A. Gorell and E. R. DeSombre: Vitam. Horm. 32, 89 (1974).
36 B. S. Katzenellenbogen and J. Gorski: Biochemical Action of Hormones, Vol. III, p. 187. Ed. G. Litwack. Academic Press, New York, London 1975.
37 R. J. B. King and W. I. P. Mainwaring: Steroid Cell Interaction, Chaps. 7, 9. Baltimore University (1974).
38 G. C. Mueller, B. Vonderhaar, U. H. Kim and M. LeMahien: Rec. Prog. Horm. Res. 28, 1 (1972).
39 B. W. O'Malley and A. R. Means: In the Cell Nucleus 3, 379 (1974).
40 G. Raspe: Adv. Biosci. 7, 1 (1971).

41 H. G. Williams-Ashman and A. H. Reddi: A. Rev. Physiol. 33, 31 (1971).
42 E. V. Jensen and E. R. DeSombre: Science 182, 126 (1973).
43 B. M. O'Malley, W. T. Schrader and T. C. Spelsberg: Adv. exp. Med. Biol. 36, 174 (1973).
44 E. V. Jensen and H. I. Jacobson: Biological Activities of Steroids in Relation to Cancer, p. 161. Eds G. Pincus and E. P. Vollmer (1960).
45 E. Milgram, M. Atger and E. E. Baulieu: Biochim. biophys. Acta 320, 267 (1973).
46 E. E. Baulieu: Adv. exp. med. Biol. 36, 80 (1973).
47 P. Feherty, D. M. Robertson, H. B. Waynsforth and A. E. Kellie: Biochem. J. 120, 837 (1970).
48 G. Giannopoulos and J. Gorski: J. biol. Chem. 246, 2530 (1971).
49 G. Shyamala and J. Gorski: J. biol. Chem. 244, 1097 (1969).
50 D. Toft, G. Shyamala and J. Gorski: Proc. natn. Acad. Sci. (USA) 57, 1740 (1967).
51 T. Erdos: Biochem. biophys. Res. Commun. 32, 338 (1968).
52 H. Rochefort and E. E. Baulieu: Acad. Sci. Ser. D. 267, 662 (1968).
53 E. V. Jensen: Dev. Biol. 3, 151 (1969).
54 E. V. Jensen, T. Suzuki, M. Numata, S. Smith and E. R. DeSombre: Steroids 13, 417 (1969).
55 A. C. Notides and S. Nielson: J. biol. Chem. 249, 1866 (1974).
56 S. Green, P. Walter, V. Kumar, A. Krust, J. Bornert, P. Argos and P. Chambon: Nature 320, 134 (1986).
57 A. Ullrich, L. Cousseue, J. S. Hayflick, T. J. Dull, A. Gray, A. W. Tam, J. Lee, Y. Yarden, T. A. Libermann, J. Schlessinger, J. Downward, E. L. V. Mayes, N. Whittle, M. D. Waterfield and P. H. Seeburg: Nature 309, 418 (1984).
58 S. D. Holmes, N. T. Van, S. Stevens and R. G. Smith: Endocrinology 109, 670 (1981).
59 R. G. Smith, M. Istria and N. T. Vau: Biochemistry 20, 5557 (1981).
60 C. Levy, P. Robel, J. P. Gautray, J. DeBrux, U. Verma, B. Descomps, E. E. Baulieu and B. Eychenne: Amer. J. Obstet. Gynec. 136, 646 (1980).
61 G. Teutsch: Proceeding of the joint symposium organized by the Indian Council of Medical Research and World Health Organization at Banglore, India, 14th Jan. 1985.
62 S. D. Holmes and R. G. Smith: Biochemistry 22, 1729 (1983).
63 J. Delettre, J. P. Mornon, T. Ojasoo and J. P. Raynaud: Perspective in Steroid Receptor Research, p. 1. Ed. F. Bresciani. Raven Press, New York 1980.
64 E. V. Jensen, P. I. Brecher, M. Numata, S. Mohla and E. R. DeSombre: Enzyme Reg. 11, 1 (1973).
65 K. R. Yamamato: J. biol. Chem. 249, 7068 (1974).
66 A. Migliaccio, A. Rotondi and F. Auricchio: Proc. natn. Acad. Sci. (USA) 81, 5921 (1984).
67 A. M. Fogelman, J. Seager, M. E. Haberland, M. Hokom, R. Tanaka and P. A. Edwards: Proc. natn. Acad. Sci. (USA) 79, 922 (1982).
68 W. T. Schrader and B. W. O'Malley: J. biol. Chem. 247, 51 (1972).
69 J. Monod, J. Wyman and J. P. Chaugeux: J. molec. Biol. 12, 88 (1965).
70 D. E. Koshland and K. E. Neet: A. Rev. Biochem. 37, 359 (1968).
71 E. V. Jensen, E. R. DeSombre, P. W. Jungblut, W. E. Stump and L. J. Roth: Autoradiography of Diffusible Substances, p. 89. Eds L. J. Roth and W. E. Stumpf (1969).
72 W. D. Noteborm and J. Gorski: Proc. natn. Acad. Sci. (USA) 60, 250 (1963).
73 D. Williams and J. Gorski: Proc. natn. Acad. Sci. (USA) 69, 3464 (1972).

74 P. Brecher, R. Vigersky, H. S. Wotiz and H. H. Wotiz: Steroids 10, 635 (1967).
75 H. R. Maurer and G. R. Chalkley: J. molec. Biol. 27, 431 (1967).
76 C. S. Teng and T. H. Hamilton: Proc. natn. Acad. Sci. (USA) 60, 1410 (1968).
77 R. J. B. King, J. Gordon, D. M. Cowan and D. R. Inman: J. Endocr. 36, 139 (1966).
78 T. C. Spelsberg, A. W. Steggles and B. W. O'Malley: J. biol. Chem. 246, 4188 (1971).
79 G. S. Harris: Nature New Biol. 231, 246 (1971).
80 R. J. B. King and J. Gordon: Nature New Biol. 240, 185 (1972).
81 D. O. Toft: J. Steroid Biochem. 3, 515 (1972).
82 T. Liang and S. Liao: J. biol. Chem. 249, 4671 (1974).
83 S. Liao, T. Liang and J. L. Tymoczko: Nature New Biol. 241, 211 (1973).
84 G. C. Mueller: Biological Activities of Steroids in Relation to Cancer. Eds G. Pincus and E. P. Vollmer (1960).
85 G. C. Mueller, A. M. Herranen and K. F. Jervell: Rec. Prog. Horm. Res. 14, 95 (1958).
86 S. Y. Tsai, M. J. Tsai, R. Schwartz, M. Kalimi, J. H. Clark and B. W. O'Malley: Proc. natn. Acad. Sci. (USA) 72, 4228 (1975).
87 S. E. Harris, A. R. Meaus, W. M. Mitchell and B. W. O'Malley: Proc. natn. Acad. Sci. (USA) 70, 3776 (1973).
88 R. Palaeios, D. Sullivan and R. T. Summers: J. biol. Chem. 248, 7530 (1973).
89 K. R. Yamamoto and B. Alberts: Cell 4, 301 (1975).
90 K. R. Yamamoto and B. Alberts: J. biol. Chem. 249, 7076 (1974).
91 K. R. Yamamoto and B. Alberts: A. Rev. Biochem. 45, 722 (1976).
92 A. F. Wilks, P. J. Cozens, I. W. Mattaj and J. P. Jost: Proc. natn. Acad. Sci. (USA) 79, 4252 (1982).
93 F. C. P. W. Meijliuk, J. N. J. Philipsen, M. Gruber and G. Ab: Nucleic Acids Res. 11, 1361 (1983).
94 M. Geiser, A. F. Wilks, M. Seldran and J. P. Jost: J. Biol. Chem. 258, 9024 (1983).
95 J. E. Burch and H. Weintraub: Cell 33, 65 (1983).
96 J. P. Jost, M. Seldran and M. Geiser: Proc. natn. Acad. Sci. (USA) 81, 429 (1984).
97 E. R. Mulvihill, J. P. LePennee and P. Chambon: Cell 28, 621 (1982).
98 J. G. Compton, W. T. Schrader and B. W. O'Malley: Proc. natn. Acad. Sci. (USA) 80, 16 (1983).
99 J. Anderson, J. H. Clark and E. J. Peek: Biochem. J. 126, 561 (1972).
100 T. C. Spelsberg, B. A. Littlefield, R. Seelke, G. M. Daui, H. Toyoda, P. Leinen, C. Thrall and O. L. Kon: Rec. Prog. Horm. Res. 39, 463 (1983).
101 C. M. Szego: Rec. Prog. Hormone Res. 30, 171 (1974).
102 J. Gorski, W. Welshons and D. Sakai: Molec. Cell Endocr. 36, 11 (1984).
103 E. R. Barrack and D. S. Caffey: Biochemical Action of Hormones, Vol. X, p. 23. Ed. G. Litwack. Academic Press, New York, London 1983.
104 B. S. Katzenellenbogen, J. A. Katzenellenbogen and D. Mordecai: Endocrinology 103, 1860 (1978).
105 D. A. Shutt and R. I. Cox: J. Endocr. 52, 299 (1972).
106 R. A. Micheli, A. N. Booth, A. L. Livingston and E. M. Bickoff: J. med. Chem. 5, 321 (1962).
107 C. Geynet, C. Millet, H. Troun and E. E. Baulieu: Gynec. Invest. 3, 2 (1972).
108 R. Hahnel, E. Twaddle and T. Ratajczak: J. Steroid Biochem. 4, 21 (1973).
109 M. Hospital, B. Busetta, R. Boucourt, H. Weintraub and E. E. Baulieu: Molec. Pharm. 8, 438 (1972).

110 M. Hospital, B. Busetta, C. Courseille and G. Precigoux: J. Steroid Biochem. 6, 221 (1975).
111 K. S. Korach, M. Metzler and J. A. Melachlan: J. biol. Chem. 254, 8963 (1979).
112 L. Terenims: Acta pharm. toxic. 31, 441 (1972).
113 J. A. Katzenellenbogen, H. M. Hsing, K. E. Carlson, W. L. McGuire, R. J. Kraay and B. S. Katzenellenbogen: Biochemistry 14, 1742 (1975).
114 B. S. Katzenellenbogen, J. A. Katzenellenbogen, E. R. Erguson and N. Krauthammer: J. biol. Chem. 253, 697 (1978).
115 C. W. Emmens: Rec. Prog. Hormone Res. 18, 415 (1962).
116 B. G. Miller and C. W. Emmens: J. Endocr. 45, 9 (1969).
117 L. Martin: Steroids 13, 1 (1969).
118 M. R. Kilbourn, A. J. Arduengo, J. T. Park and J. A. Katzenellenbogen. Molec. Pharm. 19, 388 (1981).
119 K. Miescher: Chem. Rev. 43, 367 (1948).
120 M. J. K. Harper and A. L. Walpole: J. Rep. Fert. 13, 101 (1967).
121 D. L. Dipietro, F. J. Sanders and D. A. Goss: Endocrinology 84, 1404 (1969).
122 C. W. Emmens, B. G. Miller and W. H. Ower: J. Rep. Fert. 15, 33 (1968).
123 S. Durani, A. K. Agarwal, R. Saxena, B. S. Setty, R. C. Gupta, P. L. Kole, S. Ray and N. Anand: J. Steroid Biochem. 11, 67 (1979).
124 M. Salman, S. Ray, A. K. Agarwal, S. Durani, B. S. Setty, V. P. Kamboj and N. Anand: J. med. Chem. 26, 592 (1983).
125 D. Lednicer, S. C. Lyster, B. D. Aspergren and G. W. Duncan: J. med. Chem. 9, 172 (1966).
126 M. J. K. Harper and A. L. Walpole: Nature 212, 87 (1966).
127 F. P. Palopoli, V. P. Feil, R. F. Allen and D. E. Moltkamp and A. Richardson: J. med. Chem. 10, 84 (1967).
128 L. Terenius: Acta endocr. 66, 431 (1971).
129 R. L. Sutherland, J. Mester and E. E. Baulieu: Nature 267, 434 (1977).
130 B. L. Windsor, M. R. Callantine, R. R. Humphrey, S. L. Lee, N. H. Schottin and O. P. O. Brien: Endocrinology 79, 153 (1966).
131 V. C. Jordan, M. S. Collins, L. Rowsby and G. Prestwich: J. Endocr. 75, 305 (1977).
132 R. N. Iyer and R. Gopalachari: Ind. J. Pharm. 31, 49 (1969).
133 R. Gopalachari, R. N. Iyer, V. P. Kamboj and A. B. Kar: Contraception 2, 199 (1970).
134 C. D. Jones, T. Sarez, E. H. Massey, L. J. Black and C. Tinsley: J. med. Chem. 22, 962 (1979).
135 L. J. Black, C. D. Jones and R. L. Goode: Molec Cell Endocr. 22, 95 (1981).
136 J. Delettre, J. P. Mornon, G. Lepicard, T. Ojasoo and J. P. Raynaud: J. Steroid Biochem. 13, 45 (1980).
137 E. Sakiz and G. Azadian-Boulanger: Exerpta med. 219, 865 (1971).
138 G. Azadian-Boulanger, J. Secchi and E. Sakiz: Exerpta med. 278, 129 (1973).
139 E. Sakiz, G. Azadian-Boulanger and J. P. Raynaud: Exerpta med. 273, 988 (1974).
140 J. Secchi and D. Lecaque: Cell Tiss. Res. 238, 247 (1984).
141 J. P. Raynaud, G. Azadian-Boulanger and R. Bucourt: J. Pharmac. 5, 27 (1974).
142 G. Lepicard, J. P. Mornon, J. DeLettre, T. Ojasoo and J. P. Raynaud: J. Steroid Biochem. 9, 830 (1978).
143 J. P. Raynaud, T. Ojasoo and F. Labrie: Mechanisms of Steroid Action, p. 145. Eds G. P. Lewis and M. Ginsburg. MacMillan Press, London 1981.

144 G. Teutsch: Adrenal Steroid Antagonism, p. 43, Ed. M. K. Agarwal, Walter de Gryter, Berlin 1984.
145 A. Belanger, D. Philibert and G. Teutsch: Steroids 37, 361 (1981).
146 D. Philibert, R. Deraedt, G. Teutsch, C. Tournemine and E. Sakiz: Presented at the 64th Annual Meeting of the United States Endocrine Society, San Francisco (16th June 1982) [Abst. No. 668].
147 D. L. Healy, E. E. Baulieu and G. D. Hodgen: Fert. Steril. 40, 253 (1983).
148 G. Sehaison, M. George, N. LeStrat, A. Reinberg and E. E. Baulieu: J. clin. Endocr. Metab. 61, 484 (1985).
149 L. Proulx-Ferland, J. Cote, D. Philibert and R. Deraedt: J. Steroid Biochem. 17, 80 (1982).
150 I. Jung-Testas and E. E. Baulieu: Exp. Cell Res. 147, 177 (1983).
151 D. Philibert, R. Deraedt and G. Teutsch: 8th International Congress of Pharmacology, Vol. 668, Tokyo (1981) [Abst. No. 1463].
152 R. C. Gaillard, A. Riondel, A. F. Muller, W. Herrmann and E. E. Baulieu: Proc. natn. Acad. Sci. (USA) 81, 3829 (1984).
153 M. Moguilewski and D. Philibert: J. Steroid Biochem. 20, 271 (1984).
154 A. Ulmann: J. Steroid Biochem. 27, 1009 (1987).
155 A. Gravanis, G. Schaison, M. George, J. deBrux, P. G. Stayaswaroop, E. E. Baulieu and P. Robel; J. Cin. Endocr. Metab. 60, 156 (1985).
156 E. E. Baulieu and S. J. Segal: The Antiprogestin Steroid RU 486 and Human Fertility Control. Plenum Press, New York 1985.
157 R. Wiechert and G. Neef: J. Steroid Biochem. 27 (4–6), 851 (1987).
158 LHRH and Its Analogs – Contraceptive and Therapeutic Applications. Eds B. H. Vickery, J. J. Nestor, Jr. and E. S. F. Hafez. MTP Press Ltd., Boston 1984.
159 LHRH and Its Analogs – Contraceptive and Therapeutic Applications, part II. Eds B. H. Vickery and J. J. Nestor. Jr. MTP Press Ltd., Boston 1987.
160 A. S. Dutta and B. J. A. Furr: A. Rep. med. Chem. 21, 203 (1985).
161 M. J. Karten and J. E. Rivier: Endocr. Rev. 7, 44 (1986).
162 J. J. Nestor, Jr.: Third SCI-RSC Medicinal Chemistry Symposium, p. 362. Eds R. W. Lambert. The Royal Soc. of Chemistry, London 1986 [Special Pub. No. 55].
163 B. H. Vickrey and J. J. Nestor, Jr.: Sem. Rep. Endocr. 5, 353 (1987).
164 J. Waxman, A. Man, W. F. Hendry, H. N. Whitfield, G. M. Besser, R. C. Tiptaft, A. M. Paris and R. T. Oliver: Br. med. J. 291, 1387 (1985).
165 H. M. Fraser and J. Sandow: J. clin. Endocr. Metab. 60, 579 (1985).
166 B. H. Vickery, G. I. McRae and J. C. Goodpasture: LHRH and Its Analogs – Contraceptive and Therapeutic Applications, part II, p. 517. Eds B. H. Vickery and J. J. Nestor, Jr. MTP Press Ltd., Boston 1987.
167 P. J. Dewart, A. S. McNeilly, S. K. Smith, J. Sandow, S. G. Hillier and H. M. Fraser: Acta Endocr. 114, 185 (1987).
168 R. M. Evans, G. C. Doelle, J. Lindner, V. Bradley and D. Rabin: J. clin. Invest. 73, 262 (1984).
169 S. Bhasin, R. Robinson, M. Peterson, B. S. Stein, D. Handelman, J. Raifer, D. Heber and R. S. Swerdloff: Fert. Steril. 42, 318 (1984).
170 D. R. Meldrum, Z. Tsao, S. E. Monroe, G. D. Braumstein, J. Sladek, J. K. H. Lu, W. Vale, J. Rivier, H. L. Judd and R. J. Chang: J. clin. Endocr. Metab. 58, 755 (1984).
171 N. Lahlou, M. Roger, J. L. Chaussian, M. C. Feinstein, C. Sultan, J. E. Toublane, A. V. Schally and R. Scholler: J. clin. Endocr. Metab. 65, 946 (1987).
172 M. R. A. Lalloz, A. Detta and R. N. Clayton: Endocrinology 122 (4), 1681 (1988).

173 R. Geiger: LHRH and Its Analogs – Basic and Clinical Aspects, p. 36. Eds F. Labrie, A. Belanger and A. Dupont. Elsevier Science, New York 1984.
174 R. Schwyzer: Peptides 1986, p. 7. Ed. D. Theodoropoulos. New York 1987.
175 B. Horsthemke, H. Knisatschek, J. Rivier, J. Sandow and K. Bauer: Biochem. biophys. Res. Commun. 100, 753 (1981).
176 N. Benuck and N. Marks: Life Sci. 19, 1271 (1976).
177 E. Loumaye, Z. Naor and K. J. Catt: Endocrinology 111, 730 (1982).
178 J. A. Gudmundssen, S. J. Nillius and C. Bergquist: Contraception 30, 107 (1984).
179 P. E. Breuner, D. Shoupe and D. R. Mishell, Jr.: Contraception 32, 531 (1985).
180 M. Schmidt-Gollwitzer, W. Hardt and K. Schmidt-Gollwitzer: LHRH and Its Analogs – Contraceptive and Therapeutic Applications, p. 243. Eds B. H. Vickery, J. J. Nestor, Jr. and E. S. E. Hafez. MTP Press Ltd., Boston 1984.
181 A. Lemay and N. Faure: LHRH and Its Analogs – Contraceptive and Therapeutic Applications, part II, p. 411. Eds B. H. Vickery and J. J. Nestor, Jr. MTP Press Ltd., Boston 1987.
182 K. J. Catt, E. Loumaye, P. C. Wynn, M. Iwashita, K. Hirota, R. O. Morgan and J. P. Chang: J. Steroid Biochem. 23, 677 (1985).
183 H. M. Fraser: LHRH and Its Analogs – Contraceptive and Therapeutic Applications, part II, p. 227. Eds B. H. Vickery and J. J. Nestor, Jr. MTP Press Ltd., Boston 1987.
184 P. E. Belchetz, T. M. Plant, Y. Nakai, E. J. Keogh and E. Knobil: Science 202, 631 (1978).
185 F. Schmidt, K. Sundaram, R. B. Thau and C. W. Bardin: Contraception 29, 283 (1984).
186 J. Foreman and C. Jordan: Agents Actions 13, 105 (1983).
187 R. W. Roeske, N. C. Chaturvedi, J. Rivier, W. Vale, J. Porter and M. Perrin: Peptides; Structure and Function, p. 561. Eds C. M. Deber, V. J. Hruby and K. D. Kopple. Rockford, IL 1985.
188 J. E. Rivier, J. Porter, C. L. Rivier, M. Perrin, A. Corrgan, W. A. Hook, R. P. Siraganian and W. Vale: J. med. Chem. 29, 1846 (1986).
189 K. Folkers, C. Bowers, X. Shao-bo, P. F. L. Tang and M. Kubota: Biochem. biophys. Res. Commun. 137, 709 (1986).
190 R. W. Roeske, N. C. Chaturvedi, T. Hrinyo-Pavlina and M. Kowalezuk: LHRH and Its Analogs – Contraceptive and Therapeutic Applications part II, p. 17. Eds B. H. Vickery and J. J. Nestor, Jr. MTP Press Ltd., Boston 1987.
191 K. Folkers, C. Bowers, P. F. L. Tang, D. Feng, T. Okamoto, Y. Zhang and A. Ljungquist: LHRH and Its Analogs – Contraceptive and Therapeutic Applications, part II, p. 17. Eds B. H. Vickery and J. J. Nestor, Jr. MTP Press Ltd. Boston 1987.
192 S. J. Hocart, M. V. Nekola and D. H. Coy: J. med. Chem. 30, 739 (1987).
193 S. J. Hocart, M. V. Nekola and D. H. Coy: J. med. Chem. 30, 1910 (1987).
194 A. Sydbom and L. Terenius: Agents Actions 16, 269 (1985).
195 J. J. Nestor, Jr., R. Tahilramani, T. L. Ho, G. I. McRae and B. H. Vickery: J. med. Chem. 31, 65 (1988).
196 The Work on Post-Coital, Once-a-month and Mense-regulating Agents, Report presented by Dr. B. N. Saxena, ICMR, New Delhi.
197 A. Omodei-Sale, P. Consoni, G. Galliani and L. Lerner: German Pat. 2, 819, 372 (1978) [C. A. 90, 72206c (1979)].
198 S. K. Saxena and R. R. Chaudhary: Ind. J. med. Res. 58 (3), 374 (1970).

199 A. D. Horace and D. W. Roger: US Pat. 3, 156, 698 (1964) [C. A. 62, 5258d (1964)].
200 R. R. Chaudhary: Ind. J. med. Res. 56 (II), 1720 (1968).
201 M. R. Bell: US Pat. 402, 162 (1976) [C. A. 85, 46425a (1976)].
202 M. Vincent and G. Remond: German Pat. 2, 519, 077 (1975) [C. A. 84, 59451k (1976)].
203 M. R. Bell and A. W. Zalay: US Pat. 3, 779, 943 (1974) [C. A. 80, 133246h (1974)].
204 B. S. Setty, K. V. B. Rao, R. N. Iyer and J. D. Dhar: Indian Pat. 148, 416 (1981) [C. A. 95, 168979m (1981)].
205 J. Paget Charles Jr.: US Pat. 4, 275, 210 (1981) [C. A. 95, 150641k (1981)].
206 Lilly Eli and Co.: Isr. Pat. 50, 413 (1980) [C. A. 94, 83931u (1981)].
207 W. Huang, Z. Yang and S. Peng: Nanjing Yaoxueyan Xucbao 16(2), 75 (1985).
208 V. K. Pandey, N. Raj and Y. Saxena: Biol. Mem. 11(5), 201 (1985).
209 A. Omodei-Sale, E. Toja, G. Galliani and L. J. Lerner: Pat. Specif. (Aust.) 516, 343 [C. A. 96, 162693n (1982)].
210 R. V. Coombs: US Pat. 4, 042, 704 (1977) [C. A. 87, 184493p (1977)].
211 A. Omodei-Sale, E. Toja, G. Galliani and L. J. Lerner: German Pat. 2, 551, 868 (1974) [C. A. 85, 192731p (1976)].
212 D. A. Habeck and W. J. Houlihan: US Pat. 3, 932, 430 (1976) [C. A. 84, 121821m (1976)].
213 Gruppo Lepetit S. P. A., German Pat. 2, 424, 670 (1974) [C. A. 83, 206286v (1975)].
214 M. H. Rosen: US Pat. 3, 781, 293 (1973) [C. A. 80, 82633x (1974)].
215 Council of Scientific and Industrial Research, Indian Pat. 157, 242 (1986) [C. A. 106, 4998m (1987)].
216 W. J. Houlihan: US Pat. 3, 816, 437 (1974) [C. A. 81, 105516j (1974)].
217 W. J. Houlihan: US Pat. 3, 816, 438 (1974) [C. A. 81, 91519m (1974)].
218 M. Yamamoto, T. Hirohashi, S. Inaba and H. Yamamoto: Japan Pat. 7, 376, 886 (1973) [C. A. 80, 48050v (1974)].
219 Loozen Hubert Jan Jozef, Eur. Pat. Appl. E. P. 50, 387 (1982) [C. A. 97, 127622n (1982)].
220 B. V. Rao and V. V. Somayajulu: J. Ind. chem. Soc. 57 (8), 837 (1980).
221 Ho Chih Yung: Pat Int. Appl. W. O. 8, 504, 554 (1985) [C. A. 104, 224922f (1986)].
222 Ho Chih Yung: Eur. Pat. Appl. E. P. 219, 292 (1987) [C. A. 107, 33198b (1987)].
223 A. Omodei-Sale, P. Consoni and G. Galliani: J. med. Chem. 26(8), 1187 (1983).
224 G. Galliani, F. Luzzani, G. Colombo, A. Conz, L. Mistrello, D. Barone, G. C. Lancini and A. Assandri: Contraception 33, 263 (1986).
225 Neelima: Ph. D. Thesis, Division of Medicinal Chemistry, Central Drug Research Institute, Lucknow, India, 1983.
226 P. K. Mehrotra, Neelima, A. P. Bhaduri and V. P. Kamboj: Ind. J. med. Res. 83, 614 (1986).
227 P. K. Mehrotra, Neelima, A. P. Bhaduri and V. P. Kamboj: Ind. J. med. Res., 86, 256 (1987).
228 M. S. Akhtar: Ph. D. Thesis, Division of Medicinal Chemistry, Central Drug Research Institute, Lucknow, India, 1987.
229 A. Mukherjee, M. S. Akhtar, V. L. Sharma, M. Seth, A. P. Bhaduri, A. Agnihotri, P. K. Mehrotra and V. P. Kamboj: J. med. Chem. (In press) (1989).
230 R. M. Richart, K. F. Darabi: Frontiers in Reproduction and Fertilit Control, pt. II, p. 178. Eds R. O. Greep and M. A. Koblinski. MIT Press, Cambridge, London 1977.

231 Ranjan P. Srivastava: Ph. D. Thesis, Division of Medicinal Chemistry, Central Drug Research Institute, Lucknow, India 1988.
232 M. M. Singh, P. K. Mehrotra, A. Agnihotri, Ranjan P. Srivastava, M. Seth, A. P. Bhaduri and V. P. Kamboj: Contraception 36, 239 (1987).
233 K. C. Joshi, V. P. Pathak and R. K. Chaturvedi: Pharmazie 41, 634 (1986).
234 V. K. Pandey, N. Raj and U. K. Srivastava: Indian Drugs 24(2), 87 (1986).
235 V. K. Pandey, N. Raj and U. K. Srivastava: Acta pharm. Jugosl. 36(3), 281 (1986).
236 Inhibit, FSH and Spermatogenesis, J. Rep. Fert. Suppl. Vol. 26. Eds B. P. Setchell and B. J. Weir (1979).
237 Wld Hlth Org.: Wld Hlth Org. Techn. Rep. Ser. 520 (1973).
238 U. J. Koch, F. Lorenz, K. Danehl, R. Ericsson, S. H. Hasan, D. Kcyserling, K. Lubke, M. Mehring, A. Rommler, U. Schwartz and J. Hammerstein: Contraception 14, 117 (1976).
239 S. Roy, S. Chatterjee, M. R. N. Prasad, A. K. Poddar and M. A. Pandey: Contraception 14, 403 (1976).
240 C. Wang and K. M. Yeung: Contraception 21, 245 (1980).
241 M. Foegh, C. S. Corker, W. M. Hunter, H. McLean, J. Philip, G. Schou and N. E. Skakkeback: Acta endocr. 91, 545 (1979).
242 B. Torre (de la), S. Noren, M. Hedman and E. Diczfalusy: Contraception 20, 377 (1979).
243 L. Moltz, A. Rommler, K. Post, U. Schwartz and J. Hammerstein: Contraception 21, 393 (1980).
244 C. A. Paulsen: Male Contraception – Advances and Future Prospects, p. 300. Eds G. I. Zatuchni, A. Goldsmith, J. M. Spieler and J. J. Sciarra. Harper and Row, Philadelphia 1986.
245 H. Y Lee, S. I. Kim and E. H. Kwon: Seoul J. Med. 20, 1 (1979).
246 K. E. Friedl, S. R. Plymate and C. A. Paulsen: Contraception 31, 409 (1985).
247 J. Bain, V. Rachlies, E. Robert and Z. Khait: Contraception 21, 365 (1980).
248 M. Foegh, F. Damgaard-Pedersen, J. Gormsen, J. B. Knudsen and G. Schou: Contraception 21, 381 (1980).
249 M. Foegh, M. Nichol, I. B. Petersen and G. Schou: Contraception 21, 631 (1980).
250 G. F. Weinbauer, G. R. Marshall and E. Nieschlag: Acta endocr. 113, 128 (1986).
251 B. S. Setty: Endocrinology 74, 100 (1979).
252 A. Srivastava, J. P. Maikuri and B. S. Setty: Contraception 36, 253 (1987).
253 A. Srivastava, J. P. Maikuri and B. S. Setty: Int. J. Fert. 34 (In press) (1988).
254 National Coordinating Group on Male Antifertility Agents, Chinese med. J. 91, 417 (1978).
255 M. R. N. Prasad and E. Diczfalusy: Fertility and Sterility, p. 255, Ed. R. F. Harrison, J. Bonnar and W. Thompson, MTP Press, Lancaster 1984.
256 S. J. Qian: Proceedings of the WHO-ICMR Symposium on Methods for Regulation of Male Fertility, p. 58. Eds T. C. Anand Kumar and G. M. H. Waites. ICMR, New Delhi 1985.
257 S. A. Matlin, R. Zhou, G. Bialy, R. P. Blye, R. H. Naqvi and M. C. Lindberg: Contraception 31, 141 (1985).
258 S. Z. Qian, C. Q. Zhong and Y. Xu: Contraception 33, 105 (1986).
259 R. W. James, R. Heywood, J. Colley and B. Hunter: Toxicology 11, 235 (1978).
260 W. C. L. Ford and G. M. H. Waites: Contraception 24, 577 (1981).
261 W. C. L. Ford: Contraception 25, 535 (1982).

262 T. H. Schurmeyer, U. A. Koneth, C. W. Frieschmen, J. Sandow, F. B. Akhtar and E. Nieschlag: J. clin. Endocr. Metab. 59, 19 (1984).
263 S. Bhasin, B. S. Steiner and R. S. Swerdloff: LHRH and Its Analogs – Contraceptive and Therapeutic Applications, part II, p. 427. Eds B. H. Vickery and J. J. Nestor, Jr. MTP Press Ltd, Boston 1987.
264 S. N. Pavlou, G. B. Wakefield and W. J. Kovacs: LHRH and Its Analogs – Contraceptive and Therapeutic Applications, part II, p. 245. Eds B. H. Vickery and J. J. Nestor, Jr. MTP Press Ltd, Boston 1987.
265 B. H. Vickery, G. I. McRae, D. J. Donahue, B. B. Roberts and A. C. Worden: J. Androl. 6, 48 (1985).
266 E. Nieschlag, F. B. Akhtar, T. Schurmeyer and G. Weinbauer: LHRH and Its Analogs, p. 277. Eds F. Labrie, A. Belanger and A. Dupont, Elsevier Science Publishers BV, Amsterdam 1984.
267 A. S. Laffont and G. Theron: Bull. Soc. Obstet. et Gynec. 23, 207 (1934).
268 A. Tyler and D. Bishop: Mechanisms Concerned with Conception, p. 397. Ed. C. G. Hartman. Macmilan Co., New York 1963.
269 H. M. Beier: Biochim. biophys. Acta 160, 289 (1968).
270 M. S. Dabrowolski: Bull. Int. Acad. Sci. Cracovix 5, 256 (1903).
271 B. C. Seegal, M. W. Hasson, E. C. Gayor and M. S. Rottenberg: Exp. Med. 102, 789 (1955).
272 V. C. Stevens, J. E. Powell and S. J. Sparks: Fourth Annual Meeting of the Society for the Study of Reproduction, Boston 1971.
273 S. J. Behrman and Y. Amano: Contraception 5, 357 (1972).
274 O. P. Bahl: J. biol. Chem. 244, 567 (1969).
275 V. C. Stevens, B. Cinader, J. E. Powell, A. C. Lee and S. W. Koh: Am. J. Rep. Immun. 1, 307 (1981).
276 A. C. Lee, J. E. Powell, G. W. Tregear, H. D. Niall and V. C. Stevens: Molec. Immun. 17, 749 (1980).
277 J. E. Powell, A. C. Lee, G. W. Tregear, H. D. Niall and V. C. Stevens: J. Rep. Immun. 2, 1 (1980).
278 V. C. Stevens, B. Cinader, J. E. Powell, A. C. Lee and S. W. Koh: Am. J. Rep. Immun. 1, 315 (1981).
279 V. C. Stevens, J. E. Powell, A. C. Lee and D. Griffin: Fert. Steril. 36, 98 (1981).
280 C. A. Shivers, A. B. Dulkiewicz, L. E. Franklin and E. N. Fussell: Science 178, 1211 (1972).
281 C. Das: Proceeding of the WHO-ICMR Symposium on Post-Coital, Once-a-Month and Menses-Inducing Agents in the Control of Female Fertility, p. 85. Eds B. N. Saxena, S. Mokkapati and G. S. Toteja. ICMR, New Delhi 1985.
282 C. Das, S. K. Gupta and G. P. Talwar: J. Steroid Biochem. 23, 803 (1985).
283 E. Goldberg, T. E. Wheat, J. E. Powell and V. C. Stevens: Fert. Steril. 35, 214 (1981).

Chemotherapy for systemic mycoses

By Paul D. Hoeprich
Section of Medical Mycology, Department of Internal Medicine, School of Medicine, University of California, Davis, CA 95616, USA

1	Introduction	318
2	Polyene – amphotericin B	318
3	Pyrimidine – flucytosine	330
4	Azoles	334
4.1	Miconazole	336
4.2	Ketoconazole	338
4.3	Itraconazole	340
4.4	Fluconazole	341
5	Lipopeptide – cilofungin	342
6	Allylamine – terbinafine	344
7	Combinations	345
7.1	Amphotericin B plus flucytosine	345
7.2	Amphotericin B plus azoles	346
7.3	Flucytosine plus azoles	346
7.4	Other	346
7.4.1	Amphotericin B plus either cilofungin or terbinafine	346
7.4.2	Azoles plus terbinafine	346
7.4.3	Antifungal antimicrobics plus surgery	346
	References	347

1 Introduction

The treatment of the systemic mycoses began with the discovery of amphotericin B in 1953 [1]. Thereafter, the pace of discovery of additional antifungal agents has, in comparison with other antiinfectives, been slow, indeed: flucytosine in 1957 [2]; azole derivatives in the 1960s [3], with new triazole derivatives [4,5] presently under clinical evaluation; polypeptides first reported in 1977 [6] and 1980 [7], with a semisynthetic lipopeptide [8] now nearing clinical trial; allylamines in 1978 [9], with a perorally administrable derivative under clinical study [10].

The development of new antifungal antimicrobics has occurred in the research laboratories of pharmaceutical firms, acknowledging that the need for such drugs has mounted to the point of commercial merit. There is no doubt that therapeutic successes in many other areas of medicine has preserved many patients in a state of vulnerability to infectious agents – for example, diabetes mellitus, cancer chemotherapy, extensive surgery, organ transplantation. In addition, there is nearly uniform occurrence of mycoses in patients with the acquired immunodeficiency syndrome (AIDS). The very effectiveness of our vast array of antibacterial agents has cast into relief the shortcomings of available antifungal agents, as, for example, the death of a neutropenic patient from general aspergillosis after cure of staphylococcal sepsis. Hence, activity aimed at the development of new antifungal agents is most welcome, as is an opportunity to take stock of the current situation in chemotherapy for systemic mycoses.

2 Polyene – amphotericin B

Source and physicochemical properties

While many polyenes are active against fungi, amphotericin B (AmB) is the only polyenic agent in use for the systemic therapy of mycoses. Elaborated by strains of *Streptomyces nodosus* selected for production of AmB, the compound is virtually insoluble in water if the pH is above 2 and below 11. Unstable at the extremes of pH, and degraded by light, heat, and oxygen, AmB is a heptaenic macrolide with the structure shown in I. Its deep yellow color is contributed by the rigid, heptaenic, planar, hydrophobic portion of the molecule that is linked through an oxygen bridge to the flexible, polyhydroxylated, lipophobic part of the molecule to which is attached a microlactone ring sub-

Formula I

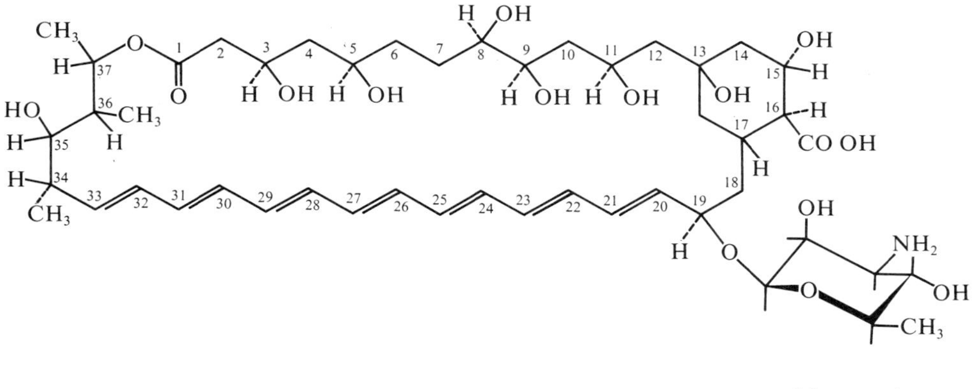

Amphotericin B
m.w. 924.10

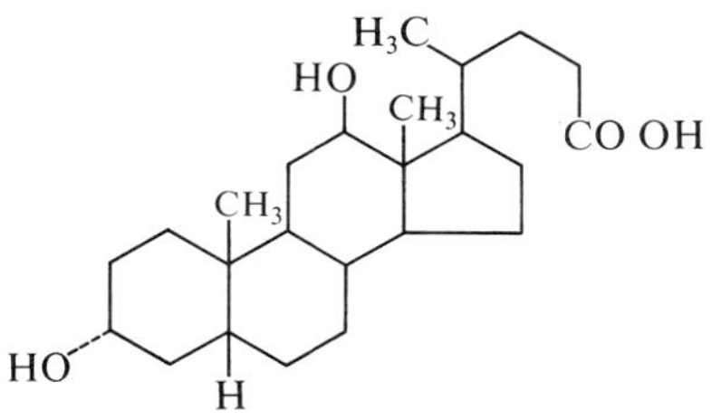

Mycosamine

Deoxycholic acid
m. w. 392.59

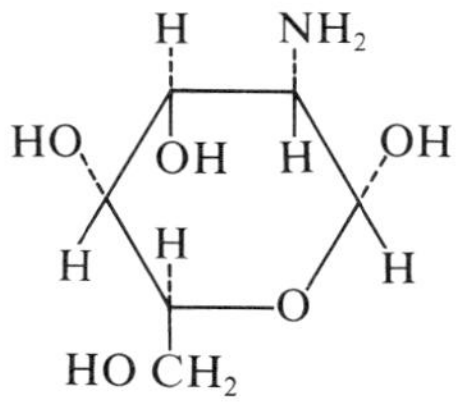

2-Aminoglucose
m. w. 179.17

tending at C-16 the one carboxyl grouping of the compound [11,12]. Mycosamine, an aminosugar also found in other antifungal polyenes, is glycosidically linked to the hydroxyl at C-19 on the macrolactone ring. Amphotericity is contributed by the carboxyl and the primary amine; an intact primary amine appears to be essential to antifungal activity, whereas the carboxyl is not [13], as was attested also by the clinical efficacy of amphotericin B methyl ester (AME), the only chemical derivative of AmB to come to clinical trial [14,15]. Greatly reduced in nephrotoxicity and better tolerated than the parent AmB, AME may be more neurotoxic, a matter fundamentally confused by the fact that all of the preparations used in therapy were mixtures containing 35–67 % AME, 2–8 % AmB, and various amounts of six to seven multimethylated derivatives [15]. Further evaluation of *pure* AME should be carried out not only as an antifungal, but also as an antiviral drug in view of its remarkable activity against the human immunodeficiency virus in H–9 cell cultures [16].

Antifungal spectrum

When tested in vitro using the AmB-deoxycholate complex (AmB-DC) that is injected intravenously (IV) in treating patients, many of the fungi that cause systemic mycoses are inhibited from growth by concentrations ≤ those attained in the blood during therapy. That is, the minimal inhibitory concentrations (MICs) of AmB-DC for clinical isolates of *Aspergillus* spp., *Blastomyces* spp., *Candida* spp., *Coccidioides immitis, Cryptococcus neoformans, Histoplasma* spp., *Sporothrix schenckki,* and the Zygomycetes are usually ≤ 1.5 ug/ml. However, there is variation in the susceptibility of clinical isolates. Many strains of *Aspergillus* spp. are not susceptible; *C. lusitaniae* are generally resistant [17], whereas the other pathogenic *Candida* spp. are usually susceptible; and *Pseudallescheria boydii, Geotrichum* spp., and many of the dematiaceous fungi that cause disease in humans are often resistant. Although AmB-DC is usually fungicidal at concentrations above the MIC when tested in vitro, it is doubtful that lethal concentrations are safely attainable in vivo.

Mechanisms of action

AmB and the other antifungal polyenes interact with sterols in the cell membranes of eukaryotic cells [18,19] to cause destabilization that is manifest as leakage of cations with minimal, reversible injury [20]; with more severe injury, loss of larger molecules may culminate in lethal damage from loss of nucleoproteins. Cholesterol, the membrane stabilizing steroid characteristic of human cells (and the cells of other mammals, helminths, protozoa, some mycoplasmas, certain viruses enveloped with constituents of the outer membrane of a mammalian host cell) may serve as a ligand for AmB. However, binding by ergosterol, the cell membrane sterol characteristic of fungi, appears to be stronger – to an extent permitting therapy. Rarely, resistance to AmB may develop during therapy [21], an occurrence marked by the disappearance of ergosterol from the fungal cell membrane, a reduction in the binding of AmB by the fungal cell membrane, and a decrease in the pathogenicity of the resistant fungi as tested in hypercorticoid mice [22].

Routes of administration

Intravenous

AmB is not absorbed from the gut after ingestion. Intravenous injection is the usual route of parenteral administration of AmB (the drug is too irritating to be given by subcutaneous, intramuscular, or intraperitoneal injection). A central venous catheter that terminates in the superior vena cava greatly facilitates IV therapy: 1) thrombophlebitis, as is commonly engendered when peripheral veins are used, is avoided; and 2) rapid mixing in a large volume of blood is assured, enabling use of higher concentrations, i. e., smaller volumes for administration.

Intrathecal

Bloodborne AmB does not attain therapeutic concentrations in the central nervous system (CNS), including the cerebrospinal fluid (CSF) [23]. Accordingly, in treating mycotic meningitis, intrathecal (IT) injection may be necessary to supplement IV therapy. Four sites are used: the lumbar space, the cisterna magna, the cervical space, and the lateral ventricles; all are hazardous. In addition, IT injection is really topical therapy; its value is determined by the mechanics of access of CSF containing AmB-DC to subarachnoid sites of injection. Lumbar injection is easiest but delivers AmB-DC into a limited volume for dilution in a region of normally sluggish flow of CSF that is distant from the usual basilar site of meningitis. So-called hyperbaric therapy [24] was designed to secure delivery to the base of the brain by injecting AmB-DC in 10 % glucose solution into the lumbar sac followed immediately by tilting the patient 15–20 % head down – a maneuver of doubtful therapeutic value. Both cisternal and lateral cervical injection enable delivery of AmB-DC close to the base of the brain, but both require a high degree of skill. Moreover, as IT therapy is continued, difficulty in needle puncture (caused by fibrosis in reaction to the AmB-DC) increases at all sites, but is particularly foreboding about the brainstem.

Surgical placement of reservoir-catheter device [25] under the scalp provides access to a lateral ventricle; because mycotic ventriculitis is almost never the problem, therapeutic value depends on normal circulation of the CSF to carry AmB-DC to the sites of infection.

Other

Subconjunctival and intravitreal injection into the eye are specialized ophthalmologic procedures [26]. Instillation into serous cavities, joint spaces, and the urinary bladder may also be carried out, and simply require direct needle puncture or urethral catheterization.

Preparations
Amphotericin B, deoxycholate complex

For parenteral administration, the limitation of insolubility in water at physiologic pH led to preparation of AmB-DC [27]. The commercial preparation is a dry powder consisting of 50 mg of AmB, 41 mg of sodium deoxycholate, and 25,2 mg of sodium phosphates; it is supplied in glass, rubber-stoppered vials sealed under nitrogen. The addition of 10 ml of sterile water for injection yields, with vigorous shaking, a clear, yellow, stable, colloidal dispersion of AmB-DC. As is indicated in I, 2-aminoglucose may also be used to stabilize AmB in aqueous colloidal form.

Amphotericin B, liposomal

Because of the toxicity of AmB-DC, many alternative preparations have been investigated. Currently under evaluation is liposomal-AmB (L-AmB), a generic designation connoting a novel carrier system for an old drug [28,29]. In principle, AmB (without DC) is incorporated in liposomes during the preparation of these biocompatible vesicles from phospholipids (usually, phosphotidylcholines and/or phosphotidylglycerols), without or with sterols (either cholesterol or ergosterol). The preparation of liposomes is simple, but many variables affect size, extent of layering of walls, and efficiency of incorporation of AmB. Toxicity appears to increase with increase in vasicle size, and the influence of sterols on toxicity is controversial. Pharmacokinetics and efficacy also appear to vary with liposomal size and constituents. There appears to be some difficulty in batch-to-batch replication of preparations of L-AmB.

Treatment
Amphotericin B, deoxycholate complex

Intravenous

About 30 min before each IV injection of a dose of AmB-DC, the patient should receive acetaminophen or aspirin (5–10 mg/kg body wt, PO; dose repeated twice at 3–4-hour intervals) and either prochlorperazine (10–20 mg, PO or IV; dose repeated once after 4 hours), metaclopramide (2mg/kg body wt, PO or IV; dose repeated once after 4 hours), or meperidine (50–75 mg, PO). The initial dose of AmB-DC is 0.25 mg/kg of body wt in 5 % glucose solution for injection. By means of an infusion pump, a quantity calculated to deliver 5–10 mg of this dose is delivered over a period of 15 min; the infusion is stopped for 15 min while the patient is observed for signs of hypotension. If none occur, the remainder of the dose is injected over 15–20 min. (Asyptomatic hypotension generally responds to elevation of the foot of the bed, with, or without the bolus injection of crystalloid; symptomatic, severe hypotension mandates termination of treatment. While therapy may again be attempted – next day, beginning with 1 mg – the probability of safe use of AmB-DC is low.) Twenty-four hours after uneventful commencement of treatment, a second dose of 0.5 mg/kg body wt is run in over 30–45 min – again with close observation for signs of hypotension. On the third day, the dose is increased to 0.75 mg/kg body wt, given over 45–60 min. The regimen of therapy is then shifted to alternate day administration of 0.75 mg/kg of body wt/dose. Blood should be obtained for assay of the concentrations of AmB in serum-bracketing the third dose of 0.75 mg/kg (7th day of therapy), obtaining specimens immediately pre-dose, and two hours post-dose; the dose per injection should be adjusted to obtain values of ≤ 1.0 µg/ml and 2.0–2.5 µg/ml, respectively. Children, adolescents, and young adults often require larger doses, e.g., 1.5–2.0 mg/kg body wt/injection, whereas 0.75 mg/kg body wt every other day is likely to be sufficient in the elderly. When a regimen is shown to be appropriate by clinical observations and measurements of concentrations in the serum, outpatient therapy may be undertaken as thrice weekly (Monday, Wednesday, Friday) injections. The advantages of periods of injection of ≤ 60 min are: 1) provocation of but one bout of rigors and emesis – typically, coming on 45 min after starting IV injection; 2) securing a gradient favoring passage from blood to tissue sites of infection; and 3) economy of pa-

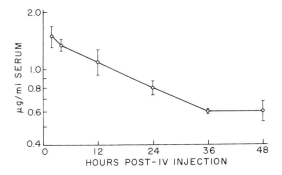

Figure 1
Plot of concentrations of amphotericin B in serum against time after completion of intravenous injection. The points are the arithmetic means, plus/minus standard errors, of concentrations in specimens obtained after three separate doses of 0.5 mg/kg body wt. The half-life was 21.5 hours.

tient care resources. Alternate day therapy is entirely suitable because the half-life of AmB in the blood is about 24 hours (Fig. 1); moreover, the adverse reactions of each dose are better tolerated with a day of respite between doses. The duration and total dosage of systemic therapy with amphotericin B must be individually determined. The clinical course and the state of renal function are of major importance, as are the nature of the causative fungus and the anatomic locus of the disease.

Intrathecal

About 30 min before IT injection is carried out, the patient should be given codeine (30–45 mg, PO; repeated once after 4 hours), and either prochlorperazine (10–20 mg PO or IV; repeated once after 4 hours) or metaclopramide (2 mg/kg body wt, PO or IV; repeated once after 4 hours). All injections IT should be preceded by removal of a volume of CSF equal to the volume of AmB-DC in 5 % glucose solution to be injected. The concentration of AmB should be 0.25 mg/ml. The first dose should be 0.025 mg (0.10 ml); with daily treatment, the dose is increased every 2 days as tolerated by the patient: 0.05 mg (0.20 ml); 0.10 mg (0.40 ml); 0.20 mg (0.80 ml); 0.30 mg (1.2 ml); 0.40 mg (1.6 ml); 0.50 mg (2.0 ml). Usually, the maximal dose that is tolerated is 0.25–0.50 mg; when the maximum tolerated dose is ascertained, it is given 2–3 times/week until evidence of active meningitis has disappeared; maintenance therapy, 1–2 injections/week, is continued for periods that vary with the causative fungus, e.g., at least for one year in patients with coccidioidal meningitis.

Other

Premedication is not necessary before instillation of AmB-DC into serous cavities, joint spaces, or the urinary bladder. A final concentration of 100–250 µg/ml (only the final dilution may be made in 0.9 % NaCl solution to avoid precipitation of the AmB) is usually tolerated.

Amphotericin B, liposomal

The protocols for the investigational use of preparations of L-AmB have varied with the investigator, as have the preparations administered to patients. The route of administration has been by IV injection, usually with daily administration of doses of 0.4–5.0 mg/kg body wt every 24, 48 or 72 hours for periods of weeks to months [30.31].

Distribution
Amphotericin B, deoxycholate

The distribution of AmB was studied in the rhesus monkey using ^{3}H-AmB complexed with ^{14}C-DC [32.33]. When observed 24 hours after the IV injection of a dose of 1 mg/kg body wt, there had been dissociation of the complex; in descending order, the concentrations of AmB were: kidneys, liver, spleen, adrenals, lungs, thyroid, heart, somatic muscle, pancreas, brain, and bone. AmB was detected in quite meager concentrations in the CSF, eye and aqueous humor, and the urine. In another study using normal sheep with surgically placed catheters that drained the pulmonary lymph, the kinetics of AmB, given IV as AmB-DC in a dose of 1 mg/kg body wt, was followed for a period of 24 hours post-dose [34]. Bioassays of concurrently collected pulmonary lymph and venous blood revealed that AmB appeared promptly in pulmonary lymph, disappearing at approximately exponential rates from both pulmonary lymph and venous blood. Apparently, neither the colloidal state of the injected AmB-DC nor binding by plasma proteins interfered with passage of AmB from the blood into the interstitium of the lungs, and thence into the pulmonary lymph.

Amphotericin B, liposomal

There are no published data describing the distribution of preparations of L-AmB after IV injection into humans. In non-human mammals, according to bioassays for AmB, the organ distribution after injection of preparations of L-AmB is primarily to the major organs of the reticuloendothelial system – liver, spleen, kidneys, and lungs – much as occurs when AmB-DC is injected [29].

Elimination
Amphotericin B, deoxycholate complex

In humans, about 5 % of the dose of AmB (injected IV as AmB-DC) is excreted as antifungally active drug in the urine during the 24 hours post-dose [23]. In rhesus monkeys, urinary excretion was about the same as in humans [35], and the major route of excretion was the bile [33].

Amphotericin B, liposomal

There are no published data regarding the elimination of L-AmB by humans. In experimental animals, elimination of AmB from the body is markedly slower after IV injection as preparations of L-AmB (particularly if a preparation rich in small, positively charged vesicles high in content of lipid is given) than when injected as AmB-DC [28, 29].

Adverse reactions
Amphotericin B, deoxycholate complex

Intravenous
Adverse reactions to AmB-DC are mainly the result of interaction of AmB with cholesterol in the membranes of cells causing perturbations of regulatory function [36]. DC, by itself, deranges the integrity of cell membranes through a detergent effect [37,38] that is probably of greatest import when AmB-DC is injected into an anatomically confined region such as the subarachnoid space. The reversible adverse reactions to systemic therapy with AmB-DC may be diminished somewhat by securing rapid mixing on injection, continuing treatment (some tolerance develop), and alternate day (or thrice weekly) treatment.

1. Chills, fever, headache, nausea, and vomiting may sometimes be decreased by pretherapy and intratherapy medications; of the many that have been suggested, possibly prochlorpromazine [39] is more effective in combating nausea than are other medications. 2. Hypotension is uncommon, usually asymptomatic, and not often engendered after the first or second dose; typically, it responds promptly to raising the foot of the bed and administering 0.9 % NaCl solution IV. Rarely, however, a patient will develop such severe hypotension that treatment must be halted. 3. Anorexia and malaise are related to the other adverse reactions, but to some extent they respond to alternate day therapy and premedication. 4. Anemia – a development in at least 75 % of patients, sometimes with thrombocytopenia – results primarily from direct suppression of erythropoiesis (and platelet formation) and also from renal failure. Hemolysis from direct interaction between erythrocytes and AmB is unlikely to be contributory because much higher concentrations than are attained in therapy are necessary by testing *in vitro* [36]. The hematocrit generally stabilizes at around 24–30 %. Treatment with iron is of no value, transfusion is only transiently beneficial, and therapy with erythropoietin has yet to be evaluated; for the present, stopping AmB appears to be critical to the correction of the anemia. 5. Nephropathy is virtually always produced by amphotericin B, but is variable in severity from patient to patient [36,40,43]. Within minutes after beginning IV injection of amphotericin B, the renal blood flow is reduced and the production of urine falls – effects that occur despite maintenance of the systemic blood pressure, and occur with every dose [44.45]. Such renal hypoperfusion appears to have particular impact in the relatively poorly vascularized medulla. Here, oxygen-dependent cation transport enzymes are pushed to maximal activity to conserve sodium because the direct toxic effect of AmB on renal tubular cell membranes is expressed by loss of cations including sodium. Anoxic necrosis of renal tubular epithelium may result [46]. Nephrotoxicity may be reduced by making certain the patient is eunatremic [42] before amphotericin B is given (stopping treatment with naturetics; easing dietary restriction of intake of sodium), and giving AmB only on alternate days. Maneuvers such as alkalinization of the urine by giving sodium bicarbonate [47,48] may help by assuring eunatremia, whereas induction of diuresis by giving mannitol is not of protective value and may be harmful [49,50]. Limiting the exposure of the kidneys to amphotericin B is of critical importance – see Fig. 2. If renal function was

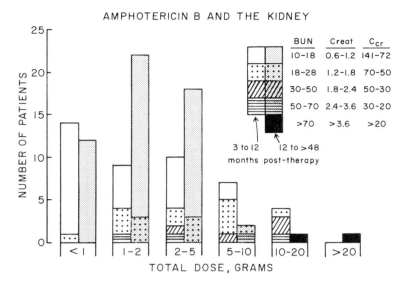

Figure 2
Relationship of total dose of intravenously administered amphotericin B deoxycholate complex to renal function; all of the patients were treated for coccidioidomycosis. The data are segregated according to duration of follow-up. [Reproduced with permission from Fig. 39–3; Hoeprich P. D., and Rinaldi, M. G., Candidosis, in: *Infectious Diseases*, 3rd edn, p. 446. Ed. P. D. Hoeprich. Harper & Row, New York]

normal at the outset of treatment and the total dose was ≤ 2 g (about 30 mg/kg body wt in an adult), fewer than 15 % of patients will have detectable, not crippling, but permanent renal damage. If the total dose was ≤ 5 g (about 75 mg/kg body wt in an adult), approximately 80 % of patients will have permanent, severe damage that is maximal in the tubules (hypokalemia, hyposthenuria, and diminished capacity to excrete acid) but also involves the glomeruli (decreased creatinine clearance, with or without azotemia [43]). Necrosis and calcification of the renal tubules is the histologic hallmark of kidneys rendered non-functional by AmB [49,51,52]. 6. The neurotoxic potential of amphotericin B is well documented. Intravenous injection has been associated with hyperthermia, hypotension, confusion, incoherence, delerium, depression, obtundation, psychotic behavior, tremors, convulsions, blurring of vision, loss of hearing, flaccid quadriparesis with degeneration of the myelin in the brachial plexus, akinetic mutism, and diffuse cerebral leukoencephalopathy [15,36,53] (Fig. 3).

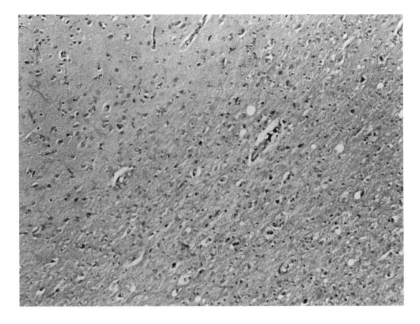

Figure 3
Diffuse cerebral leukoencephalopathy was found at autopsy of a 23-year-old man who developed akinetic mutism after receiving 5.67 g of amphotericin B deoxycholate complex by intravenous injection over a period of 4.5 months. There is almost total loss of myelin, with marked vacuolation (approximately × 100). Photomicrograph kindly supplied by David A. Rottenberg, M. D.

Amphotericin B, liposomal

As compared with AmB, the IV administration of L-AmB causes less severe immediate and short-term adverse reactions [31]. Long-term evaluation of toxicity has yet to be carried out.

Intrathecal
Intrathecal injection of AmB-DC may cause radiculitis, arachnoiditis, pareses (mono-, para-), hypesthesias, paresthesias, urinary retention, fever, impairment of vision, loss of hearing, and delerium [15,36].

Intracavitary
Injection of AmB-DC into serous cavities, joint spaces, or the urinary bladder is usually without adverse reaction.

Formula II

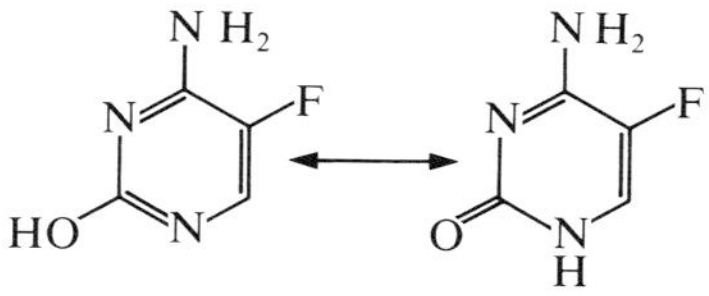

Flucytosine
m.w. 129.1

3 Pyrimidine, flucytosine

Source and physicochemical properties
Flucytosine (5-fluorocytosine, 5FC) was one of several mock pyrimidines synthesized as candidate anticancer agents [54]. It was found to have antifungal activity when a favorable effect was noted in experimental murine candidosis and cryptococcosis [2]. A stable white powder that is moderately soluble in water (about 15 g/l), 5 FC exists in tautomeric form at physiologic pH (II).

Antifungal spectrum
Initially, the results of susceptibility testing in vitro indicated lack of activity, in contrast to effectiveness in vivo in experimental animals and in humans. The paradox was resolved when it was shown that conventional, complex, and undefined culture media contained pyrimidines antagonistic to 5FC [55,56]. Native resistance (growth in $\geq$ 25 µg/ml) to 5 FC may vary in frequency according to geographic region, and appears to be increasing in frequency – about 10–15 % of isolates of *Candida albicans* [55,57–64] and about 3–5 % of isolates of *Cryptococcus neoformans* [55,57,62,65,66]. With *C. albicans,* there is a distinct relationship to serogroups in that about 15 % of Group A and about 95 % of Group B strains are natively resistant [67], whereas with *C. neoformans,* no serogroup relationship to resistance has been found [68]. Variation in susceptibility may be even greater among non-*albicans Candida* spp. [64,69]. Fungistasis is the usual effect [70]; however, most isolates of *Torulopsis glabrata* are killed at clinically relevant concen-

trations [70], with native resistance in about 5 % of isolates [60,64]. The pathogenic molds are generally resistant to 5FC [71].

Mechanisms of action

There is no evidence that 5FC is itself cytotoxic. However, some bacteria *(Salmonella typhimurium* [72] and presumably other genera of Enterobacteriaceae) and fungi *(Saccharomyces cerevisae* [73,74] and *C. albicans* [75]) internalize 5FC and then, through the action of cytosine deaminase, convert it into 5-fluorouracil (5FU). 5FU is potentially lethal as fluorouridine monophosphate, an inhibitor of thymidylate synthase, and through the formation of either fluorouridine triphosphate or fluorocytidine triphosphate [73,76] which are then incorporated into abnormal RNA, leading to faulty or non-synthesis of proteins. The

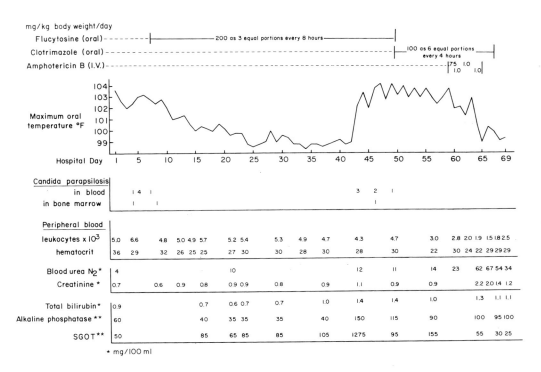

Figure 4
Course of a patient with infective endocarditis caused by *Candida parapsilosis* who was treated with flucytosine. Resistance developed during therapy. [P. D. Hoeprich, J. L. Ingraham, E. Kleker and M. J. Winship: J. infect. Dis. *130*, 112 (1974). Copyright by The University of Chicago Press. All rights reserved.]

safety of treatment with 5FC reflects lack of cytosine deaminase in humans [77,78]; however, the microflora in the colons of humans almost surely abounds in organisms that are capable of deaminating 5FC; no formal studies of this question have been reported.

Native resistance to 5FC has not been studied systematically either in fungi or bacteria, but it is thought that resistant organisms may not take up 5FC [74], lack cytosine deaminase [79], have an altered capacity to phosphorylate uridine monophosphate [75,80], or have an overproduction of pyrimidines [74,75]. Development of resistance during monotherapy with 5FC in a patient with infective endocarditis caused by *Candida parapsilosis* [79] was shown to have resulted from mutational deletion of cytosine deaminase (Fig. 4).

Routes of administration

5FC is usually administered perorally. Intravenous injection has been used in a few patients, apparently with success.

Preparations

Tablets containing either 0.25 or 0.50 g are commercially available. The investigational solution for IV administration contains 1.0 % 5 FC in 0.9 % NaCl solution.

Treatment

In Europe, the usual dose of 5FC is 200 mg/kg body wt/day, PO, given in four equal portions, 6-hourly [69,81]; in the United States, the usual recommended dose is 150 mg/kg body wt/day, PO, given in four equal portions, 6-hourly [82]. Either period of dosing is reasonable in view of a half-life in humans with normal renal function of about three hours [83–85]. The proper dose is the dose that is shown by measurement of the concentrations in the serum to yield a pre-dose (trough) concentration of 25–35 µg/ml, and a 2-hour post-dose (peak) concentration of 80–100 µg/ml. These considerations are important because there is evidence that: 1) the development of resistance may be favored when the concentrations in the blood are $\leq$ 25 µg/ml [70,84]; and 2) the ferquency of occurrence of adverse reactions increases when the concentrations in the blood exceed 120 µg/ml [81,86].

If there is renal failure, the dose must be reduced to compensate for reduced excretory capacity [83,87–89]. As a first approximation, the total daily dose should be reduced in direct proportion to the reduction in creatinine clearance, retaining an 8-hourly regimen of administration. However, in all patients with impaired renal function who must be given 5FC, the appropriateness of the dose selected must be gauged by measurement of trough/peak concentrations in the blood.

Distribution

As a small molecule, 5FC is present in all body water, encountering no barriers. The differences in concentrations, as referred to contemporaneous blood, probably reflect a temporal lag in equilibration: peritoneal fluid, 100 % [88]; CSF, 75 % [90], synovial fluid 60 % [91], aqueous humor, 20 % [92], and bronchial secretions 76 % [93].

Elimination

Flucytosine and its metabolites are virtually entirely eliminated in the urine. About 90 % of a dose given PO appears as antifungally active drug in the urine, attaining concentrations 10–100 times those in the serum [83,94,95]. Of probable relevance to toxicity in humans was the finding of about 1 % of a PO dose of ^{14}C-labelled 5FC in the urine in the form of alpha-fluoro-beta-ureido-propionic acid [85], a metabolite of 5FU most likely derived from deamination of 5FC by enteric microflora, followed by absorption and catabolism.

Adverse reactions

Adverse reactions to 5FC are not only uncommon, but also are usually of no great morbid potential – anorexia, nausea, vomiting, diarrhea, abdominal pain in about 6 %; elevations of transaminases, alkaline phosphatase (and quite rarely, hyperbilirubinemia with or without hepatomegaly) in about 5 %. Potentially more serious is depression of the bone marrow with leukopenia, thrombocytopenia (and rarely, agranulocytosis, anemia, or pancytopenia) in about 5 % [81]. The cited values are approximations because the patients to whom 5FC is often given have either underlying diseases that have already compromised the function of the gut, liver, and bone marrow, or have been treated with drugs that injure these organs; moreover, such patients are often under treatment with nephrotoxic antimicrobics, frequently including AmB. Disturbed enteric motility, especially slowing, permits a shift in the colonic microflora mouthward [100], greatly increasing the probability that intraenteric deamination of 5FC to yield 5FU will occur. If there is at the same time injury (e.g., erosion) to the enteric mucosa, the

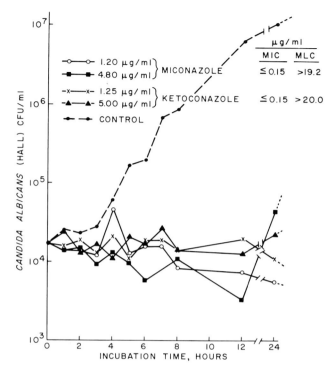

Figure 5
Plot of colony-forming units of *Candida albicans* inoculated into synthetic, defined culture medium without (control) and with miconazole or ketoconazole at two concentrations above the minimal inhibitory concentrations. Both drugs were fungistatic. [Reproduced with permission from P. D. Hoeprich and M. G. Rinaldi: Candidosis, in: *Infectious Diseases*, 4th edn. Eds P. D. Hoeprich and M. C. Jordan. Harper & Row, New York (1988) in press]

normally poor absorption of 5FU may be obviated. The net effect is increased toxicity.

4 Azoles

Nearly two decades have passed since the antifungal activity of certain azole derivatives was described [3]. They have proved to be less toxic than AmB, are active against many kinds of fungi, and the newer derivatives have excellent pharmacologic properties. Unfortunately, none of the antifungal azoles is lethal in action against any of the pathogenic fungi at concentrations relevant to therapy (Fig. 5). This is a distinct limitation of all of the antifungal azoles, licensed and investigational.

Formula III

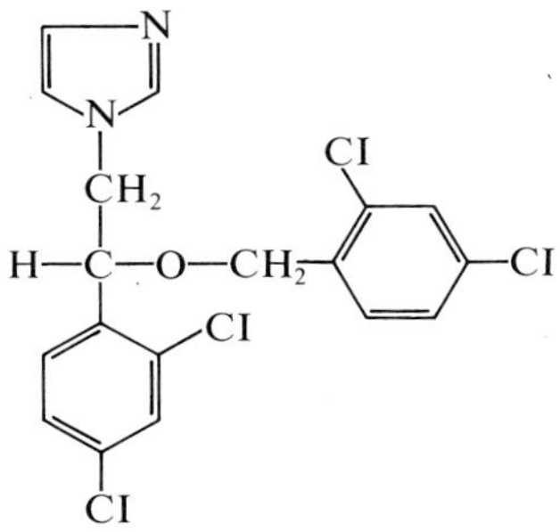

Miconazole
m.w. 416.2

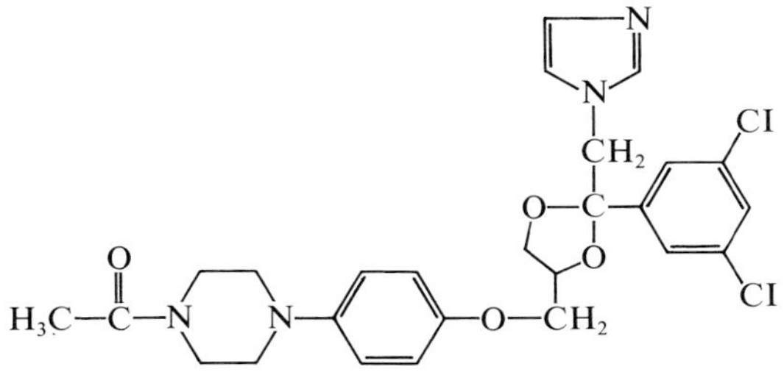

Ketoconazole
m.w. 531.4

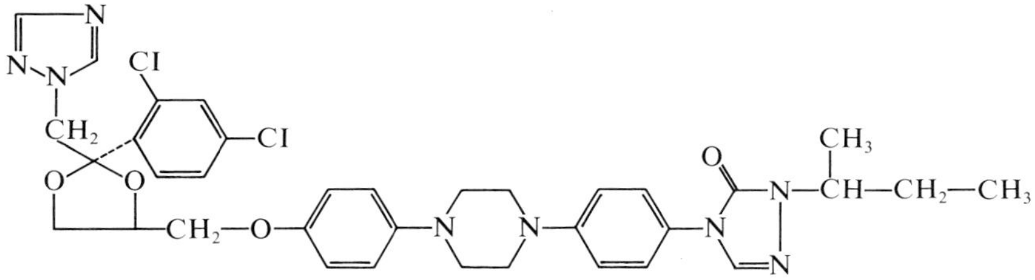

Itraconazole
m.w. 705.64

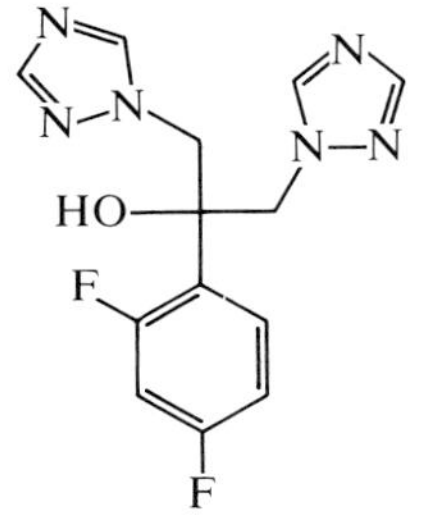

Fluconazole
m.w. 305.27

Because most systemic fungal infections are consequent on diminution of the defenses of the host, optimal therapy for systemic mycoses requires antimicrobics capable of eliminating fungal pathogens primarily through fungicidal drug action.

Source
All of the antifungal azole derivatives are the products of chemical syntheses. Of the many such compounds that have been synthesized, four have been selected for discussion (III): the two imidazoles are licensed in the United States – miconazole and ketoconazole; the two triazoles are investigational drugs – itraconazole and fluconazole.

Mechanisms of action
The antifungal azoles act through inhibition of the microsomal cytochrome P-450-dependent lanosterol 14-alpha-demethylase system [95,97]. To an extent that is dose-dependent, the biosynthesis of ergosterol is decreased and 14-methylated intermediary sterols accumulate. Both the liquidity and the permeability of the fungal cell walls suffer, the activities of surface enzymes (chitin synthetase, lipid metabolism, oxidative enzymes) are disarrayed, and there is retention of metabolites (glucose). Disordered chitin synthetase activity is of particular interest because foci of growth are affected most dramatically – sites of abcission of buds in yeast forms, and hyphal tips and septa in mycelial forms. The fungi survive, but the damage they sustain may heighten their susceptibility to destruction by phagocytes. Because depletion of ergosterol in fungi exposed to the antifungal azoles should lead to a decrease in cell membrane sites for the interaction of polyeneic antifungal agents, it is reasonable to expect that the azoles might antagonize AmB. This is indeed the case, *in vivo,* as well as *in vitro* [98–101].

4.1 Miconazole

Physicochemical properties
One of a series of beta-substituted 1-phenethylimidazoles synthesized in the late 1960s [3], miconazole (MON) is a white powder that is insoluble in water, but is soluble in a variety of organic solvents. In the form of a colloidal suspension in water, it is stable for at least a week at 37°C [102].

Antifungal spectrum

Many evaluations of the susceptibility of clinical isolates of fungi to MON are uninterpretable because they were carried out in conventional, complex, undefined culture media which antagonized the antifungal activity of MON [103]. When testing was carried out in a nonantagonistic, defined, synthetic medium, most isolates of *Candida* spp., *T. glabrata, C. neoformans, Coccidioides immitis*, and about one-third of *Aspergillus* spp. were susceptible [104]. In addition, the dermatophytes, *Sporothrix schenckii, Blastomyces dermatitidis, Histoplasma capsulatum*, and *Pseudallescheria boydii* were susceptible [111]. MON is fungistatic (Fig. 5).

Routes of administration

For treatment of systemic mycoses, MON must be injected IV. It has also been injected IT in the treatment of fungal meningitis.

Preparations

Each ml of the commercial preparation for IV injection consists of 10 mg of a colloidal suspension of MON, stabilized by 0.115 ml of Cremaphor El[R] – polyethoxylated castor oil (a mixture of ricinoleic acid polyglycol ester, glycerol polyglycol ethers, and polyglycols), 1.62 mg of methylparaben, and 0.18 mg of propylparaben.

Treatment

Doses of 25–50 mg/kg body wt/day, IV, given in two or three equal portions, either 12- or 8-hourly, have been applied in the treatment of a variety of systemic mycoses [106,107]. Intrathecal injection of doses as large as 20 mg, given at 3–7 days intervals, have been reported [108]. However, parenterally administered MON is now of little use in therapy because of the high frequency of relapse after stopping therapy, the high frequency of adverse reactions, and the advent of derivatives with more favorable properties [107,108].

Distribution

The half-life of MON after IV injection in humans is about 24 hours [109]. There is clinically insignificant entry into the urine, CSF, or sputum [106,109,110–114], although MON penetrates into pus and synovial fluid [112].

Elimination

Only about 12 % of the dose of MON can be detected in antifungally active form in the feces [109]. Most of the drug is degraded in the liver (without inducing augmented catabolic capacity as therapy is continued) to yield metabolites that appear in the urine [109,113].

Adverse reactions

The Cremaphor EL[R] component of the formulation may cause the frequent thrombophlebitis, rouleauxing of erythrocytes, and hyperlipidemia, and the uncommon anaphylaxis [112,115]. Pruritus (36 %), anemia (44 %), thrombocytosis (31 %), hyponatremia (46 %), and nausea (46 %) are toxicities probably attributable to the miconazole itself [106,110–112].

4.2 Ketoconazole

Physicochemical properties

Ketoconazole (KET) is insoluble in water at physiological pH, but does dissolve in 0.1 N HCl.

Antifungal spectrum

The antifungal spectrum of KET is virtually identical with that of MON.

Routes of administration/preparations

For systemic therapy, KET is available only in 200 mg tablets for PO administration. Absorption requires the presence of normal gastric acid; patients with achlorhydria, and those rendered functionally hypochlorhydric or achlorhydric by treatment with H_2 blockers or antacids, may not respond to PO therapy with KET. In such patients, each dose may be dissolved in 0.1 N HCL and taken through a glass tube or plastic straw; H_2 blockers should be withheld for 6–12 hours prior to administration of a dose of KET; and antacid therapy should also be withheld.

Treatment

Doses of 400 mg 12-hourly for 6–9 months may be used in the initial treatment of non-life-threatening blastomycosis [116]. For histoplasmosis, 400 mg daily for ≤ 6 months is usually sufficient [117]. Favorable clinical responses have been reported in non-life-threatening coccidioidomycosis after treatment with a single daily dose of 400–600 mg/day for one year [106]. Similar regimens have been reported to be effective in paracoccidioidomycosis [118–120]. Clinical isolates of *C. neoformans* are uniformly susceptible to KET; however, the drug should not be used to treat cryptococcal meningitis, but may be of value in isolated cutaneous cryptococcosis. While KET is recommended as the primary therapy for superficial candidosis, prolonged therapy may be necessary to maintain a response, particularly in chronic mucocutaneous candidosis [121]. Indeed, relapse of disease after discontinuing therapy is a major problem in the treatment of any systemic mycosis with KET; perhaps relapse reflects the fact of the solely fungistatic action (Fig. 5) of KET. Although the dermatophytes are susceptible to KET, in most patients with dermatophytosis, KET has no place in therapy because of the availability of other therapies with less risk of adverse reactions.

Distribution

Peak concentrations in the blood are attained 2–4 hours after ingestion of a dose; the half-life in the blood is 6–10 hours [106]. The peak concentrations in the blood range from 2–20 μg/ml after doses of 200–1200 mg [106]. KET does not attain effective concentrations in the urine, CSF (except with meningitis when the concentration may reach 15 % of the contemporaneous blood), or eye [122].

Elimination

Following non-inducible hepatic catabolism, the inactive metabolites of KET are excreted primarily in the bile, and the remainder in the urine [122].

Adverse reactions

The severity of nausea and vomiting increases in direct proportion to increases in doses, imposing a limitation in dosage [106]. Also directly related to dosage is the occurrence of endocrinopathies manifested in males as gynecomastia, loss of libido, and oligospermia, and in females as irregular menses and amenorrhea [106,107]. Hepatitis may occur and is usually reversible, although fatalities have been reported.

4.3 Itraconazole

Physicochemical properties
Insoluble in water, itraconazole is a lipophilic triazole.

Antifungal spectrum
Although the reproducibility of in vitro tests of susceptibility to itraconazole (ITR) has not been entirely satisfactory [107], it appears that some *Aspergillis* spp. are susceptible, as well as the range of pathogenic fungi listed for MON.

Routes of administration
ITR has been given to patients solely PO. The absorption of ITR is favored by ingestion with meals; it appears that gastric acid is necessary for absorption.

Preparations
For the treatment of systemic mycoses, capsules containing 50 mg of ITR have been used.

Treatment
In reported studies, doses have varied from 100 to 600 mg/day, usually given as two equal portions, 12-hourly, for periods of 6 or more months [124].
Although periods of follow-up are too short to be conclusive, it appears that ITR may be beneficial in blastomycosis, histoplasmosis, chromoblastomycosis, phaeohyphomycosis, coccidioidomycosis, sporotrichosis, and even aspergillosis; ITR may prove to be the agent of choice for the treatment of paracoccidioidomycosis.

Distribution
The half-life of ITR in the blood is about 17 hours [125]. Entry into the CSF, tracheobronchial secretions, and sputum is too poor to have a therapeutic effect. Similarly, $\leq 1\%$ of ITR is excreted in antifungally active form in the urine [108,126,127].

Elimination
In rats, ITR undergoes non-inducible catabolism by the liver yielding inactive metabolites that appear primarily in the feces, and to a lesser extent in the urine [107,127].

Adverse reactions
Preliminary experience indicates less nausea and vomiting from ITR than from KET [107,128]. Suppression of steroidogenesis in the testes or adrenal cortex my occur with high dosage [128a, 128b].

4.4 Fluconazole

Physicochemical properties
Fluconazole (FLU) is a water-soluble triazole.

Antifungal spectrum
Evaluation of the antifungal activity of FLU in vitro appears to have even less relevance to effectiveness than such studies with ITR [107,129]. However, in experimental animals, FLU appears to have much the same antifungal spectrum as ITR.

Routes of administration
Preliminary work with FLU has been carried out using PO administration.

Preparations
Capsules containing 50 mg of FLU have been used for PO administration in clinical studies. A preparation for IV administration is under evaluation.

Treatment
Single daily doses of 50–400 mg of FLU have been employed [107,133,133a,133b] in the treatment of cryptococcosis, coccidioidomycosis, candidosis and histoplasmosis. The periods of treatment have been limited, and the duration of follow-up has been inadequate for firm evaluation of efficacy. However, FLU may be the agent of choice for the treatment of cryptococcal meningitis, especially in patients with AIDS.

Distribution
Absorption of FLU from the gut appears to be independent of gastric acidity, and is virtually quantitative [107]. After PO administration, the half-life in the blood of humans is 22–24 hours. Unlike other antifungal azoles, FLU does not undergo hepatic catabolism, and antifungal-

ly active drug is present in both urine and feces. There is excellent entry into the CSF even in the absence of meningeal inflammation, with concentrations, ≤ 60 % of contemporaneous concentrations in the blood [107].

Elimination
In excess of 60 % of FLU is excreted in unchanged form in the urine, with most of the remainder of the drug appearing in the feces [107].

Adverse reaction
From limited experience, FLU has caused nausea and asymptomatic rises in hepatocellular enzymes in ≤ 5 % of patients. Thus far, there has been no evidence of interference with steroidogenesis [107].

5   **Lipopeptide – cilofungin**

Source and physicochemical properties
Cilofungin (CIL) is a semisynthetic, amphiphilic lipopeptide (IV) derived from echinocandin B [8]. It was synthesized in the Lilly Research Laboratories, Eli Lilly and Company, Indianapolis, IN, USA. The white powder is stable if stored in the absence of water.

Formula IV

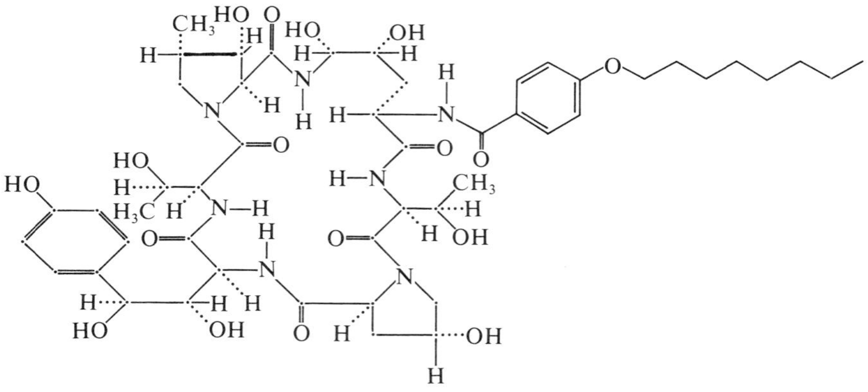

Cilofungin
m.w. 1030.15

Antifungal spectrum

Only *Candida* spp. among pathogenic fungi appear to usefully susceptible to CIL (background information supplied by Eli Lilly & Company, Indianapolis, IN). In side-by-side testing, CIL was as active as AmB against *C. albicans* and *Candida tropicalis*; *C. parapsilosis* and *Cryptococcus* spp. are resistant [8,133c].

Mechanisms of action

Actively growing *Candida* spp. are lysed, apparently as a consequence of inhibition of synthesis of the beta 1,3 glucan component of the cell wall by CIL [8]. The action is fungicidal in effect.

Routes of administration

Cilofungin is not absorbed from the gut. It has been injected IV in tests in vivo in experimental mycoses, and in volunteers.

Preparations

In work with experimental animals, CIL at a final concentration of 25 mg/ml was injected in a vehicle consisting of 33 % polyethylene glycol (v/v) in distilled water containing per 100 ml 1.53 gm Na_2HPO_4, 1.58 gm Na_2HPO_4, and 1.6 gm NaCl; the pH was set at 6.80 [8].

Treatment

For evaluation of toxicity in dogs, doses of 10–100 mg/kg wt/injection were given 5 days/week for 3 months [8].

Distribution

After IV injection, cilofungin does not enter the CSF or the aqueous humor [8].

Elimination

No data have been published.

Adverse reactions

At 100 mg/kg body wt/dose, the highest doses tested in dogs, CIL caused immediate reactions suggestive of release of histamine [8]. In addition, there were elevations in the concentrations in the serum of hepatocyte-origin enzymes, with hepatomegaly and centrilobular fatty vacuolation. Thrombophlebitis was present at the sites of IV injection [8].

Formula V

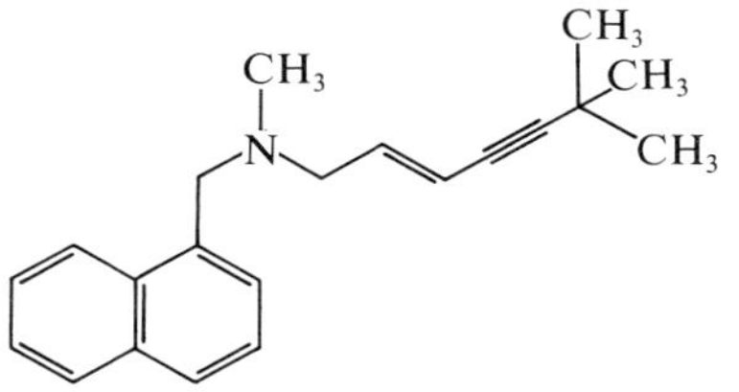

Terbinafine
m.w. 291.44

6 Allylamine – terbinafine

Source and physicochemical properties
Allylamine antifungal agents are synthetic compounds derived from heterocyclic spironaphthalenones [9,10]. From evaluation of many derivatives, it was shown that the allylamine double bond must have the *trans* orientation for good activity [7]. The most active allylamine derivative reported to date is terbinafine (TER), a white powder. A unique feature of TER is the presence of an acetylene function conjugated to the allyamine portion of the molecule (V).

Antifungal spectrum
By testing in vitro, the dermatophytes are exquisitely susceptible to TER. Among other fungi, *Aspergillis* spp. *(fumigatus, flavus, niger)*, *Candida* spp., and *S. schenckii* appear to be susceptible [8,10,134, 135,135a].

Mechanisms of action
Terbinafine inhibits squalene epoxidation causing failure of synthesis of ergosterol [10,137]. Fungi that are susceptible are killed [136]. Mammalian epoxidases appear to be 3–4 orders of magnitude less sensitive than the fungal enzymes to inhibition by terbinafine [137].

Routes of administration
Only PO administration of TER has been reported [138,138a].

Preparations
The investigational preparations of TER used for PO administration were 250 mg tablets.

Treatment
Dermatophytosis in 47 patients was treated by the administration of TER in a dose of 250 mg, PO, 12-hourly for four weeks [138].

Distribution
Peroral doses of 500 mg of TER yielded peak concentrations in the serum of 2 μg/ml; the half-life was 11.3 hours (138a). No data regarding distribution in the body have been published.

Elimination
No data have been published regarding the elimination of TER.

Adverse reactions
In the limited trials using TER thus far reported, there were no adverse reactions of consequence [138].

7   Combinations

It may be rational to combine antifungal antimicrobics in therapy if: 1) each drug of a combination is active against the fungal pathogen; 2) the mechanisms of action of the drugs differ and are non-interfering; and 3) the sites of infection are accessible to the antifungal agents. Both laboratory and clinical experience are supportive of these constructions.

7.1   Amphotericin B plus flucytosine

It was hoped that combination therapy using AmB plus 5 FC might: 1) reduce toxicity from the AmB if synergy, or simply an additive effect, permitted use of a lower dose of AmB (per injection and in total dose); and 2) prevent or delay the development of resistance to 5FC. Two clinical trials of AmB plus 5FC in the treatment of cryptococcal meningitis have been reported [139,140]. Combination therapy was not significantly more effective than treatment with AmB alone. Moreover, 30–40 % of the patients who received the combination suffered adverse reactions to the 5FC, principally suppression of bone marrow function. There are no data from controlled clinical trials to support the use of AmB plus 5 FC in the treatment of other systemic mycoses. Whenever the combination is used, surveillance of therapy by measurement of concentrations of 5FC in the blood is important because nephrotoxicity from amphotericin B may lead to accumulation of flucytosine, increasing the probability of toxicity.

7.2 Amphotericin B plus azoles

If an antifungal azole derivative is successful in blocking fungal synthesis of ergosterol, the site for the antifungal action of AmB will be absent. Hence, the combination should be antagonistic – as was demonstrated in animal models of candidosis [99,101], aspergillosis [106], and with some strains of *C. neoformans* [101].

7.3 Flucytosine plus azoles

Although 5FC might be expected to be usefully combined with FLU in the treatment of mycotic infections of the central nervous system, the interaction was, at best, one of indifference in experimental murine candidosis, cryptococcosis, and phaeohyphomycosis caused by *Wangiella dermatitidis* [101].

7.4 Other

7.4.1 Because AmB differs greatly in mechanism of action from either cilofungin or terbinafine, it is reasonable to suppose that AmB plus either agent might be non-antagonistic in effect on susceptible fungi. There are no published data on such combinations.

7.4.2 Terbinafine in acting earlier in the biosynthetic sequence of reactions leading to ergosterol than the azoles, might be additive, or even synergistic, with them. Again, there are no published data regarding such combinations.

7.4.3 Surgical and medical treatments are often essential combinations. In patients with normal host defenses correction of mechanical predisposing factors may be all that is necessary for cure of candidosis

– for example, the removal of an intravenous catheter that has been colonized by *Candida* spp. at its tip. Surgical excision or drainage of localized fungal infections should always be carried out, not only to improve prospects for cure, but also to eliminate potential sources for dissemination; preoperative, intraoperative, and postoperative administration of appropriate antifungal agents is necessary. For example, cure of candidal endocarditis appears to depend primarily on surgical excision of the infected site; antifungal antimicrobics alone are rarely, if ever, curative, but may prevent postoperative recurrences.

References (through 1 September, 1988)

1 J. D. Dutcher: Dis. Chest. *54* (Suppl.), 296 (1968).
2 E. Grunberg, E. Titsworth and M. Bennett: Antimicrob. Agents Chemother. *3*, 566 (1964).
3 E. F. Godefroi, J. Heeres, J. M. Van Cutsem and P. A. J. Janssen: J. med. Chem. *12*, 784 (1969).
4 J. Heeres, L. J. J. Backx and J. Van Cutsem: J. med. Chem. *27*, 894 (1984).
5 J. F. Ryley and R. G. Wilson: *1982* Program and Abstracts of the 22nd Interscience Conference on Antimicrobial Agents and Chemotherapy. American Society for Microbiology, Washington, DC.
6 R. E. Mason: Bull Brit. Mycol. Soc. *11*, 144 (1977).
7 M. Miyata, J. Kitamura and H. Miyata: Arch. Microbiol. *127*, 11 (1980).
8 R. S. Gordee, D. J. Zeckner, L. F. Ellis, A. L. Thakkar and L. C. Howard: J. Antibiot. *37*, 1054 (1984).
9 D. Berney and K. Schuh: Helv. chim. Acta *61*, 1262 (1978).
10 G. Petranyi, N. S. Ryder and A. Stutz: Science *224*, 1239 (1984).
11 W. Mechlinski, C. P. Schaffner, P. Ganis and G. Avitabile: Tetrahedron Lett. *44*, 3873 (1970).
12 J. O. Lampen: Am. J. clin. Path. *52*, 138 (1969).
13 C. P. Schaffner: Recent Trends in the Discovery, Development and Evaluation of Antifungal Agents. J. R. Prous Publishers, S. A., Barcelona 1987.
14 P. D. Hoeprich: Scand. J. infect. Dis. *16* (Suppl), 74 (1978).
15 P. D. Hoeprich, N. M. Flynn, M. M. Kawachi, K. K. Lee, R. M. Lawrence, L. K. Heath and C. P. Schaffner: Ann. N. Y. Acad. Sci. *544*, 517 (1988).
16 C. P. Schaffner, O. J. Plescia, D. Potani, D. Sun, A. Thornton, R. C. Pandey and P. S. Sarin: Biochem. Pharmac. *35*, 4110 (1986).
17 T. L. Hadfield, M. B. Smith, R. E. Winn and M. G. Rinaldi: Rev. infect. Dis. *9*, 1006 (1987).
18 S. C. Kinsky: Antibiotics I, Mechanisms of Action. Springer Verlag, New York 1967.
19 A. W. Norman, S. M. Spielvogel and R. G. Wong: Adv. Lipid Res. *14*, 127 (1976).
20 E. R. Block, J. E. Bennett, L. G. Livoti, W. J. Klein, R. R. MacGregor and L. Henderson: Ann intern. Med. *80*, 613 (1974).
21 D. J. Drutz and R. I. Lehrer: Ann. intern. Med.
22 R. A. Woods, M. Bard, I. E. Jackson and D. J. Drutz: J. infect. Dis. *129*, 53 (1974).
23 D. B. Louria: Antibiot. Med. clin. Ther. *5*, 295 (1958).

92 A. B. Richards, B. R. Jones, J. Withwell and Y. Clayton: Trans. Ophthalm. Soc. *89*, 867 (1969).
93 J. E. Pennington, E. R. Block and H. Y. Reynolds: Antimicrob. Agents Chemother. *6*, 324 (1974).
94 S. M. Finegold: Calif. Med. *110*, 455 (1969).
95 H. Vanden Bossche, G. Willemsens, P. Marichal, W. Cools and W. Lauwers: Mode of Action of Antifungal Agents. Cambridge University Press, Cambridge, 1984.
96 H. Vanden Bossche: Curr. Top. Med. Mycol. *1*, 313 (1985).
97 H. O. Sisler and N. N. Ragsdale: Mode of Action of Antifungal Agents. Cambridge University Press, Cambridge 1984.
98 E. P. de Fernandez, M. M. Patino, J. R. Graybill and M. H. Tarbit: Chemotherapy *18*, 261 (1986).
99 A. Polak, H. J. Scholer and M. Wall: Chemotherapy *28*, 461 (1982).
100 A. Schaffner and P. G. Frick: J. infect. Dis. *151*, 902 (1985).
101 A. Polak: Chemotherapy *33*, 381 (1987).
102 P. D. Hoeprich and A. C. Huston: J. infect. Dis. *137*, 87 (1978).
103 P. D. Hoeprich and A. C. Huston: J. infect. Dis. *134*, 336 (1976).
104 M. A. Saubolle and P. D. Hoeprich: Antimicrob. Agents Chemother. *14*, 517 (1978).
105 D. L. Calhoun and J. N. Galgiani: The Antimicrobial Agents Annual/1. Elsevier Science Publishers, Amsterdam 1986.
106 J. P. Sung, J. G. Grendahl and H. B. Levine: West. J. Med. *126*, 5 (1977).
107 J. R. Graybill: Sem. resp. Dis. *1*, 53 (1986).
108 M. S. Saag and W. E. Dismukes: Antimicrob. Agents Chemother. *32*, 1 (1988).
109 P. J. Lewi, J. Boelaert, R. Daneels, R. DeMeyere, H. Van Landuyt, J. J. P. Heykants, J. Symoens and J. Wynants: Eur. J. clin. Pharmac. *10*, 49 (1976).
110 D. A. Stevens, H. B. Levine and S. C. Deresinski: Am. J. Med. *60*, 191 (1976).
111 P. D. Hoeprich and E. Goldstein: J. Am. med. Ass. *230*, 1153 (1974).
112 P. D. Hoeprich and E. Goldstein: Clin. Res. *23*, 133A (1975).
113 J. Brugmans, J. Van Cutsem, J. Heykants, V. Scheurmans and D. Thienpont: Eur. J. clin. Pharmac. *5*, 93 (1972).
114 A. G. Bagnarello, L. A. Lewis, M. C. McHenry, A. J. Weinstein, H. K. Naito, A. J. McCullough, R. J. Lederman and T. L. Gavin: New Engl. J. Med. *296*, 497 (1977).
115 H. B. Nield: New Engl. J. Med. *296*, 1479 (1977).
116 R. W. Bradsher, D. C. Rice and R. S. Abernathy: Ann. intern. Med. *103*, 872 (1985).
117 National Institute of Allergy and Infectious Diseases Mycoses Study Group (1985). Ann. intern. Med. *103*, 861 (1985).
118 A. Restrepo, I. Gomez, L. E. Cano, M. D. Arango, F. Guiterrez, A. Saninn and M. A. Robledo: Am J. Med. *74*, (1B), 48 (1983).
119 R. Negroni, A. M. Robles, A. Arechavala, M. A. Tuculet and R. Galimberti: Rev. infect. Dis *2*, 643 (1980).
120 A. Restrepo, I. Gomez, L. E. Cano, M. D. Arango, F. Gutierrez, A. Saninn and M. A. Robledo: Am. J. Med. *74*, (1B), 53 (1983).
121 E. A. Peterson, D. W. Alling and C. H. Kirkpatrick: Ann. intern. Med. *93*, 791 (1980).
122 First International Symposium on Ketoconazole. Rev. infect. Dis. *2*, 520 (1980).
123 J. W. Rippon: Archs Derm. *122*, 399 (1986).
124 First International Symposium on Itraconazole. Rev. infect. Dis. *9*, (Suppl. 1), S1 (1987).
125 H. Van Cutsem, J. Heykants, R. DeCoster and G. Cauwenbergh: Rev. infect. Dis. *9*, (Suppl. 1), S43 (1987).

126 J. M. Heykants, W. Michiels, W. Meuldermans, J. Monbaliu, K. Lavrijsen, A. Van Peer, J. C. Levron, R. Woestenborghs and G. Cauwenbergh: Recent Trends in the Discovery, Development and Evaluation of Antifungal Agents. Telesymposia Proceedings, Barcelona 1987.
127 J. R. Perfect and D. T. Durack: J. Antimicrob. Chemother. *16*, 81 (1985).
128 P. Phillips, J. R. Graybill, R. Fetchick and J. F. Dunn: Antimicrob. Agents Chemother. *31*, 647 (1987).
128a M. Saag, R. Bradsher, S. Chapman, C. Kauffman, W. Girard, D. Stevens, C. Bowles-Patton, W. Dismukes and the NIAID Mycoses Study Group: *1988 Programm and Abstracts of the 28th Interscience Conference on Antimicrobial Agents and Chemotherapy.* American Society for Microbiology, Washington, DC.
128b P. K. Sharkey, M. G. Rinaldi, C. J. Lerner, R. J. Fetchick, J. F. Dunn and J. R. Graybill: *1988 Program and Abstracts of the 28th Interscience Conference on Antimicrobial Agents and Chemotherapy.* American Society for Microbiology, Washington, DC.
129 T. E. Rogers and J. N. Galgiani: Antimicrob. Agents Chemother. *30*, 418 (1986).
130 J. R. Graybill, E. Palou and J. Ahrens: Am. Rev. resp. Dis. *134*, 768 (1986).
131 G. S. Kobayashi, S. Travis and G. Medoff: Antimicrob. Agents Chemother. *29*, 660 (1986).
132 D. M. Dixon and A. Polak: Chemother. *33*, 129 (1987).
133 W. E. Dismukes: Ann. intern. Med. *109*, 177 (1988).
133a M. G. Rinaldi, P. A. Robinson, J. R. Graybill, J. J. Stern, A. M. Sugar, B. J. Hartman and D. M. Hilligoss: *1988 Program and Abstracts of the 28th Interscience Conference on Antimicrobial Agents and Chemotherapy.* American Society for Microbiology, Washington, DC.
133b P. A. Robinson and A. K. Knirsch: *1988 Program and Abstracts of the 28th Interscience Conference on Antimicrobial Agents and Chemotherapy.* American Society for Microbiology, Washington, DC.
133c G. S. Hall, C. Myles, K. J. Pratt and J. A. Washington: Antimicrob. Agents Chemother. *32*, 1331 (1988).
134 M. Hobbs, J. Perfect and D. Durack: Eur. J. clin. Microbiol. infect. Dis. *7*, 77 (1988).
135 E. D. Spitzer, S. J. Travis and G. S. Kobayashi: Eur. J. clin. Microbiol. infect. Dis. *7*, 80 (1988).
135a H. J. Schmitt, E. M. Bernard, J. Andrade, F. Edwards, B. Schmitt and D. Armstrong: Antimicrob. Agents Chemother. *32*, 780 (1988).
136 G. Petranyi, J. G. Meingassner and H. Mieth: Antimicrob. Agents Chemother. *31*, 1365 (1987).
137 N. S. Ryder, I. Frank and M-C. DuPont: Antimicrob. Agents Chemother. *29*, 858 (1986).
138 U. Ganzinger, A. Stutz, G. Petranyi and A. Stephen: Acta derm.-vener. suppl. 121, 155 (1986).
138a A. Stutz and G. Petranyi: J. med. Chem. *27*, 1539 (1984).
139 J. P. Utz, I. L. Garriques, M. A. Sande, J. F. Warner, G. L. Mandell, R. F. McGehee, R. J. Duma and S. Shadomy: J. infect. Dis. *32*, 368 (1975).
140 J. E. Bennett, W. E. Dismukes, R. J. Duma, G. Medoff, M. A. Sande, H. Gallis, J. Leonard, B. T. Fields, M. Bradshaw, H. Haywood, Z. A. McGee, T. R. Cate, C. J. Cobbs, J. F. Warner and D. W. Alling: New Engl. J. Med. *301*, 126 (1979).

Effects of drugs on calmodulin-mediated enzymatic actions

By Judit Ovádi
Institute of Enzymology, Biological Research Center, Hungarian Academy of Sciences, P. O. B. 7, H– 1502 Budapest HUNGARY

1	Introduction	354
2	Fine structure of calmodulin	355
3	Importance of Ca^{2+} binding	356
4	Enzymes stimulated by calmodulin	358
5	Functional domains of calmodulin for target enzymes	361
6	Interaction between calmodulin and drugs	366
7	Potency and specificity of anti-calmodulin drugs	375
8	Mechanism of action of anti-calmodulin drugs	380
9	Drugs liberating enzymatic activity from calmodulin inhibition	385
10	Past and future	387
	References	390

Abbreviations:

CaM	calmodulin
PDE	cyclic nucleotide phosphodiesterase [EC 3.1.4.17]
MLCK	myosin light chain kinase [EC 3.6.1.3]
PFK	phosphofructokinase [EC 2.7.1.11]
GAPD	glyceraldehyde-3-phosphate dehydrogenase [EC 1.2.1.12]
ELISA	enzyme-linked immunosorbent assay
EGTA	ethylene glycol bis(β-aminoethylether)-N,N' tetraacetic acid
TFP	trifluoperazine

1 Introduction

The Ca^{2+}-dependent regulatory protein, CaM, has been shown to be involved in the regulation of numerous Ca^{2+}-mediated events [1–7]. This ubiquitous protein appears to be primarily located in the soluble fraction of those cells in which its subcellular distribution has been examined [8]. However, it is partly also an integral component of supramolecular structures in the highly organized interior of the cell.

CaM has no inherent enzymatic activity; however a number of biochemical activities of CaM have been established, such as the stimulation of enzymatic activities, protein binding activities or regulation of metabolic processes [6–9 and references therein]. Recently it has been found that CaM can mediate inhibition of some new target enzymes in a Ca^{2+}-dependent manner [10–14]. Therefore, its role in signal transmission appears evident; whenever the free Ca^{2+} concentration changes in the cell, the concentration of the different Ca-CaM complexes varies, thus modulating the activity of target enzymes, which eventually determine the cellular response.

Although the effect of CaM on isolated enzymes is well established, neither the precise mode of its stimulating/inhibitory effects nor the mechanisms regulating metabolic processes have been clearly defined. Anti-CaM drugs of strikingly heterogeneous chemical structure prevent or modify the interactions of CaM with target enzymes. The use of these drugs is widely accepted, on the one hand, to gain information on the mechanism of this action in a large variety of biological systems, and on the other hand, to develop available tools for the pharmacological modification of CaM-mediated cellular events.

Several recent reviews on CaM have appeared [3, 5–9). Nevertheless, the accumulation of information in this field is explosive and the topic has so many different aspects. My intention is not to provide a comprehensive review of all CaM-mediated processes which can be controlled by drugs, rather special attention will be paid to the mechanism of action of anti-CaM drugs by which they exert their effects on the enzymatic processes. I will attempt to give a view of the specificity problem; why the effects of these drugs are not associated with selective functional effects. In order to facilitate the overview of these questions, a brief synthesis of the present knowledge of the structure and activitiy of CaM will also be presented.

2 Fine structure of CaM

One of the important contributions to the elucidation of the mechanism of action of CaM comes from the three-dimensional structure of this protein. The determination of the crystal structure of CaM to 0.3-

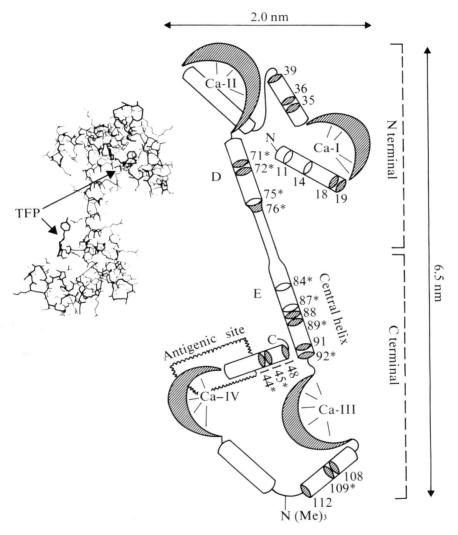

Fig. 1
Schematic representation of the structure of calmodulin by its stereoview showing the proposed binding sites for TFP [16]. N and C terminals are as indicated. Tubular regions represent α-helical segments. Full and open circles represent the hydrophobic and ionic amino acid residues of the functional domains of CaM, respectively.

nm resolution [15] revealed that the molecule looks like a dumb-bell with the globular N and C halves denoted as N- and C-terminal domains, which are linked by a 2.0-nm long α-helical stretch. The schematic representation of the structure of CaM is shown in Fig. 1. The molecule with 63 % α-helix structure is approximately 6.5 nm in length and each of the two non-interacting lobes in roughly 2.5 × 2.0 × 2.0 nm. Both lobes contain two Ca^{2+}-binding sites. The D helix runs continuously to join the E helix through the D/E-helical linker forming a long central helix connecting the two domains [16]. This central helix may be buried in the absence of Ca^{2+} and exposed when the ions are bound. This finding would be consistent with proteolytic studies of CaM, showing that in the presence of Ca^{2+} the amino acid sequence corresponding to this central helix is very sensitive to proteolytic cleavage [17, 18].

CaM belongs to the family of Ca^{2+}-binding proteins, including troponin C, which exhibits a common structural motif of helix-loop-helix, frequently referred to as the EF hand [19]. The most accurate predicted molecular model of CaM is based on the X-ray crystallographic coordinates of the highly refined structure of skeletal muscle troponin C [16].

CaM can be subdivided into four domains of similar amino acid sequences, which correspond to Ca^{2+}-binding domains. The Ca^{2+}-binding domain is a structural motif comprising a 12-residue Ca^{2+}-binding loop [cf. Fig. 1]. The primary structure of this protein has been found to be highly conserved throughout the phylogenetic scale. This "finely tooled" protein molecule is designed to mesh with a number of other macromolecules to regulate various cellular processes [7].

3   Importance of Ca^{2+} binding

There is consensus as to the number and affinity of Ca^{2+}-binding sites. However, the nature of the Ca^{2+} binding sites has not been precisely defined because the binding profile strongly depends on the environmental factors [8]. For example, CaM contains six auxiliary ion-binding sites which modulate the affinity of the Ca^{2+} binding sites [20]. Microscopic dissociation constants range for Ca^{2+} binding to CaM between $1\mu M$ and $100\mu M$ [7]. Under physiological conditions the maximum extent of saturation is 3 or 4 moles of Ca^{2+} per mole of CaM [8].

Ca^{2+} binding transforms CaM from a less-ordered asymmetrical structure into a compact symmetrical one [21]. The Ca^{2+}-free CaM has a constrained conformation due to electrostatic interactions between three pairs of the free carboxylic groups (residues between 75 and 90) and appropriate binding groups [20]. When Ca^{2+} is bound to CaM, it is accompanied by conformational changes due to the neutralization of certain negative charges. These conformational alterations include an increased α-helical content and the exposure of a hydrophobic domain on the surface of CaM, which is recognized by the target proteins as well as by the CaM antagonist drugs [22, 23]. Mg^{2+} competes with Ca^{2+} for the high affinity sites in the C-terminal of CaM; however, their bindings induce different conformations which results in the exposure of various interfaces with a potential to bind target enzymes [24]. Although it was suggested [25] that binding of two Ca^{2+} ions per CaM is necessary and sufficient to expose the hydrophobic patches, Cox et al. [26] recently presented evidence that the Ca_3-CaM and Ca_4-CaM complexes are the active species for activating a large number of CaM-dependent processes. Indeed, full exposure of the two binding sites for hydrophobic fluorescent probes requires the saturation of the four Ca^{2+}-binding sites. Therefore, the hypothesis that the various Ca^{2+}-induced conformers of CaM provide a mechanism for selective activation of target enzymes does not seem to be supported. Thus, the activation of CaM by Ca^{2+} – as a result of Ca^{2+} binding – seems to be an all-or-none process: when CaM has 3 or 4 Ca^{2+} ions bound it is active, when it has less than three it is inactive [26].

For the interaction of CaM with Ca^{2+}, two mutually exclusive binding models have been proposed: 1) four nearly independent binding sites with very similar affinities. This model is supported by direct binding and microcalorimetric studies [26, 27]; 2) two pairs of sites with pronounced positive cooperativity within each pair and a marked difference in the affinity between the pairs. Kinetic data and some conformational studies favor the latter model [28]. The apparent homotropic cooperativity in Ca^{2+} binding to the CaM-enzyme complex may be due to the transition of the complex from its low affinity form to its high affinity one in the third Ca^{2+} binding step. This concurs with the dramatic increase in the extent of Ca^{2+} binding when TFP and related drugs are bound to CaM [29]. In addition, mastoparan, a 14-residue peptide, has been also shown to increase the Ca^{2+}-binding affinity when complexed to CaM [24]. Therefore, the Ca^{2+}-binding properties of CaM are markedly influenced by the target enzyme and/or drugs with an increase of the binding constants and the appearance of quite strong, positive cooperativity [20]. It has to be added that the Ca^{2+}-dependent CaM stimulation of an enzyme activity does not necessarily reflect a physiological function of CaM. It is possible that some of the *in vitro* activities attributed to CaM may be examples of CaM's ability to substitute for an endogenous Ca^{2+}-binding protein. However, some *in vitro* activities clearly reflect the physiological roles af CaM.

4   Enzymes stimulated by CaM

During the last few years, an increasing number of enzymes have been found to be modulated by CaM. Due to the large number of these reports, I will give a short insight into the function of CaM-target enzymes. I have selected the enzymes PDE, adenylate cyclase, Ca^{2+}-Mg^{2+}-ATPase and MLCK because the effects of anti-CaM drugs have been studied with them extensively.

In the Ca^{2+} and cAMP second messenger systems important activities have been attributed to the metabolic processes catalyzed by CaM-dependent enzymes. A hypothetic model of how CaM might control the enzymes which catalyze the synthesis as well as the degradation of cAMP can be described briefly as follows: PDE and adenylate cyclase are involved in the metabolism of cAMP and both are activated by CaM. PDE es a cAMP-degrading enzyme, adenylate cyclase catalyzes the synthesis of this second messenger. CaM may function so elegantly because the adenylate cyclase is membrane-bound, in contrast to PDE; thus, Ca^{2+} penetrating the membrane first binds to membrane-bound CaM, thereby activating adenylate cyclase. PDE is activated when Ca^{2+} ions reach cytoplasmic CaM. Therefore, the balanced activation of both enzymes, enhancement or decrease of the cAMP level, is determined by their differing sensitivities to the Ca^{2+} concentration [8]. Furthermore, the mediator of cAMP action, the cAMP-dependent protein kinase can alter the function of, for example, MLCK (at least *in vitro*). Thus, these systems can influence and coordinate the regulation of intermediary metabolism and control the contractility.

PDE exists in multiple forms; the native form seems to be dimeric [30, 31]. It has been proposed that the monomer is active, whereas the dimer is inactive in the absence of CaM. Regarding the molecular mechanism of the interaction between brain PDE and CaM, it has been suggested that the active CaM demands the binding of 3 or 4 Ca^{2+}, and that one molecule of CaM per monomer of PDE is required for activity [8]. However, there are some very recent indications for the cooperative binding of PDE molecules to CaM. An exciting observation is that the initial rate of proteolysis of PDE is greater in the presence of Ca^{2+} and CaM than with EDTA, indicating that the association of PDE with CaM alters the conformation of PDE, exposing a new site with increased susceptibility to proteolytic attack. CaM can remain associated with the proteolyzed PDE and in this complex the proteolytic site is inaccessible. Whether CaM physically protects this site or induces a conformation in which the site is hidden is not known. The proteolyzed forms of PDE were fully activated and no difference in the enzymatic property was detected. It appears that a region of the protein exerts a self-inhibitory influence which is reversibly relieved upon activation by Ca^{2+} and CaM. Proteolytic activation destroys the function of this "inhibitory domain" and leads to irreversible activation of the enzyme [32]. Much less information has been accumulated about adenylate cyclase as a CaM-dependent enzyme. CaM is proposed to bind to the catalytic unit of adenylate cyclase and increases the activity of cyclase [33, 34]. Further details of the interaction remain to be established. The transport of Ca^{2+} by the Ca^{2+}-Mg^{2+}-ATPase is driven by the hydrolysis of ATP which transiently phosphorylates the ATPase. The 140-kDa plasma membrane Ca^{2+}-ATPase is the only ion-transport ATPase known to be regulated by CaM [35, 36]. Limited proteolysis and chemical modification studies suggest that the human erythrocyte enzyme contains discrete domains involved in membrane association (30–35 kDa), catalysis (76–81 kDa), and CaM-regulation (8–10 kDa). Maximal activation of the purified Ca^{2+}-ATPase could be observed at one molecule of CaM per one molecule of ATPase. Similarly to PDE, it has been found that the removal of CaM-binding domain of Ca^{2+}-ATPase induces an activation of the enzyme similar to that observed with Ca-CaM complex [37, 38]. Moreover, one clone encoding a C-terminal portion of the enzyme that appears to contain an inhibitory domain of the Ca^{2+}-ATPase has been recently isolated [36].

MLCK can be phosphorylated by the cAMP-dependent protein kinase [39], the phosphorylated form binds to CaM with a lower affinity than the unphosphorylated enzyme does [40]. Since the dissociation constant of CaM-MLCK is in the order of nM, and the intracellular concentration of CaM is in the μmolar range, the phosphorylation of MLCK would have little effect on the function of MLCK. However,

the formation of calcium-dependent hydrophobic surfaces of CaM. The importance of Met residues, 109, 124, 144 and 145, in the C-terminal binding domain was also proposed by Faust et al. [56], namely, the modifications of these methionine results in a reduced activation of both PDE and adenylate cyclase [53]. Carboxymethylation of PDE reduces the magnitude of its activation by Ca-CaM complex [57], but, carboxymethylation of CaM produces only a slight reduction of its ability to activate PDE [58, 59]. Carboxymethylation of histidine residue, nitration of tyrosil residues produced no significant changes either in W-7 binding to CaM or in the ability of CaM to stimulate PDE [52].

The genetic approach to CaM function has been attempted by site-directed mutagenesis of cloned and expressed CaM genes. Site-specific mutation of Glu 82, 83, 84 to Lys residues resulted in the expression of a CaM mutant with a 70 % decrease in activation of MLCK [60]. Putkey et al. [61] have produced a mutant CaM with 16 amino acid substitution which fully activates PDE and MLCK but only half-maximally activates calcineurin and CaM-dependent protein kinase. These results suggest that the interaction of CaM with various target proteins may involve different functional domains.

Some difference in the conservative amino acid sequence of CaMs can result in different interactions with various enzymes [62]. One of the posttranslational modifications of CaM is lysine methylation at position 115 [63]. The degree of methylation is probably not necessary for the activation of CaM-dependent enzymes [64]. However, it was shown that unmethylated CaM was less effective than the methylated counterpart in the activation of plant NAD kinase [62]. Similar results have been obtained with *Chlamydomonas* CaM, which also has an unmethylated lysine at residue 115 and may have a C-terminal extension [65]. The covalent modification of not only CaM but of the target enzymes would result in either an increase or a decrease in the concentrations of active CaM-enzyme complexes. The physiological relevance of these posttranslational modifications like phosphorylation [66], and their effects on CaM's interaction with its target proteins have not yet been established. Nevertheless, this observation suggest that certain CaM-dependent enzymes are selective in their interactions.

More insight into the molecular details governing the interaction of CaM with its target proteins was provided by using CaM fragments [18, 67]. These results show that fragments 1–77 and 78–148, each containing two Ca^{2+} binding sites and one hydrophobic domain, retain their ability to interact with several proteins [68]. PDE and cAMP-dependent phosphorylase kinase and calcineurin interact with fragment 78–148; fragment 1–77 interacts only with PDE and phosphorylase kinase; therefore the latter seems to be more selective. Other data suggest that MLCK and PDE interact with the N-terminal portion of the central helix [69, 70] while calcineurin [68] and Ca^{2+}-pumping ATPase [70] preferentially recognize the domain III region of CaM. Although the fragments can activate the enzymes, the affinity of the fragments to the target enzymes is always reduced, as compared to the native molecule [18], probably because the long central helix of CaM plays a role in the process [6]. Although these studies with fragments must be interpreted with caution, they are likely to reflect structural requirements for the activation which may differ among target enzymes. Thus, CaM might interact with different proteins at different sites. Moreover, hydrophobic forces are necessary for the interaction with and activation of enzymes, however, they are not enough to enable a fragment to activate all enzymes [7]. Although many studies offer conflicting reports about various CaM fragments activating CaM-mediated processes [18], highly purified fragments of CaM (1–77, 1–90, 1–106, 78–148, 107–148) were recently reported to be incapable of activating target proteins [71].

Since there is a single central helix in the CaM structure one would expect that the enzymes can exhaustively interact with a "unique, functional" domain of CaM. However, the data indicate that CaM has multiple binding domains and different binding sites are responsible for the binding of different target enzymes. Nevertheless, no ternary complex of CaM with two different target proteins has yet been detected [14].

Additional evidence for the heterogeneity of enzyme-binding domains on CaM comes from experiments using antibodies against CaM. The simultaneous binding of a target enzyme and an anti-CaM immunoglobulin to CaM has recently been detected [Zuklys and Ovádi, unpublished result]. Within the C-terminal domain of CaM a linear segment of 7 amino acid residues (137–143) (cf. Fig. 1) seems to be responsible for its full immunoreactivity [72]. The use of the indirect ELISA technique allowed us to investigate the accessibility of the functional domains of CaM in a CaM-antibody immune complex [73]. It has been found that the antibodies were able to bind to CaM in the presence of

PDE [74], and the binding of the two proteins to CaM exhibited positive cooperation [Zuklys and Ovádi, unpublished result]. In addition, there are some indications that a monospecific anti-CaM immuno-

Table 1
Quantitative data for drugs with anti-calmodulin activity

Compounds	Binding sites with High affinity		Low affinity		Me-thod	I_{50} (μM) PDE	ATPase	MLCK	Ref.
	K_d (μM)	number	K_d (μM)	number					
Phenothiazines									
Trifluoperazine	1.5	2	500	24–27	A				74
	5				B		7.4*		75
						10 (0.0118)			74
						1.6–2 (0.005)			76
								13–200	80
Chlorpromazine	5	3	130	17					74
		5–6							81
						42			77
						47			82
Thioridazine						18			83
						11	18		84
Naphthalenesulfonamides									
	11	3	200	9					87
W-5						240			88
W-7	4–11				B		45*		75
						26			88
						67 (0.0118)			82
								15–300	80
	1–10	2–3	>100						82, 89
TI-233	the same as W-7								
Imidazolium									
Calmidazolium	0.003				B				92
(R-24571)						0.01 (0.004)			76
								2.5–22	80
						0.3			97
Ro 22-4839						20		3.1	94
CGS 9343 B						3.3			95
Peptides									
β-endorphin	4.6								98
	4.6	2							99
Mastoparan	nanomolar								100
Melittin	nanomolar								101
Synthetic peptides	210–390 pmol								102

Compounds	Binding sites with				Me-thod	I_{50} (μM)			Ref.
	High affinity		Low affinity			PDE	ATPase	MLCK	
	K_d (μM)	number	K_d (μM)	number					
Arylalkylamines									
Bepridil	18								110
						18	30		84
						37			97
Fendiline	0.7				B			4.0*	75
						31	40		84
						18			97
Prenylamine	0.7				B			0.7*	75
						20	28		84
						18			77
						14			97
Pimozide	0.83	1				7			74
Verapamil	30				B				75
						>300	>300		84
						340			110
							>1000		112
						150			97
Dihydropyridines									
Felodipine	2.8				B				75
		2							113
						11			114
	9					3.7		12.6	105
p-chloro-felodipine	0.3					1.5			105
Nimodipine						2			110
Nicardipine						8			
Nifedipine						≥1000			112
						90			114
						15			97
Others									
Flunarizine (piperazine derivative)						23		97	
Diltiazem (benzothiazepine derivative)									
	80				B				75
						>300	>300		84
							>1000		112
						310			79
D-600	30				B				93
Vinblastine	2		10						123
						10–20 (0.023)			124, 125
Adriamycin								50–85	126
Triton X-100						14 (0.024)			127
48/80	<1					0.5 (0.03)			128
						0.015–0.025 (0.005)			21
Haloperidol	9	2	–	–					74
Butyrophenone						60			77

The numbers in brackets are the CaM concentrations used. Symbols A and B represent equilibrium dialysis and fluorescent intensity of covalently labelled CaM techniques, respectively. * smooth muscle contraction was measured.

globulin can be bound to the CaM-activated human red cell inside-out-vesicle calcium pump [73]. The basic Ca^{2+} transport was not affected by anti-CaM antibody, but the CaM activation of Ca^{2+} transport was inhibited, depending on the concentrations of Ca^{2+} and antibody. Therefore, it seems probable that the antibody does not compete with the target enzymes for CaM binding (cf. Table 3) since 1) the antigenic site of CaM is distinct from the enzyme-binding domain; 2) the immunoreactive site is accessible when CaM is part of a supramolecular complex.

6 Interaction between CaM and drugs

This review of the binding of drugs to CaM focuses on the multiple and cooperative characters of the interactions; nevertheless, a list of drugs with anti-CaM activity is presented together with their quantitative data (Table 1). Special attention will be paid to TFP because this compound was discovered first as a CaM antagonist and the mode of its binding to CaM has been analyzed most extensively. From the data some generalization can be drawn for the interaction of CaM with other drugs of different chemical structures (cf. Fig. 2). The methods used include fluorescence spectroscopy, affinity labelling, limited proteolysis, affinity chromatography, NMR, spectroscopic distance measurements, covalent modifications, etc.

Two sets of binding sites for TFP were revealed (cf. Table 1): two high affinity, Ca^{2+}-dependent sites with an apparent K_d of 1.5 μM and 24–27 low-affinity Ca^{2+}-independent sites with an apparent K_d of about 500 μM [74]. Structure-activity relationship investigations reveal that mainly hydrophobic but also electrostatic interactions play a role in the binding. Weiss and his coworkers [77] proposed the classical model for CaM inhibitors: they ought to have a bulky hydrophobic region, formed by two aromatic rings for interaction with non-polar regions within the CaM molecule, and a positively charged group, essentially a basic nitrogen atom charged at physiological pH, at a distance of not less then three carbon atoms from the hydrophobic rings (cf. Fig. 2). Phenothiazine binding to two distinct sites in different CaM domains has interesting implications for a cooperative mode of interaction between CaM and its regulator [8]. It has been suggested that the stronger high affinity TFP-binding site is located in the carboxy terminal half [78], and its affinity is at least one order of magnitude higher than that of the N-terminal domain [79].

Considerable progress has been made in the past few years in localizing the sites of interaction of CaM with its antagonists. The involvement of various amino acid residues in the CaM-drug interactions has been suggested. Ionic interactions between phenothiazines and their binding site(s) on CaM may occur between the positive charges of the amino groups of drugs and the glutamic acid rich area of CaM containing residues 82, 83, 84, 87 (cf. Fig. 1). This ionic interaction with the flexible piperazine-arm of TFP (cf. Fig. 2) may account for the greater ability of the piperazine-containing phenothiazines to increase the α-helical content, as compared to chorpromazine which lacks this ring [85] (cf. Fig. 2)..The potency of drugs increases with the increasing length of amino side chain in both phenothiazines.

Acetylation studies have shown that Lys 75 and Lys 148 may be involved in the TFP binding [86]. Recently, Faust et al. [56] have also reported the importance of these lysyl residues using labelled CaM with a chemically reactive phenothiazine. They also suggested that the amino acid residues responsible for TFP binding in the C-terminal domain are Phe 92, Met 144, Met 145, Lys 148.

Due to the functional and structural similarities of CaM and troponin C, the refined crystal structure of skeletal troponin C provided an excellent basis for the construction of a model for CaM. A site for TFP-binding on troponin C was proposed by Gariépy and Hodges [85]: a small helican region of about ten amino acids, rich in hydrophobic (aliphatic and aromatic) side chains. This sequence corresponds to amino acid residues 92–102 of rabbit skeletal Tn C, and probably to residues 82–92 in the homologous sequence of CaM [Fig. 1]. Strynadka and James [16] made a prediction for residues within the N- and C-terminal domains of CaM, close to the hinge region, which are responsible for the TFP binding: residues 84, 87, 88, 91, 92.

On the basis of the data presented in the literature, a tentative molecular model for the binding of drugs and enzymes may be derived (cf. Fig. 1): The hydrophobic patches of CaM play an important role in the interaction with drugs. These hydrophobic clefts of the N and C domains face each other on opposite sides of the central helix. Along the entrance of both apolar cavities are a number of charged acidic residues. The hydrophobic phenothiazine rings stack against the hydrophobic patch. The positively charged N of the piperazine moiety inter-

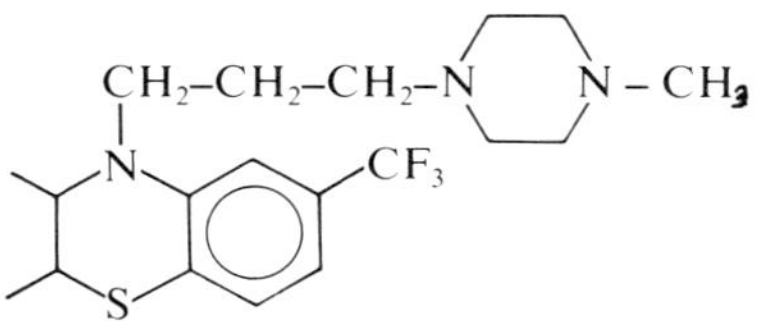

Trifluoperazine

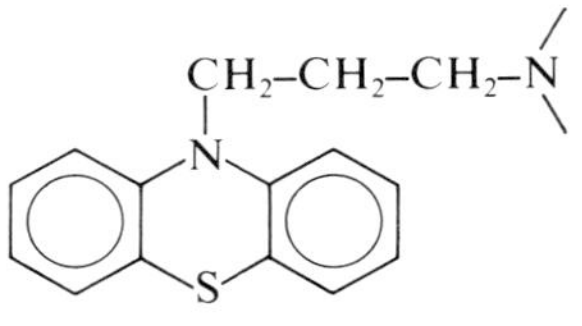

Chlorpromazine

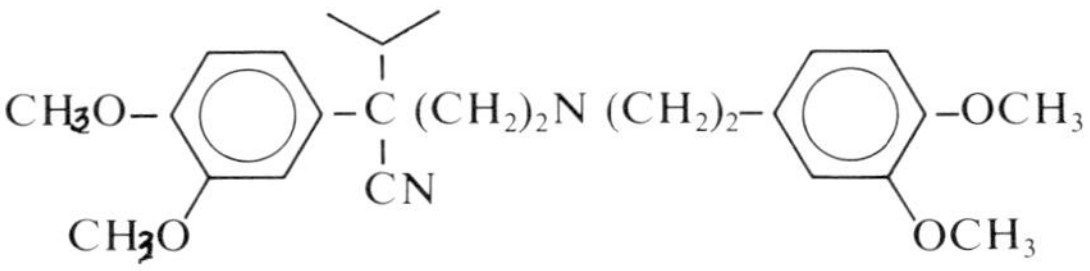

Verapamil

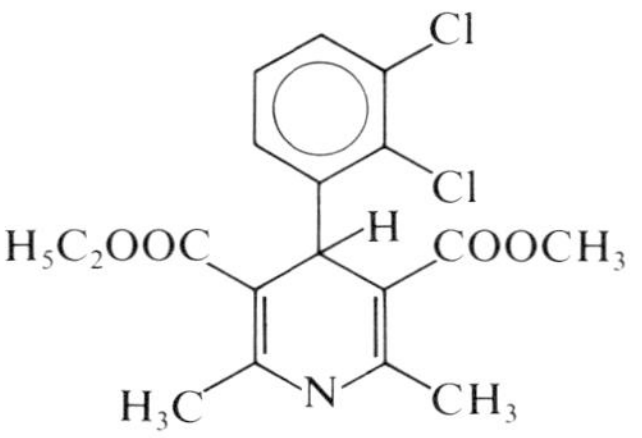

Felodipine

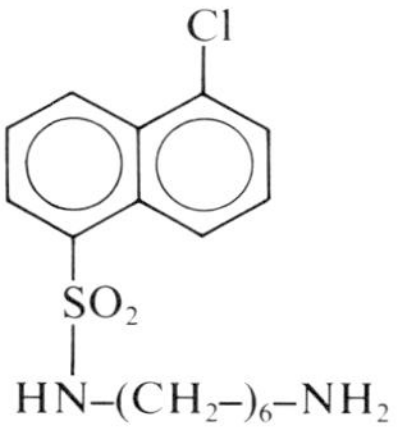

W-7

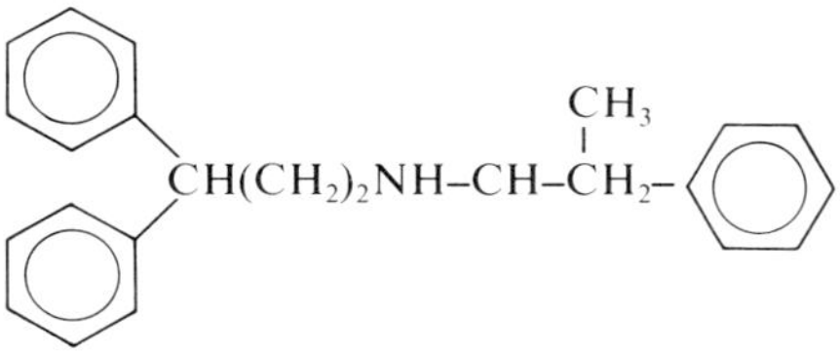

Prenylamine

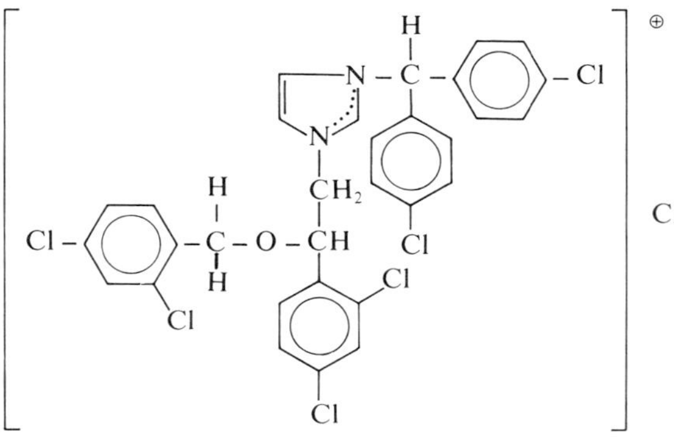

Calmidazolium

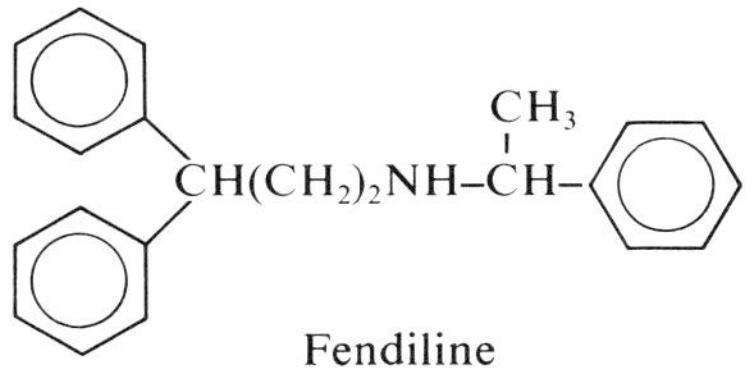

Fendiline

[Structure: diphenyl-CH(CH₂)₂NH-CH(CH₃)-phenyl]

Fig. 2
Structural formulas of some selected anti-calmodulin drugs.

acts with the negatively charged residues framing the outer rim of the hydrophobic pocket. In the C-terminal domain the phenothiazine rings of TFP nicely fit into a groove formed by the side chains of Ala 88, Val 91, Phe 92, Val 108, Met 109, Leu 112, Met 144, Met 145. The positively charged N of the piperazine ring can then easily interact with one or both of the acidic side chains of Glu 84 and Glu 87. Outside the hydrophobic patch Lys 148 runs along side the trifluoro group at position 2 of the phenothiazine ring. Similarly to the C terminal, an analogous sequence was found to be involved in TFP binding in the N-terminal domain: Ala 15, Leu 18, Phe 19, Val 35, Met 36, Leu 39, Met 71, Met 72 and for ionic interaction Glu 11 and Glu 14 and Lys 75.

A second group of compounds with proven CaM antagonistic properties includes the naphthalene-sulphonamides, which have been intensively investigated by Hidaka and his coworkers [87–89]. These compounds exhibit binding characteristics to CaM qualitatively comparable to those of the phenothiazines (cf. Table 2). W-7 has been shown to enter the cytoplasm [90] and to possess good specificity for interacting with CaM.

Calmidazolium (cf. Fig. 2), which is a highly lipophilic imidazolium derivative demands considerable interest. The affinity of this imidazolium molecule to CaM is extremely high (cf. Table 1). It seems that it is the most potent and most selective CaM antagonist until now [91]. For other imidazolium derivatives similar structure-activity relationships have been established [87, 92].

New molecules have been synthesized which seem to be potent and selective inhibitors of CaM-mediated events: Ro 22-4839 [94] and CGS 9343B [95] (cf. Table 1). Hydrophobic fluorescent probes showed that

Ro 22-4839, like TFP and W-7 binds to the hydrophobic region of CaM. However, its precise binding site to CaM is different from that of the other two drugs as suggested by differing I_{50} values of these compounds against the probes [94]. CGS 9343B has a chemical structure which corresponds to the criterion of a "good" CaM antagonist [92], it also has a hydrophobic domain consisting of a N phenyl pyrrole, separated from a basic nitrogen by at least two carbon atoms. Additionally, the 4-substituent on the piperidine in this molecule is identical to that found in pimozide, one of the most potent neuroleptic inhibitors of CaM [96] (cf. Table 1).

Peptides, depending on their chemical character, can bind to CaM predominantly via hydrophobic or ionic interactions. β-endorphin has been suggested to interact with the same hydrophobic region of CaM as phenothiazines (cf. Table 2). β-endorphin binding is antagonized by

Table 2
The role of chemical character of drugs in binding to calmodulin

Chemical character	Binding site on related to TFP binding sites	
	Identical	Different
Amphiphilic	TFP calmidazolium W-7	fendiline prenylamin* verapamil
Hydrophobic	β-endorphin	felodipine
Ionic		melittin

For details and references see text. * Data are controversal.

phenothiazines; e. g. by chlorpromazine [98] or by TFP [99] in a Ca^{2+}-dependent manner. The role of hydrophobic forces in the interaction between CaM and this peptide is also supported by cross-linking experiments [99]. Melittin (and a number of other peptides found in insect venom) interacts with CaM predominantly *via* basic, positively charged groups. Acetylation markedly reduced the binding of this group of peptides to CaM, however, it did not completely eliminate their inhibitory actions [101]. Because of the importance of positively charged groups in the interaction, it has been suggested that the interaction occurs at sites other than the phenothiazine binding site (cf. Table 2). Strynadka and James [16] proposed on the basis of their computer modeling data that melittin, as well as mastoparan, a basic amphiphilic peptide, interacts with a cluster of glutamic acids surrounding the C-terminal hydrophobic patch of CaM.

New synthetic peptides with high affinity for CaM are novel tools for studying the mechanism of action of CaM. However, their specificities have not yet been established, and their ability to enter cells is questionable [102].

It is well known that "Ca^{2+} antagonists" play a crucial role in a number of enzymatic processes. Subclassification of Ca^{2+} antagonists is presented in Fig. 3. Recently a number of Ca^{2+} channel blockers have been shown to have CaM antagonistic properties in addition to their original activities. This group of compounds includes both arylalkylamines like prenylamine [103], fendiline [104], bepridil [106], and dihydropyridines like felodipine and nimodipine [107]. Fluorescence studies suggested that bepridil binds to more sites on CaM than the other CaM inhibitors, like W-7 (five sites for bepridil versus one to three sites for other inhibitors) [108]. Binding constants for fendiline and prenylamine were found to be in the micromolar range (cf. Table 1). Weaker binding of verapamil and diltiazem to CaM was reported; they inhibit CaM in a very high concentration range, but another arylalkylamine, pimozide, binds to CaM and inhibits enzyme activation at very low concentrations, which overlaps the range at which calcium entry is blocked [106, 109]. Dihydropyridines do not easily conform to the structure-activity relationship proposed by Prozialeck and Weiss [92]. According to their different structures, they seem to have binding sites on CaM different from those of other CaM antagonists. Since dihydropyridines are neutral molecules (cf. Fig. 2) and bind with high affinity to CaM (cf. Table 1), it seems that a positively charged group is not a prerequisite for the interaction of these drugs with CaM [8, 115] (cf. Table 2).

According to Vogel [116], dihydropyridines such as felodipine, bind preferentially to the amino-terminal domain, in contrast to other drugs (e. g. TFP), which prefers to bind to the carboxy-terminal. They also propounded on the basis of Cd- 113 NMR measurements that the hinge region between the two halves of CaM (65–92) is important for felodipine-binding. This problem was studied in considerable detail by Johnson and his coworkers. They suggested that residues in the region of 75–90 are necessary for felodipine binding [117]. The probability that the binding sites of arylalkylamines differ from those of other CaM antagonists, is further substantiated by data from various laboratories. It has been suggested that the binding of felodipine to CaM is strongly enhanced in the presence of some other antagonists. Calmid-

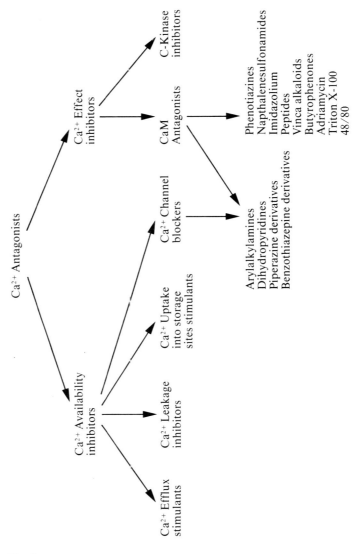

Fig. 3
Subclassification of "Ca antagonists". Only drugs which belong to Ca^{2+} channel blockers and/or CaM antagonists are listed.

azolium, prenylamine and diltiazem act as allosterically potentiating agents for felodipine binding [118]. Johnson [119] and Mills et al. [98] also found positive cooperativity in the binding of these drugs on CaM, namely, calmidazolium and prenylamine increased the binding of felodipine, suggesting allosteric interactions between the binding sites of these drugs (cf. Fig. 2). At low concentration they stimulate felodipine binding, while at higher concentrations they displace felodipine from CaM [6]. Compound 48/80 has the same effect, but it is less effective [6]. Further analysis of these phenomena revealed that calmidazolium and prenylamine could bind to one site on CaM, and increase the affinity of CaM for felodipine at its remaining site twenty- fold [118]. A fluorescent CaM-felodipine derivative exhibited a fluorescence increase on the addition of the potentiating drugs, calmidazolium, prenylamine and melittin with half-maximal fluorescence values 9×10^{-8} M, 4×10^{-6} M and 2.6×10^{-8} M, respectively [120]. Felodipine binding to CaM in the presence of other Ca^{2+} antagonists, such as prenylamine and diltiazem, indicates that various CaM antagonists bind to distinct sites and cooperative allosteric interactions can occur between these sites [115, 119]. The fact that most Ca^{2+} antagonists interact with CaM weakly or not at all, suggests that these compounds probably do not interact with CaM at pharmacologically relevant concentrations. Nevertheless, the use of these drugs led to the discovery of a positive cooperation in the binding of anti-CaM drugs.

Evidences for the cooperative binding of TFP and other anti-CaM drugs have been also presented. Newton et al. [121] have shown that the rate of covalent modification of CaM by a hydrophobic fluorescent dye is enhanced in the presence of TFP. Inagaki and Hidaka [122] suggested that the binding site of prenylamine on CaM is different from that of TFP. Distinct binding sites for TFP and fendiline have been recently reported [13] (cf. Table 2, Fig. 6).

It is an exciting problem how the accessibility of the antigenic sites is influenced by drugs. It has been found that CaM iodinated at tyrosil-99 interacts with phenothiazine and binds to an antibody. Moreover, it binds the antibody in the presence of chlorpromazine and Ca^{2+} [73 a]. This result suggests that the phenothiazine-binding domains on CaM are distinct from the immunoreactive site (residue 137–143) (cf. Fig. 1). We chose drugs from different chemical classes to study the relationship of binding sites of drugs and antibody on CaM. Anti-CaM sera were produced in rabbits against dinitrophenylated bovine CaM. The sera did not form precipitin lines in double diffusion with CaM indicating that only one major antigenic site is present on the CaM molecule. Anti-bovine CaM obtained by CaM-Sepharose-4B affinity chromatography was used for indirect ELISA. Subsaturating concentrations of antibody were added at constant concentrations of CaM in

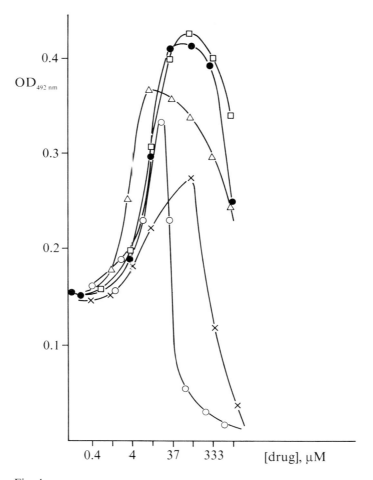

Fig. 4
Effect of drugs on the formation of calmodulin-antibody immunecomplex. Anti-human CaM raised in rabbits was used for indirect ELISA. Plates were coated with human CaM (coating concentration: 1.5 µg/ml). The subsaturating concentration of antibody was 2.5 µg/ml. The plates were read at 492 nm after partial hydrolysis of o-phenylenediamine catalyzed by goat anti-rabbit globulin conjugated with horseradish peroxidase in the presence of peroxide. Drugs added: o, calmidazolium; x, TFP; △, verapamil; o, fendiline; □, vinblastine.

plate coating. The plates were red at 492 nm for detection of immunecomplex. As shown in Fig. 4, the binding of antibody to CaM could be observed even in the presence of drugs. However, the formation of the immunecomplex depended on the chemical structure and the concentration of the drugs (Fig. 4). Moreover, at low concentrations the drugs stimulate antibody binding while at higher concentrations TFP and calmidazolium displace the antibody from CaM. The other drugs, such as vinblastine, fendiline or verapamil, although they bind to CaM in the presence of antibody, apparently do not inhibit formation of the immunecomplex. These results suggest that 1) the structurally heterogeneous CaM inhibitors are bound outside of the antigenic domain; 2) probably different binding sites of CaM are responsible for the binding; 3) positive cooperation may exist between drug and antibody binding sites.

7 Potency and specificity of anti-CaM drugs

There are contradictory data concerning the selectivity of anti-CaM drugs. Obviously, the effect depends on the dose and varies with the complexity of the system. The agents which share an amphipathic (hydrophobic and cationic) character antagonize CaM-activated enzyme and cellular activities with a potency which generally correlates the overall hydrophobicity of the molecules [7]. The compounds of lower potency might interfere with CaM activation by mechanisms other than direct interaction with CaM. The structural specificity of CaM antagonists, in addition to their hydrophobicity, might be related to their ionic region. In fact, the binding of phenothiazines to CaM is reduced when the pH is changed from 7.5 to 8.5 as well as below pH 5 [129]. This probably reflects an ionic contribution to binding of the ammonium head of TFP, in addition to the hydrophobic interaction. Aromatic rings of phenothiazines were substituted with chloro-, thio-, trifluoromethyl and hydroxyl groups in various positions [93], and it has been concluded that the hydrophobicity of the aromatic ring is the major determinant of the potency. This potency increases as the distance between the amino group and the phenothiazine nucleus is increased from 2 to 4 atoms [129]. Similar findings were obtained with 1-naphthalenesulfone derivatives.

It is possible that some drugs, even the "good" CaM antagonists, can inhibit the action of CaM by interacting with the effector enzyme [130]. Since Ca^{2+} channel blockers bind to CaM with greater or lesser affinity [93] (cf. Table 1), CaM might be an intracellular target of the Ca^{2+} channel blockers. Some of the Ca^{2+} channel blockers may be bound directly to the target enzymes and may not act via CaM. For example, the inhibition of both unactivated and CaM-activated PDE by DHP was observed, which suggests that these compounds are PDE inhibitors rather that CaM inhibitors. Similarly, according to Meltzer and Kassir [131] the inhibition of CaM activation of erythrocyte Ca-Mg-ATPase by the β-blockers, propranolol and nadolol is probably based on the interaction of the antagonist with the target enzyme. In the case of Ca-Mg-ATPase, for example, phenothiazines inhibit the basal enzyme activity and lower its Ca^{2+} sensitivity [132, 132a].

Among the Ca^{2+} antagonists the binding of felodipine to CaM is similar to that of TFP, while verapamil and diltiazem are less potent agents [93]. Verapamil and diltiazem, but not nifedipine, cannot interfere with CaM activation of the red cell Ca^{2+} pump ATPase [112]. This observation is in excellent agreement with studies by Lamers et al. [133] who found the same for CaM-activated PDE and CaM-activated cardiac sarcolemmal Ca^{2+}-ATPase. While the inhibition kinetics for verapamil and diltiazem indicate both V_{max} and K_M changes, it is not clear whether these agents inhibit stimulation of the CaM-dependent processes by interacting with the enzyme or with CaM or both [112]. The millimolar concentrations used for inhibition (cf. Table 1) are much higher than the typical dose ranges for verapamil or diltiazem (0.01 to 10μM) and nifedipine (0.001 to 1 μM). Moreover, since high concentrations of these drugs are needed to inhibit the process, it seems that the binding site of these drugs on CaM is not involved in modulation of Ca^{2+}-ATPase [112]. CaM-dependent enzymes, such as PDE, Ca^{2+}-Mg^+-ATPase and phospholipase A_2 are specifically inhibited by W-7 [134]. In addition, W-7 at concentrations in the CaM antagonistic range enhanced the noradrenaline release [135]. The facilitation of stimulation-evoked release may have several mechanisms.

Many processes where CaM is not involved are also affected by CaM antagonists at concentrations comparable to those at which they are effective as anti-CaM agents (even below micromolar concentrations). For example, dexamethasone binding to the glucocorticoid-receptor is significantly inhibited by 0.5 μM TFP [136]. Due to the lipophilic character of the drugs it should be expected that most processes occurring in a lipid environment would be affected somehow. In addition, phenothiazines, as well as imidazolium derivatives have been found to inhibit the activity of the mitochondrial pyruvate dehydrogenase complex [137] as assayed in isolated and intact mitochondria. W-7 was found to be a competitive inhibitor with respect to coenzyme A and an uncompetitive inhibitor with respect to NAD and pyruvate. Binding of drugs directly to glutamate dehydrogenase has been reported as well [138]. The affinity constants of phenothiazines to this enzyme are similar both to the relative inhibitory effects by these agents on dehydrogenase catalysis and to their relative pharmacological potencies. This may suggest that the dehydrogenase could play some role in the *in vivo* mechanism of action of these drugs. In addition, the inhibition of the NADPH-oxidase activation by W-7 in a cell-free system is not due to an inhibition of Ca^{2+}-CaM but rather to the blocking of interaction of arachidonate, an intracellular activator, with the NADPH-oxidase system. A similar effect was observed with TFP. In both cases the inhibitory effect seems to be independent of a Ca-CaM pathway, because the activation of NADPH-oxidase occurred in the presence of EDTA, and the addition of CaM had no effect [139]. Although the mechanism of CaM-independent action of these drugs has not been established [140], the inhibition of enzymes by CaM antagonists should serve to emphasize the care necessary in the interpretation of such data.

It has been suggested that Ca^{2+} and CaM play a role in the regulation of microtubule polymerization and CaM is a target of anti-cancer agents, like Vinca alkaloids [124]. The affinity of vinblastine to CaM is comparable with that of TFP (cf. Fig. 1). However, the concentration of Vinca alkaloids necessary to block CaM-dependent activities is considerably higher than that required to affect microtubule depolymerization [123]. Vinblastine at μM concentrations binds to purified human erythrocyte Ca-Mg-ATPase as well but it does not inhibit its basal ATPase activity at μM concentrations [123].

Similarly to CaM, the Ca-dependent exposure of hydrophobic regions was found in the related Ca^{2+}-binding proteins, like troponin C [141] and S-100 [142], which may be responsible partly for the lack of selectivity of CaM antagonists. Phenothiazines interact with a wide variety of Ca^{2+}-binding proteins at concentrations used for CaM antagonism [143–148].

CaM is not unique in the sense that several other molecules or processes may stimulate Ca^{2+}-CaM-dependent enzymes to the same extent as CaM does. For example, trypsin-activated ATPase is inhibited by TFP and related phenothiazines, at concentration comparable with those required to inhibit Ca^{2+} activation [149]. Differences in the mechanism of action of anti-CaM drugs is indicated by the fact that trypsin-activated ATPase is inhibited by calmidazolium as well, but this drug does not inhibit the trypsin-activated PDE. Ca-dependent enzymes can be stimulated by molecules with amphipathic character similar to CaM. ATPase and CaM-deficient PDE are activated by unsaturated fatty acids [150, 151], acid phospholipids [151, 152] or acidic proteins [153]. Activation by these agents is inhibited by phenothiazines and other CaM antagonists in a way similar to the inhibition of CaM activation [124]. However, the mechanism of inhibition of CaM- or lipophilic molecule-activated enzymes by anti-CaM drugs is different:

For the CaM-activated system:

$$E + C \underset{}{\overset{K_1}{\rightleftarrows}} E^*\text{-}C$$
basal activity → activated

$$E^*\text{-}C \underset{}{\overset{K_2}{\rightleftarrows}} E + C\text{-}D \quad \text{scheme 1}$$
activated → basal activity

For the lipophilic molecule-activated system:

$$E \underset{L}{\overset{K_3}{\rightleftarrows}} E^{\#}\text{-}L \underset{D}{\overset{K_4}{\rightleftarrows}} E^{\#\#}\text{-}L\text{-}D \quad \text{scheme 2}$$

basal activity — activated — basal or lower activity

where E, E^*, $E^{\#}$ and $E^{\#\#}$ are the symbols of the different forms of enzymes. C, D and L indicate CaM, drugs and lipophilic molecules, respectively. K are dissociation constants of the respective equilibria. Scheme 1 shows that the anti-CaM drug suspends the activating effect of CaM due to the competition of the enzyme and the drug for CaM binding. In the case of scheme 2 the enzyme is activated by lipophilic molecule and the drug can bind exhaustively to a hydrophobic surface exposed on the target enzyme. Within this ternary complex the enzyme has merely its basal or even lower activity. Since the effectivity of a drug at a given concentration is related to its affinity to the target enzyme, the concentration of drug bound to the enzyme is described by the following equations:

For the CaM-activated system:

$$[C-D] = [D]_{bound} = \frac{[D][E^*-C][K_1]}{[E][K_2]}$$

For the lipophilic molecule-activated system:

$$[E\#\#-L-D]_{bound} = \frac{[E][L][D]}{[K_3][K_4]}$$

Therefore, in both cases the concentration of bound drug depends on the concentrations of each component of the system as well as on the value of the dissociation constants; however, in the case of the CaM-activated system the stronger the interaction between CaM and enzyme the smaller the $[D]_{bound}$, while in the lipophilic molecule-activated system the stronger the binding of L to enzyme the more ternary complex can be formed.

Obviously, one of the most important questions is the correlation between the inhibition of CaM activity by the various compounds and their pharmacological functions. Such a correlation was first postulated for phenothiazines as antipsychotic agents [154]. However, there is evidence against this assumption. Roufogalis [130] observed that structure-activity relationships for the CaM inhibition of chlorpromazine derivatives differed from their antipsychotic potency. Moreover, there are potent antipsychotic agents which cannot bind to CaM probably due to their poor lipophilicity. The potency of the CaM antagonism is approximately 1000-fold lower that the anti-psychotic activity [155]. Therefore, the binding of antipsychotics to CaM is irrelevant to their clinical efficiency, though some side effects of these phenothiazines could be explained by this biochemical mechanism [8].

Another important problem about the correlation between the inhibition of CaM activity and pharmacological function is the distribution of the CaM antagonists within the cell. Penetration of these drugs from the extracellular into the intracellular phase is determined by their hydrophobicity and charge. However, it is also dependent on the fine structure of the membrane [156], since the cationic drugs prefer to distribute at hydrophobic surface [154]. A number of anti-CaM drugs have high affinity to receptors located at the external cell surface and influence the uptake and release of neurotransmitters at concentrations similar to or lower than those used for CaM antagonism. Neuropeptides also inhibit activation of PDE by CaM [118]. Since their physiological concentrations are lower than that required for CaM inhibition, their pharmacological effect on CaM is unlikely.

The specificity of CaM antagonists is poor, which limits the usefulness of the compounds as pharmacological antagonists. At the doses used for Ca^{2+} entry blockade the interaction between Ca^{2+} channel blockers and CaM can not contribute significantly to the pharmacological effects. Phenothiazines also inhibit other CaM-dependent enzymes, including the widely distributed Ca^{2+}-dependent protein kinase C [157]. Even more selective CaM antagonists, including W-7 and calmidazolium, interact with the Ca-dependent protein kinase C [158, 159]. In addition, phenothiazines displace Ca^{2+} from membranes, thus influencing cellular Ca^{2+} level [160], and block both passive and active Ca^{2+} transport [161–163]. Furthermore, no correlation between CaM antagonism and the clinical effects of the drugs has been established in the case of anti-CaM agents. However, a direct correlation of anti-CaM activity with the anti-diarrheal activity has been shown for a series of compounds, including phenothiazines [for details see 124]. As an antimycotic agent, calmidazolium, which is a potent and selective CaM antagonist for erythrocyte Ca^{2+}-ATPase, is of considerable interest [8]. CaM antagonistic local anesthetics, like lidocaine, inhibit several Ca^{2+}-dependent cellular processes: Ca^{2+} transport, smooth muscle contraction, non-muscle cell motility and cytoskeletal organization. It seems that their mechanism of action cannot be attributed simply to CaM antagonism. At present it is unclear whether there is any relationship between the binding of drugs to CaM *in vivo* and their pharmacological properties.

8 Mechanism of action of anti-CaM drugs

Considerable progress has been made in the past few years in understanding the mechanism of action of drugs in CaM-mediated processes. It is generally accepted that anti-CaM drugs affect CaM activation/inhibition of various enzymes by a competition between drugs and target enzymes for the hydrophobic region on CaM exposed by Ca^{2+}. Alternatively, however, it cannot be excluded that drugs bind to sites on CaM different from those of the target enzymes. CaM being quite a small protein, drug binding may alter its conformation which prevents CaM from activating the enzymes [7]. Nevertheless, the potency of inhibition by CaM antagonists correlates with the hydrophobicity of the antagonists [164, 165]. The Ca^{2+}-dependence of the high affinity binding of antagonists to CaM appears to be associated with a conformational change in CaM which exposes hydrophobic sites [22, 23]. The significance of the hydrophobic region exposed by Ca^{2+} is well illustrated by the correlation between the low potency of *Tetrahymena*-derived CaM, which is a less hydrophobic molecule than the others, to activate PDE [166]. Similarly, Tanaka et al. [165] have demonstrated that the activity of PDE and the binding of the fluorescent probe 2-toluidinylnaphthalene-6-sulfonate [23] or naphthalene-sulfonamide (W-7) are activated in a parallel fashion by 1–10 μM Ca^{2+}. Beside the hydrophobic forces, ionic forces are also involved in the interaction of CaM with drugs. Tricyclic CaM inhibitors [140] have been used to study the degree of activation of PDE in intact cells as well [167]. These anti-CaM agents bind to the Ca-CaM complex when this complex is not associated with the target enzyme.

The effectivity of drugs on CaM-stimulated events can be quantitatively characterized by inhibition constants, usually by I_{50} values, which are the concentrations at which the drugs produce half-maximal inhibition. Because of the competitive character of the drugs, however, I_{50} values have no meaning unless the concentration of CaM is specified. The reason of the inter-laboratory variation of I_{50} values is probably due to differences in CaM concentrations used in the experiments. Table 1 summarizes I_{50} values together with the applied CaM concentrations when available in the literature.

Two mechanisms were suggested for the activation of target enzymes: 1) Binding of regulatory ligand to CaM produces a conformer of CaM which is distinct from that produced by Ca^{2+} alone [6]; 2) Ligand-induced conformational change in a target protein would alter its affinity for CaM. For example, CaM binding to MLCK increases its affinity for its substrate, MLC. For thermodynamic reasons, substrate binding to MLCK should increase its affinity for CaM [87]. Obviously, the drugs binding to CaM may act as regulatory ligands and may enhance CaM activation of target proteins due to their binding to CaM. It was suggested [104], for example, that prenylamine depending on its concentration can enhance or inhibit the velocity of contraction of smooth muscle. Since the contraction is triggered by the interaction of CaM and MLCK, the possible explanation for the phenomenon is that prenylamine could allosterically regulate the binding of CaM to MLCK over a concentration range. Metal ions, like lanthanum [168,

169], may also allosterically affect CaM activation of PDE, in an analogous manner as anti-CaM agents. A similar mechanism may operate in the case of other regulatory ligands. Therefore, one of the possible mechanisms for a selective drug effect is ligand binding to the allosteric sites on CaM. In this way a regulatory ligand might allosterically potentiate its interaction with one protein or another, resulting in selective activation. Another possible mechanism for drug selectivity is drug-induced conformational change in a target protein, which would alter its affinity for CaM.

A covalent adduct of CaM with a norchlorpromazine derivative was synthesized, which was a potent competitive inhibitor of PDE activation by CaM [121]. This drug forms a one-to-one, Ca^{2+}-dependent complex with CaM. This complex can interact with PDE, but fails to activate PDE. The phenothiazine-CaM adduct binds also to MLCK but can no longer stimulate this enzyme [170]. The possible explanation for this phenomenon is that there are two hydrophobic sites on CaM; to one of them binds the fluorescent drug derivative, while the other is free to interact with PDE although with lower affinity and without "efficiency" [121]. In cardiac sarcolemmal vesicles, Ca-ATPase is inhibited by calmidazolium by increasing the K_M for Ca^{2+} ions. However, the effect of calmidazolium seems non-competitive in constrast to what is observed with PDE and CaM-dependent protein kinase. Since exogenously added Ca^{2+} is able to reverse calmidazolium inhibition of the ATPase, it is possible that calmidazolium binds to a site on the Ca^{2+}-CaM complex bound to Ca^{2+}-ATPase which is different from the drug-binding site on the Ca-CaM complex PDE or protein kinase.

The most promising CaM antagonist is calmidazolium. Although it does not inhibit neurotransmitter binding at concentrations lower than those required for antagonism of CaM-dependent processes like TFP, and it is an effective inhibitor of CaM-stimulated Ca^{2+} transport, it together with TFP drastically influences membrane permeability to Ca^{2+}, which seriously curbs its usefulness [171]. Phenothiazines and high concentrations of W-7 can inhibit Ca^{2+} influx into cells. Nevertheless, the binding of CaM to Ca^{2+} transporting systems *in vivo* remains an hypothesis [172]. Fendiline binds to the Ca^{2+} channel and to CaM with similar affinities. Moreover, it exerts a typical Ca^{2+}/CaM antagonistic action [97]. Pharmacokinetics reveal slow onset of action and a long half-life. Direct comparison with other Ca^{2+} antagonists shows that its potency is at least equal to that of nifedipine but, in contrast to nifedipine, verapamil and diltiazem its anti-anginal action increases in time, reaching a steady state of action after 2 to 3 weeks. Fendiline affects indirectly the binding of dihydropyridine derivatives to the membrane fragments of heart, and its binding site on the receptor may interact in a cooperative manner with different Ca^{2+} channel antagonists [173].

These are only few reports on the use of anti-CaM drugs in intact cells. These agents were shown to deplete ATP pools and increase lactate output in certain cells [174]. On the other hand, experiments with C-6-glioma cells suggest that CaM antagonists decrease the level of cAMP, probably by inhibiting the membrane-associated CaM-dependent adenylate cyclase activity [175].

Differences in the binding of dihydropyridines to sites involved in the regulation of smooth muscle contraction may explain why certain substitutions of felodipine result in dissimilar pharmacological potency and capacity or inhibit CaM-dependent processes. Very recently the effect of felodipine and its analogues (p-chloro, oxidized and t-butyl) have been examined on the activities of CaM-dependent enzymes [105]. Felodipine and the p-chloro analogue inhibited both basal and CaM-stimulated activities of PDE, suggesting that these dihydropyridine derivatives may act directly on PDE, as well as through CaM. In addition, these drugs inhibited the actin-activated Mg^{2+}-ATPase activity of smooth muscle myosin (I_{50}: 25.1 μM), as well as myosin filament assembly induced by low concentrations of CaM. This inhibition could be reserved by raising the CaM concentration. The inhibition of enzymatic processes with K_i values in the 1–10 μM range occurs at felodipine concentrations which may exist intracellularly *in vivo*. The *in vitro* inhibitory effects of felodipine are specific and quantitatively resemble the pharmacological effect on the contractile apparatus [105]. Structural alterations, however, result in a significant loss of pharmacological potency, but no loss of capacity to inhibit CaM-dependent enzymes and processes. The reason of this discrepancy has not yet been established; it may well be that CaM and Ca^{2+}-regulated enzymes represent only secondary pharmacological sites for the action of felodipine.

Further biochemical proof for the CaM antagonistic properties of anti-CaM drugs and related compounds has been derived from investigations with CaM-regulated enzyme systems. TFP, calmidazolium, fendiline, bepridil, prenylamine, and flunarizine inhibit PDE activation with K_i values between 10 μM and 10 mM, as well as MLCK activation at about 10-fold higher concentrations [176]. Verapamil, diltiazem inhibit PDE activation with a K_i values of 10 μM, but do not inhibit MLCK activation. Inhibition by each of these compounds is competitive with CaM. Dihydropyridines inhibit PDE activation non-competitively, with K_i values of 1–7 μM but do not inhibit MLCK activation. On the basis of these observations Bayer and Mannhold [97] suggested that the inhibition of PDE requires occupancy of a "high" affinity site, while the inhibition of MLCK needs the occupancy of a "low" and probably also a "high" affinity site. Again, it might be concluded that CaM interacts with different hydrophobic sites on PDE and MLCK. At present it is uncertain how anti-CaM drugs interfere with the interaction of CaM and target enzymes. The various drugs have characteristic orders of inhibition of Ca^{2+}-activated events [94]: for TFP it is PDE $>$ adenylate cyclase $>$ Ca^{2+}-Mg^{2+}-ATPase $>$ MLCK; for W-7 it is MLCK $>$ PDE $=$ adenylate cyclase $>$ Ca^{2+}-Mg^{2+}-ATPase; for Ro 22-4839 in the order MLCK $>$ PDE $>$ adenylate cyclase $>>$ Ca^{2+}-Mg^{2+}-ATPase. These orders indicate that conformational changes induced by various CaM antagonists are not the same and these agents show a great deal of selectivity toward CaM-regulated enzymes. The fact that the inhibitory effects of various CaM antagonists are different (sometimes it depends on the isoform of the target enzyme as well) supports the idea that the binding characteristic of one CaM antagonist to a certain enzyme is not always characteristic of another. Recently, Mills and his co-workers [176] claim that the allosteric interactions among sites of drugs and/or proteins on CaM might provide a mechanism for selectivity, directing CaM to specific target proteins. Significant differences have been found in affinity for CaM in various CaM-regulated systems, thus, a hierarchy of interactions may be responsible for the activation of all CaM-dependent processes [177]. Nevertheless, it seems probable that intact CaM is required for enzyme activation, its long central helix is important. This helix can represent at least a portion of the domain where antagonist drugs bind and alter the interaction of CaM with its target proteins in a multifarious manner.

9 Drugs liberating enzymatic activities from CaM inhibition

Recently, some new target enzymes of anti-CaM drugs have been recognized among the glycolytic enzymes and the mechanism of action of drugs of different chemical structure has been investigated [10, 13]. Glycolytic enzymes are present in high concentrations in the cytosolic fraction of the cell [178]. Due to the binding of CaM to the glycolytic enzymes, PFK and aldolase, their activities are inhibited. The effects of drugs on the interactions between CaM and target enzymes have been analyzed by means of covalently attached fluorescent probes

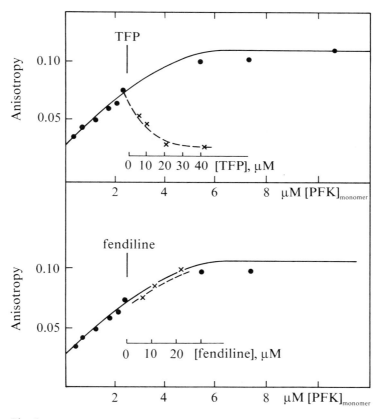

Fig. 5
Effect of drugs on the anisotropy of dansyl-calmodulin (4.5 μM) complexed with phosphofructokinase. The solid line is computer fitting with an apparent dissociation constant of 0.3 μM for the protein complex. The arrows indicate the concentration of enzyme at which TFP (A) or fendiline (B) is added to the system. Dashed line shows the change of anisotropy as a function of drug concentration.

(Fig. 5), ELISA and by enzyme kinetic approaches (Table 3). A new *in vitro* functional test was elaborated to screen molecules for anti-CaM activity as well as to study the mechanism of action of drugs [13].

Table 3
Binding and functional tests of drugs for anti-calmodulin activity

Target enzyme	Drugs	Antibody against CaM	Binding of proteins to CaM	K_d	Functional effect
1.5 μM MLCK			yes		
1.5 μM MLCK	20 μM TFP		no		
1.5 μM MLCK	20 μM fendiline		yes		
3 μM PDE			yes		
3 μM PDE	20 μM TFP		no		
3 μM PDE	20 μM verapamil		yes		
3 μM PDE	20 μM fendiline		yes		
–		0.01 μM Ab	yes*		
3 μM PDE		0.01 μM Ab	yes*		partial activation
μM ATPase					activation
μM ATPhase		0.01 μM Ab			partial activation
3 μM aldolase			yes	4.6 μM	inhibition
3 μM aldolase	20 μM TFP		no		no inhibition
3 μM aldolase	20 μM verapamil		yes		
3 μM aldolase	20 μM fendiline		yes		
3 μM aldolase		0.01 μM Ab	yes*		
0.3 μM GAPD			yes	0.5 μM	no effect
0.3 μM GAPD	40 μM TFP		no		
0.3 μM GAPD	40 μM VLB		no		
0.3 μM GAPD + +3.0 μM aldolase			no		
0.3 μM GAPD		0.01 μM Ab	no*		
0.6 μM PFK			yes	0.3 μM	inhibition
0.6 μM PFK + +3.0 μM aldolase			no		no inhibition
0.6 μM PFK	20 μM TFP		no		no inhibition
0.6 μM PFK	20 μM fendiline		yes		no inhibition
0.6 μM PFK	20 μM fendiline + 20 μM TFP		partial		no inhibition

The binding of enzymes to CaM was detected by fluorescence anisotropy measurements and indirect ELISA (*). This table contains our recent data which are partly published in 13, 14 and 73.

Table 3 summarizes the results of binding and kinetic experiments. We have found that the drugs exhibit very different effects on CaM-mediated enzymatic processes. Their actions depend both on the chemical structure of drugs and on the system used. We have demonstrated that TFP prevented or eliminated complex formation, studied in a reconstituted enzyme system, using a covalently attached fluorescent probe (Fig. 5). The Ca^{2+} channel blockers had no effect. The functional consequences of drugs on CaM-PFK interaction were investigated [13]. While TFP suspended the Ca^{2+}-mediated hysteretic inactivation of PFK, Ca^{2+} channel blockers (verapamil and nifedipine) were inactive. Fendiline, as a negative inotropic drug, seems to act as a functional CaM antagonist. Its binding to CaM does not prevent the complex formation of CaM-PFK (Fig. 5), but within this ternary complex PFK preserves or recovers its original activity measured in the absence of CaM.

The possible explanation for this finding is that the fendiline binding site on CaM differs from the TFP binding site at least in the CaM-fendiline-PFK complex [13] (Fig. 6). Fendiline does not compete directly with enzymes in binding to CaM. However, the interaction of fendiline with CaM probably induces an alteration in the tertiary structure of CaM, at least at the enzyme-binding surface. This observation resembles and extends the finding of Klee and his co-workers [18] according to which a phenothiazine-CaM adduct binds to PDE or MLCK but can no longer stimulate these enzymes. Moreover, recent evidence indicates that CaM substituted by gramicidin S, a cyclic decapeptide, activates PDE and displays the same activity as in the absence of gramicidin S [179].

It is intriguing how CaM can stimulate the basal activity of enzymes while in some cases it produces just the opposite, i. e. has an inhibitory effect. If drugs suspend the activation/inhibition of enzymes by CaM, this can be regarded as indicative of the direct effect of drugs on CaM.

10   Past and future

The alteration of CaM level within the cell may represent an important regulatory mechanism of CaM-mediated processes. In addition to the CaM-regulated enzymes, CaM has been shown to bind to a variety of structural proteins, which have no intracellular functions or these have not yet been established [6–9]. CaM binds to these proteins in a Ca-dependent or independent manner. The interaction of CaM with the "inhibitory" CaM-binding proteins may have important implications for CaM localization and regulation of multiple cellular processes. Thus, CaM could be "compartmentalized" in the cell by virtue of its interaction with particulate-binding proteins or target enzymes.

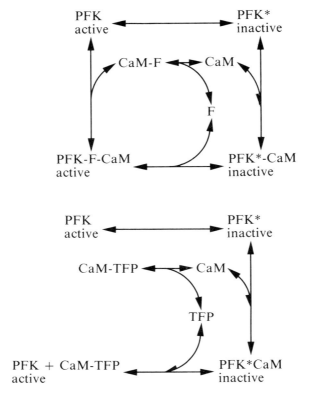

Fig. 6
Molecular models for the mechanism of action of anti-calmodulin drugs. PFK, phosphofructokinase; CaM, calmodulin; TFP, trifluorperazin; F, fendiline. For details see the text and ref. 13.

The central helix of CaM may be the structural feature required for the interaction with some target enzymes. Drugs interact with CaM *via* hydrophobic and/or ionic forces. The hydrophobic interactions alone are mainly non-specific, since such interactions may occur with the hydrophobic regions of proteins or the lipid membrane. The hydrophobic moiety may be responsible for the effectivity of the interaction. The positively charged group of the anti-CaM drugs which fit to the ionic amino acid residues of CaM, plays an important role in the specificity of the binding. The inhibitory potencies of drugs seem to differ from each other, they probably depend on the orientation of groups relative to the hydrophobic moiety. There are evidences that the various CaM-dependent enzymes do not interact with CaM in the same way [18, 67] and that drugs are bound to CaM *via* multiple sites in a complex manner.

Although the relevant drugs have widely different structures, some general conclusions can be drawn for their binding to CaM or CaM-target enzymes: 1) they bind to CaM with high affinity in a Ca-dependent way; 2) anti-CaM drugs inhibit Ca-mediated events; 3) two binding sites with high affinities are located in both halves of the CaM molecule; one of them, which seems to be the stronger one, is located in the C-terminal half of the protein [180].

CaM inhibitors exhibit a wide spectrum of activity since CaM is involved in the modulation of a number of enzymatic activities and metabolic processes. Clearly, a target enzyme with higher affinity is activated by CaM before other enzymes with lower affinity. The concentration of active Ca-CaM-enzyme ternary complexes may increase or decrease due to the effect of drugs. This mechanism would adjust the amount of CaM available for other enzymatic processes, thereby securing the intracellular regulation of multi-enzyme systems.

The specificity of commonly used CaM antagonists is rather restricted because they bind to other Ca^{2+}-sensitive proteins and inhibit CaM-independent actions at higher concentrations. Despite the lack of selectivity of most CaM antagonists, these agents are very useful in studying: 1) the Ca-induced exposure of drug-binding sites on CaM; 2) the identity of CaM-dependent processes; 3) the mechanism of activation/inhibition of target enzymes; 4) the structure-activity relationship of drugs which may provide insight into how CaM's structural domains correspond to functional domains.

Although limited amounts of data are available regarding the pharmacological significance of drugs, various drugs have proven to be useful for the future development of CaM-specific antagonists. Further research must be aimed at developing specific drugs which antagonize CaM at concentrations low enough that other metabolic activities are not to be affected. Additional criteria for CaM antagonists are that the inhibitory potency must parallel the anti-CaM potency. Therefore, the research may be directed to design and synthesize compounds which interact with the CaM binding domain of enzymes and block CaM-mediated metabolic processes.

Acknowledgments

Several parts of the work described in this chapter are the results obtained with my colleagues F. Orosz, T. Y. Christova and K. Zuklys. T. Y. C. and K. Z. were supported by ITC fellowships. Experiments with antibody against calmodulin were carried out in collaboration with J. Kramer and M. Magócsi of the National Institute of Haematology and Blood Transfusion, H–1113 Budapest. Research from the author's laboratory was supported by a grant OTKA 315/87. The author is very grateful to Dr. B. Sarkadi of the National Institute of Haematology and Blood Transfusion, Budapest, for helpful discussions and for critical reviewing of the manuscript.

References

1. W. Y. Cheung: Science 207, 19 (1980).
2. A. R. Means and J. R. Dedman: Nature (Lond.) 285, 73 (1980).
3. C. B. Klee, T. H. Crouch and P. G. Richman: A. Rev. Biochem. 49, 489 (1980).
4. C. O. Boström and D. J. Wolf: Biochem. Pharmac. 30, 1395 (1981).
5. H. Van Belle: Advance in Cyclic Nucleotide and Protein Phosphorylation Research (Greengard et al. eds) vol. 17, p. 557. Raven Press, New York (1984).
6. J. D. Johnson and J. S. Mills: Med. Res. Rev. 6, 341 (1986).
7. J. C. Stoclet, D. Gérard, M. C. Kilhoffer, C. Lucnier, R. Miller and P. Schaeffer: Prog. Neurobiol. (Oxf.) 29, 321 (1987).
8. R. Mannhold: Drugs of future 9, 677 (1984).
9. L. J. Van Eldik and D. M. Watterson: Calcium and Cell Physiology (D. Marmé ed.) p. 105. Springer-Verlag, Berlin-Heidelberg-New York-Tokyo (1984).
10. G. W. Mayr: Eur. J. Biochem 143, 513 (1984).
11. G. W. Mayr: Eur. J. Biochem 143, 521 (1984).
12. B. Buschmeier, H. E. Meyer and G. W. Mayr: J. biol. Chem. 262, 9454 (1987).
13. F. Orosz, T. Y. Christova and J. Ovádi: Molec. Pharmac. 33, 678 (1988).
14. F. Orosz, T. Y. Christova and J. Ovádi: Biochem. biophys. Acta, in press.
15. Y. S. Babu, J. S. Sack, J. J. Greenhough, C. E. Bugg, A. R. Means and W. J. Cook: Nature (Lond.) 315, 37 (1985).
16. N. C. J. Strynadka and M. N. G. James: Proteins: Structure, Function and Genetics 3, 1 (1988).
17. W. Drabikowski, H. Brzeska and S. Y. Venyaminow: J. biol. Chem. 254, 11584–1159(1982).
18. D. L. Newton, M. D. Oldewurtel, M. H. Krinks, J. Shiloach and C. B. Klee: J. biol. Chem. 259, 4419 (1984).
19. R. H. Kretzinger: CRC Crit. Rev. Biochem. 8, 119 (1980).
20. J. A. Cox: Biochem. J. 249, 621 (1988).
21. I. C. Y. Kuo and C. J. Coffee: J. biol. Chem. 251, 1603 (1976).
22. D. C. LaPorte, B. M. Weiman and D. R. Storm: Biochemistry 19, 3814 (1980).

23 T. Tanaka and H. Hidaka: J. biol. Chem. *255,* 11078 (1980).
24 J. E. Van Eyk and R. S. Hodges: Biochem. Cell. Biol. *65,* 982 (1988).
25 D. Burger, J. A. Cox, M. Comte and E. A. Stein: Biochemistry *23,* 1966 (1984).
26 J. A. Cox, M. Comte, A. Malnoe, D. Burger and E. A. Stein: Metal Ions in Biological System (H. Sigel, ed.) vol. 17, p. 215. Marcel Dekker, New York (1984).
27 S. Ilda and J. D. Potter: J. Biochem. (Tokyo) *99,* 1765 (1986).
28 T. H. Crouch and C. B. Klee: Biochemistry *19,* 3692 (1980).
29 C. H. Keller, B. B. Olwen, D. C. LaPorte and D. R. Storm: Biochemistry *21,* 156 (1982).
30 C. B. Klee, T. H. Crouch and M. H. Krinks: Biochemistry *18,* 722 (1979).
31 M. E. Morrill, S. T. Thompson and E. Stellwagen: J. biol. Chem. *254,* 4371 (1979).
32 R. L. Kincaid and M. Vaughan: Calcium and Cell Function (W. Y. Cheung ed.) vol. 6, p. 44 (1986).
33 C. O. Broström, Y. C. Huang, B. M. Breckenridge and D. J. Wolff: Proc. natl Acad. Sci USA *72,* 64 (1975).
34 W. Y. Cheung, L. S. Bradham, T. J. Lynch, Y. M. Lin and E. A. Tallant: Biochem. biophys. Res. Commun. *66,* 1055 (1975).
35 H. W. Jarrett and J. T. Penniston: Biochem. biophys. Res. Commun. *77,* 1210 (1977).
36 P. Brandt, M. Zurini, R. L. Neve, R. E. Rhoads and T. C. Vanaman: Proc. natl Acad. Sci. USA *85,* 2914 (1988).
37 M. Zurini, J. Krebs, J. T. Penniston and E. Carafoli: J. biol. Chem. *259,* 618 (1984).
38 B. Sarkadi, A. Enyedi, Z. Foldes-Papp and G. Gárdos: J. biol. Chem. *261,* 9552 (1986).
39 R. S. Adelstein, M. A. Cont, D. R. Hathaway and C. B. Klee: J. biol. Chem. *253,* 834 (1978).
40 M. A. Conti and R. S. Adelstein: J. biol. Chem. *256,* 3178 (1981).
41 L. Gregori, P. M. Gillevet, P. Doan and V. Chan: Curr. Top. Cell. Reg. *27,* 447 (1985).
42 S. R. Anderson and D. A. Malencik: Calcium and Cell Function (W. Y. Cheung ed.) vol. 6, p. 2 (1986).
43 J. A. Cox, M. Comte, J. E. Fitton and W. F. DeGrada: J. biol. Chem. *260,* 2527 (1985).
44 D. A. Malencik and S. R. Anderson: Biochemistry *21,* 3480 (1982).
45 D. A. Malencik and S. R. Anderson: Biochemistry *22,* 1995 (1983).
46 D. K. Blumental, K. Takio, A. M. Edelman, H. Charbonneau, K. Titani, K. A. Walsh and E. G. Krebs: Proc. natl. Acad. Sci. USA *82,* 3187 (1985).
47 W. F. DeGrado, F. G. Prendergast, H. R. Wolfe and J. A. Cox: J. Cell Bioochem. *29,* 83 (1985).
48 R. M. Hanly, A. R. Means, B. E. Kemp and S. Shenolikar: Biochem. biophys. Res. Comm. *152,* 122 (1988).
49 T. J. Lukas, W. H. Burgess, F. G. Prendergast, W. Lau and D. M. Watterson: Biochemistry *25,* 1458 (1986).
50 K. C. Wang, H. Y. Wong, J. H. Wang and H. Y. P. Lam: J. biol. Chem. *258,* 12110 (1983).
51 B. E. Kemp, R. B. Pearson, V. C. Guerriero, I. C. Bagchi and A. R. Means: J. biol. Chem. *262,* 2542 (1987).
52 T. Tanaka, T. Ohmura and H. Hidaka: Pharmacology *26,* 249 (1983).
53 M. Walsh and F. C. Stevens: Biochemistry *16,* 2742 (1977).
54 M. Walsh and F. C. Stevens: Biochemistry *17,* 3924 (1978).
55 J. Krebs and E. Carafoli: Eur. J. Biochem. *124,* 619 (1982).
56 F. Faust, M. Slisz and M. Jarrett: J. biol. Chem. *262,* 1938 (1987).

57 M. L. Billingsley, D. Kuhn, P. A. Velletri, R. Kincaid and W. Lovenberg: J. biol. Chem. *259*, 6630 (1984).
58 C. Gagnon, S. Kelley, V. Mangeniello, M. Vaughan, C. Odya, W. Stritmatter, A. Hoffman and F. Hirta: Nature (Lond.) *291*, 515 (1981).
59 M. L. Billingsley, P. A. Velletri, R. H. Roth and R. J. Delorenzo: J. biol. Chem. *258*, 5352 (1983).
60 T. A. Craig, D. M. Watterson, F. G. Prendergast, J. Haiech and D. M. Roberts: J. biol. Chem. *262*, 3278 (1987).
61 J. A. Putkey, G. F. Draelta, G. R. Slaughter, C. B. Klee, P. Choen, J. T. Stull and A. R. Means: J. biol. Chem. *261*, 9896 (1986).
62 D. M. Roberts, W. H. Burgess and D. M. Watterson: Plant Physiol. *75*, 796 (1984).
63 H. Runte, L. Jurgensmeier, C. Unger and H. D. Soling: FEBS Lett. *147*, 125 (1982).
64 J. A. Cox, C. Ferraz, J. G. Demaille, R. O. Perez, D. VanTuinen and D. Marmé: J. biol. Chem. *257*, 10694 (1982).
65 M. Schleicher, T. J. Lukas and D. M. Watterson: Archs Biochem. Biophys. *228*, (1984).
66 Y. D. Plancke and E. Lazarides: Molec. cell. Biol. *3*, 1412 (1983).
67 D. Guerini, J. Krebs and E. Carafoli: J. biol. Chem. *259*, 15172 (1984).
68 W. C. Ni and C. B. Klee: J. biol. Chem. *260*, 6974 (1985).
69 W. Szyja, A. Wrzosek, H. Brzeska and M. G. Sarzaia: Cell Calcium *7*, 73 (1986).
70 D. Mann and T. C. Vanaman: Meth. Enzymol. *139*, 417 (1987).
71 H. J. Vogel, L. Lindahl and E. Thulin: FEBS Lett. *157*, 241 (1983).
72 L. J. Van Eldik and D. M. Watterson: J. biol. Chem. *256*, 4205 (1984).
73 K. Zuklys, J. Kramer, M. Magócsi and J. Ovádi: submitted for publication.
73a L. J. Van Eldik and D. M. Watterson: Calcium Physiology (D. Marmé ed.) p. 210. Springer, Berlin (1984).
74 R. M. Levin and B. Weiss: Pharmac. exp. Ther. *208*, 454 (1979).
75 J. D. Johnson and D. A. Fugman: J. Pharmac. exp. Ther. *226*, 330 (1983).
76 H. Van Belle: Cell Calcium *2*, 483 (1981).
77 B. Weiss, W. C. Prozialeck and T. L. Wallace: Biochem. Pharmac. *31*, 2217 (1982).
78 A. Anderson, S. Forsen, E. Thulin and H. J. Vogel: Biochemistry *22*, 2309 (1983).
79 D. L. Dalgarno, R. E. Klevit, B. A. Levine, G. M. H. Scott, R. J. P. Williams, J. Gergely, Z. Grabarek, P. C. Leavis, R. J. A. Grand and W. Drabikowski: Biochem. biophys. acta *791*, 164 (1984).
80 M. Zimmer and F. Hofmann: Eur. J. Biochem. *142*, 393 (1984).
81 D. R. Marshak, T. J. Lukas and D. M. Watterson: Biochemistry, *24*, 144 (1985).
82 H. Hidaka, T. Yamaki, M. Naka, T. Tanaka, H. Hayahashi and R. Kobaya shi: Molec. Pharmac. *17*, 66 (1980).
83 R. M. Levin and B. Weiss: Molec Pharmac. *12*, 581, (1976).
84 G. Gigl, D. Hartweg, E. Sanchez-Delgado, G. Metz and K. Gietzen: Cell Calcium *8*, 327 (1987).
85 J. Gariépy and R. S. Hodges: Biochemistry *22*, 1586 (1983).
86 D. P. Giedroc, S. K. Sinha, K. Brew and D. Puett: J. biol. Chem. *260*, 13406 (1985).
87 H. Hidaka, M. Asano and T. Tanaka: Molec. Pharmac. *20*, 571, (1981).
88 T. Tanaka, T. Ohmura and H. Hidaka: Molec. Pharmac. *22*, 403 (1982).
89 T. Tanaka, M. Inagaki and H. Hidaka: Archs Biochem. Biophys. *220*, 188 (1983).
90 H. Hidaka, Y. Sasaki, T. Tanaka, T. Endo, S. Ohno, Y. Fujii and T. Nagata: Proc. natl Acad. Sci USA *78*, 4354 (1981).

91 K. Gietzen, A. Wüthrich and H. R. Bader: Biochem. biophys. Res. Commun. *101*, 418 (1981).
92 W. C. Prozialeck and B. Weiss: J. Pharmac. exp. Ther. *222*, 509 (1982).
93 J. D. Johnson and L. Wittenauer: Biochem. J. *211*, 473 (1983).
94 T. Nakajima and A. Katoh: Molec. Pharmac. *32*, 140 (1987).
95 J. A. Norman, J. Ansell, G. A. Stone, L. P. Wennogle and J. W. F. Wasley: Molec. Pharmac. *31*, 535 (1987).
96 J. A. Norman, A. H. Drummond and P. Moser: Molec. Pharmac. *16*, 1089 (1979).
97 R. Bayer and R. Mannhold: Pharmatherapeutica *5*, 103 (1987).
98 M. Sellinger-Barnette and B. Weiss: Molec. Pharmac. *21*, 86 (1982).
99 D. P. Giedroc, D. Puett, N. Ling and J. V. Staros: J. biol. Chem. *258*, 16 (1983).
100 D. A. Malencik and S. R. Anderson: Biochem. biophys. Res. Commun. *114*, 50 (1983).
101 Y. Maulet and J. A. Cox: Biochemistry *22*, 5680 (1983).
102 S. R. Anderson and D. A. Malencik: Calcium and Cell Function (W. Y. Cheung ed.) vol. 6, p. 2 (1986).
103 M. Asano, Y. Suzuki and H. Hidaka: J. Pharmac. exp. Ther. *220*, 191 (1982).
104 H. Metzger, H. O. Sterm, G. Pfitzer and J. C. Rüegg: Arzneimittel-Forsch. *32*, suppl. 11, 1425 (1982).
105 M. P. Walsh, C. Sutherland and G. C. Scott-Woo: Biochem. Pharmac. *37*, 1569 (1988).
106 C. Lugnier, A. Follenius, D. Gereard and J. C. Stoclet: Eur. J. Pharmac. *98*, 157 (1984).
107 S. L. Boström, B. Ljung, S. Mardh, S. Forsén and E. Thulin: Nature (Lond.) *292*, 777 (1981).
108 H. Itoh, T. Tanaka, Y. Mitani and H. Hidaka: Biochem. Pharmac. *35*, 217 (1986).
109 J. C. Stoclet, C. Lugnier, A. Follenius, J. M. Schefter and D. Gerard: Calcium Entry Blockers and Tissue Protection (T. Gogfraind ed.) p. 31. Raven Press, New York (1985).
110 H. Itoh, I. Tomahito and H. Hidaka: J. Pharmac. exp. Therap. *230*, 737 (1984).
111 P. M. Epstein, K. Fiss, R. Hachisu and D. M. Andrenyak: Biochem. biophys. Res. Commun. *105*, 1142 (1982).
112 H. C. Kim and B. U. Raess: Biochem. Pharmac. *37*, 917 (1988).
113 T. S. Mills, B. C. Bailey and F. D. Johnson: Biochemistry *24*, 4897 (1985).
114 A. M. Minocherhomje and B. D. Roufogalis: Cell Calcium *5*, 57 (1984).
115 R. Mannhold, R. Rodenkirchen, R. Bayer and W. Haas: Arzneimittel-Forsch. *34*, 407 (1984).
116 H. J. Vogel: Calcium in Drug Action (P. F. Baker ed.) Springer Verlag (1987).
117 J. D. Johnson, L. A. Wittenauer, E. Thulin, S. Forsén and H. J. Vogel: Biochemistry *25*, 2226 (1986).
118 J. S. Mills, B. L. Bailey and J. D. Johnson: Biochemistry *24*, 4897 (1985).
119 J. D. Johnson: Biochem. biophys. Res. Commun. *112*, 787 (1983).
120 J. D. Johnson, J. S. Mills and J. H. Collins: Biochemistry *25*, 2228 (1988).
121 D. L. Newton, T. R. Burke, K. C. Rice and C. B. Klee: Biochemistry *22*, 5412 (1983).
122 M. Inagaki and H. Hidaka: Pharmacology *29*, 75 (1984).
123 K. Gietzen, A. Wütrich and H. Bader: Molec. Pharmac. *22*, 413 (1982).
124 K. Watanabe and W. L. West: Fed Proc. *41*, 2292 (1982).
125 K. Watanabe and W. L. West: Experientia *35*, 1487 (1987).
126 N. Katoh, B. C. Wise, R. W. Wrenn and J. F. Kou: Biochem. J. *198*, 199 (1981).

127 R. K. Sharma and J. H. Wang: Biochem. biophys. Res. Commun. *100*, 710 (1981).
128 K. Gietzen, P. Adamczyk-Engelmann, A. Wütrich, A. Konstantinova and H. Bader: Biochem. biophys. Acta *736*, 109 (1983).
129 B. Weiss, W. C. Prozialeck and T. L. Wallance: Biochem. Pharmacol. *31*, 2217 (1982).
130 B. D. Roufogalis: Calcium and Cell Physiology (D. Marmé ed.) p. 148. Springer-Verlag, Berlin (1984).
131 H. I. Meltzer and S. Kassir: Biochem. biophys. Acta *755*, 452 (1983).
132 E. K. Rooney and A. G. Lee: Biochem. biophys. Acta *732*, 428 (1983).
132a F. F. Vincenzi: Ann. N. Y. Acad. Sci. *402*, 368 (1982).
133 J. M. J. Lamers, K. J. Cysouw and P. D. Verdouw: Biochem. Pharmac. *34*, 3837 (1985).
134 M. Asano and H. Hidaka: Calcium and Cell Function (W. Y. Cheung, ed.) p. 123. Academic Press (1984).
135 W. Reimann, U. Köllhofer and B. Wagner: Eur. J. Pharmac. *147*, 481 (1988).
136 V. Bohemen and G. G. Rousseau: FEBS Lett. *143*, 21 (1982).
137 J. A. Miernyk, T. K. Fang and D. D. Randall: J. biol. Chem. *262*, 15338 (1987).
138 F. M. Veronese, R. Bevilacqua and I. M. Chaiken: Molec. Pharmac. *15*, 313 (1979).
139 A. Sakata, E. Ida, M. Tominaga and K. Onone: Biochem. biophys. Res. Commun. *148*, 112 (1987).
140 S. H. Snyder, S. P. Banerjee, H. I. Yamamura and D. Greenberg: Science *184*, 1243 (1974).
141 T. Tanaka and H. Hidaka: Biochem. Int. *2*, 71 (1981).
142 D. R. Marshak, D. M. Watterson and L. J. Van Eldik: Proc. natl Acad. Sci. USA *78*, 6793 (1981).
143 P. B. Moore and J. R. Dedman: J. biol. Chem. *257*, 9663 (1982).
144 J. F. Head, S. Spielberg and B. Kaminer: Biochem. J. *209*, 797 (1983).
145 S. Nagao, S. Kudo and Y. Nozowas: Biochem. Pharmac. *30*, 2709 (1981).
146 R. M. Levin and B. Weiss: Biochem. biophys. Acta *540*, 197 (1978).
147 L. J. Van Eldik, G. Piperno and D. M. Watterson: Proc. natl Acad. Sci. USA *77*, 4779 (1980).
148 J. P. McManus: FEBS Lett. *126*, 245 (1981).
149 B. Sarkadi, A. Enyedi, A. Nyers and G. Gárdos: Ann. N. Y. Acad. Sci. *402*, 329 (1982).
150 D. R. Taverna and J. D. Hanahan: Biochem. biophys. Res. Commun. *94*, 652 (1980).
151 A. Al-Jobore and B. D. Roufogalis: Can. J. Biochem. *59*, 880 (1981).
152 V. Niggli, E. S. Adunyah and E. Carafoli: J. biol. Chem. *256*, 8588 (1981).
153 A. M. Minocherhomjee and B. D. Roufogalis: Biochem. J. *206*, 517 (1982).
154 B. Weiss, R. Fertel, R. Figlin and P. Uzunov: Molec Pharmac. *10*, 615 (1974).
155 P. Seeman: Pharmac. Rev. *32*, 229 (1980).
156 M. P. Sheetz and S. J. Singer: Proc. natl Acad. Sci. USA *71*, 4457 (1974).
157 Y. Takai, A. Kishimoto, Y. Iwasa, Y. Kawahara, T. Mori and Y. Nishizuka: J. biol. Chem. *254*, 3692 (1979).
158 R. C. Schatzman, R. L. Raynor and J. F. Kuo: Biochem. biophys. Acta *755*, 144 (1983).
159 B. C. Wise and J. F. Kuo: Biochem. Pharmac. *32*, 1259 (1983).
160 P. Seeman: Pharmac. Rev. *24*, 583 (1972).
161 Y. Larndry, M. Amellal and M. Ruckstuhl: Biochem. Pharmac. *30*, 2031 (1981).
162 B. U. Raess and F. F. Vincenzi: Molec. Pharmac. *18*, 253 (1980).

163 R. M. Levin and B. Weiss: Neuropharmacology *19*, 169 (1980).
164 J. A. Norman, A. H. Drummond and P. Moser: Molec. Pharmac. *16*, 1089 (1979).
165 T. Tanaka, T. Ohmura and H. Hidaka: Molec. Pharmac. *22*, 403 (1982).
166 M. Ignagaki, M. Naka, Y. Nozawa and H. Hidaka: FEBS Lett. *151*, 67 (1983).
167 C. H. Reynolds and P. T. J. Claxton: Biochem. Pharmac. *31*, 419 (1982).
168 S. H. Chao, Y. Suzuku, J. R. Zysk and W. Y. Cheung: Molec. Pharmac. *26*, 75 (1984).
169 J. L. Mills and J. D. Johnson: J. biol. Chem. *260*, 15100 (1985).
170 B. L. Newton, C. B. Klee, J. Woodgett and P. Cohen: Biochem. biophys. Acta *845*, 533 (1985).
171 P. A. Lucchesi and C. R. Scheid: Cell Calcium *9*, 87 (1988).
172 S. F. Flaim, M. D. Brannan. S. C. Swigart, M. M. Gleason and L. D. Muschek: Proc. natl Acad. Sci. USA *82*, 1237 (1985).
173 F. Orosz, E. Fejér, J. Gaál, J. Ovádi: J. molec. cell. Card. *19*, (suppl. III) 6, (1987).
174 A. N. Corps, T. R. Hesketh and J. C. Metcalfe: FEBS Lett. *138*, 280 (1982).
175 J. A. Norman and M. Staehelin: Molec. Pharmac. *22*, 395 (1982).
176 J. S. Mills, B. L. Bailey and J. D. Johnson: Biochemistry *24*, 4897 (1985).
177 T. Tanaka, E. Yamada, T. Sone and H. Hidaka: Biochemistry *22*, 1030 (1983).
178 P. A. Srere: Science *158*, 936 (1967).
179 J. A. Cox, M. Milos and M. Comte: Biochem. J. *246*, 495 (1987).
180 J. Buerkler, J. Krebs and E. Carafoli: Cell Calcium *8*, 123 (1987).

1 Introduction

The foundations of the science of genetics were laid last century by the Augustinian monk Gregor Mendel [1]. As a result of his experiments in cross-breeding peas and beans, he observed that the inheritance of single traits can be explained by genetic factors which are passed on from one generation to the next. At the suggestion of Johannsen (1908) these fundamental units of heredity were named genes. Modern recombinant DNA techniques allow the genes in the complex human genome to be identified and mapped, and permits the analysis of their fine structure. Modern molecular human genetics is also known as "new genetics" [2]. Recombinant DNA tech-

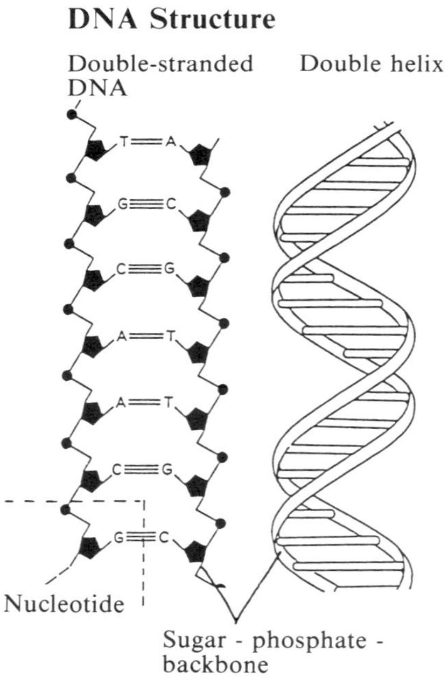

DNA Structure

Figure 1
The DNA molecule is a very long double-stranded chain. Each strand is made up of millions of nucleotides, which are composed of a sugar, a phosphate group and a base. The sugar (pentagon) and the phosphate group (circlet) of each nucleotide contribute to the backbone of the DNA strand. The four bases in DNA are adenine (A), guanine (G), thymine (T) and cytosine (C). They form the rungs. Whenever an adenine occurs in one strand, a thymine must occur opposite in the other strand. Guanine always pairs with cytosine. Hydrogen bonds hold the two complimentary bases together. The backbones of the two strands wind around each other to form a double helix.

niques have led to a remarkable extension of biotechnology in pharmaceutical industries.

2 Basic concepts of human genetics and gene technology

By inheritance we understand the transmission of traits from one generation to the next. What in fact is transmitted is genetic information. This, together with environmental influences, determines our physical and mental characteristics.

2.1 Chromosomes and DNA; structure and function

The fertilization of an ovum by a sperm brings together the information containing the genetic plans for a new individual with all his characteristics and capabilities. Divided among 46 individual chromosomes (23 pairs), this information is stored almost exclusively in the cell nucleus. During cell division, the chromosomes can be identified and assessed by light microscopy. They consist of proteins and desoxyribonucleic acid (DNA) which was discovered in 1869 by the Basle chemist Friedrich Miescher [3]. The thread-like DNA molecule, whose basic building elements are nucleotides linked together in a chain, has the form of a double helix resembling an evenly twisted rope ladder (Fig. 1) [4]. It is the chemical carrier of the genetic information. The fundamental units of heredity, the genes, are arranged in a linear fashion along the DNA molecule, which reaches a length of more than 1.5 metres in each cell. Each gene occupies a sequence of 800 to as many as 2,000,000 base pairs [5]. It contains the information for the synthesis of a specific protein (Fig. 2). Proteins are complex organic compounds composed of 20 different amino acids. The incorporation of an individual amino acid is coded by a sequence of three adjacent nucleotides (triplet). Thus, the myosin or actin gene ensures that the types of protein that make muscle contraction possible are formed in the muscles. Other genes direct the formation of the component proteins of haemoglobin (the oxygen-carrying molecule in red blood cells), or the production of collagen, a component of skin, bone and connective tissue. Other proteins promote immune defence in the form of antibodies, or act as enzymes that regulate complex metabolic processes. Thus, proteins perform countless vital tasks in our body.

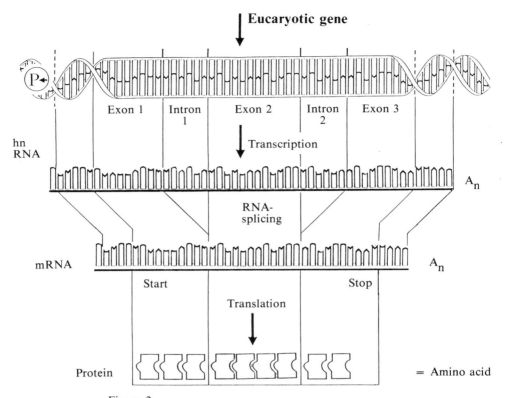

Figure 2
A gene consists of a sequence of DNA containing sufficient nucleotide triplets (only reduced number shown) to code for the appropriate amino acids of the gene product (protein). The coding sequences are interrupted by introns (intervening sequences = IVS). Genes also have sequences of varying lengths at both ends. The P region (promoter) is of particular importance for its regulation.

2.2 What are genetic diseases?

The manner in which each disease is manifested, and its accessibility to treatment, depend on the genetically determined constitution of the patient. True genetic diseases are defined as those caused by a defect in the genome. Clinically relevant mutations occur at the chromosome level where they lead to well-known syndromes such as Down's syndrome (trisomy 21 or mongolism) and at the gene level where they are responsible for the monogenic or Mendelian disorders that concern us here [6]. Diseases with a so-called multifactorial or polygenic inheritance that are of great importance in medical practice result from an interaction between one or more disadvantageous genes and environmental influences. They include obesity, essential hypertension, allergies, most forms of cancer, various endogenous psychiatric disorders and most congenital malformations.

On the whole, genetic diseases are common. Affected persons often do not survive to puberty or are unable to reproduce as a result of their infirmity. Because mutations continually occur during gametogenesis, severe hereditary disorders do not become extinct in our population.

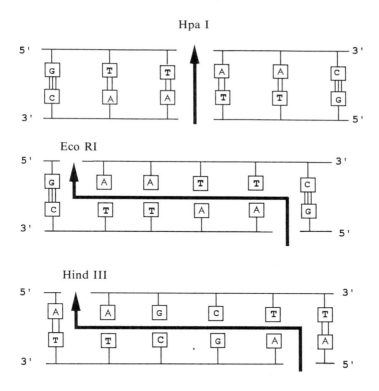

Figure 3
Examples of DNA sequences (restriction sites) recognized and cut by three widely used restriction nucleases. These are obtained from various species of bacteria: Hpa I is from *Haemophilus parainfluenzae*, Eco RI is from *Escherichia coli* and Hind III is from *Haemophilus influenzae*.

2.3 The instruments of recombinant DNA techniques: restriction enzymes and DNA probes

Probably the most important requirement for the analysis and recombination of DNA is the availability of the restriction endonucleases (type II) occurring in bacteria, that recognize a given nucleotide sequence of DNA and cut both strands with a high degree of specificity (Fig. 3). For work which contributed towards the discovery of the restriction endonucleases and their use in molecular genetics, the Swiss scientist, Werner Arber [7], and two American researchers, Hamilton O. Smith and Daniel Nathans [8], were awarded the 1978 Nobel Prize for medicine. By means of restriction endonucleases obtained from bacteria, the DNA under study can be cut in a test tube *(in vitro)*. More than 100 different restriction endonucleases are already commercially available.

A further important step was the discovery of antibiotic resistance in bacteria. Besides their own genome, bacteria possess small, circular DNA-duplex plasmids in which antibiotic resistance is located. These plasmids can easily be transferred from one bacterial cell to another. The efficient DNA analysis methods developed by Walter Gilbert [9] and Frederick Sanger [10] also contributed to the rapid development of gene technology. These enable the sequence of base pairs in a stretch of DNA to be determined. Finally, nucleotides can already be successfully assembled in sequence in a test tube.

3 The use of gene technology in medicine today

Gene technology is more widely used in present-day medicine than generally known. It permits the production of valuable proteins for diagnosis, prophylaxis and therapy. It also allows better diagnosis of monogenic disorders and the reliable identification of their clinically healthy gene carriers, the so-called heterozygotes. It offers an opportunity to detect and categorize infecting microorganisms with DNA probes. Gene transfer into the genome of body cells is a potential form of therapy.

3.1 Production of proteins using gene technology

Recombinant DNA technology has made possible new processes for the production of drugs [11]. Active substances endogenous to the human body such as insulin [12, 13] or interleukins are now available for the treatment of diabetes and cancer patients. For their production, the genetic factor responsible for their synthesis is separated from the complex human genome and inserted into the genome of bacteria or cells in culture [14]. The living organisms or cells into which the foreign gene has been inserted then produce the desired human protein as well as their own. Such protein can be extracted only in insufficient quantities, if at all, from a cadaver or from human tissue, and total chemical synthesis is not possible due to the complex structure of these substances.

Table 1
Recombinant Proteins (Adopted from M. Steinmetz)

Protein	Company	Application
Insulin	Eli Lilly	Diabetes
Growth hormone (GH)	Genentech	Growth disorders
Tissue plasminogen activator (t-PA)	Genentech	Stroke
alpha-Interferon	Roche/Schering-Plough	Leukaemias, Kaposi sarcoma
Granulocyte colony stimulating factor	Amgen/Roche	Leukaemia, side effects of chemotherapy, AIDS
Erythropoetin	Amgen/Integrated Genetics	Anaemia
gamma-Interferon	Amgen/Biogen/Genentech/Interferon Sciences	Cancer, infectious diseases, rheumatic arthritis, condylomata, sclerodermia
Interleukin-2	Amgen/Biogen/Cetus/Collaborative Research/Roche	Cancer
Tumour necrosis factor (TNF)	Biogen/Cetus/Genentech	Cancer
Clotting factor VIII	Genetics Institute	Haemophilia A
Epidermal growth factor	Chiron	Wound healing
Hepatitis B vaccine	Merck	Prevention of hepatitis
Circum-sporozoite-protein	Roche/Smith Kline French/Chiron	Prevention of malaria
gp160/gp120	Bristol-Myers/Chiron/MicroGeneSys/Oncogen	Prevention of AIDS
HIV antigenes	Roche	HIV diagnosis

Today, primarily "natural" proteins made by recombinant DNA technology are on the market or undergoing clinical trials (Table 1). These substances make it possible to treat patients who cannot produce these proteins or can do so only in insufficient quantities, or who require them in particularly large amounts in critical situations of their lives. Here, we are referring to naturally occurring substances, but protein engineering can also be used to modify active sites, to build compound drugs, to increase pharmacological effectiveness (e.g. by increasing stability, reducing side effects etc.) or to envisage basically new therapeutic strategies.

3.2 Diagnostic reagents for microorganisms and recombinant subunit vaccines

The DNA diagnosis of pathogenic microorganisms and parasites is probably superior in a number of respects to conventional serological, biochemical or culture diagnosis methods, particularly with regard to a more specific classification of these pathogens or their identification in infected tissue or body fluids [15]. The proteins on the surface of infectious agents (e.g. envelope proteins of a virus) are recognized as antigenes by our immune system, and specific antibodies are produced to neutralise them. The application of recombinant subunit vaccines can eliminate many limitations and dangers of conventional types of vaccines (attenuated live vaccines or killed vaccines) which contain their own genome [16].

3.3 Avoiding the difficulties of large-scale genetically engineered production: transgenic domestic animals instead of transgenic bacteria

Seen as a whole, the potential of recombinant DNA techniques for the production of proteins and protein derivatives is far from being fully exploited. Large-scale production by these methods has not, however, developed as quickly as was expected a few years ago. The industrial cultivation of genetically modified microorganisms presents all kinds of difficulties, and it is a disadvantage that the design of their genes differs considerably from that of our own. The bacteria are unable to produce complex human proteins on their own. Cultivated animal and human cells do not have this disadvantage as producers of protein. Attempts are also being made to substitute genetically modified domestic animals [17]. For instance, transgenic domestic animals are being bred whose milk contain the human clotting factor IX which is lacking in haemophilia. The ability to produce transgenic mice that are predisposed to specific malignancies and also to other diseases should accelerate our understanding of the underlying pathological mechanisms and the designing and testing of new therapies.

3.4 DNA diagnostics and genetic counselling

Given the large number and the heterogeneity of genetic disorders [18], counselling is possible only if a precise diagnosis of the disorder has been established. Until now the physician depended almost entirely on the evaluation of the clinical symptoms in the patient and his relatives, and could occasionally draw upon biochemical and/or cell biological aids. The DNA technique allows the genetic mutations in the genome to be identified and followed in a family [19]. The reliable identification of clinically normal carriers, the so-called heterozygotes, is one of the main advantages of the new genetics. In a steadily growing number of hereditary diseases, persons seeking advice can be given reliable information with regard to their own health and that of their offspring. Genetic counselling has undergone a decisive improvement as a result of more precise information. Experience shows that it is often easier to live with an unfavourable result than with the inevitable uncertainty associated with conventional statistical risk assessments. Most of the affected persons, although not all of them, are therefore strongly in favour of these genetic investigations.

DNA diagnosis is carried out by means of the above-mentioned restriction endonucleases and DNA probes, with which it is possible to distinguish the DNA fragment(s) associated with the normal or defective gene(s) (Fig. 4). For DNA diagnosis, the DNA sequences flanking the gene or located within it (the latter being known as "introns") are frequently investigated. There the human genome contains DNA sequence variants – variants in the base sequence – which apparently are biologically neutral and do not specify the primary structure of the gene product. As a result of the variability of nucleotide sequences between unrelated individuals, the restriction endo-

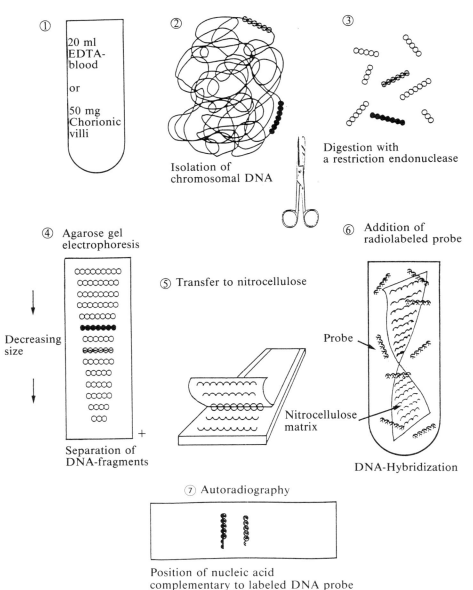

Figure 4
A specific restriction enzyme cuts genomic DNA ②. Resulting DNA fragments are separated by size, by means of agarose gel electrophoresis ④. The fragments are denaturated (transferred to a matrix such as nitrocellulose), while the special separation of the fragments is preserved (Southern transfer) ⑤. Hybridization is performed on the matrix with a radioactively labelled, single stranded DNA segment called DNA probe, which is complementary to the DNA fragments of interest ⑥, ⑦.

nucleases cuts the DNA at different points; this is the cause of what is known as restriction fragment length polymorphism (RFLP) [20–22].

3.5 Prenatal diagnosis of monogenic disorders

Since the entire genome is present in most human cells, defective genes can be identified irrespectively of the available cell sample, i.e. irrespectively of its cell-specific expression, a fact which widens the range of monogenic disorders accessible to prenatal diagnosis [23]. Cells that have a genome identical to that of the fetus can be obtained by amniocentesis from the amniotic fluid into which they are desquamated, by chorionic villus sampling from the chorion (early development stage of the fetal part of the placenta), or by fetoscopy and fetal blood sampling. The first two methods are of practical relevance. Chorionic villus sampling enables prenatal diagnosis in the 8th to 10th week of pregnancy, whereas an amniocentesis cannot be performed before the 16th week.

If we examine the indications for the prenatal genetic examinations carried out in Switzerland over the last 20 years, a striking feature is that most of them (2/3) were performed because of advanced maternal age. After the age of 35 the risk of a new numeric chromosome aberration that can lead to well-known conditions such as Down's syndrome (mongolism) rises to above 1%. Conditions detectable only by DNA diagnosis do not account for even 1% of all indications. Since such severe monogenic diseases are quite rare in our society, they are unlikely to lead to a substantial rise in the number of prenatal examinations carried out in the next few years. This situation is different in subpopulations in which specific hereditary diseases are relatively common, e.g. Tay-Sachs disease in Ashkenazi Jews or thalassaemia in Sardinia and in Cyprus.

3.6 Presymptomatic/preclinical diagnosis of genetic diseases

Genetic diseases are not necessarily congenital. They may become manifest only during the patients lifetime, although the defective gene is present in all cells from the outset. The presymptomatic diagnosis of a genetic disease is appropriate if the prognosis of the disease is likely to be influenced by medical intervention or if the reliable detection of the predisposition is of importance for family planning. Between 1965 and 1978, without methods based on recombinant DNA technology, 1,673,168 neonates were examined under the Swiss newborn babies screening programme for genetic diseases that result in an inability to tolerate specific nutrients (amino acids or sugar). As a result, 470 children were detected in whom the severe genetic disease could be prevented from becoming manifest by means of a diet.

Gene technology will become increasingly important for the presymptomatic diagnosis (diagnosis before they become manifest) of diseases that appear only in adult life. Such conditions include familial hypercholesterolaemia [24], which leads to cardiovascular diseases if left untreated, and malignancies of the gastrointestinal tract, endocrine organs, skin (including melanoma) and eventually breast cancer, whose incidence is elevated in families with an inherited susceptibility. These conditions often produce no clinical symptoms for a long time. A carcinoma of the colon can be present for up to 10 years before it extends beyond the limits of the gut and/or metastasizes to other organs. Members of such families worry about their own health because they observe the fate of their relatives and are reminded of a genetic disposition for the disease. In presymptomatic diagnosis a suspected gene defect can be detected in a person at risk. Preventive measures can then be systematically applied in order to diagnose and remove a malignancy in good time or to protect the individual from a specific environmental factor to which he/she is particularly sensitive.

3.7 Gene therapy

Most genetic diseases still cannot be treated effectively. This accounts for the endeavour to find new means of therapy [25, 26]. Genetic engineering could permit interventions in the genome of human body cells that would give a patient lifelong freedom from his hereditary disease.

When evaluating the possibilities and limitations of gene therapy in humans, it is necessary to determine 1) how this intervention is to be made and 2) where, i.e. on what cells, it is to be carried out. Among the various possible approaches to gene therapy, the insertion of a normal gene at a non-determinable locus in the human genome is one that can now be considered. The techniques for the locus-specific replacement of a defective gene with a good one are not yet efficient enough, and involve the risk of new mutations.

In animal studies of germ line gene therapy the fertilized ovum often does not survive gene transfer. The insertion of the foreign gene is not regularly successful. Furthermore, even couples with a heavy genetic burden have a large chance of producing healthy offspring. Germ cells or fertilized ova (zygotes) that bear the normal gene cannot be distinguished from those with the defective one. Nor can the success of gene therapy be assessed before the embryo is transferred into the uterine cavity, and the prospects of a pregnancy following embryo transfer are small. Virtually all ethical committees dealing with questions of gene therapy recommend that it should not be carried out on human zygotes and embryos, not only for such medico-technical reasons, but also for reasons of more far-reaching significance.

Somatic cell gene therapy

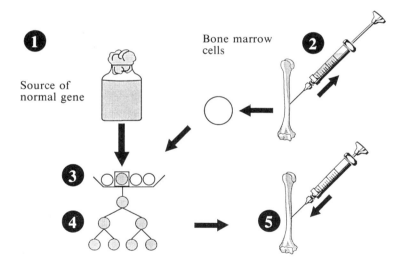

Figure 5
Procedure for gene transfer into bone marrow cells: 1. Gene cloning: production of multiple copies of the normal gene. 2. Removal of bone marrow from the patient and isolation of marrow cells. 3. Insertion of the normal (foreign) gene into the bone marrow cells, for instance by transfection with a retroviral vector carrying it. 4. Selection and multiplication of successfully healed cells. 5. Infusion of these cells into the patient whose blood-forming system has previously been destroyed.

For a few genetic diseases, the gene therapy of body cells (somatic cells) is likely to become a possibility in the foreseeable future. Here an attempt will be made to insert a foreign, functioning gene only into body cells which are already differentiated and cannot correctly perform an important function due to a gene defect (Fig. 5). For this purpose, however, such cells would first have to be grown in culture and subjected to treatment by gene technology, after which those in which the poorly controllable gene transfer had been successful would have to be fished out of the millions in the culture. Only such successfully treated cells would be returned to the patient, who would then be cured of his genetic disease for the rest of his life.

The limitations of somatic gene therapy are evident if one bears in mind how few types of differentiated cells survive outside the human body and multiply in vitro, i.e. under tissue culture conditions. Thus, there can be no question of interventions in brain cells, which are very frequently involved in genetic diseases. Finally, it is always too late for the gene therapy of somatic cells if irreversible defects are already present at birth.

Furthermore, gene therapy cannot be carried out unless the precise gene defect is known and the normal gene has been isolated and cloned. In the present state of the art it would therefore be negligence bordering on recklessness to hold out hopes of a cure through gene therapy to patients with genetic diseases and their relatives. Bioethically, this form of treatment differs little from the generally accepted procedures of organ, tissue or cell transplantation. The transferred gene dies with the individual whose cells have been treated.

4 The "vitreous" individual

Gene technology has become the greatest challenge in modern medicine; it will decisively influence the development of medical science by giving a progressively better understanding of the genetic plans for the structure and function of our body [27, 28]. This will enable us to obtain a fundamentally new insight into the cause and nature of common chronic disorders occurring in adulthood. If it is possible to determine the individual genes that regularly or occasionally predispose to these conditions – diseases of multifactorial origin such as cardiovascular diseases, allergies, endogenous psychoses or malignancies – then there is a justified hope that the environmental factors involved in the causation of the diseases can also be identified and classified. New possibilities of effective disease prevention can also be derived from such knowledge.

To map our genes is a long-standing aim of research into human genetics [29]. Very slow progress has been made by family studies and cell biological investigations. Only gene technology has brought a real breakthrough in gene mapping. The "genome project", the biggest-ever biological research scheme, has been started with the aim of deciphering the entire human genome by means of coordinated efforts. Its goal is understanding the normal and abnormal structure and function of human DNA. So far, we know the full DNA sequence of the mitochondrion, a cytoplasmic particle involved in cellular respiration [30]. But there is no need to fear that one day every one of us can simply be genetically "X-rayed". Our genome is based on two sets (one set from the mother and one from the father) of 3×10^9 base pairs. In order to recognize our individual genetic identity, 6 000 million base pairs would have to be identified and registered. Needless to say, the necessary laboratory space and computer capacity will never be available. The future use of gene technology will therefore have to remain restricted to relevant genetic problems in individual persons and families.

5 Fears and misgivings

As a result of the rapid development of methods using gene technology for diagnosis and therapy, we have increasingly come to discuss whether we really want the possibilities that have been created and whether they are ethically acceptable. Weighing benefits against possible hazards is not always a simple matter. If the new methods based on gene technology can be used to help persons who need important proteins that can be produced only by these techniques, or who can be protected by the early detection of serious diseases, then undisputed advances will have been achieved. On the other hand genetic examinations will also touch on important points of moral concern, although it should be borne in mind that these problems have long existed but have again become a focus of attention as a result of gene technology:

1. Respect for human life, also before birth.
2. Valuation of life, or rather the quality of life.
3. Autonomy of the individual, right of self-determination (independence of social constraints).

It is widely feared that a new age of inhumanly applied eugenics could come upon us if, for instance, prenatal diagnosis were to become mandatory under the pressure of the explosion in health care costs. Fears are also aroused by speculation over what a tyrannical regime could do with genetic engineering. As a result of the perversed racial hygiene theories propagated under the Nazi-regime, we in the German-speaking countries find it particularly hard to come to terms with genetic questions. In a pluralistic society, there is unlikely ever to be complete agreement on the applications of gene technology and the restriction of its use. But it is to be hoped that through openness, understanding and tolerance we shall be able to find solutions that are acceptable not only to the community at large but also to the affected individuals themselves.

References

1 G. Mendel: Verh. Naturf. Ver. Brünn *14*, 3 (1865).
2 D. J. Weatherall: The New Genetics and Clinical Practice. Oxford Univ. Press, Oxford (1985).
3 F. Miescher: Hoppe-Seyler's med.-chem. Unters. *4*, 441 (1871).
4 F. H. C. Crick and J. D. Watson: Proc. Roy. Soc. (A) *223*, 80 (1954).
5 OTA-report: Mapping our genes. Genome projects: How big, how fast? p. 61. The Johns Hopkins Univ. Press, Baltimore (1988).
6 V. A. McKusick: Mendelian Inheritance in Man. Catalogues of Autosomal Dominant, Autosomal Recessive and X-linked Phenotypes (8th edition). Johns Hopkins University Press, Baltimore (1988).
7 W. Arber and S. Linn: Ann. Rev. Biochem. *38*, 467 (1969).
8 H. O. Smith and D. Nathans: J. molec. Biol. *81*, 419 (1976).
9 A. M. Maxam and W. Gilbert: Proc. natl Acad. Sci. USA *74*, 560 (1977).
10 F. Sanger, S. Nickler and A. R. Coulson: Proc. natl Acad. Sci. USA *74*, 5463 (1977).
11 D. V. Goeddel, D. G. Kleid, F. Bolivar, H. L. Hynecker et al., Proc. natl Acad. Sci. USA *76*, 106 (1979).
12 I. S. Johnson: Science *219*, 632 (1983).
13 T. Taniguchi, L. Guarente, T. Roberts, D. Kimelman, J. Douhan and M. Ptashne: Proc. natl Acad. Sci. USA *77*, 5230 (1980).
14 S. L. Berger and A. R. Kimmel: In: Methods in Enzymology, Vol. 152, Acad. Press, London (1987).
15 D. H. Gillespie: J. clin. Lab. Anal. *1*, 42 (1987).
16 F. Brown: Ann. Rev. Microbiol. *3*, 221 (1984).
17 F. Constantini and R. Jaenisch (eds): Genetic Manipulation of the Early Mammalian Embryo. 20th Banbury Report. Cold Spring Harbour Laboratory (1985).

18 H. Harris: Principles of Human Biochemical Genetics. (2nd edition). Northholland Publishing Company, Amsterdam (1975).
19 C. T. Caskey: Science *236*, 1223 (1987).
20 R. White, M. Lippert, D. T. Bishop et al., Nature *313*, 101 (1984).
21 Y. W. Kan and A. M. Dozy: Lancet *ii*, 910 (1978).
22 J. F. Gusella: Ann. Rev. Biochem. *55*, 831 (1986).
23 J. R. Gosden and C. M. Gosden: Oxford Rev. Reprod. Biol. *7*, 73 (1985).
24 Int. Symp. on Familial Hypercholesterolaemia. Arteriosclerosis 1989 (suppl. 1).
25 W. F. Anderson: Science *226*, 401 (1984).
26 Hj. Müller: Experientia *43*, 375 (1987).
27 S. H. Orkin: Cell *47*, 845 (1986).
28 A. Motulsky: Science *219*, 135 (1983).
29 V. A. McKusick and F. J. Ruddle: Genomics *1*, 103 (1987).
30 Mitochondrial DNA and genetic disease. Editorial. Lancet *i*, 250 (1989).

Immunotherapy for leprosy and tuberculosis

By J. L. Stanford
Department of Medical Microbiology, University College and Middlesex School of Medicine, London W1P 7LD, United Kingdom

1	Introduction	416
1.1	Koch's immunotherapy	416
2	The Koch phenomenon	417
2.1	Immunotherapy based on the Koch phenomenon	419
3	A thumbnail sketch of immunotherapy from Robert Koch to the advent of chemotherapy	420
3.1	Immunotherapy since the introduction of effective chemotherapy	422
3.2	Marianum Antigen	422
4	Immunotherapy today	425
4.1	Immunotherapy and protective immunity	425
4.2	Protective immunity	425
4.3	Antigens	426
4.4	Adjuvants	428
5	Choice of an immunotherapeutic	429
5.1	Mechanisms of protective immunity	429
5.2	The importance of group i, common mycobacterial, antigens	430
5.3	From protective immunity to the Koch phenomenon – and back again?	431
5.4	Suppression of the Koch response	432
5.5	Are Koch responses and protective responses evoked by the same antigens?	434
5.6	Should immunotherapy for tuberculosis and leprosy be the same?	434
6	Why *Mycobacterium vaccae*?	436
6.1	Description of *Mycobacterium vaccae*	436
7	Immunotherapy with *Mycobacterium vaccae*, an up-to-date appraisal	437
7.1	Preparations used	437
7.2	Application to a guinea pig model of immunotherapy	438
7.3	Preliminary studies with leprosy patients	440
7.4	Responses at the sites of immunotherapeutic injection	442
7.5	Preliminary studies in tuberculosis patients	443
7.6	Studies in progress	443
8	Conclusion	444
	References	445

1 Introduction

Prior to the development of effective chemotherapy for tuberculosis and leprosy, the only approach to active treatment for these diseases was to enhance the patients' own immunity. Attempts to accomplish this were an obvious goal once the infectious nature of tuberculosis became appreciated from the work of Villemin in the 1870's. The observations that many persons in close contact with either leprosy or tuberculosis did not develop disease, that the disease followed a very chronic course in a considerable proportion of patients, and that others overcame their disease, showed the potential for such an approach. It is often forgotten that prior to the 1940's and the discovery of streptomycin as many as 50 % of patients with clinically manifest tuberculosis actually recovered. It is this proclivity toward natural recovery that has made assessment of early treatment methods for tuberculosis so difficult, and there were many claims of treatments for the disease that probably depended on case selection and natural recovery for their "efficacy". The earliest attempts at planned immunotherapy for tuberculosis may have been made in Paris, but the first person to claim real success, and have experimental observations to support these claims, was Robert Koch [1].

1.1 Koch's immunotherapy

In retrospect there was tremendous logic in Koch's approach. He reasoned that those patients who did best were those who had the greatest reaction around their lesions, and that if every case could have this reactivity boosted their chances of survival would be increased [2]. In the absence of any way of killing bacilli and reducing the bacillary/antigenic load, such an approach would probably be that adopted today. Koch observed that a guinea pig previously infected with virulent tubercle bacilli was capable of overcoming a subsequent small intradermal challenge with the same bacilli by necrosing all the tissues in contact with these bacilli and sloughing the dead tissue together with the living bacilli within [1]. Thus the pig would show healing without disease at the site of the second challenge, whilst going on to die from its original infection with the same organism.

It is this fundamental observation that provides the basis for all subsequent work on immunoprophylaxis, and immunotherapy of tuberculosis and leprosy, although the way in which the phenomenon underlies modern immunotherapy is quite different from the interpretation put on it by earlier workers.

It is difficult to see how Koch could have come to different conclusions at the time when he made them, but, nonetheless, he made a fundamental error which we have still not completely thrown aside. Although Koch's phenomenon demonstrated superinfection immunity from reinfection, it is not the mechanism of protective immunity that prevents an individual from becoming infected for the first time, or the mechanism that we would wish to see following BCG vaccination.

2   The Koch phenomenon

Today we know that Koch's phenomenon is due to a combination of cellular immune mechanisms involving macrophages, T cells, gamma interferon and tumour necrotic factor (TNF) [3]. Nevertheless, we remain unsure of the parts played in such cellular responses by antibodies, which may modify cellular responses, by granulocytes and even fixed tissue cells. We know too, that the active metabolites of vitamin D may mediate towards the tissue necrosing response [3]. That the actual mechanism of cell death through the action of TNF is also becoming established, although the parts played by HLA, secreted antigens of the bacilli themselves, and different glycoforms of immunoglobulins remain obscure. The following is an attempt to put together an explanation of the mechanism of the Koch phenomenon as seen in tuberculosis.

Bacilli inside a cell secrete antigens, and perhaps substances with their own cytotoxicity, which may kill the cell. The bacilli released from the dead cell may remain extracellular, or be re-phagocytosed, but in either case they continue to secrete antigens. Secreted antigens may become attached to cell surfaces, probably including those of endothelial cells lining local capillaries. Certain cells of the macrophage series, and perhaps dendritic cells present antigens to T cells capable of recognising them, and these T cells release mediators including gamma interferon, which activate macrophages.

The macrophages may be further activated by vitamin D3, and by attachment to their surface of the fc portions of agalactosyl IgG (GO, G naught) molecules of uncertain fab specificity [4]. Other antibodies, perhaps against the exposed N acetyl glucosamine of GO, may compound the effect. Whatever the precise chemistry of the reaction, macrophages release TNF in large quantities.

TNF has both generalised and local effects. Its rapid release into the circulation can be catastrophic and lead to death from tuberculin shock, or related conditions in certain non-mycobacterial infections. Its slow release is modified, by substances of probable hepatic origin, thus tuberculosis is often worse, and frequently fatal in persons with alcoholic liver damage [5]. The modified effect may be no longer catastrophic, but still acts as cachectic factor leading to severe weight loss in tuberculosis patients and in patients with advanced malignant tumours. (It is interesting to note that patients in the latter group usually lose tuberculin skin test positivity, as do cases of advanced tuberculosis, and thereafter respond poorly to anticancer immunotherapy.)

Locally TNF kills endothelial cells and perhaps other cells, but how it is targetted remains unknown. One suggestion would be that the fab of some GO molecules, attached to macrophages by their fc fragments, is specific for mycobacterial antigen attached to the endothelial cell surface. This brings the TNF-secreting cell into close proximity with the endothelial cell whose death is achieved. Death of capillary endothelial cells eventually results in lowered oxygen tension in the tissues, with further cell death and lowered metabolic and replicating activity of the bacilli themselves. Thus the Koch reaction results in chronic abscess formation in which an anoxic caseous mass still containing viable bacilli is surrounded by a thick fibrous wall.

Patients with such fibrocaseous disease may live for many years, perhaps all their lives, without reactivation to more active disease states. Such a reaction occurring around limited disease is the phenomenon of natural healing, and all immunotherapy prior to the development of successful chemotherapy depended on its exaggeration.

2.1 Immunotherapy based on the Koch phenomenon

Careful use of immunotherapy based on augmentation of the Koch phenomenon undoubtedly prolonged the lives of many patients with both tuberculosis and leprosy, but also undoubtedly led to the premature death of other patients with tuberculosis [6]. With our present knowledge of cellular immunity and human genetics it might be possible to develop a successful immunotherapeutic regimen of this type. Death due to the acute effects of circulating TNF might be prevented by the use of inhibitors of its cytotoxic activity, but it is more difficult to see how one could prevent death from tuberculous pneumonia due to the abscesses bursting into bronchioles, or from respiratory failure due to excessive loss of lung tissue. Nonetheless, such a system may be the only way of combatting the gradual spread of infections with totally drug-resistant strains which are becoming an alarmingly frequent finding in India, Pakistan, Afghanistan and Iran. The place of the Koch phenomenon in leprosy is less easy to determine. Paucibacillary (BT/TT) leprosy is a granulomatous response to unknown group iv, species-specific antigens of the bacillus in which tissue death by TNF release is not seen. It seems more likely that the granuloma in BT/TT disease is a response of the appropriate type to kill the bacilli, but yet one that is inefficient for unknown, perhaps genetic, reasons. At first sight Koch's phenomenon does not seem to play a part in multibacillary (LL/BL) leprosy in which there appears to be no cellular immune response to the bacilli. Many patients develop reactions known as erythema nodosum leprosum (ENL) during treatment of LL/BL disease, and careful studies during the height of such reactions show tissue death, the temporary prescence of GO, and other indicators of Koch's phenomenon [7]. Such reactions appear to be due to a temporary relaxation of homeostatic control of immune response, suggesting that the Koch phenomenon underlies LL/BL disease, but its action against the antigens of *M. leprae* is actively suppressed. If this is true, then multibacillary rather than paucibacillary leprosy has some analogy with caseating tuberculosis. Koch-type skin test responses to tuberculin [8], presumably directed at the species-specific antigens of *M. tuberculosis*, quite frequently occur in patients with multibacillary disease who have received some months of chemotherapy. They occur less frequently in paucibacillary disease, and responses to tuberculin are generally smaller amongst TT/BT patients than amongst BL/LL patients.

It is difficult to see a place for immunotherapy that enhances Koch's phenomenon in leprosy, since relaxation of suppression of this response could be very damaging in this disease. Prior to chemotherapy for leprosy, a syndrome was occasionally observed in which multibacillary patients suddenly developed massive necrosis of their lesions and died within 24–48 hours.

3 A thumbnail sketch of immunotherapy from Robert Koch to the advent of effective chemotherapy

Following Koch's announcement in 1890 [1] that he could prevent the development of tuberculosis in the guinea pig by repeated injections of what became known as his "brown fluid", claims were made for the successful treatment of tuberculosis in man, especially for Lupus vulgaris, which could be easily observed [9]. The lesions of Lupus literally necrosed and fell off, and the sites healed just as the intradermal second challenge sites of guinea pigs did. This was fine if deep tuberculous lesions were few, and in places where their enlargement and subsequent encapsulation would not impinge on any vital organ. It was fine too if the patient survived the prostration, depressed respiration, and fever due to circulating TNF, that lasted 24 hours after administration of the first course of immunotherapy with the brown fluid [10].

What is now known as tuberculin shock can be fatal in experimental animals, and was soon found to be fatal in man. A combination of deaths due to tuberculin shock, to increase in size of internal lesions, or to tuberculous pneumonia after lung lesions have burst into bronchial passages, led to Koch's immunotherapy being given up within a few years of its inception. Nonetheless, there can be little doubt that Koch cured many patients of their disease, and adherents to his methods were publishing success stories until the turn of the century [11].

Koch announced the nature of his brown fluid to be a preparation that we would now call Old Tuberculin (OT), and modifications of "tuberculin therapy" were still being investigated until the discovery of streptomycin in 1947. Immunodiffusion analysis of OT, and of the subsequently developed purified protein derivative (PPD), shows that they contain a high concentration of slow grower associated, group ii antigens, and lesser concentrations of species-specific, group iv antigens and common mycobacterial, group i antigens [12]. Recent studies in Kuwait have shown that a considerable proportion of BCG-vaccinated healthy persons respond to group ii antigens in skin tests [13]. They have also shown that tuberculosis patients lack skin test responsiveness to both group i and group ii antigens. The place of cellular responsiveness to group ii antigens obviously requires further elucidation.

An interesting discovery made in the 1940's was that a very similar reaction around plaques of Lupus vulgaris followed treatment with large doses of vitamin D (a macrophage activator), and a number of cases were successfully treated [14]. However, once more death could accompany the use of this system through renal failure, and it was very much a 2-pronged weapon.

Tuberculosis serum therapy was introduced in the early years of the twentieth century, and it did seem that this approach could be successful, even in very advanced cases of disease. In fact, one of its major proponents, Spahlinger, advocated use of his serum prepared in black irish hunters (horses) for the severely afflicted, and his special suspension of tubercle bacilli for those with less severe disease [15]. The combination of his lack of formal medical training, tremendous jealousies within the profession, and so-called evidence that antibodies played no part in immunity from tuberculosis, lead to his eventual failure. Even today, most textbooks repeat the old dogma that antibodies are not important in tuberculosis. It was probably serum factors, even if not antibody, that determined life or death from tuberculin shock after Koch's immunotherapy, and there is a real need for a reappraisal of humoral factors in mycobacterial disease. With increasing knowledge of the part that antibodies (such as GO) may play in immune cellular interactions, views of their importance are likely to be revised.

Similar approaches to those used in tuberculosis were applied to the immunotherapy of leprosy, again with the same mixture of success and failure. If anything, the early work in leprosy is even more difficult to assess than that in tuberculosis. After the early attempts at immunotherapy of leprosy using Koch's system [16], many other methods were tried. Often these were based on organisms isolated from leprous tissues or the fomites of leprosy patients, not all of them mycobacteria. If they were effective, and some of them may have been, it is impossible to be certain of their modes of action, but they may have worked through the stimulation of responsiveness to group i, common mycobacterial antigens. Since *M. leprae* is not known to possess antigens of group ii specificity, it is unlikely that responses to them will prove important in the control of leprosy.

3.1 Immunotherapy since the introduction of effective chemotherapy

Other than some experimental studies in animals and the recent double blind trial showing Levamisole to be useless in the treatment of pulmonary tuberculosis in man [17], little work has been done in the field of immunotherapy for tuberculosis since the discovery of streptomycin. In leprosy an interest in immunotherapy has remained, despite the discovery of the sulphones. In part this is because dapsone has been less successful than the antituberculosis drugs, and in part it has been due to the large amount of immunological interest in leprosy.

The first major attempt at immunotherapy of leprosy after the second world war was with Marianum Antigen, and the papers written about it make fascinating reading.

3.2 Marianum Antigen

This preparation is particularly interesting because its use, which was recent, but is now historical, illustrates many aspects of immunotherapy of leprosy and teaches many lessons which have already been forgotten.

Strain Chauviré was grown on Sauton's medium from a case of leprosy, and was used as a suspension of autoclaved bacilli to induce Lepromin conversion in young, Lepromin negative, close contacts of leprosy patients in 1952. This promising beginning led to the use of what was termed "l'antigène Chauviré" in a much larger study in Cameroun, and results were reported on 6 persons without leprosy, 195 with indeterminate leprosy, 65 with tuberculoid disease, and 73 with lepromatous disease [18]. Each person was tested with Lepromin and the Mitsuda reaction was measured 3–4 weeks later; 3 injections of the Chauviré antigen were then given at monthly intervals, and 2 months after the last injection the Lepromin test was repeated. The results were striking; the 4 healthy persons and the 4 tuberculoid patients who were Mitsuda negative at the beginning of the study all became positive at the end. Of 170 Mitsuda negative indeterminate patients, 116 (68 %) converted to positive, and of 63 Mitsuda negative lepromatous patients, 32 (51 %) became positive. A small proportion of the patients who started with Mitsuda positivity became negative during the study.

In the same year (1953) strain Chauviré was described as the type strain of a new species, named *Mycobacterium marianum* in honour of the Marist order to which belonged Sister Marie-Suzanne who carried out the first studies with it [19]. Thereafter the reagent became known as Marianum Antigen. However, in 1955 *M. scrofulaceum* was described, and later studies showed that the 2 species were synonymous. Because of possible confusion with the earlier described *M. marinum,* a quite different species, the Judicial Commission on Bacterial Nomenclature, favoured *M. scrofulaceum* over *M. marianum,* despite its chronological priority. Thus, in today's terms, Marianum Antigen is a suspension of killed *M. scrofulaceum.*

Extensive studies were reported from Cameroun in 1955 [20], by which time 6 injections of Marianum Antigen at monthly intervals had been adopted instead of 3. Consideration was given to the local and general side effects of 2,638 injections. In 15–20 % of injections there was minimal or no local reaction; in 70–75 % reactions were considered to be moderate, and in 5–15 % reactions were severe. Undergoing a moderate reaction to the injection, the patient suffered a fairly severe headache and burning pain at the site of injection within a few hours, with pyrexia of 38–39° C. The local lesion ulcerated over the next 2 weeks and dried to a scab by the end of the 4th week when the next injection was due. During this time focal congestive reactions took place in some or all of the superficial lesions of the disease. The most severe 3 % of reactions commenced soon after injection with intense local irritation, and formation of a papule up to 3 cm in diameter. By the 20th day this necrosed to form an ulcer through the complete thickness of the skin, perhaps 1 cm in diameter and this crusted over to leave a permanent scar 1 cm or more across. Accompanying the local reaction was marked and painful lymphadenopathy of the local modes, a marked general reaction with intense headache for many days, a pyrexias of 39° C or more, and the patients were bedridden for several days. During this reaction leprous lesions became congested, painful, and sometimes bled.

Clinically about 20 % of the patients did badly, in about 20 % there was little change in the disease, and in 60 % there was significant improvement above that expected from dapsone treatment alone.

A report from Cebu in the Philippines [21] in 1957 reported much less significant improvement, as did a report from Nigeria in the same year [22]. However, it is of interest to note that the local and general reactions following the injections given in Nigeria were described to be less severe and prolonged than those reported from Cameroun [20]. In the following year there was a report from Fiji, also of disappointing results [23]. The last 2 papers, in 1964 [24], and 1966 [25], came from Cameroun and Korea respectively, and both support the efficacy of the Antigen. The Korean study is particularly interesting since the major part of it was a comparison between the results of monthly doses of Marianum Antigen without other treatment (2,502 patients), and with dapsone monotherapy (284 patients). The 2 treatment groups did equally well, and if anything the Antigen-treated group did best. 66.5 % patients receiving the Antigen clinically improved in comparison with 59.5 % of the dapsone group. Approximately 83 % of bacteriologically positive patients receiving Marianum Antigen had become negative by the end of the study.

Partly because of the way the results were reported, partly because of lack of adequate control groups, and partly because none of the studies were blind, it is difficult to accurately assess the effects that Marianum Antigen may have had. It is interesting that the best results were reported tor the use of the Antigen *without* the use of dapsone, and it may be that the constant slight immunomodulating effect of dapsone militated against the Antigen's efficacy. It is also possible that the Antigen was more effective in some places than in others; just as BCG vaccine is against tuberculosis [26–28].

Despite the deficiencies of the reports, Marianum Antigen seems to have caused improvement in a number of patients, just as Koch's immunotherapy did 70 years earlier. In fact Marianum Antigen must be considered to be of the same lineage as the Brown Fluid, with the necessity of repeated injections, generalised side effects and tendency for the lesions to necrose. If its action was to enhance the Koch response, then the claims for 93 % [20], 30 % [21], and 65 % [22] of Lepromin Mitsuda convertion following the use of Marianum Antigen means that this response is a correlate of the Koch phenomenon, at least under some circumstances. Thus, although investigated after the introduction of dapsone as therapy for leprosy, Marianum Antigen should be seen as belonging to an earlier era, rather than paving the way to new approaches.

4   Immunotherapy today

The modern research on immunotherapy to be described in this section has a different foundation from the Koch phenomenon. It is not designed for drug-resistant organisms, but as an adjunct to effective chemotherapy to improve therapeutic efficacy, reduce the period of drug administration, improve patient compliance with treatment, and reduce the incidence of relapse or reinfection.

Prior to the development of effective chemotherapy, immunotherapy had to be undertaken in the presence of large numbers of actively metabolising and multiplying bacilli releasing large quantities of antigens and other immunomodulating substances. Under such conditions enhancement of certain aspects of the immune mechanisms already operative was all that could be done. With our modern methods of short-course multidrug therapy for both leprosy and tuberculosis, the position is completely changed. With all but a small number of persisters killed within the first few weeks of starting treatment, and with rapidly declining antigen loads, the Koch reaction is no longer the immune state of choice. Thus immunotherapy aimed at stimulating the Koch reaction is no longer to be advocated for infections with drug-sensitive organisms.

4.1   Immunotherapy and protective immunity

Under these changed conditions of low bacterial and antigenic load, protective immunity of the kind following the use of BCG vaccine, where it is successful, might be appropriate. Reintroduction of an immune protective mechanism leading to the death of live bacilli is the basis of recent attempts at immunotherapy [8, 28, 29].

4.2   Protective immunity

Although we know that this exists, and have been using BCG vaccination to induce it for more than 60 years, the actual mechanism remains the subject of surmise. It is thought that invading mycobacteria are killed by macrophages, but it has not been possible to demonstrate reproducibly actual killing of either tubercle or leprosy bacilli by cells of human origin *in vitro*, although mouse cells can kill some mycobacterial species. It can be shown *in vitro* that human macrophages in certain states of activation can prevent replication of phagocytosed bacteria without killing them [30]. Nonetheless, *in vivo* evidence suggests that mycobacteria are killed in the tissues and the mechanism may soon be elucidated.

A major question asked over many years has been whether the mechanism of protective immunity is, or is not, the Koch phenomenon. There is little doubt that protective immunity and allergy to mycobacterial products are different. Some of the arguments for this belief are discussed below after a consideration of the bacterial contributions of antigens and adjuvants.

4.3 Antigens

The antigens of mycobacteria are many and varied, but they can be conveniently considered as belonging to 4 groups of taxonomic importance [31] (fig. 1). Group i, common mycobacterial antigens are shared by all mycobacterial species and by nocardiae [32]. Some of

Figure 1
The antigens of mycobacteria and their occurrence in related genera.

	Groups of antigens			
	I	II	III	IV (species specific)
M. tuberculosis	/////	/////		/////?
BCG	/////	/////		/////?
M. scrofulaceum	/////	/////		/////
M. intracellulare	/////	/////		/////
M. fortuitum	/////		/////	/////
M. chelonei	/////		/////	/////
M. vaccae	/////			/////?
M. leprae	/////			?/////
Nocardiae	/////		/////	
Corynebacteria	////?			
Escherichia	//?			

Group I antigens are shared by all mycobacteria and nocardiae (common mycobacterial antigens). Group II antigens are shared by all members of the slowly growing subgenus of mycobacteria (slow grower associated antigens). Group III antigens are shared by fast growing species except *M. vaccae* and by nocardiae (fast grower associated antigens). Note that *M. leprae* lacks both group II and III antigens.

them are shared with rhodococci and corynebacteria, and at least one or two are shared with such distant genera as *Escherichia*. Group ii, slow grower associated antigens are shared by slowly growing mycobacterial species such as the tubercle bacillus, *Mycobacterium scrofulaceum, M. avium* etc. Group iii, fast grower associated antigens are shared by most rapidly growing species such as *M. fortuitum* and *M. smegmatis* and by nocardiae. There are only 2 species known to lack both group ii and iii antigens, the leprosy bacillus and *M. vaccae* [33]. Group iv, species-specific antigens are only shared amongst the members of a single species, although there may be slight differences between subspecies as to which of these antigens they possess [31]. Some of the antigenic determinants of mycobacteria are the same as those of mammals [34], and it is these cross-reactive antigens that are likely to be responsible for autoimmune aspects of mycobacterial diseases.

It is tempting to think that when we have a full understanding of tissue cross-reactive mechanisms, we may also begin to understand mechanisms of tissue predilection of bacteria. So far, no one has convincingly explained why leprosy bacilli are attracted to the superficial dermis (dermatotrophism) and to peripheral nerves (neurotrophism). Antigens known to be shared with these sites may result in leprosy bacilli becoming subject to tissue organisational processes that ensure the limitation of tissues and maintain tissue integrity. If there are antigens that might be recognised by immune surveillance, what better place could there be to hide than in tissues bristling with the same antigens? Could such arguments also explain reduced tuberculin reactivity when certain vital organs such as the kidneys are attacked by the tubercle bacillus?

The antigens of mycobacteria may be proteins, peptidoglycans, polysaccharides or lipopolysaccharides, either actively secreted or occurring in the cytoplasm or cell envelope. The antigens presented by live organisms will be different in a number of respects from those of dead organisms. Live bacilli actively secrete a limited range of substances [35], and only some structural antigens are exposed on the cell surface of the intact organism. Secreted antigens will rapidly diffuse away from a dead bacillus, and as it autolyses cytoplasmic antigens will escape and envelope antigens which are not usually exposed will become so. Thus immune mechanisms directed against dead bacilli may not be useful in the eradication of live bacilli, and host tissue may be damaged by extraneous responses. A possible incidence of this occurs to the 65 kDa heat shock protein of tubercle bacilli. This protein is not thought to be secreted, and therefore only becomes available to host mechanisms after bacterial death. It has been found that the human equivalent of this protein has a number of epitopes cross-reacting with that of mycobacteria, although apparently not with that of *Escherichia coli* [36]. Rheumatoid arthritis patients, in whom cell-mediated and antibody responses to the 65 kDa antigen are exaggerated, have higher antibody levels to the *mycobacterial* heat shock protein than do tuberculosis patients [37]. The 65 kDa human protein is present in large quantities in the arthritic joint. Thus 65 kDa antigen of mycobacteria may induce immune mechanisms, which become autoimmune against joints, leading to rheumatoid changes in those with a genetic predisposition to develop them [38].

4.4   Adjuvants

Mycobacteria have been known for many years to possess remarkable abilities to make vertebrate immune systems respond strongly to their antigens, sometimes in ways that confer protection, and sometimes in ways that constitute an essential part of the pathogenetic process. Indeed dead mycobacteria incorporated in oily emulsions, known as Freund's adjuvant, can be used to stimulate antibody production to almost any antigen mixed in the emulsion. Several adjuvant substances in the cell wall of mycobacteria have been described and extensively studied [39]. Muramyl dipeptide (MDP) is a simple, water soluble substance derived from the peptidoglycan layer of the cell envelope of most bacteria. Mycobacteria also possess waxy adjuvants, such as "wax D", which may enhance particular patterns of cell-mediated response, as well as the antibody-promoting effects of MDP. Although studies have been devoted to the adjuvants of tubercle and leprosy bacilli and of saprophytes such as *M. phlei* and *M. smegmatis*, to the best of my knowledge none of the studies has specifically addressed the question of the differences between the adjuvants of pathogens versus nonpathogens. Indirect evidence suggests that such a difference exists, and may have crucial importance in the selection of organisms to be used in immunoprophylaxis and immunotherapy.

5   Choice of an immunotherapeutic

From considerations such as these on antigens and adjuvants some definitions of the requirements for an active immunotherapeutic can be based. We need to stimulate and focus immune mechanisms that lead to the annihilation or total stasis of viable mycobacteria within the body. This must be accomplished without inducing unnecessary responses to already dead bacteria, without inducing death of host tissue cells, and without inducing life-threatening responses such as the intravascular release of TNF or massive increases in extracellular fluid.

5.1   Mechanisms of protective immunity

Biopsies of human skin reactions thought to be those of protective immunity have been compared with biopsies of reactions thought to be of Koch type, and very few differences have been found between them so far [40]. The same lymphocytes and macrophages appear to be present, and extravasation of extracellular fluid appears to occur in both types of reaction. The part played by granulocytes has not been studied in the reactions. The best explanation of the different clinical appearances and significances of the two types of reaction would seem to depend upon release of different cytokines, or different modulation of responses to them.
In vitro studies of vitamin D metabolites and of gamma interferon show them to be mediators of the Koch phenomenon, so that they must either be capable of acting differently in different situations *in vivo*, or they are not the mediators of protective immunity [41].
Recent studies of chemiluminescence in the presence of PPD show differences between the neutrophil granulocytes of tuberculosis patients and those of persons with tuberculin positivity following BCG vaccination [42]. Other studies of chemiluminescence of buffy coat neutrophils from healthy blood donors in the presence of zymosan show differing effects of adding to the system sera from tuberculosis patients, tuberculin positive BCG recipients and tuberculin negative BCG recipients [43]. Studies on infiltrates of skin test sites within 24 hours of injection of tuberculin in an BCG recipient individual, who had no induration to tuberculin at 72 hours, but was in constant laboratory contact with mycobacteria, showed a heavy infiltration with

eosinophil polymorphs which rapidly degranulated. Thus it may be that substances modulating the response of cells to the same cytokines may be present in sera or released from neutrophil or eosinophil cells in the presence of eliciting antigen or adjuvant.

5.2   The importance of group i, common mycobacterial antigens

Although our ignorance of the mechanism of protective immunity is considerable, many studies have shown that responses to group i, common mycobacterial antigens are important in the induction of protective immunity:
1) The efficacy of BCG vaccine in the prevention of tuberculosis and leprosy has been shown in the United Kingdom and Uganda, respectively; its failure to do so in some other countries is irrelevant to the argument. The antigenic relationship between the leprosy bacillus, the tubercle bacillus and BCG is shown in figure 1, and it can be seen that the only the group i antigens are shared between them.
2) Of course not all antigens are shown by such a system, but skin test studies of patients with either tuberculosis or leprosy show their loss of responsiveness to antigens of group i specificity, and lymphocyte transformation studies show that tuberculosis patients also lose responsiveness to group ii antigens.
3) Similar studies of quadruple skin testing of healthy persons show an increase in category 1 responders (to group i antigens) in BCG recipients (table 1). Studies of the peripheral blood of healthy BCG recipients show that the majority of T cells recognising mycobacte-

Table 1
The categories of responders to simultaneous skin tests with 4 new tuberculins

	Tuberculin	Leprosin A	Scrofulin	Vaccin
Category 1	+	+	+	+
Category 2	−	−	−	−
Category 3	+ or −	+ or −	+ or −	+ or −
Category 4	+	+ or −	+	+ or −

Groups of antigens to which the categories respond

Category 1 responds to group I (and possibly groups II and IV as well).
Category 2 responds to none of the groups at routine test concentration.
Category 3 responds to group IV of the species they have met.
Category 4 responds to group II (and possibly group IV as well).

rial antigens respond to the common antigens, whereas similar studies of patients with mycobacterial diseases show that their T cells respond predominantly to antigens of group iv, species-specific specificity.
4) Animal studies have been published showing that protective immunity induced with one mycobacterial species operates against challenge with any other mycobacterial species.

5.3   From protective immunity to the Koch phenomenon – and back again?

There is plenty of evidence that progression occurs from cell-mediated protective immunity to the cell-mediated Koch phenomenon [8, 28]. This occurs in apparently protected individuals developing tuberculosis, in a proportion of apparently healthy individuals in contact with disease, but who do not develop its clinical manifestations, and in persons repeatedly meeting environmental species capable of inducing either type of response. From this has arisen the concept of a small amount of contact with environmental mycobacteria being a good thing, and a large amount, a bad thing. As a generalisation, this seems to be true. Thus in Iran, environmentally induced sensitisation by *M. scrofulaceum* boosts the ability to recognise contact with leprosy bacilli by the development of skin test positivity to Leprosin A (table 2). In Burma, the much larger sized skin test responses to Scrofulin (prepared from *M. scrofulaceum*) are associated with reduced ability to become Leprosin A positive after BCG [44].
Studies of skin test responses to tuberculin and subsequent development of tuberculosis have shown that those with large skin test responses (roughly correlating with the Koch phenomenon) are most likely to develop the disease. The next most likely individuals to develop clinical tuberculosis are those with negative tuberculin responses who have not previously received BCG vaccination. The best protected are those with small sized responses. Those who do not become tuberculin positive after BCG, despite their formation of a post-vaccination scar, tend to lack HLA-DR 3, and are just as well protected from tuberculosis as are those who become tuberculin positive [45, 46].

Table 2
The different effects of environmental contact with *Mycobacterium scrofulaceum* in Burma and Iran on acquisition of Leprosin A positivity, in relation to BCG vaccination, and size of response to Scrofulin.

	Scrofulin +ve	Scrofulin −ve
BCG non-vaccinated		
Iranian children	50 %	15 %
Burmese children	18 %	16 %
BCG vaccinated		
Iranian children	71 %	30 %
Burmese children	15 %	33 %
	Mean size of +ve responses to Scrofulin	
Iranian children	7,8 mm	
Burmese children	11,4 mm	

Note that among those who do not respond to Scrofulin, responses to Leprosin A are the same in Iran and Burma. The small-sized responses to Scrofulin in Iran are associated with enhanced ability to recognise Leprosin A, which is further enhanced by BCG vaccination. The larger-sized responses to Scrofulin in Burma are associated with no improvement in recognition of Leprosin A, either with or without BCG vaccination.

The trial of BCG against leprosy in Burma, although in general only producing a disappointing 18 % of protection, was much more successful in children aged less than 5 years with small-sized responses to tuberculin, in whom 65 % protection was induced [47]. It is particularly disappointing that in this study which could have told us so much, the Heaf test rather than the Mantoux test was used, and long-term investigation has been blocked by WHO and the Burmese government.

It is much more difficult to be sure whether reversion from the Koch phenomenon to cell-mediated protective immunity ever occurs naturally. Studies of patients many years after their recovery from clinical tuberculosis suggests that it may, but much more investigation is required to establish this point. Figure 2 shows diagrammatically the changes in cellular response types accompanying the development, clinical course, and recovery due to chemotherapy of a tuberculosis patient.

Figure 2
Cellular immunity and tuberculosis

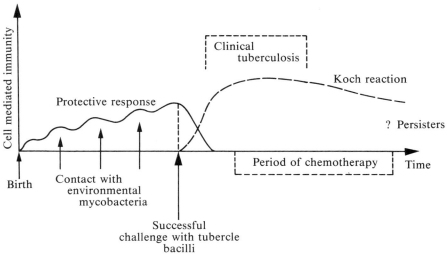

5.4 Suppression of the Koch response

There is little doubt that the Koch response can be suppressed under several sets of circumstances, but this does not necessarily mean that it has been replaced by a protective response. One situation in which the Koch response to tuberculin disappears is in severe pulmonary tuberculosis, patients with which may be completely tuberculin negative, and return to positivity with effective chemotherapy [48]. Other pulmonary tuberculosis patients, even those with only moderate disease, may be transiently tuberculin negative. Tuberculous pleurisy and renal tuberculosis are associated with a fluctuating tuberculin response quite frequently. Contact with large numbers of environmental mycobacteria can also be instumental in switching off the Koch reaction to tuberculin [8]. The mechanism by which this occurs may operate through the gut, as may the skin test allergy associated with severe pulmonary tuberculosis in which large numbers of tubercle bacilli are being swallowed.

The Koch phenomenon is a part of the disease in tuberculosis, and a person in whom a Koch response is present but in whom there is no other evidence of infection with *M. tuberculosis* might be considered as having the disease in an immunological sense. If this is so, then immunotherapy for tuberculosis is immunotherapy for the Koch phenomenon, and the ramifications of the thought are considerable. Immunotherapy designed for clinical tuberculosis would be useful for individuals who make an inappropriate, Koch-type response after BCG vaccination (perhaps the major factor underlying the vaccines failure in Burma). The same approach might be effective in the treatment of rheumatoid arthritis, in which the basic anomaly appears to be the Koch phenomenon directed against antigenic epitopes on the 65 kDa protein in joints cross-reactive with mycobacteria [38].

Moving in the same train of thought yet one more stage, of immunocorrection of the Koch phenomenon ist mechanism-dependent, rather than antigen-dependent, then the same immunotherapy could be used to prevent death due to acute release of TNF into the circulation in children with malaria, and delay in the progression in those infected with the Human Immunodeficiency Virus (replication of which has been reported to be stimulated by TNF) into the fatal AID syndrome.

5.5   Are Koch responses and protective responses evoked by the same antigens?

Koch responses are usually to group iv, species-specific antigens, but may also be to group ii, slow grower associated antigens [8]. So far, we have not detected this kind of response to group i antigens. Protective responses, although primarily to group i antigens, can also be directed to antigens of groups ii and iv.

5.6   Should immunotherapy for tuberculosis and leprosy be the same?

The answer is probably yes, since recognition of common mycobacterial antigens is as necessary for destructive recognition of leprosy bacilli as it is for tubercle bacilli, and although the Koch response is not an obvious part of the immune abnormality of lepromatous leprosy, its release from control could be catastrophic as discussed above.

The attempts to use *M. leprae* itself in immunotherapy of leprosy [49, 50] may be ill-conceived (although it would be interesting to see its effects in immunotherapy of tuberculosis). The cellular infiltrations produced by its injection could well be against the species-specific determinants that the bacillus shares with tissue, and therefore more damaging to the very tissues that it is desirous to protect, although this may be modified by the presence of live BCG given in the same injection. The reactions that it would induce or enhance would be expected to be to the antigens of dead bacilli, rather than to the group i antigens of living organisms. The experience that immunotherapy with *M. leprae* has to be repeated so many times to achieve an effect could be that it is quite difficult to break self-tolerance [50]. The same could be true for immunotherapy of tuberculosis with preparations containing the species-specific (pathology generating) antigens of the tubercle bacillus. Thus Koch's brown fluid may have killed some of those with extensive disease by respiratory insufficiency due to an increase in alveolar destruction. Too vigourous treatment with rifampicin- and isoniazid-containing regimens may kill patients with extensive lung disease due to local release of immunoreactive forms of potentially cross-reacting antigens, and perhaps an increase in circulating TNF [51]. The simultaneous use of steroids may help to prevent this. Even tiny stimuli to the Koch phenomenon such as tuberculin tests are known to precipitate clinical tuberculosis in some patients with previously subclinical lesions. Thus there is not only no rationale to the use of preparations of the infecting species in immunotherapy of these diseases, but good contra-indications for doing so. The effects of attempting immunotherapy of experimental tuberculosis of the guinea pig with injections of irradiation-killed BCG Glaxo are demonstrated by the experiment shown in table 3. Although this is only one experiment, it is interesting that it appears to bear out theoretical expectations.

What we want in an ideal immunotherapeutic is an optimal presentation of the antigens stimulating protective immunity together with regulatory activity against the mechanism of the Koch phenomenon. For the reasons discussed below preparations of an organism such as *Mycobacterium vaccae* may be ideal in its combination of a high concentration of group i, common mycobacterial, antigen, and its possible immunomodulatory effects [52]

Table 3
The disadvantageous effects of attempted immunotherapy with irradiation killed BCG in a guinea pig (GP) tuberculosis model.

	Mean extent of lesions at *post mortem* examination						
	Infection site	Local node	Opposite node	Spleen	Liver	Lungs	Total
Control GPs	1.9	1.8	0.3	0.9	0.7	1.8	7.3
BCG × 1	1.7	2.2	0.3	1.3	1.0	1.7	8.2
BCG × 2	1.7	2.3	0	2.0	1.3	3.0	10.0
M. vaccae × 1	2.2	1.9	0.1	0.7	0.6	1.5	7.0
M. vaccae × 2	1.7	2.3	0	0.5	0.7	1.3	6.5

6 Why *Mycobacterium vaccae?*

There is a fundamental schism between those who consider that the best immunotherapeutic, or immunoprophylactic, for a mycobacterial diesease lies amongst the group iv, species-specific antigens, and those who consider induction of the relevant mechanism to group i, common mycobacterial antigen is the best approach. As a supporter of the latter view, it would be difficult for me to give a fair assessment of the efforts of the adherants to the former. Suffice it to say, that achievements in immunotherapy of leprosy with *M. leprae*, with the ICRC bacillus [53, 54], or with *Mycobacterium W* [55] are no worse, and perhaps better, than those I can report with *M. vaccae* immunotherapy, yet I suggest that their effects are via the group i antigens that they all possess.

6.1 Description of *Mycobacterium vaccae*

Mycobacterium vaccae is a fast-growing species with photochromogenic, scotochromogenic and non-pigmented variants. It has never been implicated in the aetiology of human disease, and publications claiming it as a cause of thelitis in cattle report dubious taxonomic markers and appear to embody confusion with the rejected name: *M. aquae* [56, 57]. First reported in Europe as present in the environment of cattle, *M. vaccae* has been isolated from many places, and available information suggests that it is an inhabitant of the surroundings of farm animals and a dweller in neutral or acidic moist soil. The only form of *M. vaccae* that has been used in immunotherapy in R 877 R (NCTC 11659), which is a selected rough, scotochromogenic, variant of a strain isolated originally from grass roots collected near Lake Kyoga in central Uganda [58].

7   Immunotherapy with *Mycobacterium vaccae*, an up-to-date appraisal
7.1   Preparations used

R 877 R inoculated onto Sauton's medium solidified with 1.2 % agar (table 4) is incubated at 32° C until good growth has been obtained. The surface growth is removed with a spatula and weighed. 1 mg wet weight is approximately equivalent to 10^9 bacilli. The mass of bacilli are suspended in borate buffered saline containing tween 80 (table 4), diluted to the required concentration, and killed either by irradiation or by autoclaving. The irradiation system used was exposure to 2.5 Mrads from a ^{60}Co source, and autoclaving was for 10 minutes at 115° C. Both systems achieved 100 % kill of the bacilli as assessed by culture on Loewenstein-Jensen medium.

Injections have always been of 0.1 ml given intradermally. The only additives to the suspension tested so far have been new tuberculin 0.2 µg protein/ml, and Murabutide (MDP) as discussed below.

Table 4
Preparation of Sauton's agar medium and of borate buffered saline.

Sauton's Agar			
Asparagine	6 g	Glycerol	30 ml
Citric acid	2 g	K_2HPO_4	1.5 g
$MgSO_4$	0.25 g	$FeNH_4$ citrate	0.05 g

Add distilled water to 1 l.
Adjust pH to 6.2 using concentrated ammonia.
Add 12 g agar, and autoclave at 121 °C.
When cool, add 25 ml of 40 % glucose solution, mix and pour.

Borate buffered saline (M/15 borate)			
$Na_2B_4O_7.10\,H_2O$	3.63 g	H_3BO_3	5.25 g
NaCl	6.19 g	Tween 80	1 drop

Add distilled water to 1 l, heat to disolve and autoclave at 121 °C.

Figure 3

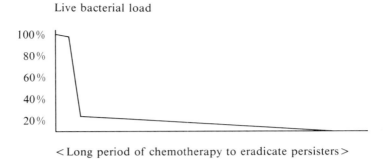

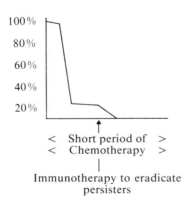

7.2   Application to a guinea pig model of immunotherapy

Both tuberculosis and leprosy are diseases acquired after a variable amount of contact with mycobacteria from the environment, and consequently in the presence of a variable amount of the different elements of immunity. Thus to prepare an animal model system that might parallel the human situation, some mycobacterial experience must be given to the animals prior to experimental infection. Similarly, any form of treatment should not start until signs are shown by the animal, which if occurring in a human would induce him to seek medical advice. Immunotherapy should then be given at a time after starting chemotherapy when most bacilli should have been killed (fig. 3). Treatment thereafter has to be foreshortened in order that the experiment can be finished within a reasonable time scale. The guinea pig model shown diagrammatically in figure 4 has been designed to fullfil these requirements [59]. Some of the results obtained with the model are shown in table 3. In the upper part of the table the disadvantageous effects of immunotherapy with killed BCG are shown, and are referred to above. In the lower part of the table the advantageous effects of immunotherapy with killed *M. vaccae* are illustrated. Using this model we have been able to show that live *M. vaccae* is ineffective, killed *M. chitae* is ineffective, and that

Figure 4
Guinea pig model of immunotherapy of tuberculosis.

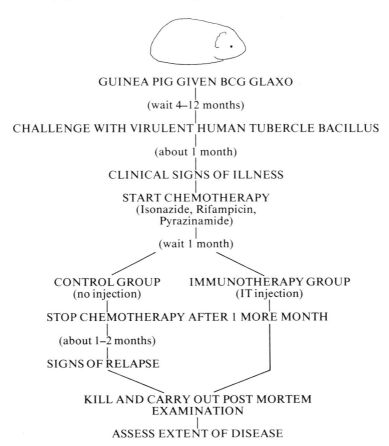

M. vaccae is more effective when killed by autoclaving than when killed by irradiation [59]. We have also shown that immunotherapy with *M. vaccae* is ineffective if given on the first day of chemotherapy when the bacterial and antigenic loads of the animals are high. Although some effects of immunotherapy and some of its constraints can be measured by an animal system, the short life span of most laboratory animals and the impossibility of infecting them by a natural means, seriously reduces their value. Thus we have made slow progress by directly working with human patients.

7.3   Preliminary studies with leprosy patients

Patients who have had multibacillary forms of leprosy maintain many of their abnormal aspects of immunity for some years after they are bacteriologically cured. In such patients the effects on their immune state of various forms of immunotherapy can be investigated. What we hoped to find was a return to the pattern of immunity shown by healthy individuals living in the same surroundings.
128 patients who had received some years of chemotherapy for multibacillary leprosy in Sanatorio Fontilles in Spain, gave their informed consent to participate in immunotherapy studies [29]. 86 patients have received injections of *M. vaccae* and 44 have remained as controls. All have continued with their prescribed chemotherapy throughout the study.
Over a period of 8 years the following potential immunotherapeutic preparations were tested:

10^7 *M. vaccae* irradiated
10^8 *M. vaccae* irradiated
10^9 *M. vaccae* irradiated
10^9 *M. vaccae* irradiated plus Tuberculin (1/10th of the routine skin test concentration).

Each consenting patient was skin tested with Tuberculin, Leprosin A, Scrofulin and Vaccin prior to injection of the immunotherapeutic, and retested with the same reagents 1 and/or 2 years later. Immunotherapy was considered successful if Leprosin A positivity was induced. The 2 low doses of *M. vaccae* had little effect, whereas both preparations containing 10^9 bacilli induced significant Leprosin A

Table 5
Results of the preliminary trial of immunotherapy with *Mycobacterium vaccae* on leprosy patients in Spain.

Results obtained 1–2 years after immunotherapy

5a Recognition of Leprosin A

1) Control group	1/47	2 %
2) 10^7 *M. vaccae*	3/29	10 % } dose too
3) 10^8 *M. vaccae*	0/6	0 % } low
4) 10^9 *M. vaccae*	13/34	38 % } better than control
5) 10^9 *M. v.* + Tuberculin	14/27	52 % } $p < 0.00001$

5b Results of quadruple skin testing

	No IT 144 LL/BL	IT received 26 failures	IT received 32 successes	No IT 39 BT/TT
Tuberculin	62 %	62 %	91 %	77 %
Leprosin A	3 %	0 %	100 %	62 %
Scrofulin	17 %	19 %	50 %	56 %
Vaccin	15 %	12 %	69 %	62 %

convertion (table 5a), and the addition of Tuberculin also appeared to be beneficial. When the full set of follow-up skin test results were compared for the successful and failed groups, there were marked differences (table 5b). The failures were not different from the whole group prior to immunotherapy, whereas the successes had responses similar to those of patients with treated paucibacillary disease, and with healthy control persons [60].

Must of our patients had been HLA typed by the Department of Immunohaematology in Leiden, Holland, and the immunotherapy results were analysed in relation to HLA types. There were found to be negative correlations between success of immunotherapy and DR or DQw markers. HLA-DR 2 is known to be associated with tuberculoid disease in family studies [61], and is known to be protective for rheumatoid arthritis [62]. Very recent data shows a strong correlation of DR 2 and DQw 1 with smear positive pulmonary tuberculosis in Indonesians [63].

Thus the fact that only about half of our leprosy patients became Leprosin A positive, and in whom immunotherapy is thought to have been successful, may reflect genetic differences. Whether there is a method of immunotherapy that would work in all patients remains uncertain. It is interesting to note that the reports on immunotherapy with Marianum Antigen, the ICRC bacillus and with BCG plus *M. leprae*, show that they only work in about half of patients. It seems likely, therefore, that HLA restricts for success whichever the immunotherapeutic.

7.4 Responses at the sites of immunotherapeutic injection

One of the major reasons why Marianum Antigen was not investigated further was the severity of local responses and the degree of unsightly scarring that followed its injection [20–22].
Local responses to injections of killed *M. leprae* plus live BCG can be considerable, and probably depend on immune mechanisms to tubercle bacilli which are not always reflected by the tuberculin test. Nonetheless if the benefits of such immunotherapy can be proved, its local effects would be warranted. Local responses to the ICRC bacillus and *Mycobacterium W* are reported to be similar to those following BCG plus *M. leprae*.
The responses following 7 days and 1 year after injection of *M. vaccae* into our patients are shown in table 5c. It can also be seen from

5c	Local reaction to immunotherapy and Leprosin A conversion			
	Immunotherapy successful Leprosin A positive			Immunotherapy failed Leprosin A negative
10^9 *M. vaccae* 7 days	12.4 ± 2.1		p < 0.03	8.1 ± 5.4 mm
scar at 1 year	4.5 ± 3.2		ns	3.2 ± 3.3 mm
10^9 *M. v.* + T 7 days	7.0 ± 2.9		p < 0.05	4.8 ± 2.9 mm
scar at 1 year	4.0 ± 3.0		ns	1.9 ± 1.9 mm

The reactions to 10^9 *M. v.* + T at 7 days are significantly smaller than those after 10^9 *M. vaccae* alone.

this table that local responses to injection at 7 days are related to success in Leprosin A conversion. The worst generalised effects of injection have been slight malaise and feverishness in a few patients during the night following, nothing like the prostration that could follow Koch's immunotherapy, or even Marianum Antigen.
It would be interesting to test our patients with Lepromin, rather than Leprosin A, since all were negative at earlier times in their dis-

ease. However, since Lepromin is a suspension of killed M. leprae, it might interfere with the results, and, therefore, has not been carried out.

7.5   Preliminary experiments in tuberculosis patients

This small study was performed on 10 patients with bacteriologically proven pulmonary tuberculosis attending the Middlesex Hospital [64]. All gave their informed consent prior to the study, and all had received at least 3 months of multiple drug therapy.

Each patient was skin tested with Tuberculin, Leprosin A, Scrofulin and Vaccin, and the character of the response to Tuberculin was qualitatively assessed to be of Koch type. Two of the patients were given an intradermal injection of saline, and 8 received an intradermal injection of 10^9 M. vaccae. One month later, the skin tests were repeated, and the type of response to Tuberculin assessed again. This was repeated on 4 of the patients a year later.

The injections of saline made no difference to the responses which remained of Koch type, whereas in not a single case was the response of this type after the injection of M. vaccae (see table 6). The local responses were no worse than those seen in leprosy patients, and no complaints of other side effects were made.

Table 6
Qualitative changes in reactions to new Tuberculin induced in tuberculosis patients by injections of saline or of 10^9 M. vaccae. Middlesex Hospital study.

	1/12 after IT	1 year after IT
	9.8 ± 2.5 (8)	19.7 ± 5.6 (3)
Before IT		
20.6 ± 5.5 K(10)		
	1/12 after saline	1 year after saline
	24.5 ± 3.5 K(2)	40.0 K(1)

K = Koch response; Numbers of patients tested are shown in parentheses.

7.6   Studies in progress

As a result of these preliminary investigations it was decided that preparations of M. vaccae were suitable for further investigation as immunotherapeutics for both leprosy and tuberculosis, and a number of studies are progressing. Early results obtained by a team in

Kuwait, lead by Dr. G. M. Bahr, have shown that 10^9 irradiated *M. vaccae* plus tuberculin plus Murabutide (MDP), or 10^9 autoclaved *M. vaccae* alone, can restore lymphocyte transformation to group i, common mycobacterial antigen. It has also been possible to show that these preparations of *M. vaccae* can specifically increase IgG levels to the 30 kDa antigen of *M. tuberculosis*. This antigen is amongst the proteins actively secreted by tubercle bacilli, and as a fibronectin binding protein [65, 66], may play an important role in the pathogenesis of tuberculosis. Although we do not know yet whether antibody to this protein has significance in immune control of tuberculosis, it is tempting to think that it may have.

A double blind study of immunotherapy given 6 weeks after the start of chemotherapy, is almost completed under Dr. T. Corrah in the Gambia. In this study 10^9 irradiated *M. vaccae* plus tuberculin is being compared with an injection of saline. Patients who do not comply with treatment and escape from control will be actively followed to their villages and relapse rates compared for the 2 groups.

Two small studies are being carried out in India to investigate the possible role of immunotherapy with *M. vaccae* in the control of leprosy reactions. The only data we have so far is that the injection does not make the patients' reaction worse. Further information is required before any beneficial effects can be assessed.

8   Conclusion

Although one of the first scientifically designed systems for treating both leprosy and tuberculosis, immunotherapy has still to prove itself. The position is very different today from that facing its originators in the 1890's. No longer does immunotherapy have to be a complete treatment, no longer does it have to be given repeatedly, and no longer does it have to be given in the presence of large bacterial and antigenic loads. As an adjunct to effective chemotherapy to shorten the course of treatment and to harness immune mechanisms to eradicate persisting bacilli that are metabolising too slowly for drugs to be effective is its potential for the future. It is difficult to see an alternative to immunotherapy in this role, and it is difficult to see what part the pure proteins of the so-called second generation vaccines are going to play in the fight against these diseases that remains so far from over.

References

1. R. Koch: An address on bacteriological research delivered before the International Medical Congress, held in Berlin, August 1980. Br. med. J. 2, 380–383 (1890).
2. R. Koch: A further communication on a remedy for tuberculosis. (translation from the german) Br. med. J. 2, 1193–1195 (1890).
3. G. A. W. Rook, J. Taverne, C. Leveton and J. Steele: The role of gamma interferon, vitamin D_3 metabolites and tumour necrosis factor in the pathogenesis of tuberculosis. Immunology 62, 229–234 (1987).
4. R. B. Parekh, R. A. Dwek, B. J. Sutton, D. L. Fernandez, A. Leung, D. R. Stanworth and T. W. Rademacher: Association of rheumatoid arthritis and primary osteoarthritis with changes in the glycosylation pattern of total serum IgG. Nature 316, 452 (1985).
5. G. A. W. Rook: Role of activated macrophages in the immunopathology of tuberculosis. Br. med. Bull. 44, 611–623 (1988).
6. Anonymous: Professor Koch's remedy for tuberculosis; (report from) Austria. Br. med. J. 2, 1490 (1890).
7. E. Filley, A. Andreoli, J. Steele, M. Waters, D. Wagner, D. Nelson, K. Tung, T. Rademacher, R. Dwek and G. A. W. Rook: A transient rise in agalactosyl IgG correlating with free interleukin 2 receptors, during episodes of erythema nodosum leprosum. Clin. exp. Immun. 72, 343–347 (1989).
8. J. L. Stanford, M. J. Shield and G. A. W. Rook: How environmental mycobacteria may predetermine the protective efficacy of BCG. Tubercle 62, 55–62 (1981).
9. R. Koch: Further Communications on a remedy for tuberculosis. Lancet 2, 1085–1086 (1890).
10. R. Koch: A further communication on a remedy for tuberculosis. (translated from the German) Br. med. J. 1, 125–127 (1891).
11. J. G. Sinclair Coghill: Sequel of a case treated by Koch's tuberculin, with the results of the necropsy. Lancet 2, 1219–1220 (1895).
12. J. L. Stanford: Immunity to mycobacterial infections – a few new ideas and a re-evaluation of some old ones. Hjärta Kärl Lungor Kvartalsskrift 77, 83–92 (1982).
13. I. C. McManus, D. N. J. Lockwood, J. L. Stanford, M. A. Shaaban, M. Abdul Ati and G. M. Bahr: Recognition of a category of responders to group ii, slow-grower associated antigens amongst senior school children using a statistical model. Tubercle 69, 275–281 (1988).
14. G. S. Dowling and E. W. Prosser-Thomas: Treatment of lupus vulgaris with calciferol. Lancet 1, 919–924 (1946).
15. L. MacAssey and C. W. Saleeby: Eds Spahlinger contra tuberculosis 1908–1934. John Bale, sons and Danielsson Ltd. London (1934).
16. H. Spahlinger: Note on the treatment of tuberculosis. Lancet 1, 5–8 (1922).
17. C. Abou Zeid, I. Smith, J. M. Grange, J. Steele and G. A. W. Rook: Subdivision of daughter strains of Bacille Calmette Guerin (BCG) according to secreted proteins patterns. J. gen. Microbiol. 132, 3047–3053 (1986)
18. D. B. Young, A. Mehlert, V. Bal P. Mendez-Samperio, J. Ivanyi and J. R. Lamb: Stress proteins and the immune response to mycobacteria – antigens as virulence factors? Antonie Van Leeuwenhoek 54, 431–439 (1988).
19. P. T. Abraham: Leprosy and Professor Koch's treatment. Letter to the Lancet 2, 1300 (1890).
20. O. B. Swai: A double-blind study of explore the role of levamisole in short-course chemotherapy for pulmonary tuberculosis. Proceedings of the XXVIth IUAT World Conference, Singapore, November 1986 (in press 1987).

21 M. Blanc, M.-T. Prost and Marie-Suzanne Soeur: Influence de l'injection d'une *Mycobacterium* isole d'un cas de lepre (souche chauvire) sur la reaction de Mitsuda. Bulletin de la Societe de Path. exot. 46, 1009–1014 (1953).

22 G. Penso: Criteri generali par determinare la posizione sistematica di un micobatterio. Symposium V, *Actinomycetales,* morphology, biology and systematics, Proceedings of the VIth International Microbiology Congress. Instituto Superiore di Sanita, Rome (1953).

23 M. Blanc, M.-T. Prost, Micheal, Lemaire, E. Kuna, J. Essele and J.-M. Nkoa: Clinical and therapeutic study of an antigen prepared with *Mycobacterium marianum* applied to 457 leprosy patients. Int. J. Leprosy 23, 23–31 (1955).

24 J. G. Tolentino: Results of six months supplementary treatment of lepromatous leprosy patients with *Mycobacterium marianum* vaccine. Int. J. Leprosy 25, 351–355 (1957).

25 A. L. Relvich: Experience with Antigen *Marianum* in the treatment of leprosy. Leprosy Rev. 28, 150–156 (1957).

26 D. W. Beckett: A trial of Antigen *Marianum* as an adjunct of DDS in the treatment of lepromatous leprosy. Leprosy Rev. 29, 209–214 (1958).

27 P. Ondoua, M. T. Prost and Sister M de la Trinite: Clinical and immunological results obtained with the *Marianum* antigen after more than ten years of therapeutic use. Leprosy Rev. 35, 297–303 (1964).

28 A. Bagalawis, E. Oh und M. Whang: Result of *Marianum* antigen in the treatment of leprosy. Leprosy Rev. 37, 51–55 (1966).

29 Medical Research Council: BCG and vole bacillus vaccines in the prevention of tuberculosis in adolescents. First report. Br. med. J. 1, 413–427 (1956).

30 Tuberculosis Prevention Trial, Madras. Trial of BCG vaccines in South India for tuberculosis prevention. Ind. J. med. Res. 70, 349–363 (1979).

31 G. A. W. Rook, G. M. Bahr and J. L. Stanford: The effect of two distinct forms of cell-mediated response to mycobacteria on the protective efficacy of BCG. Tubercle 62, 63–68 (1981).

32 J. L. Stanford, J. Terencio de las Aguas, P. Torres, B. O. Gervasioni and R. Ravioli: Studies on the effects of a potential immunotherapeutic agent in leprosy patients. Proceedings of the IVth European Leprosy Symposium, Santa Margherita Ligure, Italy, October 1986. Hlth Coop. Pap. 7, 201–206 (1987).

33 G. A. W. Rook, J. Steele, L. Fraher, S. Barker, R. Karmali, J. O'Riordan and J. L. Stanford: Vitamin D_3, gamma interferon, and control of proliferation of *Mycobacterium tuberculosis* by human monocytes. Immunology 57, 159–163 (1986).

34 J. L. Stanford and J. M. Grange: The meaning and structure of species as applied to mycobacteria. Tubercle 55, 143–152 (1974).

35 J. L. Stanford and J. K. Wong: A study of the relationship between *Nocardia* and *Mycobacterium diernhoferi* – a typical fast growing *Mycobacterium.* Br. J. exp. Path. 55, 291–295 (1974).

36 J. L. Stanford, G. A. W. Rook, J. Convit, T. Godal, G. Kronvall, R. J. W. Rees and G. P. Walsh: Preliminary taxonomic studies on the leprosy bacillus. Br. J. exp. Path. 56, 579–585 (1975).

37 C. J. Thorns and J. A. Morris: Common epitopes between mycobacterial and certain host tissue antigens. Clin. exp. Immun. 61, 323–328 (1985).

38 G. M. Bahr, G. A. W. Rook, M. Al-Saffar, J. van Emden, J. L. Stanford and K. Behbehani: Antibody levels to mycobacteria in relation to HLA type: evidence for non-HLA-linked high levels of antibody to the 65kD heat shock protein of *M. bovis* in rheumatoid arthritis. Clin. exp. Immun. 74, 211–215 (1988).

39 J. Holoshitz, F. Koning, J.E. Coligan, J. de Bruyn and S. Strober: Isolation of CD4-DC8-mycobacteria-reactive T lymphocyte clones from rheumatoid arthritis synovial fluid. Nature *339*, 226–229 (1989).
40 D. Stewart Tull: Immunologically important constituents of mycobacteria: Adjuvants. in: The biology of the mycobacteria, vol. 2. Eds C. Ratledge and J.L. Stanford, Academic Press, London (1983).
41 J.S. Beck, S.M. Morley, J.H. Gibbs, R.C. Potts, M.I. Ilias, T. Kardjito, J.M. Grange, J.L. Stanford and R.A. Brown: The cellular responses of tuberculosis and leprosy patients and of healthy controls in skin tests to 'new tuberculin' and leprosin A. Clin. exp. Immun. *64*, 484–494 (1986).
42 E.M. Lilius: Personal communication (1988).
43 E.M. Lilius: Personal communication (1988).
44 M.J. Shield and J.L. Stanford: The epidemiological evaluation, in Burma, of the skin test reagent LRA6; a cell-free extract from armadillo-derived *Mycobacterium leprae*. Part 2: Close contacts and non-contacts of bacilliferous leprosy patients. Int. J. Leprosy *50*, 446–454 (1982).
45 Medical Research Council: BCG and vole bacillus vaccines in the prevention of tuberculosis in adolescents. Third report. Br. med. J. *1*, 973–978 (1963).
46 W. van Eden, R.R. de Vries, J.L. Stanford and G.A.W. Rook: HLA-DR3 associated genetic control of response to multiple skin tests with new tuberculins. Clin exp. Immun. *52*, 287–292 (1983).
47 L.M. Bechelli, K. Lwin, P. Gallego Garbajosa, M.M. Gyi, K. Uemura, R. Sundaresan, C. Tamondong, M. Matejka, H. Sansarricq and J. Walter: BCG vaccination of children against leprosy: nine year findings of the controlled WHO trial in Burma. Bull. WHO *51*, 93–99 (1974).
48 G.A.W. Rook, J.W. Carswell, J.L. Stanford: Preliminary evidence for the trapping of antigen-specific lymphocytes in the lymphoid tissue of 'anergic' tuberculosis patients. Clin. exp. Immun. *26*, 129–132 (1976).
49 D.H. Raulet: Antigens for gamma-delta T cells. Nature *339*, 342–343 (1989).
50 K.A. Clouse, D. Powell, I. Washington, G. Poli, K. Strebel, W. Farrar, P. Barstad, J. Kovacs, A.S. Fauci and T.M. Folks: Monokine regulation of Human Immunodeficiency Virus-1 expression in a chronically infected human T cell clone. Immun. *142*, 431–438 (1989).
51 M. Ito, M. Baba, A. Sato, K. Hirabayashi, F. Tanabe, S. Shigeta and E. de Clercq: Tumour necrosis factor enhances replication of human immunodeficiency virus (HIV) *in vitro*. Biochem. Biophys. Res. Comm. *158*, 307–312 (1989).
52 T.A. Folks, K.A. Clouse, J. Justement, A. Rabson, E. Duh, J.H. Kehrl and A.S. Fauci: Proc. nat. Acad. Sci. USA *86*, 2365–2368 (1989).
53 D. Aboulafia, S.A. Miles, S.R. Saks and R.T. Mitsuyasu: Intravenous recombinant tumour necrosis factor in the treatment of AIDS-related Kaposi's sarcoma. J. AIDS *2*, 54–58 (1989).
54 J. Convit, M.E. Pinardi, G. Rodriguez Ochoa, M. Ulrich, J.L. Avila and M. Goihman: Elimination of *Mycobacterium leprae* subsequent to local *in vivo* activation of macrophages in lepromatous leprosy by other mycobacteria. Clin. exp. Immun. *17*, 261–265 (1974).
55 J. Convit, N. Ananzazu, M. Ulrich, M. Pinardi, O. Reyes and J. Alvarado: Immunotherapy with a mixture of *M.leprae* and BCG in different forms of leprosy and in Mitsuda-negative contacts. Int. J. Leprosy *50*, 415–424 (1982).
56 J.K. Onwubalili, G.M. Scott and H. Smith: Acute respiratory distress related to chemotherapy of advanced pulmonary tuberculosis: A study of two cases and review of the literature. Q. J. Med. *230*, 599–610 (1986).

57 P. M. Nye, J. L. Stanford, G. A. W. Rook, P. Lawton, M. MacGregor, C. Reily, D. Humber, P. Orege, C. R. Revankar, J. Terencio de las Aguas and P. Torres: Suppressor determinants of mycobacteria and their potential relevance to leprosy. Leprosy Rev. *57,* 147–157 (1986).

58 C. V. Bapat, K. J. Ranadive and V. Khanolkar: In vitro cultivation of an acid fast mycobacterium isolated from human lepromatous leprosy. Ind. J. Path. Bact. *1,* 156–159 (1958).

59 M. G. Deo, C. V. Bapat, V. Bhalerao, R. M. Chaturvedi, W. S. Bhatki and R. G. Chulawala: Antileprosy potentials of ICRC vaccine. A study in patients and healthy volunteers. Int. J. Leprosy *51,* 540–549 (1983).

60 S. Chaudhuri, A. Fotedar und G. P. Talwar: Lepromin conversion in repeatedly Lepromin negative BL/LL patients after immunisation with autoclaved *Mycobacterium w*. Int. J. Leprosy *51,* 159–168 (1983).

61 J. Oudar, L. Joubert and L. Valette: Repercussion de la thélite nodulaire mycobacterien sur le dépistage allergique de la tuberculose bovine. Soc. Sci. vet. med. comp., Lyon *70,* 121–126 (1968).

62 K. Shimizu, S. Masaki and T. Hirose: Studies on mycobacterial infection of cows in Hokkaido. Jap. J. vet. Sci. *43,* 13–19 (1981).

63 J. L. Stanford and R. C. Paul: A preliminary report on some studies of environmental mycobacteria. Annls Soc. Belge Med. trop. *53,* 389–393 (1973).

64 J. L. Stanford, S. Lucas, G. Cordess, G. A. W. Rook and S. Barnass: Immunotherapy of tuberculosis in mice and guinea pigs. Proceedings of the XXVIth IUAT World Conference, Singapore, November 1986 (in press, 1987).

65 P. Torres, J. L. Stanford, J. Terencio, J. Gozalbez, B. Gervasioni and R. Fernandez: The potential use of killed *Mycobacterium vaccae* in the treatment of multibacillary leprosy. Int. J. Leprosy (submitted, 1989).

66 W. van Eden, R. R. P. de Vries, N. K. Mehra, M. C. Vaidya, J. D'Amaro and J. J. van Rood: HLA segregation of tuberculoid leprosy: confirmation of the DR2 marker. J. infect. Dis. *141,* 693–701 (1980).

67 P. Stastny, E. J. Ball, M. A. Khan, N. J. Olsen, T. Pincus and X. Gao: HLA-DR4 and other genetic markers in rheumatoid arthritis. Br. J. Rheumat. *27,* (suppl. 2) 132–138 (1989).

68 G. H. Bothamley, J. S. Beck, G. M. T. Schreuder, J. D'Amaro, R. R. P. de Vries, T. Kardjito and J. Ivanyi: Association of tuberculosis and *M. tuberculosis* specific antibody levels with HLA. J. infect. Dis. (1989)

69 A. Pozniak, J. L. Stanford, G. A. W. Rook and N. McI. Johnson: A tentative introduction of immunotherapy into the treatment of tuberculosis. Proceedings of the XXVIth IUAT World Conference, Singapore, November 1986 (in press, 1987).

70 T. L. Ratliff, J. A. McGarr, C. Abou-Zeid, G. A. W. Rook, J. L. Stanford, J. Aslanzadeh and E. J. Brown: Attachment of mycobacteria to fibronectin-coated surfaces. Infect. Immun. *134,* 1307–1313 (1988).

71 C. Abou-Zeid, T. L. Ratliff, H. G. Wiker, M. Harboe, J. Bennedsen and G. A. W. Rook: Characterisation of fibronectin-binding antigens released by *M. tuberculosis* and *M. bovis* BCG. Infect. Immun. *56,* 3046–3051 (1988).

Index Vol. 33

The references of the Subject Index are given in the language of the respective contribution.
Die Stichworte des Sachregisters sind in der jeweiligen Sprache der einzelnen Beiträge aufgeführt.
Les termes repris dans la Table des matières sont donnés selon la langue dans laquelle l'ouvrage est écrit.

Abdominal pain 35
Acetaminophen 323
Acetylcholine 30, 147
ACTH 164
Actin 64
Acyclovir 110
Adenine 398
Adenylate cyclase 156, 161, 358
Adrenal cortex 19
Adrenal cortical cells 75
Adrenal medulla 19
Adrenaline 19, 152
Adrenergic receptors 151
Adriamycin 240
AIDS 111, 114, 318, 403
AIDS virus 101
AIDS, prevention of 403
Albumine 163
Allenoic acid 281
Allergic reactions 72
Allergies 401, 410
Allylamines 318
Alopecia 195
Alpha-interferon 403
Alprazolam 141
Alveolar macrophages 64
Alzheimer's disease 33
Amantadine 106
Amaurosis fugax 52
Amethopterin 179
Amino acids 20
2-Aminoglucose 319
Aminoglycosides 84
Aminopterin 179
Amphotericin B 318
Ampicillin 88
Analgesia 21
Analgesic properties 56
Analgesics 139
Anemia 327, 232, 403
Anesthetics 143
Angina 49

Angina pectoris 48
Angioplasty 50
Anorexia 327
Antazoline 145
Anthelminthics 172
Anthriscus 219
Anti-catabolic 77
Anti-FMD vaccines 118
Anti-idiotypic antibodies 115
Anti-inflammatory agents 64, 77
Anti-inflammatory properties 56
Anti-rheumatic drugs 63
Antiandrogens 300
Antianxiety agents 140
Antiarrhythmics 136
Antibacterial agents 318
Antibodies 399
Antibody formation 78
Anticancer agents 144
Anticonvulsants 141
Antiemetic activities 142
Antiemetics 138
Antiestrogens 280
Antifungal agents 318
Antiglucocorticoid 287
Antimanics 138
Antimicrobials 83
Antiplatelet drugs 51
Antiprogesterone 297
Antiprogestins 280, 284
Antipsychotics 138
Antischistosomal drugs 146
Antithemostatic properties 44
Antitumor agents 146, 165
Antitumor therapy 113
Antiviral agents 110
Antiviral vaccines 115
Anxiety 23
Apopicropodophyllotoxins 222
Arachidonate metabolism 50
Arachidonic acid 46, 64, 157, 163
Argentinian hemorrhagic fevers 110

Arginine 273, 361
Artery by-passes 51
Arthritis 44, 403
Ascites 233
Aspergillosis 318
Aspergillus spp. 320, 340
Aspirin 43, 323
Astrocytoma 239
Atherosclerosis 48, 59
Avian paramyxoviruses 101
Azoles 318, 334
Azospermia 303
Azotemia 328

B-lymphocytes 78
Bacillus subtilis 229
Basophils 73
Bepridil 371, 384
Beta-endorphin 22
Blastomyces spp. 320
Blastomycosis 338
ß-Blockers 152, 376
Blood platelets 44
Brest cancer 240
Bronchial asthma 165
Brown Fluid 424
Brown-Pearce carcinoma 218
Brucella abortus 86
Brucella spp. 85, 91
Bunyaviruses 101
Bupivacaine 143
Buspirone 140

Calcineurin 362
Calcium 72
Calmidazolium 368, 383
Calmodulin 353
Calpactins 64
Calvaria 72
Cancer 401, 403
Cancer chemotherapy 318
Candida albicans 330
Candida lusitaniae 320
Candida neoformans 330
Candida parapsilosis 332
Candida spp. 320
Candidosis 330, 339
Cardiovascular diseases 410
Catecholamines 19, 24, 152, 155
Catecholamine receptors 152
Catharactics 172
Cefamandole 87
Cefazolin 88
Cefotaxime 87
Centochroman 281
Cephalexin 87
Cerebral ischemia 49

Cerebrovascular disease 51
Chaviré antigen 422
Chlamydia species 85
Chlamydomonas 362
Chloramphenicol 86, 88
Chlormadinone 268
Chlorpromazine 148, 368
Cholesterol 320
Cholinergic agents 148
Cholinergic antagonists 148
Chondrocytes 70
Choriomeningitis viruses 101
Chromatin 248
Cibenzoline 136
Cilofungin 342
Cingestrol 269
Ciprofloxacin 86, 88
Circum-sporozoite-protein 403
Cisplatin 144, 240
Clindamycin 86, 88
Clomiphene 281
Clotting factor 403
Cocaine 11, 37
Coccidioides immitis 320
Coccidiomycosiis 339
Codeine 139
Colchicine 165, 172, 175, 179, 241, 254
Colitis 232
Colitis ulcerosa 196
Collagen 399
Collagen fibers 72
Condylomata acuminata 172
Contraceptives 267
Coronary disease 152
Cortex 30
Corticosterone 20, 75
Corticotropin 286
Cortisol 20, 286
Cotinine 16
Cotinine N-oxide 16
Coxsackievirus B4 117
Coxsackievirus 101
Cryptococcosis 330, 339
Cryptococcus neoformans 32
Cyclohexamide 68
Cyclohexane carboxylic acid 282
Cycloheximide 164
Cyclooxygenase 55, 72, 157, 163
Cyclophospamide 227
Cyclosporin A 241
Cysteine 273
Cytokines 78
Cytoplasm 272
Cytosine 398
Cytosine deaminase 332
Cytosol 163

Dapsone 422, 424
3-Deazaguanine 111
Decidual cells 293
Dehydropodophyllotoxin 172
Demethylenepodophyllotoxin 206
Demethylpodophyllotoxin 251
Dengue 101
Deoxycholic acid 319
Depression 23
Dermatophytosis 345
Desoxypicrosikkimotoxin 208
Desoxypodophyllinic acid 174, 221
Desoxypodophyllotoxin 172, 205
Dexamethasone 64, 74, 286, 287
Dextrosulpiride 138
Diabetes 78, 403
Diabetes mellitus 318
Diacylglycerol 157
Diacylglycerol lipase 157
1,2-Diamines 135
Diampromid 139
Diarrhoea 35, 193, 195
Diarylethylenes 280, 282
Diazepam 141
Dibucaine 143
Dichlorflavan 107
Diethylstilbestrol 283
Diltiazem 376, 383
Dimethyl sulfoxide 191
Dipyridamole 51, 54, 58
Diseases, genetic 400
Disoxaril 107
Diuresis 286
DNA 397, 399
Doisynolic acid 280, 283
Down's syndrome 400, 409
Doxasozin 137
Drosophila C 101

Edema 70
Ehrlich ascites 226
Ehrlich ascites tumor 192
Emesis 193
Encephalitis 116
Encephalomyelitis 101, 196, 228
Endarterectomy 55
Endocytosing cells 89
Endocytosis 85
Endometrium 271
Endonuclease 278, 402
Endothelial cells 70, 75
Enterobacteriaceae 331
Enterovirus 101
Enviroxime 112
Epidermal growth factor 403
Epipodophyllotoxins 171

Epoxygenase 157
Ergosterol 320
Ergot alkaloids 152
Erythroid precursors 72
Erythromycin 86, 88
Erythropoietin 72, 327, 403
Escherichia coli 428
Estradiol 273, 281
Estrogen 268
Estrogen receptor 273, 275
Estrogens 280
Estrone 273
Ethambutol 86, 88
Ethynodiol 269
Etidocaine 143
Etoposide 169, 176, 211, 224, 240

Fatigue 35
Fatty acids 19, 157
Felodipine 368
Fendiline 369
Fertility 268
Fibrolasts 64
Flecainide 136
Fluconazole 335
Flucytosine 318, 330, 345
Flunarizine 384
5-Fluorouracil 331
Foot-and-mouth disease
 virus 101
Fungistasis 330
Furans 294

G proteins 153
Gamma-interferon 403, 417, 429
Gastrointestinal agents 142
Gastrointestinal bleeding 44
Gene 399
Genome 400
Gentamycin 87
Geotrichum spp. 320
Gestational choriocarcinoma 240
Gestogens 268
Gestrinone 284
Glucagon 162
Glucocorticoid therapy 73
Glucocorticoids 19, 63, 153, 163, 165
Gluconeogensis 20
Glucose 19
Glucosidase 189
Glutamate dehydrogenase 377
Glutaramic acid 142
Glycogen 20
Glycogenolysis 160
Gonadal peptides 300
Gonadotropins 271, 286, 300, 301

Gossypol 302
Gout 179
Granulopoiesis 78
Growth disorders 403
Growth factor 51
Growth hormone 403
Guanidine 111
Guanine 398
Guillain-Barre syndrome 116

Haemoglobin 399
Haemophilia A 403
Haemophilus influenzae 85, 401
Haemophilus parainfluenzae 401
Headache 35
Heaf test 432
Hemagglutinin H3 105
Hemodialysis 55
Hemopoiesis 72
Hemopoietin-1 72
Hemostasis 45
Hepatic-receptors 159
Hepatitis A 101, 116
Hepatitis B 101
Hepatitis B vaccine 118, 403
Hepatitis, prevention of 403
Hepatocellular carcinoma 240
Hepatocytes 77, 161
Heroin 11
Herpes simplex 229
Hexestrol 281
Hippocampus 30
Histamine 73
Histamine release 72
Histidine 274
Histoplasma spp. 320
Histoplasmosis 338
HIV antigenes 403
Hycanthone 146
Hydroxycotinine 16
Hydroxyindoleacetic acid 24
Hydroxytamoxifen 283
Hypertension 152, 155
Hypokalemia 302, 328
Hyposthenuria 328
Hypotension 327
Hypotensives 137
Hypothalamus 75

Imipenem 87
Imipramine 148
Immune functions 165
Immune systems 165
Immunodeficiency virus (HIV) 105
Immunoglobulin 75
Immunopharmacology 73

Immunosuppression 228
Indoleamines 24
Indomethacin 71
Inflammation 70
Influenza A virus 99
Influenza virus 101, 116
Inosin monophosphate
  dehydrogenase 110
Inositol 162
Insulin 403
Interceptives 289
Interferon 165
Interferon-2 77
Interferons 112
Interleukin 164
Interleukin-2 403
Ischemic events 47
Isodoproterenol 162
Isodesoxypodophyllotoxin 223
Isoniazid 86, 88
Isoproterenol 160
Isoxazoles 294
Itraconazole 335
Juniperus silicicola 219

Kaposi's sarcoma 240, 403
Ketoconazole 335
Ketogenesis 19
Koch phenomenon 417
Kupfer cells 164

Lassa fever 110
Legionella spp. 85
Lepromatous disease 422
Lepromin 422
Leprosin A 431
Leprosy 415, 419
Leucine 273
Leukemia 174, 181, 195, 226, 227, 240, 403
Leukopenia 193, 230, 332
Leukotrienes 64
Levamisole 147, 422
Levosulpiride 138
Lewis acids 187
Lewis lung carcinoma 227
Lidocaine 136, 143
Lincomycin 86
Lipocortin 163
Lipopolysaccharides 68, 70, 79, 427
Lipotropin 287
Listeria 91
Listeria monocytogenes 85
Local anesthetics 143
Lucanthone 146
Lung cancer 240

Lupus vulgaris 420
Lymphocytes 70
Lymphomas 239
Lymphotoxin 72
Lynestrenol 269
Lyphomas 240
Lysergic acid 146
Lysine 273, 361
Lysosomal enzymes 280
Lysosomes 89

Macrophages 64, 78, 83, 84, 88, 417
Magnesium influx 156
Male contraceptives 300
Malformations, congenital 401
Mannitol 327
Mantoux test 432
Marianum Antigen 423
Mastocytoma 200, 207, 226
Measles 114, 116
Measles virus 101
Mebrofenin 146
Mechlorethamine 179
Megestrol 268
Melanoma 227
Melittin 163, 360, 370
Mendelian disorders 400
Meningitis 84, 324
Menses inducers 291
Meperidine 323
Mepivacaine 143
Mepridine 139
6-Mercaptopurine 193
Metaclopramide 323
Metalloproteinase 71
Methionine 274
Methionines 361
Methotrexate 165
Methylparaben 337
Metronidazole 86
Miconazole 335
Micrococcus pyogenes 229
Mitopodoside 193
Mitosis 184
Mitsuda reaction 422
Mongolism 400, 407
Monocytes 68, 70
Morphine 139
Moxestrol 284
Multiple sclerosis 196
Mumps 114
Muramyl-dipeptide 304
Murine hepatitis 101
Murine leukemia 192
Murray valley encephalitis 101
Mutation frequency 96

Mutation rate 96
Mutations, genetic 405
Mycobacterium leprae 419, 435
Mycobacteria spp. 85
Mycobacterium aquae 436
Mycobacterium avium 427
Mycobacterium marianum 423
Mycobacterium phlei 423
Mycobacterium scrofulaceum 423, 431
Mycobacterium smegmatis 427
Mycobacterium tuberculosis 419, 434
Mycobacterium vaccae 436
Mycosamine 319
Mycoses 317
Myelin 328
Myocardial infarction 45, 47, 48

Nadolol 376
Nafarelin 289
Nafoxidine 281
Naphthalene-sulphonamides 369
Nausea 35, 195
Neisseria gonorrhoea 85
Neisseria meningitidis 85
Neomycin 98
Neopodophyllotoxin 203, 208
Nephropathy 327
Nephrotoxicity 327
Neuroaminidase 99
Neuroblastoma 240
Neuroleptics 138
Neutrophils 70, 83
Nicotinamide 162
Nicotine 9
Nicotine N-oxide 16
Nifedipine 376, 382
Nimodipine 371
Noradrenaline 19
Norepinephrine 147
Norethisterone 269
Norgestrel 269
Nuclease 277
Nucleotides 44, 51

Obesity 401
Ofloxacin 88
Old Tuberculin 420
Oligodeoxynucleotides 113
Organ cultures 83, 89
Organ transplantation 318
Orthopramides 138
Ovarian cancer 240
Ovulation 268
Oxamniquine 147
Oxotremorine 145

Pain 70
Palpitation 35
Panencephalitis virus 101
Parathyroid hormone 72
Pefloxacin 88
Peltatin 172, 178
Peltatinic acid 198
Penicillin 87, 89
Penicillin G 88
Peptic ulcers 142
Peptide modulators 148
Peptide transmitters 148
Peptidoglycans 427
Peptidomimetic agents 148
Periodontal disease 69, 72
Peripheral vascular disease 53
Peritonitis 233
Pertusis toxin 156, 162
Phagocytes 84
Phenampromid 139
Phenothiazine derivatives 146
Phenetolamine 137
Phenylephrine 160
Phenylpyrroles 139
Phenytoin 141
Phosphatase 76
Phosphodiesterase 156
Phosphoenolpyruvate carboxykinase 76
Phospholipase 64, 70, 156, 161
Phospholipase A2 75
Phospholipids 47, 64, 322
Phosphorylase 160, 162
Phosphorylase kinase 361
Phosphorylation 159
Picornaviruses 107, 109, 112
Picropodophyllinic acid 198, 206
Picropodophyllotoxin 173, 178, 204, 221
Picrosikkimtoxin 207
Pituitary cells 75
Plasminogen 71, 403
Platelet adhesion 46
PMN Leucocytes 86
Pneumonia, tuberculous 420
Podophyllic acid 206
Podophyllin 172
Podophyllinic acid 173, 198, 203
Podophyllinic acid hydrazides 196
Podophyllol 223
Podophyllotoxin 172, 177, 204
Podophyllotoxin carbamates 217
Podophyllotoxin glucoside 169
Podophyllotoxone 221
Podophyllum peltatum L. 172, 219
Podophyllum sikkimensis 172
Podophyllum emodi Wall. 172
Podorhizol 174

Pokeweed mitogen 75
Polynucleotides 113
Polyomyelitis 114, 116
Polyovirus 101
Polyvirus 3D 98
Polypeptides 318
Polyphosphoinositides 156
Polypeptide hormones 73
Polysaccharides 427
Polysorbate 80 189
Post-ovulatory events 268
Prazosin 137, 155
Prenylamine 368, 371, 384
Preproencephalin 78
Prilocaine 143
Procainamide 136
Prochlorperazine 323, 327
Procollagenase 71, 75
Progesterone receptors 271, 275
Progesterone 293
Prokinetic activities 142
Proline 273
Promethazine 146, 148
Promonocytes 64
Promonocytic cell 74
Propanolol 138, 152, 376
Propiram 139
Propylparaben 337
Proresid 173, 191, 196
Proresipar 191
Prostacyclin 70
Prostaglandin D2 73
Prostaglandin formation 70
Prostaglandins 46, 47, 52, 64, 71, 271
Protein kinase 157, 158, 163, 164, 358
Proteins 427
Proteoglycan 69, 71
Pseudallescheria boydii 320
Pseudomonas 226
Psoriasis 196
Psychiatric disorders 401
Psychoses, endogenous 410
Pyrilamine 145
Pyrrolase 286

Quingestanol 269

RA synovia 68
Rabies 116
Raclopride 138
Receptor blockers 160
Recombinant TNF 72
Reinfarction 49
Remoxipride 138
Reovirus 101
Reserpine 137

Respiratory diseases 165
Respiratory syncytial virus 116
Restriction enzyme 277
Retrovir 111
Retroviruses 99
Rheumatic diseases 403
Rheumatoid arthritis 65, 75, 195, 196, 228
Rhinovirus 109
Ribavirin 111, 114
Ribonuclease 67, 74, 97
Ribonucleoprotein 277
Ribosyltransferases 162
Rickettsia spp. 85
Rifampin 86, 88, 89
Rimantadine 106
RNA genomes 94
RNA polymerases 98
RNA viruses 100
Rous sarcoma virus 101
Roxithromycin 86
Rubella 116

Salicylates 44
Salmonella typhi 86
Salmonella typhimurium 331
Sandimmune 176
Sarcina lutea 229
Sarcoma 181, 194, 206, 226
Schistomiasis 147
Sclerodermia 403
Scrofulin 431
Serine 158
Serine proteinase 71
Serotonin 24, 51, 146
Serum 161
Sikkimotoxin 172, 206
Silicicolin 178, 219, 223
Sindbis 101
Sindbis virus 98, 110
Skin cancers 179
Smallpox 114
Sociability 24
Sodium salicylate 44
Spermatogenesis 300
Spermidine 147
Spermine 147
Sphingosine 147
Sporothrix spp. 320
St. Louis encephalitis 101
Staphylococcal sepsis 318
Steroid hormones 270
Steroids 65, 280
Stilbenes 280
Stomatitis virus 101
Streptococcus aronson 229

Streptococcus haemolyticus 229
Streptokinase 58
Streptomyces nodosus 318
Streptomycin 88, 416
Stroke 403
Subendothelium 51
Sulfamethoxazole 86
Sulphones 422
Sulpiride 138
Surgery 318
Sweating 35
Sympatho-adrenal system 165
Synoviocytes 75

T-cells 417
T-lymphocytes 78
TAME esterase 73
Tamoxifen 281
Tay-Sachs disease 407
Teniposide 171, 211, 223, 239
Terbinafine 344
Testicular cancer 240
Testosterone 301
Tetracycline 86, 88
Thalamus 30
Threonine 158
Thrombin 46
Thrombocytopenia 232, 327
Thromboembolism 54
Thrombolytic agents 58
Thrombophlebitis 321
Thrombosis 45, 47, 48, 59
Thrombotic disease 45
Thromboxane 47, 49, 52, 57
Thymidine 165, 193, 194, 245, 251
Thymine 398
Thymolysis 286
Thymus 232
Thyroxine 153
Tobacco mosaic virus 101
Tocainide 136
Topoisomerase 175, 246
Torulopsis glabrata 330
Trans-doisynolic acid 281
Tremor 35
Triarylethylenes 280, 283
Triarylpropiones 280
Triazoles 294
Trifluoperazine 368
Trimethoprim 86
Trioxyphen 284
Tripelennamine 145
Tromboxane 64
Trophoblastic cells 293
Troponin 356, 377
Trypterigium wilfordii 302

Tryptophan pyrrolase 286
Tuberculin shock 418
Tuberculosis 415
Tubulin 179
Tumor necrosis factor 165, 403, 417
Tween 191
Tyrosin kinase 64
Tyrosine 275
Tyrosine aminotransferase 76

Uremia 55
Uridine 245
Uteroglobin 303

Vaccines 114
Vascular mortality 49
Vasoconstriction 45
Vasoconstrictors 52
Vasodilation 45
Vasodilators 70
Vasopressin 163
Vasospasm 48, 52
Venezuelan equine encephalomyelitis 10
Venous thrombosis 54
VePesid 176
Verapamil 368, 376, 383
Vesicular stomatitis virus 98
Vinblastine 165, 377

Vinca alkaloids 179, 377
Vincaleukoblastine 181, 194
Vincristine 165
Viral disease 93
Viral enzymes 111
Viral polymerases 110
Viroids 101
Virus evolution 93
Virus variation 116
Vitamin D metabolites 429
Vitamin D3 418
Vitellogenin 278
Vomiting 35, 195
Vumon 176

Walker carcinosarcoma 217, 226

Yeast Killer elements 101
Yellow fever 101
Yohimbine 137
Yoshida sarcoma 181, 192, 217

Zacopride 142
Zygomycetes 320
Zymosan 429

Index of Titles
Verzeichnis der Titel
Index des titres
Vol. 1–33 (1959–1989)

Acetylen-Verbindungen als Arzneistoffe, natürliche und synthetische
 14, 387 (1970)
Adenosine receptors: Clinical implications and biochemical mechanisms
 32, 195 (1988)
Adipose tissue, the role of in the distribution and storage of drugs
 28, 273 (1984)
β-Adrenergic blocking agents
 20, 27 (1976)
β-Adrenergic blocking agents, pharmacology and structure-activity
 10, 46 (1966)
β-Adrenergic blocking drugs, pharmacology
 15, 103 (1971)
Adrenergic receptor research, recent developments
 33, 151 (1989)
Adverse reactions of sugar polymers in animals and man
 23, 27 (1979)
Allergy, pharmacological approach
 3, 409 (1961)
Alzheimer's disease, implications of immunomodulant therapy
 32, 21 (1988)
Amebic disease, pathogenesis of
 18, 225 (1974)
Amidinstruktur in der Arzneistofforschung
 11, 356 (1968)
Amines, biogenic and drug research
 28, 9 (1984)
Amino- und Nitroderivate (aromatische), biologische Oxydation und Reduktion
 8, 195 (1965)
Aminonucleosid-Nephrose
 7, 341 (1964)

Amoebiasis, chemotherapy
 8, 11 (1965)
Amoebiasis, surgical
 18, 77 (1974)
Amoebicidal drugs, comparative evaluation of
 18, 353 (1974)
Anabolic steroids
 2, 71 (1960)
Analgesia and addiction
 5, 155 (1963)
Analgesics and their antagonists
 22, 149 (1978)
Ancylostomiasis in children, trial of bitoscanate
 19, 2 (1975)
Androgenic-anabolic steroids and glucocorticoids, interactions
 14, 139 (1970)
Anthelmintic action, mechanisms of
 19, 147 (1975)
Anthelminticaforschung, neuere Aspekte
 1, 243 (1959)
Anthelmintics, comparative efficacy
 19, 166 (1975)
Anthelmintics, laboratory methods in the screening of
 19, 48 (1975)
Anthelmintics, structure-activity
 3, 75 (1961)
Anthelmintics, human and veterinary
 17, 110 (1973)
Antiarrhythmic compounds
 12, 292 (1968)
Antiarrhythmic drugs, recent advances in electrophysiology of
 17, 34 (1973)
Antibacterial agents of the nalidixic acid type
 21, 9 (1977)

Antibiotics, structure and biogenesis
  2, 591 (1960)
Antibiotic activities, in vitro models for the study of
  31, 349 (1987)
Antibiotika, krebswirksame
  3, 451 (1961)
Antibody titres, relationship to resistance to experimental human infection
  19, 542 (1975)
Anticancer agents, metabolism of
  17, 320 (1973)
Antidiabetika, orale
  30, 281 (1986)
Antifertility substances, development
  7, 133 (1964)
Antitumor antibiotics, the chemistry of DNA modification
  32, 411 (1988)
Anti-filariasis campaign: its history and future prospects
  18, 259 (1974)
Antifungal agents
  22, 93 (1978)
Antihypertensive agents
  4, 295 (1962), 13, 101 (1969)
Antihypertensive agents
  20, 197 (1976)
Antihypertensive agents 1969–1981
  25, 9 (1981)
Anti-inflammatory agents, nonsteroid
  10, 139 (1966)
Anti-inflammatory drugs, biochemical and pharmacological properties
  8, 321 (1965)
Antikoagulantien, orale
  11, 226 (1968)
Antimalarials, 8-aminoquinolines
  28, 197 (1984)
Antimetabolites, revolution in pharmacology
  2, 613 (1960)
Antimicrobials, penetration of human cells
  33, 83 (1989)
Antituberculous compunds with special reference to the effect of combined treatment, experimental evaluation of
  18, 211 (1974)
Antiviral agents
  22, 267 (1978)
Antiviral agents
  28, 127 (1984)
Art and science of contemporary drug development
  16, 194 (1972)

Arterial pressure by drugs
  26, 353 (1982)
Arzneimittel, neue
  1, 531 (1959), 2, 251 (1960), 3, 369 (1961), 6, 347 (1963), 10, 360 (1966)
Arzneimittel, Wert und Bewertung
  10, 90 (1966)
Arzneimittelwirkung, Einfluss der Formgebung
  10, 204 (1966)
Arzneimittelwirkung, galenische Formgebung
  14, 269 (1970)
Aspirin as an antithrombotic agent
  33, 43 (1989)
Asthma, drug treatment of
  28, 111 (1984)
Atherosclerosis, cholesterol and its relation to
  1, 127 (1959)
Axoplasmic transport, pharmacology and toxicology
  28, 53 (1984)
Ayurveda
  26, 55 (1982)
Ayurvedic medicine
  15, 11 (1971)

Bacterial cell surface and antimicrobial resistance
  32, 149 (149)
Bacterial vaccines, approaches to the rational design
  32, 375 (1988)
Bacteria and phagocytic cells, surface interaction between
  32, 137 (1988)
Basic research, in the US pharmaceutical industry
  15, 204 (1971)
Benzimidazole anthelmintics chemistry and biological activity
  27, 85 (1983)
Benzodiazepine story
  22, 229 (1978)
Beta blockade in myocardial infarction
  30, 71 (1986)
Bewertung eines neuen Antibiotikums
  22, 327 (1978)
Biliary excretion of drugs and other xenobiotics
  25, 361 (1981)
Biochemical acyl hydroxylations
  16, 229 (1972)
Biological activity, stereochemical factors
  1, 455 (1959)

Biological response quantification in toxicology, pharmacology and pharmacodynamics
  21, 105 (1977)
Bitoscanate, a field trial in India
  19, 81 (1975)
Bitoscanate, clinical experience
  19, 96 (1975)
Bitoscanate, experience in the treatment of adults
  19, 90 (1975)

Calmodulin-mediated enzymatic actions, effects of drugs on
  33, 353 (1989)
Cancer chemotherapy
  8, 431 (1965), 20, 465 (1976)
Cancer chemotherapy
  25, 275 (1981)
Cancerostatic drugs
  20, 251 (1976)
Carcinogenecity testing of drugs
  29, 155 (1985)
Carcinogens, molecular geometry and mechanism of action
  4, 407 (1962)
Cardiovascular drug interactions, clinical importance of
  25, 133 (1981)
Cardiovascular drug interactions
  29, 10 (1985)
Central dopamine receptors, agents acting on
  21, 409 (1977)
Central nervous system drugs, biochemical effects
  8, 53 (1965)
Cestode infections, chemotherapy of
  24, 217 (1980)
Chemical carcinogens, metabolic activation of
  26, 143 (1982)
Chemotherapy of schistosomiasis, recent developments
  16, 11 (1972)
Cholera infection (experimental) and local immunity
  19, 471 (1975)
Cholera in Hyderabad, epidemiology of
  19, 578 (1975)
Cholera in non-endemic regions
  19, 594 (1975)
Cholera, pandemic, and bacteriology
  19, 513 (1975)
Cholera pathophysiology and therapeutics, advances
  19, 563 (1975)
Cholera, researches in India on the control and treatment of
  19, 503 (1975)
Cholera toxin induced fluid, effect of drugs on
  19, 519 (1975)
Cholera toxoid research in the United States
  19, 602 (1975)
Cholera vaccines in volunteers, antibody response to
  19, 554 (1975)
Cholera vibrios, interbiotype conversions by actions of mutagens
  19, 466 (1975)
Cholesterol, relation to atherosclerosis
  1, 127 (1959)
Cholinergic mechanism-monoamines relation in certain brain structures
  6, 334 (1972)
Cholinergic neurotransmitter system, behavioral correlates of presynaptic events
  32, 43 (1988)
Clostridium tetani, growth in vivo
  19, 384 (1975)
Communicable diseases, some often neglected factors in the control and prevention of
  18, 277 (1974)
Conformation analysis, molecular graphics
  30, 91 (1986)
Contraception
  21, 293 (1977)
Contraceptive agents, development of
  33, 261 (1989)
Convulsant drugs – relationships between structure and function
  24, 57 (1980)
Cooperative effects in drug-DNA interactions
  31, 193 (1987)
Cyclopropane compounds
  15, 227 (1971)

Deworming of preschool community in national nutrition programmes
  19, 136 (1975)
1,2-Diamine functionality, medicinal agents
  33, 135 (1989)

Diarrhoea (acute) in children, management of
   19, 527 (1975)
Diarrhoeal diseases (acute) in children
   19, 570 (1975)
3,4-Dihydroxyphenylalanine and related compounds
   9, 223 (1966)
Diphtheria, epidemiological observations in Bombay
   19, 423 (1975)
Diphtheria, epidemiology of
   19, 336 (1975)
Diphtheria in Bombay
   19, 277 (1975)
Diphtheria in Bombay, age profile of
   19, 417 (1975)
Diphtheria in Bombay, studies on
   19, 241 (1975)
Diphtheria, pertussis and tetanus, clinical study
   19, 356 (1975)
Diphtheria, pertussis and tetanus vaccines
   19, 229 (1975)
Diphtheria toxin production and iron
   19, 283 (1975)
Disease control in Asia and Africa, implementation of
   18, 43 (1974)
Disease-modifying antirheumatic drugs, recent developments in
   24, 101 (1980)
Diuretics
   2, 9 (1960)
DNA technology, significance in medicine
   33, 397 (1989)
Dopamine agonists, structure-activity relationships
   29, 303 (1985)
Drug action and assay by microbial kinetics
   15, 271 (1971)
Drug action, basic mechanisms
   7, 11 (1964)
Drug combination, reduction of drug action
   14, 11 (1970)
Drug discovery, organizing for
   32, 329 (1988)
Drug discovery, serendipity and structural research
   30, 189 (1986)
Drug in biological cells
   20, 261 (1976)
Drug latentiation
   4, 221 (1962)
Drug-macromolecular interactions, implications for pharmacological activity
   14, 59 (1970)
Drug metabolism
   13, 136 (1969)
Drug metabolism (microsomal), enhancement and inhibition of
   17, 12 (1973)
Drug-metabolizing enzymes, perinatal development of
   25, 189 (1981)
Drug potency
   15, 123 (1971)
Drug research
   10, 11 (1966)
Drug research and development
   20, 159 (1976)
Drugs, biliary excretion and enterohepatic circulation
   9, 299 (1966)
Drugs, structures, properties and disposition of
   29, 67 (1985)

Egg-white, reactivity of rat and man
   13, 340 (1969)
Endocrinology, twenty years of research
   12, 137 (1968)
Endotoxin and the pathogenesis of fever
   19, 402 (1975)
Enterobacterial infections, chemotherapy of
   12, 370 (1968)
Enzyme inhibitors of the renin-angiotensin system
   31, 161 (1987)
Estrogens, oral contraceptives and breast cancer
   25, 159 (1981)
Excitation and depression
   26, 225 (1982)
Experimental biologist and medical scientist in the pharmaceutical industry
   24, 83 (1980)

Fifteen years of structural-modifications in the field of antifungal monocyclic 1-substituted 1H-azoles
   27, 253 (1983)
Filarial infection, immuno-diagnosis
   19, 128 (1975)

Filariasis, chemotherapy
  9, 191 (1966)
Filariasis in India
  18, 173 (1974)
Filariasis, in four villages near Bombay, epidemiological and biochemical studies in
  18, 269 (1974)
Filariasis, malaria and leprosy, new perspectives on the chemotherapy of
  18, 99 (1974)
Fluor, dérivés organiques d'intérêt pharmacologique
  3, 9 (1961)
Fundamental structures in drug research Part I
  20, 385 (1976)
Fundamental structures in drug research Part II
  22, 27 (1978)
Further developments in research on the chemistry and pharmacology of synthetic quinuclidine derivatives
  27, 9 (1983)

GABA-Drug Interactions
  31., 223 (1987)
Galenische Formgebung und Arzneimittelwirkung
  10, 204 (1966), 14, 269 (1970)
Ganglienblocker
  2, 297 (1960)
Glucocorticoids: anti-inflammatory and immuno-suppressive effects
  33, 63 (1989)

Heilmittel, Entwicklung
  10, 33 (1966)
Helminthiasis (intestinal), chemotherapy of
  19, 158 (1975)
Helminth infections, progress in the experimental chemotherapy of
  17, 241 (1973)
Helminthic infections, immunodiagnosis of
  19, 119 (1975)
Helminth parasites, treatment and control
  30, 473 (1986)
High resolution nuclear magnetic resonance spectroscopy of biological samples as an aid to drug development
  31, 427 (1987)

Homologous series, pharmacology
  7, 305 (1964)
Hookworm anaemia and intestinal malabsorption
  19, 108 (1975)
Hookworm disease and trichuriasis, experience with bitoscanate
  19, 23 (1975)
Hookworm disease, bitoscanate in the treatment of children with
  19, 6 (1975)
Hookworm disease, comparative study of drugs
  19, 70 (1975)
Hookworm disease, effect on the structure and function of the small bowel
  19, 44 (1975)
Hookworm infection, a comparative study of drugs
  19, 86 (1975)
Hookworm infections, chemotherapy of
  26, 9 (1982)
Human sleep
  22, 355 (1978)
Hydatid disease
  19, 75 (1975)
Hydrocortisone, effects of structural alteration on the antiinflammatory properties
  5, 11 (1963)
5-Hydroxytryptamine and related indolealkylamines
  3, 151 (1961)
5-Hydroxytryptamine receptor agonists and antagonists
  30, 365 (1986)
Hypertension and brain neurotransmitters
  30, 127 (1986)
Hypertension, recent advances in drugs against
  29, 215 (1985)
Hypertension: Relating drug therapy to pathogenic mechanisms
  32, 175 (1988)
Hypolipidemic agents
  13, 217 (1969)

Immune system, the pharmacology of
  28, 83 (1984)
Immunization, host factors in the response to
  19, 263 (1975)
Immunization of a village, a new approach to herd immunity
  19, 252 (1975)

Immunization, progress in
  19, 274 (1975)
Immunology
  20, 573 (1976)
Immunology in drug research
  28, 233 (1984)
Immunostimulation with peptidoglycan or its synthetic derivatives
  32, 305 (1988)
Immunopharmacology and brain disorders
  30, 345 (1986)
Immunosuppression agents, procedures, speculations and prognosis
  16, 67 (1972)
Immunotherapy for leprosy and tuberculosis
  33, 415 (1989)
Impact of natural product research on drug discovery
  23, 51 (1979)
Indole compounds
  6, 75 (1963)
Indolstruktur, in Medizin und Biologie
  2, 227 (1960)
Industrial drug research
  20, 143 (1976)
Influenza virus, functional significance of the various components of
  18, 253 (1974)
Interaction of drug research
  20, 181 (1976)
Intestinal nematodes, chemotherapy of
  16, 157 (1972)
Ion and water transport in renal tubular cells
  26, 87 (1982)
Ionenaustauscher, Anwendung in Pharmazie und Medizin
  1, 11 (1959)
Isotope, Anwendung in der pharmazeutischen Forschung
  7, 59 (1964)

Ketoconazole, a new step in the management of fungal disease
  27, 63 (1983)

Leishmaniases
  18, 289 (1974)
Leprosy, some neuropathologic and cellular aspects of
  18, 53 (1974)
Leprosy in the Indian context, some practical problems of the epidemiology of
  18, 25 (1974)
Leprosy, malaria and filariasis, new perspectives on the chemotherapy of
  18, 99 (1974)
Levamisole
  20, 347 (1976)
Light and dark as a "drug"
  31, 383 (1987)
Lipophilicity and drug activity
  23, 97 (1979)
Lokalanästhetika, Konstitution und Wirksamkeit
  4, 353 (1962)
Lysostaphin: model for a specific enzymatic approach to infectious disease
  16, 309 (1972)

Malaria, advances in chemotherapy
  30, 221 (1986)
Malaria chemotherapy, repository antimalarial drugs
  13, 170 (1969)
Malaria chemotherapy, antibiotics in
  26, 167 (1982)
Malaria, eradication in India, problems of
  18, 245 (1974)
Malaria, filariasis and leprosy, new perspectives on the chemotherapy of
  18, 99 (1974)
Mast cell secretion, drug inhibition of
  29, 277 (1985)
Mass spectrometry in pharmaceutical research, recent applications of
  18, 399 (1974)
Mechanism of action of anxiolytic drugs
  31, 315 (1987)
Medical practice and medical pharmaceutical research
  20, 491 (1976)
Medicinal chemistry, contribution to medicine
  12, 11 (1968)
Medicinal research: Retrospectives and perspectives
  29, 97 (1985)
Medicinal science
  20, 9 (1976)
Membrane drug receptors
  20, 323 (1976)
Mescaline, and related compounds
  11, 11 (1968)
Metabolism of drugs, enzymatic mechanisms
  6, 11 (1963)

Metabolism (oxidative) of drugs and
other foreign compounds
   17, 488 (1973)
Metronidazol-therapie, Trichomonasis
   9, 361 (1966)
Molecular pharmacology
   20, 101 (1976)
Molecular pharmacology, basis for drug design
   10, 429 (1966)
Monitoring adverse reactions to drugs
   21, 231 (1977)
Monoaminoxydase-Hemmer
   2, 417 (1960)
Monoamine oxidase, inhibitors of
   30, 205 (1986)
Mycoses, chemotherapy for
   33, 317 (1989)

Narcotic antagonists
   8, 261 (1965), 20, 45 (1976)
Necator americanus infection, clinical field trial of bitoscanate
   19, 64 (1975)
Nematoide infections (intestinal) in Latin America
   19, 28 (1975)
Nicotine: an addictive substance or a therapeutic agent
   33, 9 (1989)
Nitroimidazoles as chemotherapeutic agents
   27, 163 (1983)
Noise analysis and channels at the postsynaptic membrane of skeletal muscle
   24, 9 (1980)

Ophthalmic drug preparations, methods for elucidating bioavailability mechanisms of
   25. 421 (1981)

Parasitic infections in man, recent advances in the treatment of
   18, 191 (1974)
Parasitosis (intestinal), analysis of symptoms and signs
   19, 10 (1975)
Pertussis agglutinins and complement fixing antibodies in whooping cough
   19, 178 (1975)
Pertussis, diphtheria and tetanus, clinical study
   19, 356 (1975)
Pertussis, diphtheria and tetanus vaccines
   19, 229 (1975)
Pertussis, epidemiology of
   19, 257 (1975)
Pertussis vaccine
   19, 341 (1975)
Pertussis vaccine composition
   19, 347 (1975)
Pharmacology of the brain: the hippocampus, learning and seizures
   16, 211 (1972)
The pharmacology of caffeine
   31, 273 (1987)
Phenothiazine und Azaphenothiazine
   5, 269 (1963)
Photochemistry of drugs
   11, 48 (1968)
Podophyllotoxin glucoside – etoposide
   33, 169 (1989)
Pyrimidinones as biodynamic agents
   31, 127 (1987)
Placeboproblem
   1, 279 (1959)
Platelets and atherosclerosis
   29, 49 (1985)
Progesterone receptor binding of steroidal and nonsteroidal compounds
   30, 151 (1986)
Propellants, toxicity of
   18, 365 (1974)
Prostaglandins
   17, 410 (1973)
Protozoan and helminth parasites
   20, 433 (1976)
Psychopharmaka, Anwendung in der psychosomatischen Medizin
   10, 530 (1966)
Psychopharmaka, strukturelle Betrachtungen
   9, 129 (1966)
Psychosomatische Medizin, Anwendung von Psychopharmaka
   10, 530 (1966)
Psychotomimetic agents
   15, 68 (1971)
Pyrimidinones as biodynamic agents
   31, 127 (1987)

Quaternary ammonium salts, chemical nature and pharmacological actions
   2, 135 (1960)
Quaternary ammonium salts – advances in chemistry and pharmacology since 1960
   24, 267 (1980)
Quinazoline derivatives
   26, 259 (1982)

Quinazolones, biological activity
  14, 218 (1970)
Quinolones
  31, 243 (1987)
Quinuclidine derivatives, chemical structure and pharmacological acitivity
  13, 293 (1969)

Red blood cell membrane, as a model for targets of drug action
  17, 59 (1973)
Renin-angiotensin system
  26, 207 (1982); *31*, 161 (1987)
Reproduction in women, pharmacological control
  12, 47 (1968)
Research, preparing the ground: importance of data
  18, 239 (1974)
Rheumatherapie, Synopsis
  12, 165 (1968)
Ribonucleotide reductase inhibitors as anticancer and antiviral agents
  31, 101 (1987)
Risk assessment problems in chemical oncogenesis
  31, 257 (1987)
RNA virus evolution and the control of viral disease
  33, 93 (1989)

Schistosomiasis, recent progress in the chemotherapy of
  18, 15 (1974)
Schwefelverbindungen, therapeutisch verwendbare
  4, 9 (1962)
Shock, medical interpretation
  14, 196 (1970)
Serum electrolyte abnormalities caused by drugs
  30, 9 (1986)
Social pharmacology
  22, 9 (1978)
Spectrofluorometry, physicochemical methods in pharmaceutical chemistry
  6, 151 (1963)
Stereoselective drug metabolism and its significance in drug research
  32, 249 (1988)
Stoffwechsel von Arzneimitteln, Ursache von Wirkung, Nebenwirkung und Toxizität
  15, 147 (1971)

Strahlenempfindlichkeit von Säugetieren, Beeinflussung durch chemische Substanzen
  9, 11 (1966)
Structure-activity relationships
  23, 199 (1979)
Substruktur der Proteine, tabellarische Zusammenstellung
  16, 364 (1972)
Sulfonamide research
  12, 389 (1968)

T-cell factors, antigen-specific and drug research
  32, 9 (1988)
Teratogenic hazards, advances in prescreening
  29, 121 (1985)
Terpenoids, biological activity
  6, 279 (1963), *13*, 11 (1969)
Tetanus and its prevention
  19, 391 (1975)
Tetanus, autonomic dysfunction as a problem in the treatment of
  19, 245 (1975)
Tetanus, cephalic
  19, 443 (1975)
Tetanus, cholinesterase restoring therapy
  19, 329 (1975)
Tetanus, diphtheria and pertussis, clinical study
  19, 356 (1975)
Tetanus, general and pathophysiological aspects
  19, 314 (1975)
Tetanus in children
  19, 209 (1975)
Tetanus in Punjab and the role of muscle relaxants
  19, 288 (1975)
Tetanus, mode of death
  19, 439 (1975)
Tetanus neonatorum
  19, 189 (1975)
Tetanus, pertussis and diphtheria vaccines
  19, 229 (1975)
Tetanus, present data on the pathogenesis of
  19, 301 (1975)
Tetanus, role of beta-adrenergic blocking drug propranolol
  19, 361 (1975)
Tetanus, situational clinical trials and therapeutics
  19, 367 (1975)

Tetanus, therapeutic measurement
  19, 323 (1975)
Tetracyclines
  17, 210 (1973)
Tetrahydroisoquinolines and
β-carbolines
  29, 415 (1985)
Thymoleptika, Biochemie und
Pharmakologie
  11, 121 (1968)
Toxoplasmosis
  18, 205 (1974)
Treatment of helminth diseases –
challenges and achievements
  31, 9 (1987)
Trichomonasis, Metronidazol-Therapie
  9, 361 (1966)
Trichuriasis and hookworm disease in
Mexico, experience with bitoscanate
  19, 23 (1975)
Tropical diseases, chemotherapy of
  26, 343 (1982)
Tropical medicine, teaching
  18, 35 (1974)
Tuberculosis in rural areas of
Maharashtra, profile of
  18, 91 (1974)

Tuberkulose, antibakterielle
Chemotherapie
  7, 193 (1964)
Tumor promoters and antitumor agents
  23, 63 (1979)

Unsolved problems with vaccines
  23, 9 (1979)

Vaccines, controlled field trials of
  19, 481 (1975)
Vibrio cholerae, cell-wall antigens of
  19, 612 (1975)
Vibrio cholerae, recent studies on genetic
recombination
  19, 460 (1975)
Vibrio cholerae, virulence-enhancing
effect of ferric ammonium citrate on
  19, 564 (1975)
Vibrio parahaemolyticus in Bombay
  19, 586 (1975)
Vibrio parahaemolyticus infection in
Calcutta
  19, 490 (1975)

Wurmkrankheiten, Chemotherapie
  1, 159 (1959)

Author and Paper Index
Autoren- und Artikelindex
Index des auteurs et des articles
Vol. 1–33 (1959–1989)

Petrussis agglutinins and complement fixing antibodies in whooping cough *19,* 178 (1975)	Dr. K. C. Agarwal Dr. M. Ray Dr. N. L. Chitkara Department of Microbiology, Postgraduate Institute of Medical Education and Research, Chandigarh, India
Pharmacology of clinically useful beta-adrenergic blocking drugs *15,* 103 (1971)	Prof. Dr. R. P. Ahlquist Professor of Pharmacology, School of Medicine, Medical College of Georgia, Augusta, Georgia, USA Dr. A. M. Karow, Jr. Assistant Professor of Pharmacology, School of Medicine, Medical College of Georgia, Augusta, Georgia, USA Dr. M. W. Riley Assistant Professor of Pharmacology, School of Medicine, Medical College of Georgia, Augusta, Georgia, USA
Adrenergic beta blocking agents *20,* 27 (1976)	Prof. Dr. R. P. Ahlquist Professor of Pharmacology, Medical College of Georgia, Augusta, Georgia, USA
Trial of a new anthelmintic (bitoscanate) in ankylostomiasis in children *19,* 2 (1975)	Dr. S. H. Ahmed Dr. S. Vaishnava Department of Paediatrics, Safdarjung Hospital, New Delhi, India
Development of antibacterial agents of the nalidixic acid type *21,* 9 (1977)	Dr. R. Albrecht Department of Drug Research, Schering AG, Berlin
The mode of action of anti-rheumatic drugs. 1. Anti-inflammatory and immunosuppressive effects of glucocorticoids *33,* 63 (1989)	Anthony C. Allison and Simon W. Lee Syntex Research, 3401 Hillview Avenue, Palo Alto, CA 94002, USA

Biological activity in the quinazolone series *14*, 218 (1970)	Dr. A. H. Amin Director of Research, Alembic Chemical Works Co. Ltd., Alembic Road, Baroda 3, India Dr. D. R. Mehta Dr. S. S. Samarth Research Division, Alembic Chemical Works Co. Ltd., Alembic Road, Baroda 3, India
The pharmacology of caffeine *31*, 273 (1987)	M. J. Arnaud Nestec Ltd. Nestlé Research Centre, Vers-chez-les-Blanc, CH-1000 Lausanne 26, Switzerland, Postal address: P.O.Box 353, CH-1800 Vevey, Switzerland
Enhancement and inhibition of microsomal drug metabolism *17*, 11 (1973)	Prof. Dr. M. W. Anders Department of Pharmacology, University of Minnesota, Minneapolis, Minnesota, USA
Reactivity of rat and man to egg-white *13*, 340 (1969)	Dr. S. I. Ankier Allen & Hanburys Ltd., Research Division, Ware, Hertfordshire, England
Enzyme inhibitors of the renin-angiotensin system *31*, 161 (1987)	Michael J. Antonaccio, Ph. D. John J. Wright, Ph. D. Bristol Myers Company, Pharmaceutical Research and Development Division, 5 Research Parkway, Wallingford, CT 06492, USA
Narcotic antagonists *8*, 261 (1965)	Dr. S. Archer Assistant Director of Chemical Research, Sterling-Winthrop Research Institute, Rensselaer, New York, USA Dr. L. S. Harris Section Head in Pharmacology, Sterling-Winthrop Research Institute, Rensselaer, New York, USA
Recent developments in the chemotherapy of schistosomiasis *16*, 11 (1972)	Dr. S. Archer Associate Director of Research, Sterling-Winthrop Research Institute, Rensselaer, New York, USA Dr. A. Yarinsky Sterling-Winthrop Research Institute, Rensselaer, New York, USA
Recent progress in the chemotherapy of schistosomiasis *18*, 15 (1974)	Prof. Dr. S. Archer Professor of Medicinal Chemistry, School of Science, Department of Chemistry, Rensselaer Polytechnic Institute, Troy, N. Y. 12181, USA

Recent progress in research on narcotic antagonists 20, 45 (1976)	Prof. Dr. S. Archer Professor of Medicinal Chemistry, School of Science, Department of Chemistry, Rensselaer Polytechnic Institute, Troy, New York, USA Dr. W. F. Michne Sterling-Winthrop Research Institute, Rensselaer, New York, USA
Molecular geometry and mechanism of action of chemical carcinogens 4, 407 (1962)	Prof. Dr. J. C. Arcos Department of Medicine and Biochemistry, Tulane University, U. S. Public Health Service, New Orleans, Louisiana, USA
Molecular pharmacology, a basis for drug design 10, 429 (1966) Reduction of drug action by drug combination 14, 11 (1970)	Prof. Dr. E. J. Ariëns Institute of Pharmacology, University of Nijmegen, Nijmegen, The Netherlands
Stereoselectivity and affinity in molecular pharmacology 20, 101 (1976)	Prof. Dr. E. J. Ariëns Dr. J. F. Rodrigues de Miranda Pharmacological Institute, University of Nijmegen, Nijmegen, The Netherlands Prof. Dr. P. A. Lehmann F. Departamento de Farmacologia y Toxicologia, Centro de Investigación y Estudios Avanzados, Instituto Politécnico Nacional, México D. F., México
Recent advances in central 5-hydroxytryptamine receptor agonists and antagonists 30, 365 (1986)	Lars-Erik Arvidsson Uli Hacksell Department of Organic Pharmaceutical Chemistry, Uppsala Biomedical Center, University of Uppsala, S-751 23, Uppsala, Sweden Richard A. Glennon Department of Medicinal Chemistry, Virginia Commonwealth University, Richmond, Box 581 MCV Station, Virginia 23298, USA
Drugs affecting the renin-angiotensin system 26, 207 (1982)	Dr. R. W. Ashworth Pharmaceuticals Division, Ciba-Geigy Corporation, Summit, New Jersey, USA
Tetanus neonatorum 19, 189 (1975) Tetanus in children 19, 209 (1975)	Dr. V. B. Athavale Dr. P. N. Pai Dr. A. Fernandez Dr. P. N. Patnekar Dr. Y. S. Acharya Department of Pediatrics, L. T. M. G. Hospital, Sion, Bombay 22, India

Toxicity of propellants *18*, 365 (1974)	Prof. Dr. D. M. Aviado Professor of Pharmacology, Department of Pharmacology, School of Medicine, University of Pennsylvania, Philadelphia, USA
Neuere Aspekte der chemischen Anthelminticaforschung *1*, 243 (1959)	Dr. J. Bally Wissenschaftlicher Mitarbeiter der Sandoz AG, Basel, Schweiz
Problems in preparation, testing and use of diphtheria, pertussis and tetanus vaccines *19*, 229 (1975)	Dr. D. D. Banker Chief Bacteriologist, Glaxo Laboratories (India) Ltd., Bombay 25, India
Recent advances in electrophysiology of antiarrhythmic drugs *17*, 33 (1973)	Prof. Dr. A. L. Bassett and Dr. A. L. Wit College of Physicians and Surgeons of Columbia University, Department of Pharmacology, New York, N. Y., USA
Stereochemical factors in biological activity *1*, 455 (1959)	Prof. Dr. A. H. Beckett Head of School of Pharmacy, Chelsea College of Science and Technology, Chelsea, London, England
Industrial research in the quest for new medicines *20*, 143 (1976) The experimental biologist and the medical scientist in the pharmaceutical industry *24*, 83 (1980)	Dr. B. Berde Head of Pharmaceutical Research and Development, Sandoz Ltd., Basle, Switzerland
Newer diuretics *2*, 9 (1960)	Dr. K. H. Beyer, Jr. Vice-President, Merck Sharp and Dohme Research Laboratories, West Point, Pennsylvania, USA Dr. J. E. Bear Director of Pharmacological Chemistry, Merck Institute für Therapeutic Research, West Point, Pennsylvania, USA
Recent developments in 8-aminoquinoline antimalarials *28*, 197 (1984)	Dr. A. P. Bhaduri, Scientist B. K. Bhat, M. Seth, Central Drug Research Institute, Lucknow, 226001 India
Studies on diphtheria in Bombay *19*, 241 (1975)	M. Bhaindarkar Y. S. Nimbkar Haffkine Institute, Parel, Bombay 12, India
Bitoscanate in children with hookworm disease *19*, 6 (1975)	Dr. B. Bhandari Dr. L. N. Shrimali Department of Child Health, R. N. T. Medical College, Udaipur, India

Recent studies on genetic recombination in *Vibrio cholerae* 19, 460 (1975)	Dr. K. Bhaskaran Central Drug Research Institute, Lucknow, India
Interbiotype conversion of cholera vibrios by action of mutagens 19, 466 (1975)	Dr. P. Bhattacharya Dr. S. Ray WHO International Vibrio Reference Centre, Cholera Research Centre, Calcutta 25, India
Experience with bitoscanate in hookworm disease and trichuriasis in Mexico 19, 23 (1975)	Prof. Dr. F. Biagi Departamento de Parasitología, Facultad de Medicina, Universidad Nacional Autónoma de Mexico, Mexico
Analysis of symptoms and signs related with intestinal parasitosis in 5,215 cases 19, 10 (1975)	Prof. Dr. F. Biagi Dr. R. López Dr. J. Viso Departamento de Parasitología, Facultad de Medicina, Universidad Nacional Autónoma de Mexico, Mexico
Untersuchungen zur Biochemie und Pharmakologie der Thymoleptika 11, 121 (1968)	Dr. M. H. Bickel Privatdozent, Medizinisch-Chemisches Institut der Universität Bern, Schweiz
The role of adipose tissue in the distribution and storage of drugs 28, 273 (1984)	Prof. Dr. M. H. Bickel Universität Bern, Pharmakologisches Institut, 3008 Bern, Schweiz
The β-adrenergic blocking agents, pharmacology, and structure-activity relationships 10, 46 (1966)	Dr. J. H. Biel Vice-President, Research and Development, Aldrich Chemical Company Inc., Milwaukee, Wisconsin, USA Dr. B. K. B. Lum Department of Pharmacology, Marquette University School of Medicine, Milwaukee, Wisconsin, USA
Prostaglandins 17, 410 (1973)	Dr. J. S. Bindra and Dr. R. Bindra Medical Research Laboratories, Pfizer Inc., Groton, Connecticut, USA
In vitro models for the study of Antibiotic Activities 31, 349 (1987)	J. Blaser Medizinische Poliklinik, Departement für Innere Medizin, Universitätsspital, Rämistr. 100, CH-8091 Zurich, Switzerland S. H. Zinner Department of Medicine, Brown University, Roger Williams General Hospital, Providence, Rhode Island 02908, USA
The red blood cell membrane as a model for targets of drug action 17, 59 (1973)	Prof. Dr. L. Bolis Università degli Studi di Roma, Istituto di Fisiologia Generale, Roma, Italia

Epidemiology and public health. Importance of intestinal nematode infections in Latin America *19*, 28 (1975)	Prof. Dr. D. Botero R. School of Medicine, University of Antioquia, Medellin, Colombia
Clinical importance of cardiovascular drug interactions *25*, 133 (1981)	Dr. D. C. Brater Division of Clinical Pharmacology, Departments of Pharmacology and Internal Medicine, The University of Texas, Health Science Center at Dallas, 5323 Harry Hines Boulevard, Dallas, Texas, USA
Update of cardiovascular drug interactions *29*, 9 (1985)	D. Craig Brater, M. D. Michael R. Vasko, Ph. D. Departments of Pharmacology and Internal Medicine, The University of Texas Health Science Center at Dallas and Veterans Administration Medical Center, 4500 Lancaster Road, Dallas, TX 75216
Serum electrolyte abnormalities caused by drugs *30*, 9 (1986)	D. Craig Brater, M. D. Departments of Pharmacology and Internal Medicine, The University of Texas Health Science Center, 5323 Harry Hines Boulevard, Dallas, Texas 75235, USA
Some practical problems of the epidemiology of leprosy in the indian context *18*, 25 (1974)	Dr. S. G. Browne Director, Leprosy Study Centre, 57a Wimpole Street, London, England
Brain neurotransmitters and the development and maintenance of experimental hypertension *30*, 127 (1986)	Jerry J. Buccafusco, Ph. D. Department of Pharmacology and Toxicology, and Psychiatry, Medical College of Georgia and Veterans Administration Medical Center, Augusta, Georgia 30912, USA Henry E. Brezenoff, Ph. D. Department of Pharmacology, University of Medicine and Dentistry of New Jersey, Newark, New Jersey 07103, USA
Die Ionenaustauscher und ihre Anwendung in der Pharmazie und Medizin *1*, 11 (1959) Wert und Bewertung der Arzneimittel *10*, 90 (1966)	Prof. Dr. J. Büchi Direktor des Pharmazeutischen Institutes der ETH, Zürich, Schweiz
Cyclopropane compounds of biological interest *15*, 227 (1971) The state of medicinal science *20*, 9 (1976)	Prof. Dr. A. Burger Professor Emeritus, University of Virginia, Charlottesville, Virginia, USA
Human and veterinary anthelmintics (1965–1971) *17*, 108 (1973)	Dr. R. B. Burrows Mount Holly, New Jersey, USA

The antibody basis of local immunity to experimental cholera infection in the rabbit ileal loop *19*, 471 (1975)	Dr. W. Burrows Dr. J. Kaur University of Chicago, P.O.B. 455, Cobden, Illinois, USA
Les dérivés organiques du fluor d'intérêt pharmacologique *3*, 9 (1961)	Prof. Dr. N. P. Buu-Hoï Directeur de Laboratoire à l'Institut de chimie des substances naturelles du Centre National de la Recherche Scientifique, Gif-sur-Yvette, France
Teaching tropical medicine *18*, 35 (1974)	Prof. Dr. K. M. Cahill Tropical Disease Center, 100 East 77th Street, New York City 10021, N.Y., USA
Anabolic steroids *2*, 71 (1960)	Prof. Dr. B. Camerino Director of the Chemical Research Laboratory of Farmitalia, Milan, Italy Prof. Dr. G. Sala Department of Clinical Chemistry and Director of the Department of Pharmaceutical Therapy, Farmitalia, Milan, Italy
Immunosuppression agents, procedures, speculations and prognosis *16*, 67 (1972)	Dr. G. W. Camiener Research Laboratories, The Upjohn Company, Kalamazoo, Michigan, USA Dr. W. J. Wechter Research Head, Hypersensitivity Diseases Research, The Upjohn Company, Kalamazoo, Michigan, USA
Dopamine agonists: Structure-activity relationships *29*, 303 (1985)	Joseph G. Cannon The University of Iowa, Iowa City, Iowa 52242, USA
Analgesics and their antagonists: recent developments *22*, 149 (1978)	Dr. A. F. Casy Norfolk and Norwich Hospital and University of East Anglia, Norwich, Norfolk, England
Chemical nature and pharmacological actions of quaternary ammonium salts *2*, 135 (1960)	Prof. Dr. C. J. Cavallito Professor, Medicinal Chemistry, School of Pharmacy, University of North Carolina, Chapel Hill, North Carolina, USA Dr. A. P. Gray Director of the Chemical Research Section, Neisler Laboratories Inc., Decatur, Illinois, USA
Contributions of medicinal chemistry to medicine – from 1935 *12*, 11 (1968) Quaternary ammonium salts – advances in chemistry and pharmacology since 1960 *24*, 267 (1980)	Prof. Dr. C. J. Cavallito Professor, Medicinal Chemistry, School of Pharmacy, University of North Carolina, Chapel Hill, North Carolina, USA

Changing influences on goals and incentives in drug research and development 20, 159 (1976)	Prof. Dr. C. J. Cavallito Ayerst Laboratories, Inc., New York, N. Y., USA
Über Vorkommen und Bedeutung der Indolstruktur in der Medizin und Biologie 2, 227 (1960)	Dr. A. Cerletti Direktor der medizinisch-biologischen Forschungsabteilung der Sandoz AG, Basel, Schweiz
Cholesterol and its relation to atherosclerosis 1, 127 (1959)	Prof. Dr. K. K. Chen Department of Pharmacology, University School of Medicine, Indianapolis, Indiana, USA Dr. Tsung-Min Lin Senior Pharmacologist, Division of Pharmacologic Research, Lilly Research Laboratories, Indianapolis, Indiana, USA
Effect of hookworm disease on the structure and function of small bowel 19, 44 (1975)	Prof. Dr. H. K. Chuttani Prof. Dr. R. C. Misra Maulana Azad Medical College & Associated Irwin and G. B. Pant Hospitals, New Delhi, India
The psychotomimetic agents 15, 68 (1971)	Dr. S. Cohen Director, Division of Narcotic Addiction and Drug Abuse, National Institute of Mental Health, Chevy Chase, Maryland, USA
Implementation of disease control in Asia and Africa 18, 43 (1974)	Prof. Dr. M. J. Colbourne Department of Preventive & Social Medicine, University of Hong Kong, Sassoon Road, Hong Kong
Structure-activity relationships in certain anthelmintics 3, 75 (1961)	Prof. Dr. J. C. Craig Department of Pharmaceutical Chemistry, University of California, San Francisco, California, USA Dr. M. E. Tate Post Doctoral Fellow, University of New South Wales, Department of Organic Chemistry, Kensington, N. S. W., Australia
Contribution of Haffkine to the concept and practice of controlled field trials of vaccines 19, 481 (1975)	Dr. B. Cvjetanovic Chief Medical Officer, Bacterial Diseases, Division of Communicable Diseases, WHO, Geneva, Switzerland
Antifungal agents 22, 93 (1978)	Prof. Dr. P. F. D'Arcy Dr. E. M. Scott Department of Pharmacy, The Queen's University of Belfast, Northern Ireland

Some neuropathologic and cellular aspects of leprosy *18*, 53 (1974)	Prof. Dr. D. K. Dastur Dr. Y. Ramamohan Dr. A. S. Dabholkar Neuropathology Unit, Grant Medical College and J. J. Group of Hospitals, Bombay 8, India
Autonomic dysfunction as a problem in the treatment of tetanus *19*, 245 (1975)	Prof. Dr. F. D. Dastur Dr. G. J. Bhat Dr. K. G. Nair Department of Medicine, Seth G. S. Medical College and K. E. M. Hospital, Bombay 12, India
Studies on *V. parahaemolyticus* infection in Calcutta as compared to cholera infection *19*, 490 (1975)	Dr. B. C. Deb Senior Research Officer, Cholera Research Centre, Calcutta, India
Biochemical effects of drugs acting on the central nervous system *8*, 53 (1965)	Dr. L. Decsi Specialist in Clinical Chemistry, University Medical School, Pécs, Hungary
Some reflections on the chemotherapy of tropical diseases: Past, present and future *26*, 343 (1982)	Dr. E. W. J. de Maar
Drug research – whence and whither *10*, 11 (1966)	Dr. R. G. Denkewalter Vice-President for Exploratory Research, Merck Sharp & Dohme Research Laboratories, Rahway, New Jersey, USA Dr. M. Tishler President, Merck Sharp & Dohme Research Laboratories, Rahway, New Jersey, USA
Serendipity and structured research in drug discovery *30*, 189 (1986)	George de Stevens Drew University, Madison, New Jersey 07940, USA
Hypolipidemic agents *13*, 217 (1969)	Dr. G. De Stevens Vice-President and Director of Research, CIBA Pharmaceutical Company, Summit, New Jersey, USA Dr. W. L. Bencze Research Department, CIBA Pharmaceutical Company, Summit, New Jersey, USA Dr. R. Hess CIBA Limited, Basle, Switzerland
The interface between drug research, marketing, management, and social, political and regulatory forces *20*, 181 (1976)	Dr. G. de Stevens Executive Vice President & Director of Research, Pharmaceuticals Division, CIBA-GEIGY Corporation, Summit, New Jersey, USA

Antihypertensive agents *20*, 197 (1976)	Dr. G. De Stevens Dr. M. Wilhelm Pharmaceuticals Division, CIBA-GEIGY Corporation, Summit, New Jersey, USA
Medicinal research: Retrospectives and Perspectives *29*, 97 (1985)	George DeStevens Department of Chemistry, Drew University, Madison, N.J., USA
RNA virus evolution and the control of viral disease *33*, 93 (1989)	Esteban Domingo Department of Biology, University of California San Diego, La Jolla, California 92093, USA
Transport and accumulation in biological cell systems interacting with drugs *20*, 261 (1976)	Dr. W. Dorst Dr. A. F. Bottse Department of Pharmacology, Vrije Universiteit, Amsterdam, The Netherlands Dr. G. M. Willems Biomedical Centre, Medical Faculty, Maastricht, The Netherlands
Immunization of a village, a new approach to herd immunity *19*, 252 (1975)	Prof. Dr. N. S. Deodhar Head of Department of Preventive and Social Medicine, B. J. Medical College, Poona, India
Surgical amoebiasis *18*, 77 (1974)	Dr. A. E. deSa Bombay Hospital, Bombay, India
Epidemiology of pertussis *19*, 257 (1975)	Dr. J. A. D'Sa Glaxo Laboratories (India) Limited, Worli, Bombay 25, India
Profiles of tuberculosis in rural areas of Maharashtra *18*, 91 (1974)	Prof. Dr. M. D. Deshmukh Honorary Director Dr. K. G. Kulkarni Deputy Director Dr. S. S. Virdi Senior Research Officer Dr. B. B. Yodh Memorial Tuberculosis Reference Laboratory and Research Centre, Bombay, India
The pharmacology of the immune system: Clinical and experimental perspectives *28*, 83 (1984)	Prof. Dr. Jürgen Drews, Director Sandoz Ltd., Pharmaceutical Research and Development, CH-4002 Basel, Switzerland
An overview of studies on estrogens, oral contraceptives and breast cancer *25*, 159 (1981)	Prof. Dr. V. A. Drill Department of Pharmacology, College of Medicine, University of Illinois at the Medical Center, Chicago, Ill. 60680, USA
Aminonucleosid-nephrose *7*, 341 (1964)	Dr. U. C. Dubach Privatdozent, Oberarzt an der Medizinischen Universitäts-Poliklinik Basel, Schweiz

Impact of researches in India on the control and treatment of cholera *19*, 503 (1975)	Dr. N. K. Dutta Director, Vaccine Institute, Baroda, India
The perinatal development of drugmetabolizing enzymes: What factors trigger their onset? *25*, 189 (1981)	Prof. Dr. G. J. Dutton Dr. J. E. A. Leakey Department of Biochemistry, The University Dundee, Dundee, DD1 4HN, Scotland
Laboratory methods in the screening of anthelmintics *19*, 48 (1975)	Dr. D. Düwel Helminthology Department, Farbwerke Hoechst AG, Frankfurt/Main 80, Federal Republic of Germany
Progress in immunization *19*, 274 (1975)	Prof. Dr. G. Edsall Department of Microbiology, London School of Hygiene and Tropical Medicine, London W.C.1, England
Host factors in the response to immunization *19*, 263 (1975)	Prof. Dr. G. Edsall Department of Microbiology, London School of Hygiene and Tropical Medicine, London, W.C.1, England M.A. Belsey World Health Organization, Geneva, Switzerland Dr. R. LeBlanc Tulane University School of Public Health and Tropical Medicine, New Orleans, La., USA L. Levine State Laboratory Institute, Boston, Mass., USA
Drug-macromolecular interactions: implications for pharmacological activity *14*, 59 (1970)	Dr. S. Ehrenpreis Associate Professor and Head Department of Pharmacology, New York Medical College, Fifth Avenue at 106th Street, New York, N.Y. 10029, USA
Betrachtungen zur Entwicklung von Heilmitteln *10*, 33 (1966)	Prof. Dr. G. Ehrhart Farbwerke Hoechst AG, Frankfurt a. M.-Höchst, BR Deutschland
Progress in malaria chemotherapy. Part 1. Repository antimalarial drugs *13*, 170 (1969) New perspectives on the chemotherapy of malaria, filariasis and leprosy *18*, 99 (1974)	Dr. E. F. Elslager Section Director, Chemistry Department, Parke, Davis & Company, Ann Arbor, Michigan, USA
Recent research in the field of 5-hydroxytryptamine and related indolealkylamines *3*, 151 (1961)	Prof. Dr. V. Erspamer Institute of Pharmacology, University of Parma, Parma, Italy

The chemistry of DNA modification by antitumor antibiotics 32, 411 (1988)	Jed F. Fisher and Paul A. Aristoff Cardiovascular Diseases Research, The Upjohn Company, Kalamazoo, MI 49 001, USA
Bacteriology at the periphery of the cholera pandemic 19, 513 (1975)	Dr. A. L. Furniss Public Health Laboratory, Maidstone, England
Iron and diphtheria toxin production 19, 283 (1975)	Dr. S. V. Gadre Dr. S. S. Rao Haffkine Institute, Bombay 12, India
Effect of drugs on cholera toxin induced fluid in adult rabbit ileal loop 19, 519 (1975)	Dr. B. B. Gaitonde Dr. P. H. Marker Dr. N. R. Rao Haffkine Institute, Bombay 12, India
Drug action and assay by microbial kinetics 15, 519 (1971) The pharmacokinetic bases of biological response quantification in toxicology, pharmacology and pharmacodynamics 21, 105 (1977)	Prof. Dr. E. R. Garrett Graduate Research Professor The J. Hillis Miller Health Center, College of Pharmacy, University of Florida, Gainesville, Florida, USA
The chemotherapy of enterobacterial infections 12, 370 (1968)	Prof. Dr. L. P. Garrod Department of Bacteriology, Royal Postgraduate Medical School, Hammersmith Hospital, London, England
The use of neutrophils, macrophages and organ cultures to assess the penetration of human cells by antimicrobials 33, 83 (1989)	Zell A. McGee, Gary L. Gorby and Wanda S. Updike The Center for Infectious Diseases, Diagnostic Microbiology and Immunology, University of Utah School of Medicine, Salt Lake City, Utah, 84132 USA
Metabolism of drugs and other foreign compounds by enzymatic mechanisms 6, 11 (1963)	Dr. J. R. Gillette Head, Section on Enzymes Drug Interaction, Laboratory of Chemical Pharmacology, National Heart Institute, Bethesda 14, Maryland, USA
Orale Antidiabetika 30, 281 (1986)	Heiner Glombik Rudi Weyer Hoechst AG, D-6230 Frankfurt (M) 80, Deutschland
The art and science of contemporary drug development 16, 194 (1972)	Dr. A. J. Gordon Associate Director, Department of Scientific Affairs, Pfizer Pharmaceuticals, 235 East 42nd Street, New York, USA Dr. S. G. Gilgore President, Pfizer Pharmaceuticals, 235 East 42nd Street, New York, USA

Basic mechanisms of drug action 7, 11 (1964) Isolation and characterization of membrane drug receptors 20, 323 (1976)	Prof. Dr. D. R. H. Gourley Department of Pharmacology, Eastern Virginia Medical School, Norfolk, Virginia, USA
Zusammenhänge zwischen Konstitution und Wirksamkeit bei Lokalanästhetica 4, 353 (1962)	Dr. H. Grasshof Forschungschemiker in Firma M. Woelm, Eschwege, Deutschland
Das Placeboproblem 1, 279 (1959)	Prof. Dr. H. Haas Leiter der Pharmakologischen Abteilung Knoll AG, Ludwigshafen, und Dozent an der Universität Heidelberg Dr. H. Fink und Dr. G. Härtefelder Forschungslaboratorien der Knoll AG, Ludwigshafen, Deutschland
Approaches to the rational design of bacterial vaccines 32, 377 (1988)	Peter Hambleton Stephen D. Prior and Andrew Robinson Public Health Labaratory Service, Centre for Applied Microbiology and Research, Porton Down, Salisbury, Wilts. SP 4 OJG, U. K.
Clinical field trial of bitoscanate in *Necator americanus* infection, South Thailand 19, 64 (1975)	Dr. T. Harinasuta Dr. D. Bunnag Faculty of Tropical Medicine, Mahidol University, Bangkok, Thailand
Pharmacological control of reproduction in women 12, 47 (1968) Contraception – retrospect and prospect 21, 293 (1977)	Prof. Dr. M.J.K. Harper The University of Texas, Health Science Center at San Antonio, San Antonio, Texas, USA
Drug latentiation 4, 221 (1962)	Prof. Dr. N. J. Harper Head of the Department of Pharmacy, University of Aston, Birmingham 4, England
Chemotherapy of filariasis 9, 191 (1966) Filariasis in India 18, 173 (1974)	Dr. F. Hawking Clinical Research Centre, Watford Road, Harrow, Middlesex, England
Recent studies in the field of indole compounds 6, 75 (1963)	Dr. R. V. Heinzelman Section Head, Organic Chemistry, The Upjohn Company, Kalamazoo, Michigan, USA Dr. J. Szmuszkovicz Research Chemist, The Upjohn Company, Kalamazoo, Michigan, USA

Neuere Entwicklungen auf dem Gebiete therapeutisch verwendbarer organischer Schwefelverbindungen 4, 9 (1962)	Dr. H. Herbst Forschungschemiker in den Farbwerken Hoechst, Frankfurt a.M., Deutschland
The management of acute diarrhea in children: an overview 19, 527 (1975)	Dr. N. Hirschhorn Consultant Physician and Staff Associate, Management Sciences for Health, One Broadway, Cambridge, Mass., USA
The tetracyclines 17, 210 (1973)	Dr. J. J. Hlavka and Dr. J. H. Booth Lederle Laboratories, Pearl River, N. Y., USA
Chemotherapy for systemic mycoses 33, 317 (1989)	Paul D. Hoeprich Section of Medical Mycology, Department of Internal Medicine, School of Medicine, University of California, Davis, CA 95616, USA
Relationship of induced antibody titres to resistance to experimental human infection 19, 542 (1975)	Dr. R. B. Hornick Dr. R. A. Cash Dr. J. P. Libonati The University of Maryland School of Medicine, Division of Infectious Diseases, Baltimore, Maryland, USA
Recent applications of mass spectrometry in pharmaceutical research 18, 399 (1974)	Mag. Sc. Chem. G. Horváth Research Chemist, Research Institute for Pharmaceutical Chemistry, Budapest, Hungary
Risk assessment problems in chemical oncogenesis 31, 257 (1987)	G. H. Hottendorf Medical University of South Carolina, Charleston, South Carolina 29425, USA
Recent developments in disease-modifying antirheumatic drugs 24, 101 (1980)	Dr. I. M. Hunneyball Research Department, Boots Co. Ltd., Pennyfoot Street, Nottingham, England
The pharmacology of homologous series 7, 305 (1964)	Dr. H. R. Ing Reader in Chemical Pharmacology, Oxford University, and Head of the Chemical Unit of the University Department of Pharmacology, Oxford, England
Progress in the experimental chemotherapy of helminth infections. Part 1. Trematode and cestode diseases 17, 241 (1973)	Dr. P. J. Islip The Wellcome Research Laboratories, Beckenham, Kent, England
Pharmacology of the brain: the hippocampus, learning and seizures 16, 211 (1972)	Prof. Dr. I. Izquierdo Dr. A. G. Nasello Departamento de Farmacología, Facultad de Ciencias Químicas, Universidad Nacional de Córdoba, Estafeta 32, Córdoba, Argentina

Cholinergic mechanism—monoamines relation in certain brain structures *16*, 334 (1972)	Prof. Dr. J. A. Izquierdo Department of Experimental Pharmacology, Facultad de Farmacia y Bioquímica, Buenos Aires, Argentina
The development of antifertility substances *7*, 133 (1964)	Prof. Dr. H. Jackson Head of Department of Experimental Chemotherapy, Christie Hospital and Holt Radium Institute, Paterson Laboratories, Manchester 20, England
Agents acting on central dopamine receptors *21*, 409 (1977)	Dr. P. C. Jain Dr. N. Kumar Medicinal Chemistry Division, Central Drug Research Institute, Lucknow, India
Recent advances in the treatment of parasitic infections in man *18*, 191 (1974) The levamisole story *20*, 347 (1976)	Dr. P. A. J. Janssen Director, Janssen Pharmaceutica, Research Laboratories, Beerse, Belgium
Recent developments in cancer chemotherapy *25*, 275 (1981)	Dr. K. Jewers Tropical Product Institute, 56/62, Gray's Inn Road, London, WC1X8LU, England
Search for pharmaceutically interesting quinazoline derivatives: Efforts and results (1969–1980) *26*, 259 (1982)	Dr. S. Johne Institute of Plant Biochemistry, The Academy of Sciences of the German Democratic Republic, DDR-4010 Halle (Saale), PSF 250
A review of advances in prescribing for teratogenic hazards *29*, 121 (1985)	E. Marshall Johnson, Ph. D. Daniel Baugh Institute, Jefferson College, Thomas Jefferson University, 1020 Locust Street, Philadelphia, PA 19107
A comparative study of bitoscanate, bephenium hydroxynaphthoate and tetrachlorethylene in hookworm infection *19*, 70 (1975)	Dr. S. Johnson Department of Medicine III, Christian Medical College Hospital, Vellore, Tamilnadu, India
Tetanus in Punjab with particular reference to the role of muscle relaxants in its management *19*, 288 (1975)	Prof. Dr. S. S. Jolly Dr. J. Singh Dr. S. M. Singh Department of Medicine, Medical College, Patiala, India
Virulence-enhancing effect of ferric ammonium citrate on *Vibrio cholerae* *19*, 546 (1975)	Dr. I. Joó Institute for Serobacteriological Production and Research 'HUMAN', WHO International Reference Centre for Bacterial Vaccines, Budapest, Hungary

Toxoplasmosis *18*, 205 (1974)	Prof. Dr. B. H. Kean The New York Hospital – Cornell Medical Center, 525 East 68th Street, New York, N. Y., USA
Tabellarische Zusammenstellung über die Substruktur der Proteine *16*, 364 (1972)	Dr. R. Kleine Physiologisch-Chemisches Institut der Martin-Luther-Universität, 402 Halle (Saale), DDR
Experimental evaluation of antituberculous compounds, with special reference to the effect of combined treatment *18*, 211 (1974)	Dr. F. Kradolfer Head of Infectious Diseases Research, Biological Research Laboratories, Pharmaceutical Division, Ciba-Geigy Ltd., Basle, Switzerland
The oxidative metabolism of drugs and other foreign compounds *17*, 488 (1973)	Dr. F. Kratz Medizinische Kliniken und Polikliniken, Justus-Liebig-Universität, Giessen, BR Deutschland
Die Amidinstruktur in der Arzneistofforschung *11*, 356 (1968)	Prof. Dr. A. Kreutzberger Wissenschaftlicher Abteilungsvorsteher am Institut für pharmazeutische Chemie der Westfälischen Wilhelms-Universität Münster, Münster (Westfalen), Deutschland
Present data on the pathogenesis of tetanus *19*, 301 (1975) Tetanus: general and pathophysiological aspects; achievements, failures, perspectives of elaboration of the problem *19*, 314 (1975)	Prof. Dr. G. N. Kryzhanovsky Institute of General Pathology and Pathological Physiology, AMS USSR, Moscow, USSR
Lipophilicity and drug activity *23*, 97 (1979)	Dr. H. Kubinyi Chemical Research and Development of BASF Pharma Division, Knoll AG, Ludwigshafen/Rhein, Federal Republic of Germany
Klinisch-pharmakologische Kriterien in der Bewertung eines neuen Antibiotikums. Grundlagen und methodische Gesichtspunkte *22*, 327 (1978)	Prof. Dr. H. P. Kuemmerle München/Eppstein, BR Deutschland
Adrenergic receptor research: Recent developments *33*, 151 (1989)	George Kunos Laboratory of Physiologic and Pharmacologic Studies, National Institute on Alcohol Abuse and Alcoholism, National Institutes of Health, Bethesda, MD 20892, USA

Über neue Arzneimittel *1*, 531 (1959), *2*, 251 (1960), *3*, 369 (1961), *6*, 347 (1963), *10*, 360 (1966)	Dr. W. Kunz Forschungschemiker in Firma Dr. Schwarz GmbH, Monheim (Rheinland), BR Deutschland
Die Anwendung von Psycho-pharmaka in der psychosomatischen Medizin *10*, 530 (1966)	Dr. F. Labhardt Privatdozent, stellvertretender Direktor der psychiatrischen Universitätsklinik, Basel, Schweiz
The bacterial cell surface and antimicrobial resistance *32*, 149 (1988)	Peter A. Lambert Pharmaceutical Sciences Institute, Aston University, Birmingham B4 7ET, U. K.
Therapeutic measurement in tetanus *19*, 323 (1975)	Prof. Dr. D. R. Laurence Department of Pharmacology, University College, London, and Medical Unit, University College Hospital Medical School, London, England
Physico chemical methods in pharmaceutical chemistry, 1. Spectrofluorometry *6*, 151 (1963)	Dr. H. G. Leemann Head of the Analytical Department in the Pharmaceutical Division of Sandoz Ltd, Basle, Switzerland Dr. K. Stich Specialist for Questions in Ultraviolet and Fluorescence Spectrophotometry, Analytical Department, Sandoz Ltd., Basle, Switzerland Dr. Margrit Thomas Research Chemist in the Analytical Department Research Laboratory, Sandoz Ltd., Basle, Switzerland
Biochemical acyl hydroxylations *16*, 229 (1972)	Dr. W. Lenk Pharmakologisches Institut der Universität München, Nussbaumstrasse 26, München, BR Deutschland
Cholinesterase restoring therapy in tetanus *19*, 329 (1975)	Prof. Dr. G. Leonardi Department of Medicine, St. Thomas Hospital, Portogruaro, Venice, Italy Dr. K. G. Nair Prof. Dr. F. D. Dastur Department of Medicine, Seth G. S. Medical College and K. E. M. Hospital, Bombay 12, India
Biliary excretion of drugs and other xenobiotics *25*, 361 (1981)	Prof. Dr. W. G. Levine Department of Molecular Pharmacology, Albert Einstein College of Medicine, Yeshiva University, 1300 Morris Park Avenue, Bronx, New York 10461, USA

Structures, properties and disposition of drugs *29*, 67 (1985)	Eric J. Lien Biomedicinal Chemistry, School of Pharmacy, University of Southern California, Los Angeles, Calif. 90033, USA
Ribonucleotide reductase inhibitors as anticancer and antiviral agents *31*, 101 (1987)	Eric J. Lien Section of Biomedicinal Chemistry, School of Pharmacy, Universtity of Southern California, Los Angeles, Calif. 90033, USA
Interactions between androgenic-anabolic steroids and glucocorticoids *14*, 139 (1970)	Dr. O. Linèt Sinai Hospital of Detroit, Department of Medicine, 6767 West Outer Drive, Detroit, Michigan 48235
Drug inhibition of mast cell secretion *29*, 277 (1985)	R. Ludowyke D. Lagunoff Department of Pathology, St. Louis University, School of Medicine, 1402 S. Grand Blvd. St. Louis, Mo 63104
Reactivity of bentonite flocculation, indirect haemagglutination and casoni tests in hydatid disease *19*, 75 (1975)	Dr. R. C. Mahajan Dr. N. L. Chitkara Division of Parasitology, Department of Microbiology, Postgraduate Institute of Medical Education and Research, Chandigarh, India
Epidemiology of diphtheria *19*, 336 (1975)	Dr. L. G. Marquis Glaxo Laboratories (India) Limited, Worli, Bombay 25, India
Biological activity of the terpenoids and their derivatives *6*, 279 (1963)	Dr. M. Martin-Smith Reader in Pharmaceutical Chemistry, University of Strathclyde, Department of Pharmaceutical Chemistry, Glasgow, C. 1, Scotland Dr. T. Khatoon Lecturer in Chemistry at the Eden Girls College, Dacca, East Pakistan
Biological activity of the terpenoids and their derivatives – recent advances *13*, 11 (1969)	Dr. M. Martin-Smith Reader in Pharmaceutical Chemistry, University of Strathclyde, Glasgow, C. 1, Scotland Dr. W. E. Sneader Lecturer in Pharmaceutical Chemistry, University of Strathclyde, Glasgow, C. 1, Scotland

Antihypertensive agents 1962–1968 *13*, 101 (1969) Fundamental structures in drug research – Part I *20*, 385 (1976) Fundamental structures in drug research – Part II *22*, 27 (1978) Antihypertensive agents 1969–1980 *25*, 9 (1981)	Prof. Dr. A. Marxer Dr. O. Schier Chemical Research Department, Pharmaceuticals Division, Ciba-Geigy Ltd., Basle, Switzerland
Relationships between the chemical structure and pharmacological activity in a series of synthetic quinuclidine derivatives *13*, 293 (1969)	Prof. Dr. M. D. Mashkovsky All-Union Chemical Pharmaceutical Research Institute, Moscow, USSR Dr. L. N. Yakhontov All-Union Chemical Pharmaceutical Research Institute, Moscow, USSR
Further developments in research on the chemistry and pharmacology of synthetic quinuclidine derivatives *27*, 9 (1983)	Prof. M. D. Mashkovsky Prof. L. N. Yakhontov Dr. M. E. Kaminka Dr. E. E. Mikhlina S. Ordzhonikidze All-Union, Chemical Pharmaceutical Research Institute, Moscow, USSR
On the understanding of drug potency *15*, 123 (1971) The chemotherapy of intestinal nematodes *16*, 157 (1972)	Dr. J. W. McFarland Pfizer Medical Research Laboratories, Groton, Connecticut, USA
Zur Beeinflussung der Strahlenempfindlichkeit von Säugetieren durch chemische Substanzen *9*, 11 (1966)	Dr. H.-J. Melching Privatdozent, Oberassistent am Radiologischen Institut der Universität Freiburg i.Br., Freiburg i.Br., Deutschland Dr. C. Streffer Wissenschaftlicher Mitarbeiter am Radiologischen Institut der Universität Freiburg i.Br., Freiburg i.Br., Deutschland
Analgesia and addiction *5*, 155 (1963)	Dr. L. B. Mellett Assistant Professor of Pharmacology, University of Michigan Medical School, Ann Arbor, Michigan, USA Prof. Dr. L. A. Woods Department of Pharmacology, College of Medicine, State University of Iowa, Iowa City, USA
Comparative drug metabolism *13*, 136 (1969)	Dr. L. B. Mellett Head, Pharmacology & Toxicology, Kettering-Meyer Laboratories, Southern Research Institute, Birmingham, Alabama, USA

Mechanism of action of anxiolytic drugs *31*, 315 (1987)	T. Mennini S. Caccia S. Garattini Istituto di Ricerche Farmacologiche "Mario Negri", Via Eritrea 62, 20157 Milan, Italy
Pathogenesis of amebic disease *18*, 225 (1974) Protozoan and helminth parasites – a review of current treatment *20*, 433 (1976)	Prof. Dr. M. J. Miller Tulane University, Department of Tropical Medicine, New Orleans, Louisiana, USA
Medicinal agents incorporating the 1,2-diamine functionality *33*, 135 (1989)	Erik T. Michalson and Jacob Szmuszkovicz Department of Chemistry, University of Notre Dame, Notre Dame, IN 46556, USA
Synopsis der Rheumatherapie *12*, 165 (1968)	Dr. W. Moll Spezialarzt FMH Innere Medizin – Rheumatologie, Basel, Schweiz
On the chemotherapy of cancer *8*, 431 (1965) The relationship of the metabolism of anticancer agents to their activity *17*, 320 (1973) The current status of cancer chemotherapy *20*, 465 (1976)	Dr. J. A. Montgomery Kettering-Meyer Laboratory, Southern Research Institute, Birmingham, Alabama, USA
The significance of DNA technology in medicine *33*, 397 (1989)	Hansjakob Müller Dept of Human Genetics, University Childrens Hospital, CH-4005 Basel, and Lab. of Human Genetics, Dept of Research, Kantonsspital, CH-4031 Basel, Switzerland
Der Einfluss der Formgebung auf die Wirkung eines Arzneimittels *10*, 204 (1966) Galenische Formgebung und Arzneimittelwirkung. Neue Erkenntnisse und Feststellungen *14*, 269 (1970)	Prof. Dr. K. Münzel Leiter der galenischen Forschungsabteilung der F. Hoffmann-La Roche & Co. AG, Basel, Schweiz
A field trial with bitoscanate in India *19*, 81 (1975)	Dr. G. S. Mutalik Dr. R. B. Gulati Dr. A. K. Iqbal Department of Medicine, B. J. Medical College and Sassoon General Hospital, Poona, India
Comparative study of bitoscanate, bephenium hydroxynaphthoate and tetrachlorethylene in hookworm disease *19*, 86 (1975)	Dr. G. S. Mutalik Dr. R. B. Gulati Department of Medicine, B. J. Medical College and Sassoon General Hospital, Poona, India

Ganglienblocker 2, 297 (1960)	Dr. K. Nádor o. Professor und Institutsdirektor, Chemisches Institut der Tierärztlichen Universität, Budapest, Ungarn
Nitroimidazoles as chemotherapeutic agents 27, 163 (1983)	Dr. M. D. Nair Dr. K. Nagarajan Ciba-Geigy Research Centre, Goreagon East, Bombay 400063
Recent advances in cholera pathophysiology and therapeutics 19, 563 (1975)	Prof. Dr. D. R. Nalin Johns Hopkins School of Medicine and School of Public Health. Guest Scientist, Cholera Research Hospital, Dacca, Bangladesh
Preparing the ground for research: importance of data 18, 239 (1974)	Dr. A. N. D. Nanavati Assistant Director and Head, Department of Virology, Haffkine Institute, Bombay, India
Mechanism of drugs action on ion and water transport in renal tubular cells 26, 87 (1982)	Prof. Dr. Yu. V. Natochin I. M. Sechenov Institute of Evolutionary Physiology and Biochemistry, Leningrad, USSR
Progesterone receptor binding of steroidal and nonsteroidal compounds 30, 151 (1986)	Neelima M. Seth A. P. Bhaduri Division of Medicinal Chemistry, Central Drug Research Institute, Lucknow 226001, India
Recent advances in drugs against hypertension 29, 215 (1985)	Neelima B. K. Bhat A. P. Bhaduri Central Drug Research Institute, Lucknow – 226001, India
High resolution nuclear magnetic resonance spectroscopy of biological samples as an aid to drug development 31, 427 (1987)	J. K. Nicholson Department of Chemistry, Birkbeck College, University of London, Gordon House, 29, Gordon Square, London WC1E6BT, England Ian D. Wilson Department of Safety of Medicines, ICI Pharmaceuticals Division, Mereside, Alderley Park, Macclesfield, Cheshire SK 10 4TG, England
Antibody response to two cholera vaccines in volunteers 19, 554 (1975)	Y. S. Nimbkar R. S. Karbhari S. Cherian N. G. Chanderkar R. P. Bhamaria P. S. Ranadive Dr. B. B. Gaitondé Haffkine Institute, Parel, Bombay 12, India

Surface interaction between bacteria and phagocytic cells *32*, 137 (1988)	L. Öhman G. Maluszynska K.-E. Magnusson and O. Stendahl Department of Medical Microbiology, Linsköping University, S-581 85 Linsköping, Sweden
Die Chemotherapie der Wurmkrankheiten *1*, 159 (1959)	Prof. Dr. H.-A. Oelkers Leiter der pharmakologischen und parasitologischen Abteilung der Firma C. F. Asche & Co., Hamburg-Altona, Deutschland
GABA-Drug Interactions *31*, 223 (1987)	Richard W. Olsen Department of Pharmacology, School of Medicine, and Brain Research Institute, University of California, Los Angeles, Calif. 90024, USA
Drug research and human sleep *22*, 355 (1978)	Prof. Dr. I. Oswald University Department of Psychiatry, Royal Edinburgh Hospital, Edingburgh, Scotland
Effects of drugs on calmodulin-mediated enzymatic actions *33*, 353 (1989)	Judit Ovádi Institute of Enzymology, Biological Research Center, Hungarian Academy of Sciences, P. O. B. 7, H-1502 Budapest, Hungary
An extensive community outbreak of acute diarrhoeal diseases in children *19*, 570 (1975)	Dr. S. C. Pal Dr. C. Koteswar Rao Cholera Research Centre, Calcutta, India
Drug and its action according to Ayurveda *26*, 55 (1982)	Dr. Shri Madhabendra Nath Pal
3,4-Dihydroxyphenylalanine and related compounds *9*, 223 (1966)	Dr. A. R. Patel Post-Doctoral Research Assistant. Department of Chemistry, University of Virginia, Charlottesville, Virginia, USA Prof. Dr. A. Burger Department of Chemistry, University of Virginia, Charlottesville, Virginia, USA
Mescaline and related compounds *11*, 11 (1968)	Dr. A. R. Patel Post-Doctoral Research Assistant, Department of Chemistry, University of Virginia, Charlottesville, Virginia, USA
Experience with bitoscanate in adults *19*, 90 (1975)	Dr. A. H. Patricia Dr. U. Prabakar Rao Dr. R. Subramaniam Dr. N. Madanagopalan Madras Medical College, Madras, India

Monoaminoxydase-Hemmer 2, 417 (1960)	Prof. Dr. A. Pletscher Direktor der medizinischen Forschungsabteilung F. Hoffmann-La Roche & Co. AG, Basel, und Professor für Innere Medizin an der Universität Basel Dr. K. F. Gey Medizinische Forschungsabteilung F. Hoffmann-La Roche & Co. AG, Basel Schweiz Dr. P. Zeller Chefchemiker in Firma F. Hoffmann-La Roche & Co. AG, Basel, Schweiz
What makes a good pertussis vaccine? 19, 341 (1975) Vaccine composition in relation to antigenic variation of the microbe: is pertussis unique? 19, 347 (1975) Some unsolved problems with vaccines 23, 9 (1979)	Dr. N. W. Preston Department of Bacteriology and Virology, University of Manchester, Manchester, England
Antibiotics in the chemotherapy of malaria 26, 167 (1982)	Dr. S. K. Puri Dr. G. P. Dutta Division of Microbiology, Central Drug Research Institute, Lucknow 226001, India
Clinical study of diphtheria, pertussis and tetanus 19, 356 (1975)	Dr. V. B. Raju Dr. V. R. Parvathi Institute of Child Health and Hospital for Children, Egmore, Madras 8, India
Epidemiology of cholera in Hyderabad 19, 578 (1975)	Dr. K. Rajyalakshmi Dr. P. V. Ramana Rao Institute of Preventive Medicine, Hyderabad, Andhra Pradesh, India
Adenosine receptors: Clinical implications and biochemical mechanisms 32, 195 (1988)	Vickram Ramkumar George Pierson and Gary L. Stiles Departments of Medicine and Biochemistry, Duke University Medical Center, Durham, NC 27710, USA
Problems of malaria eradication in India 18, 245 (1974)	Dr. V. N. Rao Joint Director of Health Services (Health), Maharashtra, Bombay, India
The photochemistry of drugs and related substances 11, 48 (1968)	Dr. S. T. Reid Lecturer in Chemical Pharmacology, Experimental Pharmacology Division, Institute of Physiology, The University, Glasgow, W.2, Scotland

Orale Antikoagulantien *11*, 226 (1968)	Dr. E. Renk Dr. W. G. Stoll Wissenschaftliche Laboratorien der J. R. Geigy AG, Basel, Schweiz
Mechanism-based inhibitors of monoamine oxidase *30*, 205 (1986)	Lauren E. Richards Alfred Burger Department of Chemistry, University of Virginia, Charlottesville, Virginia 22901, USA
Tetrahydroisoquinolines and β-carbolines: Putative natural substances in plants and animals *29*, 415 (1985)	H. Rommelspacher R. Susilo Department of Neuropsychopharmacology, Free University, Ulmenallee 30, D-1000 Berlin 19, F R G
Functional significance of the various components of the influenza virus *18*, 253 (1974)	Prof. Dr. R. Rott Institut für Virologie, Justus-Liebig- Universität, Giessen, Deutschland
Behavioral correlates of presynaptic events in the cholinergic neurotrans- mitter system *32*, 43 (1988)	Roger W. Russell Department of Pharmacology, University of California, Los Angeles, CA 90024-1735, USA
Role of beta-adrenergic blocking drug propranolol in severe tetanus *19*, 361 (1975)	Prof. Dr. G. S. Sainani Head, Upgraded Department of Medicine, B. J. Medical College and Sassoon General Hospitals, Poona, India Dr. K. L. Jain Prof. Dr. V. R. D. Deshpande Dr. A. B. Balsara Dr. S. A. Iyer Medical College and Hospital, Nagpur, India
Studies on *Vibrio parahaemolyticus* in Bombay *19*, 586 (1975)	Dr. F. L. Saldanha Dr. A. K. Patil Dr. M. V. Sant Haffkine Institute, Parel, Bombay 12, India
Pharmacology and toxicology of axoplasmic transport *28*, 53 (1984)	Dr. Fred Samson, Ph. D., Director Ralph L. Smith Research Center, The University of Kansas Medical Center, Department of Physiology Dr. J. Alejandro Donoso Ralph L. Smith Research Center, The University of Kansas Medical Center, Department of Neurology, Kansas City, Kansas 66103, USA
Clinical experience with bitoscanate *19*, 96 (1975)	Dr. M. R. Samuel Head of the Department of Clinical Development, Medical Division, Hoechst Pharmaceuticals Limited, Bombay, India

Tetanus: Situational clinical trials and therapeutics 19, 367 (1975)	Dr. R. K. M. Sanders Dr. M. L. Peacock Dr. B. Martyn Dr. B. D. Shende The Duncan Hospital, Raxaul, Bihar, India
Epidemiological studies on cholera in non-endemic regions with special reference to the problem of carrier state during epidemic and non-epidemic period 19, 594 (1975)	Dr. M. V. Sant W. N. Gatlewar S. K. Bhindey Haffkine Institute, Parel, Bombay 12, India
Epidemiological and biochemical studies in filariasis in four villages near Bombay 18, 269 (1974)	Dr. M. V. Sant, W. N. Gatlewar and T.U.K. Menon Department of Zoonosis and of Research Divison of Microbiology, Haffkine Institute, Bombay, India
Hookworm anaemia and intestinal malabsorption associated with hookworm infestation 19, 108 (1975)	Prof. Dr. A. K. Saraya Prof. Dr. B. N. Tandon Department of Pathology and Department of Gastroenterology, All India Institute of Medical Sciences, New Delhi, India
The effects of structural alteration on the anti-inflammatory properties of hydrocortisone 5, 11 (1963)	Dr. L. H. Sarett Director of Synthetic Organic Chemistry, Merck Sharp & Dohme Research Laboratories, Rahway, New Jersey, USA Dr. A. A. Patchett Director of the Department of Synthetic Organic Chemistry, Merck Sharp & Dohme Research Laboratories, Rahway, New Jersey, USA Dr. S. Steelman Director of Endocrinology, Merck Institute for Therapeutic Research, Rahway, New Jersey, USA
The impact of natural product research on drug discovery 23, 51 (1979)	Dr. L. H. Sarett Senior Vice-President for Science and Technology, Merck & Co., Inc., Rahway New Jersey, USA
Anti-filariasis campaign: its history and future prospects 18, 259 (1974)	Prof. Dr. M. Sasa Professor of Parasitology, Director of the Institute of Medical Science, University of Tokyo, Tokyo, Japan
Platelets and atherosclerosis 29, 49 (1985)	Robert N. Saunders, Sandoz Research Institute, East Hanover, N. J., USA

Pyrimidinones as biodynamic agents 31, 127 (1987)	Anil K. Saxena Shradha Sinha Division of Medicinal Chemistry, Central Drug Research Institute, Lucknow 226 001, India
Immuno-diagnosis of helminthic infections 19, 119 (1975)	Prof. Dr. T. Sawada Dr. K. Sato Dr. K. Takei Department of Parasitology, School of Medicine, Gunma University, Maebashi, Japan
Immuno-diagnosis in filarial infection 19, 128 (1975)	Prof. Dr. T. Sawada Dr. K. Sato Dr. K. Takei Department of Parasitology, School of Medicine, Gunma University, Maebashi, Japan Dr. M. M. Goil Department of Zoology, Bareilly College, Bareilly (U. P.), India
Quantitative structure- activity relationships 23, 199 (1979)	Dr. A. K. Saxena Dr. S. Ram Medicinal Chemistry Division, Central Drug Research Institute, Lucknow, India
Advances in chemotherapy of malaria 30, 221 (1986)	Anil K. Saxena Mridula Saxena Division of Medicinal Chemistry, Central Drug Research Institute, Lucknow 226001, India
Phenothiazine und Azaphenothiazine als Arzneimittel 5, 269 (1963)	Dr. E. Schenker Forschungschemiker in der Sandoz AG, Basel, Schweiz Dr. H. Herbst Forschungstechniker in den Farbwerken Hoechst, Frankfurt a. M., Deutschland
Antihypertensive agents 4, 295 (1962)	Dr. E. Schlittler Director of Research of CIBA Pharmaceutical Company, Summit, New Jersey, USA Dr. J. Druey Director of the Department of Synthetic Drug Research of CIBA Ltd., Basle, Switzerland Dr. A. Marxer Research Chemist of CIBA Ltd., Basle, and Lecturer at the University of Berne, Switzerland

Die Anwendung radioaktiver Isotope in der pharmazeutischen Forschung 7, 59 (1964)	Prof. Dr. K. E. Schulte Direktor des Instituts für Pharmazie und Lebensmittelchemie der Westfälischen Wilhelms-Universität Münster, Münster (Westfalen), Deutschland Dr. Ingeborg Mleinek Leiterin des Isotopen-Laboratoriums, Institut für Pharmazie und Lebensmittelchemie der Westfälischen Wilhelms-Universität Münster, Münster (Westfalen), Deutschland
Natürliche und synthetische Acetylen-Verbindungen als Arzneistoffe 14, 387 (1970)	Prof. Dr. K. E. Schulte Direktor des Instituts für pharmazeutische Chemie der Westfälischen Wilhelms-Universität Münster, Münster (Westfalen), Deutschland Dr. G. Rücker Dozent für pharmazeutische Chemie an der Westfälischen Wilhelms-Universität Münster, Münster (Westfalen), Deutschland
Central control of arterial pressure by drugs 26, 353 (1982)	Dr. A. Scriabine Dr. D. G. Taylor Miles Institute for Preclinical Pharmacology, P.O. Box 1956, New Haven, Connecticut 06509, USA Dr. E. Hong Instituto Miles de Terepeutica Experimental, A. P. 22026, Mexico 22, D. F.
The structure and biogenesis of certain antibiotics 2, 591 (1960)	Dr. W. A. Sexton Research Director of the Pharmaceuticals Division of Imperial Chemical Industries Ltd., Wilmslow, Cheshire, England
Role of periodic deworming of preschool community in national nutrition programmes 19, 136 (1975)	Prof. Dr. P. M. Shah Institute of Child Health Dr. A. R. Junnarkar Reader in Preventive and social Medicine Dr. R. D. Khare Research Assistant, Institute of Child Health, J. J. Group of Government Hospitals and Grant Medical College, Bombay, India
Quinolones 31, 243 (1987)	Dr. med. Pramod M. Shah Zentrum der Inneren Medizin, Klinikum der J.-W.-von-Goethe-Univerität, Theodor-Stern-Kai 7, D-6000 Frankfurt/Main 70
Advances in the treatment and control of tissue-dwelling helminth parasites 30, 473 (1986)	Satyavan Sharma Cogswell Laboratory, Chemistry Department, Rensselaer Polytechnic Institute, Troy, New York 12180, USA

Chemotherapy of cestode infections 24, 217 (1980)	Dr. Satyavan Sharma Dr. S. K. Dubey Dr. R. N. Iyer Medicinal Chemistry Division, Central Drug Research Institute, Lucknow 226001, India
Chemotherapy of hookworm infections 26, 9 (1982)	Dr. Satyavan Sharma Dr. Elizabeth S. Charles Medicinal Chemistry Division, Central Drug Research Institute, Lucknow 226001, India
The benzimidazole anthelmiticschemistry and biological activity 27, 85 (1983)	Dr. Satyavan Sharma Dr. Syed Abuzar Central Drug Research Institute, Lucknow 226001, India
Treatment of helminth diseases challenges and achievements 31, 9 (1987)	Satyavan Sharma Medicinal Chemistry Division, Central Drug Research Institute, Lucknow 226 001, India
Ayurvedic medicine – past and present 15, 11 (1971)	Dr. Shiv Sharma 'Baharestan', Bomanji Petit Road, Cumballa Hill, Bombay, India
Mechanisms of anthelmintic action 19, 147 (1975)	Prof. Dr. U. K. Sheth Seth G. S. Medical College and K. E. M. Hospital, Parel, Bombay 12, India
Aspirin as an antithrombotic agent 33, 43 (1989)	Melvin J. Silver and Giovanni Di Minno Cardeza Foundation of Hematologic Research, Thomas Jefferson University, Philadelphia, PA 19107, USA, and Second Medical School, Naples University, Naples, Italy
Immunopharmacological approach to the study of chronic brain disorders 30, 345 (1986)	Vijendra K. Singh and H. Hugh Fudenberg Department of Basic and Clinical Immunology and Microbiology, Medical University of South Carolina, 171 Ashley Avenue, Charleston, S. C. 29425, USA
Implications of immunomodulant therapy in Alzheimer's disease 32, 21 (1988)	Vijendra K. Singh and H. Hugh Fudenberg Department of Microbiology and Immunology, Medical University of South Carolina Charleston, SC 29425, USA
Some often neglected factors in the control and prevention of communicable diseases 18, 277 (1974)	Dr. C. E. G. Smith Dean, London School of Hygiene and Tropical Medicine, Keppel Street, London, England

Tetanus and its prevention *19*, 391 (1975)	Dr. J. W. G. Smith Epidemiological Research Laboratory, Central Public Health Laboratory, London England
Growth of *Clostridium tetani in vivo* *19*, 384 (1975)	Dr. J. W. G. Smith Epidemiological Research Laboratory, Central Public Health Laboratory, London England Dr. A. G. MacIver Department of Morbid Anatomy, Faculty of Medicine, Southampton University, Southampton, England
The biliary excretion and enterohepatic circulation of drugs and other organic compounds *9*, 299 (1966)	Dr. R. L. Smith Senior Lecturer in Biochemistry at St. Mary's Hospital Medical School (University of London), Paddington, London, W.2, England
Noninvasive pharmacodynamic and bioelectric methods for elucidating the bioavailability mechanisms of ophthalmic drug preparations *25*, 421 (1981)	Dr. V. F. Smolen President and Chief Executive Officer Pharmacontrol Corp. 661 Palisades Ave., P.O. Box 931, Englewood Cliffs, New Jersey, 07632
On the relation between chemical structure and function in certain tumor promoters and anti-tumor agents *23*, 63 (1979) Relationships between structure and function of convulsant drugs *24*, 57 (1980)	Prof. Dr. J. R. Smythies Department of Psychiatry, University of Alabama in Birmingham Medical Center, Birmingham, Alabama, USA
Gram-negative bacterial endotoxin and the pathogenesis of fever *19*, 402 (1975)	Dr. E. S. Snell Glaxo Laboratories Limited, Greenford, Middlesex, England
Emerging concepts towards the development of contraceptive agents *33*, 267 (1989)	Ranjan P. Srivastava and A. P. Bhaduri Division of Medicinal Chemistry, Central Drug Research Institute, Lucknow 226 001, India
Strukturelle Betrachtungen der Psychopharmaka: Versuch einer Korrelation von chemischer Konstitution und klinischer Wir- kung *9*, 129 (1966)	Dr. K. Stach Stellvertretender Leiter der Chemischen Forschung der C. F. Boehringer & Söhne GmbH, Mannheim-Waldhof, Deutschland Dr. W. Pöldinger Oberarzt für klinische Psychopharmakologie an der Psychiatrischen Universitätsklinik Basel, Basel, Schweiz
From podophyllotoxin glucoside to etoposide *33*, 169 (1989)	H. Stähelin and A. von Wartburg Preclinical Research, Sandoz Ltd., Basel, Switzerland

Chemotherapy of intestinal helminthiasis *19*, 158 (1975)	Dr. O. D. Standen The Welcome Research Laboratories, Beckenham, Kent, England
Immunotherapy for leprosy and tuberculosis *33*, 415 (1989)	J. L. Stanford Department of Medical Microbiology, University College and Middlesex School of Medicine, London W1P 7LD, United Kingdom
The leishmaniases *18*, 289 (1974)	Dr. E. A. Steck Department of the Army, Walter Reed Army Institute of Research, Division of Medicinal Chemistry, Washington, D.C., USA
The benzodiazepine story *22*, 229 (1978)	Dr. L. H. Sternbach Research Department, Hoffmann-La Roche Inc., Nutley, New Jersey, USA
Immunostimulation with peptidoglycan or its synthetic derivatives *32*, 305 (1988)	Duncan E. S. Stewart-Tull Department of Microbiology, University of Glasgow, Glasgow G 61 1 QH Scotland
Hypertension: Relating drug therapy to pathogenic mechanisms *32*, 175 (1988)	David H. P. Streeten and Gunnar H. Anderson Jr. State University of New York, Health Science Center, Syracuse, N.Y. 13210, USA
Progress in sulfonamide research *12*, 389 (1968) Problems of medical practice and of medical-pharmaceutical research *20*, 491 (1976)	Dr. Th. Struller Research Department, F. Hoffmann-La Roche & Co. Ltd., Basle, Switzerland
Antiviral agents *22*, 267 (1978)	Dr. D. L. Swallow Pharmaceuticals Division, Imperial Chemical Industries Limited, Alderley Park, Macclesfield, Cheshire, England
Antiviral agents 1978–1983 *28*, 127 (1984)	Dr. D. L. Swallow, M. A., B. Sc., D. Phil., F.R.S.C. Imperial Chemical Industries PLC, Pharmaceutical Division, Alderley Park, Macclesfield, Cheshire SK 10 4 TG, England
Ketoconazole, a new step in the management of fungal disease *27*, 63 (1983)	Dr. J. Symoens Dr. G. Cauwenbergh Janssen Pharmaceutica, B–2340 Beerse, Belgium

Antiarrhythmic compounds *12*, 292 (1968)	Prof. Dr. L. Szekeres Head of the Department of Pharmacology, School of Medicine, University of Szeged, Szeged, Hungary Dr. J. G. Papp Senior Lecturer, University Department of Pharmacology, Oxford, England
Practically applicable results of twenty years of research in endocrinology *12*, 137 (1968)	Prof. Dr. M. Tausk State University of Utrecht, Faculty of Medicine, Utrecht, Netherlands
Stereoselective drug metabolism and its significance in drug research *32*, 249 (1988)	Bernard Testa and Joachim M. Mayer Ecole de Pharmacie, Université de Lausanne, CH-1005, Lausanne, Switzerland
Age profile of diphtheria in Bombay *19*, 412 (1975)	Prof. Dr. N. S. Tibrewala Dr. R. D. Potdar Dr. S. B. Talathi Dr. M. A. Ramnathkar Dr. A. D. Katdare Topiwala National Medical College, BYL Nair Hospital and Kasturba Hospital for Infectious Diseases, Bombay 11, India
On conformation analysis, molecular graphics, Fentanyl and its derivatives *30*, 91 (1986)	J. P. Tollenaere H. Moereels M. Van Loon Department of Theoretical Medicinal Chemistry, Janssen Pharmaceutica Research Laboratories, 2340 Beerse, Belgium
Antibakterielle Chemotherapie der Tuberkulose *7*, 193 (1964)	Dr. F. Trendelenburg Leitender Arzt der Robert-Koch-Abteilung der Medizinischen Universitätskliniken, Homburg, Saar, Deutschland
Diphtheria *19*, 423 (1975)	Prof. Dr. P. M. Udani Dr. M. M. Kumbhat Dr. U. S. Bhat Dr. M. S. Nadkarni Dr. S. K. Bhave Dr. S. G. Ezuthachan Dr. B. Kamath The Institute of Child Health, J. J. Group of Hospitals, and Grant Medical College, Bombay 8, India

Biologische Oxydation und Reduktion am Stickstoff aromatischer Amino- und Nitroderivate und ihre Folgen für den Organismus *8,* 195 (1965) Stoffwechsel von Arzneimitteln als Ursache von Wirkungen, Nebenwirkungen und Toxizität *15,* 147 (1971)	Prof. Dr. H. Uehleke Pharmakologisches Institut der Universität Tübingen, 74 Tübingen, Deutschland
Mode of death in tetanus *19,* 439 (1975)	Prof. Dr. H. Vaishnava Dr. C. Bhawal Dr. Y. P. Munjal Department of Medicine, Maulana Azad Medical College and Associated Irwing and G. B. Pant Hospitals, New Delhi, India
Comparative evaluation of amoebicidal drugs *18,* 353 (1974) Comparative efficacy of newer anthelmintics *19,* 166 (1975)	Prof. Dr. B. J. Vakil Dr. N. J. Dalal Department of Gastroenterology, Grant Medical College and J. J. Group of Hospitals, Bombay, India
Cephalic tetanus *19,* 443 (1975)	Prof. Dr. B. J. Vakil Prof. Dr. B. S. Singhal Dr. S. S. Pandya Dr. P. F. Irani J. J. Group of Hospitals and Grant Medical College, Bombay, India
The effect and usefulness of early intravenous beta blockade in acute myocardial infarction *30,* 71 (1986)	Anders Vedin, M. D., Ph. D. Claes Wilhelmsson, M. D., Ph. D. From the Cardiac Unit and the Department of Medicine, Östra Hospital, University of Göteborg, S-41685 Göteborg, Sweden
Methods of monitoring adverse reactions to drugs *21,* 231 (1977) Aspects of social pharmacology *22,* 9 (1978)	Prof. Dr. J. Venulet Division of Clinical Pharmacology, Department of Medicine, Hôpital Cantonal and University of Geneva, Geneva, Switzerland. Formerly: Senior Project Officer, WHO Research Centre for International Monitoring of Adverse Reactions to Drugs, Geneva, Switzerland
The current status of cholera toxoid research in the United States *19,* 602 (1975)	Dr. W. F. Verwey Dr. J. C. Guckian Dr. J. Craig Dr. N. Pierce Dr. J. Peterson Dr. H. Williams, Jr. The University of Texas Medical Branch, Galveston, State University of New York Medical Center (Downstate), and Johns Hopkins University School of Medicine, USA

Cell-kinetic and pharmacokinetic aspects in the use and further development of cancerostatic drugs *20*, 521 (1976)	Prof. Dr. M. von Ardenne Forschungsinstitut Manfred von Ardenne, Dresden, GDR
The problem of diphtheria as seen in Bombay *19*, 452 (1975)	Prof. Dr. M. M. Wagle Dr. R. R. Sanzgiri Dr. Y. K. Amdekar Institute of Child Health, J. J. Group of Hospitals and Grant Medical College, Bombay 8, India
Nicotine: an addictive substance or a therapeutic agent? *33*, 9 (1989)	David M. Warburton Department of Psychology, University of Reading, Building 3, Earley Gate, Whiteknights, Reading RG6 2AL, United Kingdom
Cell-wall antigens of *V. cholerae* and their implication in cholera immunity *19*, 612 (1975)	Dr. Y. Watanabe Dr. R. Ganguly Bacterial Diseases, Division of Communicable Diseases, World Health Organization, Geneva 27, Switzerland
Antigen-specific T-cell factors and drug research *32*, 9 (1988)	David R. Webb Synthex Research, Palo Alto, CA 94303, USA
Where is immunology taking us? *20*, 573 (1976)	Dr. W. J. Wechter Dr. Barbara E. Loughman Hypersensitivity Diseases Research, The Upjohn Company, Kalamazoo, Michigan, USA
Immunology in drug research *28*, 233 (1984)	Dr. W. J. Wechter, Ph. D., Research Manager Dr. Barbara E. Loughman, Ph. D., Research Head The Upjohn Company, Kalamazoo, Michigan 49001, USA
Metabolic activation of chemical carcinogens *26*, 143 (1982)	Dr. E. K. Weisburger Division of Cancer Cause and Prevention, National Cancer Institute, Bethesda, Maryland 20205, USA
A pharmacological approach to allergy *3*, 409 (1961)	Dr. G. B. West Reader in the School of Pharmacy, Department of Pharmacology, University of London, London, England

A new approach to the medical interpretation of shock *14*, 196 (1970)	Dr. G. B. West Scientific Secretary, The British Industrial Biological Research Association, Woodmansterne Road, Carshalton, Surrey, England Dr. M. S. Starr Department of Pharmacology, St. Mary's Hospital Medical School, University of London, London, England
Adverse reactions of sugar polymers in animals and man *23*, 27 (1979)	Dr. G. B. West Department of Paramedical Sciences, North-East London Polytechnic, London, England
Biogenic amines and drug research *28*, 9 (1984)	Dr. G. B. West Department of Paramedical Sciences, North-East London Polytechnic, England
Some biochemical and pharmacological properties of anti-inflammatory drugs *8*, 321 (1965)	Dr. M. W. Whitehouse Lecturer in Biochemistry at the University of Oxford, Oxford, England
Wirksamkeit und Nebenwirkungen von Metronidazol in der Therapie der Trichomonasis *9*, 361 (1966)	Dr. K. Wiesner Tierarzt, wissenschaftlicher Mitarbeiter der Pharmawissenschaftlichen Literaturabteilung, Farbenfabriken Bayer AG, Leverkusen, Deutschland Dr. H. Fink Leiter der Pharmawissenschaftlichen Literaturabteilung, Farbenfabriken Bayer AG, Leverkusen, Deutschland
Carcinogenicity testing of drugs *29*, 155 (1985)	G. M. Williams, J. H. Weisburger Naylor Dana Institute for Desease Prevention, American Health Foundation, Valhalla, N. Y. 10595
Organizing for drug discovery *32*, 329 (1988)	Michael Williams and Gary L. Neil* Pharamaceutical Division Ciba-Geigy; Summit NJ, USA * Biotechnology and basic Research Support, The Upjohn Company, Kalamazoo, MI 49001, USA
Drug treatment of asthma *28*, 111 (1984)	Prof. Dr. Archie F. Wilson, M. D., Ph. D. University of California, Irvine Medical Center, Orange, CA 92683, USA
Cooperative effects in drug-DNA interactions *31*, 193 (1987)	W. David Wilson Department of Chemistry and Laboratory for Microbial and Biochemistry Sciences, Georgia State University, Atlanta, Georgia 30307, USA

Nonsteroid antiinflammatory agents 10, 139 (1966)	Dr. C. A. Winter Senior Investigator Pharmacology, Merck Institute for Therapeutic Research, West Point, Pennsylvania, USA
A review of the continuum of drug-induced states of excitation and depression 26, 225 (1982)	Prof. Dr. W. D. Winters Departments of Pharmacology and Internal Medicine, School of Medicine, University of California, Davis, California 95616, USA
Basic research in the US pharmaceutical industry 15, 204 (1971)	Dr. O. Wintersteiner The Squibb Institute for Medical Research, New Brunswick, New Jersey, USA
Light and dark as a "drug" 31, 383 (1987)	Anna Wirz-Justice Psychiatric University Clinic, Wilhelm Klein Strasse 27, CH-4025 Basel, Switzerland
The chemotherapy of amoebiasis 8, 11 (1965)	Dr. G. Woolfe Head of the Chemotherapy Group of the Research Department at Boots Pure Drug Company Ltd., Nottingham, England
Antimetabolites and their revolution in pharmacology 2, 613 (1960)	Dr. D. W. Woolley The Rockefeller Institute, New York, USA
Noise analysis and channels at the postsynaptic membrane of skeletal muscle 24, 9 (1980)	Dr. D. Wray Lecturer, Pharmacology Department, Royal Free Hospital School of Medicine, Pond Street, London NW3 2QG, England
Krebswirksame Antibiotika aus Actinomyceten 3, 451 (1961)	Dr. Kh. Zepf Forschungschemiker im biochemischen und mikrobiologischen Laboratorium der Farbwerke Hoechst, Frankfurt a.M., Deutschland Dr. Christa Zepf Referentin für das Chemische Zentralblatt, Kelkheim (Taunus), Deutschland
Fifteen years of structural modifications in the field of antifungal monocyclic 1-substituted 1H-azoles 27, 253 (1983)	Dr. L. Zirngibl Siegfried AG, Zofingen, Switzerland
Lysostaphin: model for a specific anzymatic approach to infectious disease 16, 309 (1972)	Dr. W. A. Zygmunt Department of Biochemistry, Mead Johnson Research Center, Evansville, Indiana, USA Dr. P. A. Tavormina Director of Biochemistry, Mead Johnson Research Center, Evansville, Indiana, USA